Advances in Intelligent Systems and Computing

Volume 465

Series editor

Janusz Kacprzyk, Polish Academy of Sciences, Warsaw, Poland
e-mail: kacprzyk@ibspan.waw.pl

About this Series

The series "Advances in Intelligent Systems and Computing" contains publications on theory, applications, and design methods of Intelligent Systems and Intelligent Computing. Virtually all disciplines such as engineering, natural sciences, computer and information science, ICT, economics, business, e-commerce, environment, healthcare, life science are covered. The list of topics spans all the areas of modern intelligent systems and computing.

The publications within "Advances in Intelligent Systems and Computing" are primarily textbooks and proceedings of important conferences, symposia and congresses. They cover significant recent developments in the field, both of a foundational and applicable character. An important characteristic feature of the series is the short publication time and world-wide distribution. This permits a rapid and broad dissemination of research results.

More information about this series at http://www.springer.com/series/11156

Radek Silhavy · Roman Senkerik
Zuzana Kominkova Oplatkova
Petr Silhavy · Zdenka Prokopova
Editors

Software Engineering Perspectives and Application in Intelligent Systems

Proceedings of the 5th Computer Science On-line Conference 2016 (CSOC2016), Vol 2

Editors
Radek Silhavy
Faculty of Applied Informatics
Tomas Bata University in Zlín
Zlín
Czech Republic

Petr Silhavy
Faculty of Applied Informatics
Tomas Bata University in Zlín
Zlín
Czech Republic

Roman Senkerik
Faculty of Applied Informatics
Tomas Bata University in Zlín
Zlín
Czech Republic

Zdenka Prokopova
Faculty of Applied Informatics
Tomas Bata University in Zlín
Zlín
Czech Republic

Zuzana Kominkova Oplatkova
Faculty of Applied Informatics
Tomas Bata University in Zlín
Zlín
Czech Republic

ISSN 2194-5357 ISSN 2194-5365 (electronic)
Advances in Intelligent Systems and Computing
ISBN 978-3-319-33620-6 ISBN 978-3-319-33622-0 (eBook)
DOI 10.1007/978-3-319-33622-0

Library of Congress Control Number: 2016937380

Printed on acid-free paper

This Springer imprint is published by Springer Nature
The registered company is Springer International Publishing AG Switzerland

Preface

This book constitutes the refereed proceedings of the Software Engineering Perspectives and Application in Intelligent Systems Section of the 5th Computer Science On-line Conference 2016 (CSOC 2016), held in April 2015.

The volume Software Engineering Perspectives and Application in Intelligent Systems brings 42 of the accepted papers. Each of them presents new approaches and methods to real-world problems and exploratory research that describes novel approaches in the field of cybernetics and automation control theory.

Particular emphasis is laid on modern trends in the selected fields of interest. New algorithms or methods in a variety of fields are also presented.

CSOC 2016 has received (all sections) 254 submissions, 136 of them were accepted for publication. More than 60 % of all accepted submissions were received from Europe, 20 % from Asia, 16 % from America and 4 % from Africa. Researchers from 32 countries participated in CSOC 2016.

CSOC 2016 intends to provide an international forum for the discussion of the latest high-quality research results in all areas related to computer science. The addressed topics are theoretical aspects and applications of computer science, artificial intelligence, cybernetics, automation control theory and software engineering.

Computer Science On-line Conference is held online and broad usage of modern communication technology improves the traditional concept of scientific conferences. It brings equal opportunity to participate to all researchers around the world.

The editors believe that readers will find the proceedings interesting and useful for their own research work.

March 2016

Radek Silhavy
Roman Senkerik
Zuzana Kominkova Oplatkova
Petr Silhavy
Zdenka Prokopova

Program Committee

Program Committee Chairs

Zdenka Prokopova, Ph.D., Associate Professor, Tomas Bata University in Zlín, Faculty of Applied Informatics, email: prokopova@fai.utb.cz

Zuzana Kominkova Oplatkova, Ph.D., Associate Professor, Tomas Bata University in Zlín, Faculty of Applied Informatics, email: kominkovaoplatkova@fai.utb.cz

Roman Senkerik, Ph.D., Associate Professor, Tomas Bata University in Zlín, Faculty of Applied Informatics, email: senkerik@fai.utb.cz

Petr Silhavy, Ph.D., Senior Lecturer, Tomas Bata University in Zlín, Faculty of Applied Informatics, email: psilhavy@fai.utb.cz

Radek Silhavy, Ph.D., Senior Lecturer, Tomas Bata University in Zlín, Faculty of Applied Informatics, email: rsilhavy@fai.utb.cz

Roman Prokop, Ph.D., Professor, Tomas Bata University in Zlín, Faculty of Applied Informatics, email: prokop@fai.utb.cz

Program Committee Chairs for Special Sections

Intelligent Information Technology, System Monitoring and Proactive Management of Complex Objects

Prof. Viacheslav Zelentsov, Doctor of Engineering Sciences, Chief Researcher of St. Petersburg Institute for Informatics and Automation of Russian Academy of Sciences (SPIIRAS)

Program Committee Members

Boguslaw Cyganek, Ph.D., D.Sc., Department of Computer Science, University of Science and Technology, Kraków, Poland

Krzysztof Okarma, Ph.D., D.Sc., Faculty of Electrical Engineering, West Pomeranian University of Technology, Szczecin, Poland

Monika Bakosova, Ph.D., Associate Professor, Institute of Information Engineering, Automation and Mathematics, Slovak University of Technology, Bratislava, Slovak Republic

Pavel Vaclavek, Ph.D., Associate Professor, Faculty of Electrical Engineering and Communication, Brno University of Technology, Brno, Czech Republic

Miroslaw Ochodek, Ph.D., Faculty of Computing, Poznań University of Technology, Poznań, Poland

Olga Brovkina, Ph.D., Global Change Research Centre Academy of Science of the Czech Republic, Brno, Czech Republic

Elarbi Badidi, Ph.D., College of Information Technology, United Arab Emirates University, Al Ain, United Arab Emirates

Luis Alberto Morales Rosales, Head of the Master Program in Computer Science, Superior Technological Institute of Misantla, Mexico

Mariana Lobato Baes, M.Sc., Research-Professor, Superior Technological of Libres, Mexico

Abdessattar Chaâri, Professor, Laboratory of Sciences and Techniques of Automatic Control and Computer engineering, University of Sfax, Tunisian Republic

Gopal Sakarkar, Shri. Ramdeobaba College of Engineering and Management, Republic of India

V.V. Krishna Maddinala, Assistant Professor, GD Rungta College of Engineering and Technology, Republic of India

Anand N. Khobragade, Scientist, Maharashtra Remote Sensing Applications Centre, Republic of India

Abdallah Handoura, Assistant Professor, Computer and Communication Laboratory, Telecom Bretagne, France

Technical Program Committee Members

Ivo Bukovsky
Miroslaw Ochodek
Bronislav Chramcov
Eric Afful Dazie

Michal Bliznak
Donald Davendra
Radim Farana
Zuzana Kominkova Oplatkova
Martin Kotyrba
Erik Kral
David Malanik
Michal Pluhacek
Zdenka Prokopova
Martin Sysel
Roman Senkerik
Petr Silhavy
Radek Silhavy
Jiri Vojtesek
Eva Volna
Janez Brest
Ales Zamuda
Roman Prokop
Boguslaw Cyganek
Krzysztof Okarma
Monika Bakosova
Pavel Vaclavek
Olga Brovkina
Elarbi Badidi

Organizing Committee Chair

Radek Silhavy, Ph.D., Tomas Bata University in Zlín, Faculty of Applied Informatics, e-mail: rsilhavy@fai.utb.cz

Conference Organizer (Production)

OpenPublish.eu s.r.o.
Web: http://www.openpublish.eu
e-mail: csoc@openpublish.eu

Conference Website, Call for Papers

http://www.openpublish.eu

Contents

The Effect of Nutrition Education System for Elementary School Students in Nutrition Knowledge

Yi-Horng Lai

Abstract The purpose of this study is to introduce the graphic presentation food safety and sanitation learning system with parent participation in element school's health and physical education curriculum. The students were divided into four groups: control group, control group with parent participation, learning system group, and learning system group with parent participation. There were three extra variables in this study: learning system, parent participation, and gender. The research data (three exams scores) was obtained before the course, in the middle of the course, and at the end of the course. The results indicate that, first, the estimate of slope of learning system is significantly correlated with parent participation; second, male elementary school students and female elementary school students were similar in the growth rate between each time points; third, the relationship between the initial exam score and the following two exam scores were not significant. Based on the results, it can be concluded that the use of learning system and parent participation were helpful for elementary school students to acquire food safety and sanitation knowledge in the health and physical education curriculum.

Keywords Computer-assisted instruction (CAI) · Parent participation · Nutrition knowledge · Latent growth model

1 Introduction

Schools play an important role in students' health promotion and disease prevention. Since elementary school students tend to form their health knowledge, attitudes and behavior through school education, it is important for schools to deliver effective health program. [1].

Y.-H. Lai (✉)
Department of Health Care Administration, Oriental Institute of Technology, New Taipei City, Taiwan
e-mail: FL006@mail.oit.edu.tw

R. Silhavy et al. (eds.), *Software Engineering Perspectives and Application in Intelligent Systems*, Advances in Intelligent Systems and Computing 465,
DOI 10.1007/978-3-319-33622-0_1

In 2000, Taiwan's Ministry of Education implemented the Nine-Year Integrated Course, and "Health and Sports" is one of the main fields of study. The integration of information technology is a new learning tool for food safety and sanitation, which is an important part of health education.

Computer tailored food safety and sanitation education may be more effective than traditional food safety and sanitation education because messages are tailored to individual behavior, needs and beliefs. Therefore, the messages are more capable of targeting at individuals and may have stronger motivational effects. Computer tailored food safety and sanitation education has been studied for different dietary behaviours, in different target populations, and in different settings. In recent years, studies [2–4] that assessed the effects of interactive technology in food safety and sanitation education were based on the behavior change theory. Computer tailored food safety and sanitation education is more likely to be read, remembered, and experienced as it is more personally relevant compared to standard materials. Furthermore, the computer tailored food safety and sanitation education also appears to have a greater effect in motivating people to change their diet, their fat intake in particular, although at present no definite conclusions can be drawn.

The purpose of this study is to explore the effects of the graphic presentation food safety and sanitation education informatics network (Computer Assist Instruction; CAI) with parent participation in elementary school's health and physical education curriculum.

1.1 The Information Technology and Nutrition Education

Computer tailored food safety and sanitation education is an innovative and promising tool to motivate people to make healthy dietary changes. It provides respondents with individualized feedback about their dietary behaviors, motivations, attitudes, norms, and skills and mimics the process of person-to-person dietary counselling. The available evidence indicates that computer tailored food safety and sanitation education is more effective in motivating people to make dietary changes than the traditional food safety and sanitation information, especially for reduction of dietary fat. The effectiveness of computer tailoring has been attributed to the fact that individualized feedback commands greater attention, is processed more intensively, contains less redundant information, and is appreciated better than more general intervention materials. Interactive technology offers good opportunities for the application of computer tailored food safety and sanitation education, and some studies of web-based computer tailoring have shown promising results. [5].

The results of Oenema, Brug, and Lechner's study [6] indicated that interactive, web-based computer-tailored food safety and sanitation education can lead to changes in determinants of behaviour. Food safety and sanitation educators are encouraged to explore the opportunities and challenges of these new technologies to enhance their work [7]. Brug, Steenhuis, Assema, and Vries's study [2] pointed out that computer-tailored food safety and sanitation information is a promising means of stimulating people to change their diet toward dietary recommendations.

Bensley, Anderson, Brusk, Mercer, and Rivas's study [8] claimed that Internet food safety and sanitation education was a viable alternative to traditional food safety and sanitation education for increasing fruit and vegetable consumption in some women, infants, and children clients. Besides, food safety and sanitation education through the telehealth service resulted in positive effects on the risk factors for metabolic syndrome, nutrient intake, and dietary habits [3]. In Tyrovolasa, Tountasb, Polychronopoulosa, and Panagiotakos's study [4], active food safety and sanitation policy and enhancement of food safety and sanitation services within the public health care system can contribute to improved health and quality of life among older populations. The food guidance system provides the basis for the food guidance presented in the American Red Cross food safety and sanitation course, "Better Eating for Better Health", and in "Dietary Guidelines and Your Diet", which is a series of bulletins developed by USDA to help consumers use the Dietary Guidelines [9].

1.2 The Parent Participation and Nutrition Education

Niemeiera, Hektnerb, and Enger's study [10] mentioned that weigh-related health interventions with parent participation could more effectively reduce body mass indexes of child and adolescent participants. In addition, longer interventions that include parent participation appear to have greater success. Suggestions for future research and related interventions are provided. Parents who participated in the intervention increased the scores on the nutrition knowledge test, and there also was a significant association between degree of family involvement, higher grain servings, and lower cholesterol intake [11].

2 Methods

The sample of this study was students of an elementary school in Taiwan. The participants came from four classes, and they were assigned in four groups respectively. The four groups were: control group, control group with parent participation, learning system group, and learning system group with parent participation. The research framework is shown in Fig. 1.

The Latent Growth Model of this study is illustrated in Fig. 2. Learning system group was the group (class) that applied the graphic presentation food safety and sanitation education system in the four-week food safety and sanitation education.

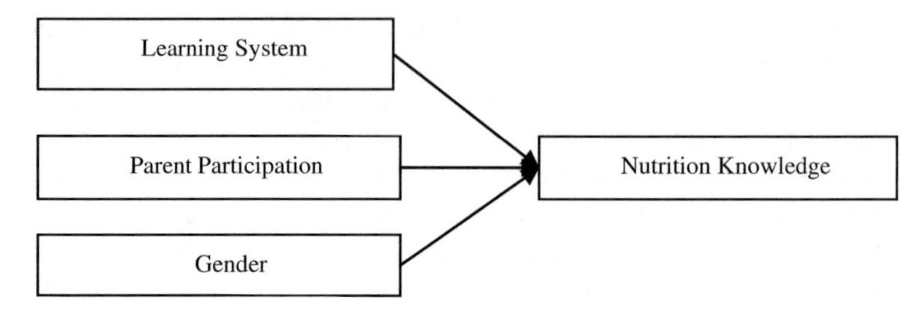

Fig. 1 Research framework of this study

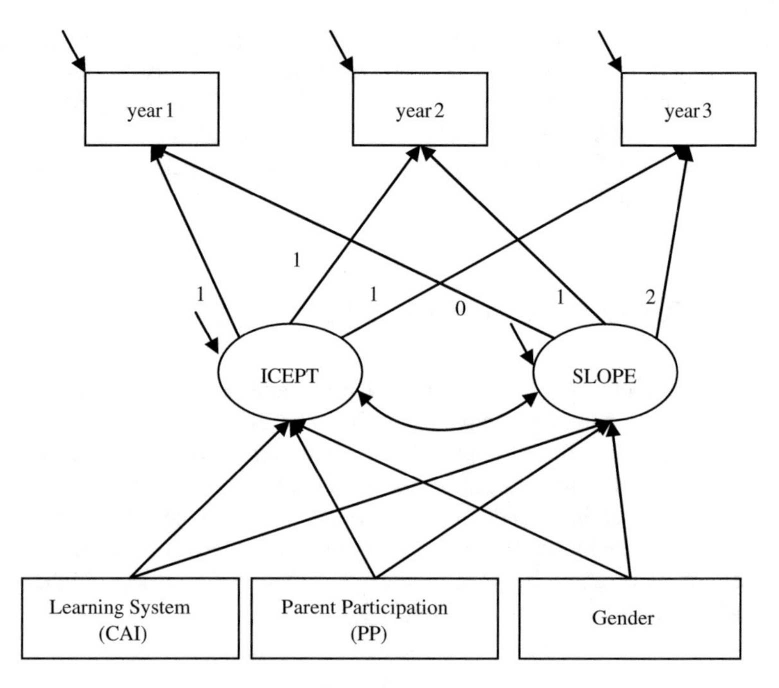

Fig. 2 The latent growth model of the effect of nutrition education system for elementary school students in nutrition knowledge

Learning system group with parent participation was the group (class) that applied the graphic presentation food safety and sanitation education system in the four-week food safety and sanitation education. Each parent was given a guide book and had the access to the feedback system for this graphic presentation food safety and sanitation education program. Students were required to do homework with the graphic presentation food safety and sanitation education system with their parents.

Control group was the group (class) that did the four-week food safety and sanitation education by traditional teaching method. This group did not involve the graphic presentation food safety and sanitation education system. Control group with parent participation was the group (class) that did the four-week food safety and sanitation education by the traditional teaching method. Each parent was given a guide book and the access to the feedback system. Students were required to do food safety and sanitation education homework by the traditional method with their parents.

The 3 exam scores (year1, year2, and year3) were obtained before the course, in the middle of the course, and at the end of the course. The questions on the three exams were based on the learning materials of the four-week food safety and sanitation education of health and physical education curriculum [12].

2.1 Research Data

124 students of an elementary school participated in this study. The data of this study was collected by Y.Y. Chu for the research: The Food Safety and Sanitation Education Information System (Research Project ID is NSC93-2516-S-034-001). This research was completed on July, 31, 2005. Data in this present study was obtained from Survey Research Data Archive (SRDA) [12] (Table 1).

There were four groups in this study: control group (learning system = 0, parent participation = 0), control group with parent participation (learning system = 0, parent participation = 1), learning system group (learning system = 1, parent participation = 0), and learning system group with parent participation (learning system = 1, parent participation = 1). The 3 exam scores on nutrition knowledge (year1, year2, and year3) were obtained at 3 time points before and after the learning system was used.

There were three extra variables in this study: learning system, parent participation, and gender, and they were all binary variables. Learning System = 1 means using the learning system; Learning System = 0 means not using the learning system. Parent Participation = 1 means learning with parent participation, and Parent Participation = 0 means learning without parent participation. Gender = 1 means male students, and Gender = 0 means female students.

There were a total of 59 female students (47.58 %) and 65 male students (52.42 %). There were 17 male students and 14 female students in the control

Table 1 Data summarize of the research

Geographic		Frequency	%
Gender	Female	59	47.58
	Male	65	52.42
Group	Control group	31	25.00
	Control group with parent participation	31	25.00
	Learning system group	31	25.00
	Learning system group with parent participation	31	25.00
Total		124	100.00

group; there were 16 male students and 15 female students in the control group with parent participation; there were 17 male students and 14 female students in the learning system group; and there were 16 male students and 15 female students in the learning system group with parent participation. Each group all had 31 students (Table 1).

2.2 *Data Analysis*

The data was analysed with latent growth modelling by using the Mplus 7 software and R 3.2.3. Latent growth modelling is a statistical technique used in the structural equation modelling framework to estimate growth trajectory. It is a longitudinal analysis technique to estimate growth over a period of time.

Latent Growth Model represents repeated measures of dependent variables as a function of time and other measures. The relative standing of an individual at a specific time point is modelled as a function of an underlying process, the parameter values of which vary randomly across individuals. Latent Growth Curve Methodology can be used to investigate systematic change, or growth, and inter individual variability in this change. A special topic of interest is the correlation among the growth parameters, the so-called initial status and growth rate, as well as their relation with time varying and time invariant covariates [13].

3 Results

The relationship between time points and the average of exam scores are shown in Table 2.

As for the model fit information of this study, Chi-Square Test of model fit was 57.296 (df = 4), RMSEA was 0.328, and CFI was 0.626. The estimate of the effect of learning system, parent participation, and gender were found in Table 3 and Table 4. The intercepts of the estimate of use of learning system, parent participation, and gender were not different. The 4 groups (control group, control group with parent participation, learning system group, and learning system group with parent participation) in this study were similar at the initial time point.

Table 2 Data summarize of the scores in three time points

	N	Mean	S.D.	Low	High
Time 1	124	11.63	1.07	9.00	14.00
Time 2	124	13.87	1.86	9.00	17.00
Time 3	124	13.80	1.91	10.00	17.00

Table 3 Fixed and random parameter estimates for latent growth curve of this study

		Estimate	S.E.	P-value
Intercept	Mean	3.060	0.166	<0.001
	Variance	0.134	0.208	0.519
Slope	Mean	0.177	0.098	0.007
	Variance	0.029	0.090	0.750

Table 4 The effect of learning system, parent participation, and gender

		Estimate	S.E.	P-value
Intercept	Learning system	0.183	0.169	0.279
	Parent participation	0.136	0.168	0.419
	Gender	0.010	0.162	0.953
Slope	Learning system	0.429	0.100	<0.001
	Parent participation	0.385	0.100	<0.001
	Gender	−0.060	0.096	0.534

The estimates of fixed and random parameter for latent growth curve of this study were displayed in Table 3. It can be found that the estimate of intercept mean and slope mean were 3.060 (p-value < 0.001) and 1.777 (p-value < 0.001). But the estimate of relationship between intercept and slope was −0.004, P-Value was 0.972. The relationship between intercept and slope was not significant. It means that the scores were higher than the previous time. Besides, no matter what the initial score is, students could all perform better in the following exams.

Table 4 and Fig. 3 shows that the relationship between the estimate of slope of learning system and parent participation was significant (P-value < 0.05). The relationship between gender and slope was not significant. The estimate of slope of

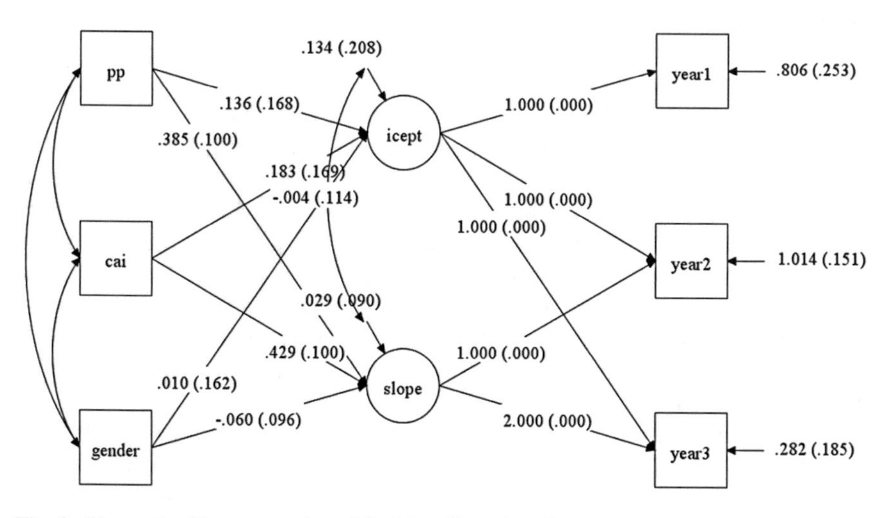

Fig. 3 The result of latent growth model of the effect of nutrition education system for elementary school students in nutrition knowledge

factor of learning system was 0.385, and the estimate of slope of factor of parent participation was 0.429. The growth rate of each exam time was positively correlated with the use of learning system and parent participation. The score of elementary school students that used food safety and sanitation education information system was higher than those who did not use food safety and sanitation education information system by 0.429 points in each growth time. The score of elementary school students with participating parents was higher than those without participating parents by 0.385 points in each growth time. Male elementary school students and female elementary school students were similar in slope (the growth rate between each time).

4 Conclusions

In recent years, computer applications have emerged as a viable means of gathering and disseminating food safety and sanitation information. Both stand-alone and on-line applications are being used to provide information on in the food safety and sanitation education for the public, paraprofessionals, and professionals. While the use of on-line communication applications such as multimedia and electronic discussion groups are just emerging as important tools among food safety and sanitation educators, the exponential growth of the Internet and the World Wide Web (WWW) are making these technologies more and more accessible. Food safety and sanitation educators are encouraged to explore the opportunities and cope with challenges of these new technologies to enhance their work.

The use of learning system was helpful for elementary school students to gain more nutrition knowledge, and this result was consistent with Brug et al. [5], Oenema et al. [6], Brug et al. [2], Park et al. [3], Tyrovolasa et al. [4]. Parent participation was helpful for elementary school students to learn more nutrition knowledge, and this result was the same as Niemeiera et al. [10] and Hopper et al. [11]. Gender made no difference in the result of nutrition knowledge acquisition. Besides, the initial score did not affect students' subsequent performances.

Finally, the learning of nutrition knowledge will be cultivated by both of the using of CAI (such as graphic presentation food safety and sanitation education system) and parent participation. Teachers could enhance the result of health education with computer learning system and parent support. CAI could be used as assistant teaching materials of food safety and sanitation education. Students could study with CAI on internet by themselves. CAI could reduce the time and expense of resource actual-food practicing. Furthermore CAI can also improve learning efficiency of school-children to achieve balance dietary behaviour.

It can be concluded that the acquisition of nutrition knowledge can be cultivated by both CAI (such as graphic presentation food safety and sanitation education system) and parent participation. Teachers can enhance the effectiveness of health education with computer learning system and parent support. CAI can supplement teaching materials of food safety, while parents' involvement can help students to achieve better in the learning system. To sum up, this paper has found that the use of CAI with parent participation can effectively improve the food safety and sanitation education in elementary school.

Acknowledgments This study is based in part on data from the Survey Research Data Archive (SRDA) provided by the Academia Sinica. The interpretation and conclusions contained herein do not represent those of Survey Research Data Archive (SRDA) or Academia Sinica.

References

1. Ubbes, V.A., Cottrell, R.R., Ausherman, J.A., Black, J.M., Wilson, P.C., Snider, J.: professional preparation of elementary teachers in ohio: status of k-6 health education. J. Sch. Health. **69**(1), 17–21 (1999)
2. Brug, J., Steenhuis, I., Assema, P.V., Vries, H.D.: The impact of a computer-tailored nutrition intervention. Prev. Med. **25**, 236–242 (1996)
3. Park, S.Y., Yang, Y.J., Kim, Y.: Effects of Nutrition education using a ubiquitous healthcare (u-health) service on metabolic syndrome in male workers. Korean J. Nutr. **44**(3), 231–242 (2011)
4. Tyrovolasa, S., Tountasb, Y., Polychronopoulosa, E., Panagiotakos, D.: A parametric model of the role of nutritional services within the health care system, in relation to cardiovascular disease risk among older individuals. Int. J. Cardiol. **155**(1), 110–114 (2012)
5. Brug, J., Oenema, A., Campbell, M.: Past, present, and future of computer-tailored nutrition education. Am. J. Clin. Nutr. **77**(4), 1028–1034 (2003)
6. Oenema, A., Brug, J., Lechner, L.: Web-based tailored nutrition education: results of a randomized controlled trial. Health Educ. Res. **16**(6), 647–660 (2001)
7. Kolasa, K.M., Miller, M.G.: New developments in nutrition education using computer technology. J. Nutr. Educ. **28**(1), 7–14 (1996)
8. Bensley, R.J., Anderson, J.V., Brusk, J.J., Mercer, N., Rivas, J.: Impact of internet vs traditional special supplemental nutrition program for women, infants, and children nutrition education on fruit and vegetable intake. J. Am. Diet. Assoc. **111**(5), 749–755 (2011)
9. Cronin, F.J., Shaw, A.M., Krebs-Smith, S.M., Marsland, P.M., Light, L.: Developing a food guidance system to implement the dietary guidelines. J. Nutr. Educ. **19**(6), 281–302 (1987)
10. Niemeiera, B.S., Hektnerb, J.M., Enger, K.B.: Parent participation in weight-related health interventions for children and adolescents: a systematic review and meta-analysis. Prev. Med. **55**(1), 3–13 (2012)
11. Hopper, C.A., Munoz, K.D., Gruber, M.B., MacConnie, S., Schonfeldt, B.S., Shunk, T.: A school-based cardiovascular exercise and nutrition program with parent participation: an evaluation study. Child. Health Care. **25**(3), 221–235 (1996)
12. Chu, Y.Y.: The graphic presentation nutrition education informatics network. The Survey Research Data Archive. https://srda.sinica.edu.tw/search/gensciitem/1025 (2007). Accessed 18 July 2012
13. Preacher, K.J., Wichman, A.J., MacCallum, R.C., Briggs, N.E.: Latent growth curve modeling. Struct. Equ. Model. Multi. J. **19**(1), 152–155 (2012)

MATP: A Multi-agent Model for the University Timetabling Problem

Houssem Eddine Nouri and Olfa Belkahla Driss

Abstract This paper proposes a multi-agent model for solving the university course timetabling problem. It is composed of cooperating agents enabling highly distributed processing of the problem and incorporating constraints that have not been considered by previous works. The aim of our model is to provide a best solution satisfying hard and soft constraints while reducing temporal complexity. To analyze the efficiency of our model, we give experimental results based on real instances of the Higher Business School of Tunis by analyzing the variation effect of the lecture and teacher numbers on the messages number and the CPU execution time, and the variation effect of the assignment priority score on the percentage of teacher's preferences satisfaction.

Keywords University course timetabling problem · Multi-agent system · Negotiation · Messaging exchange system

1 Introduction

The timetabling problem is an instance of the personal scheduling problems which has become more diffused in our real life. It is well known as an NP-complete problem. This problem is pervasive in all practical aspects of modern society. It plays a very important role in many types of organizations such as hospitals, transport companies, protection services and emergency and universities. In our case, we focus more precisely on the problem of the university timetabling problem.

H.E. Nouri (✉) · O.B. Driss
Stratégies d'Optimisation et Informatique intelligentE,
Institut Supérieur de Gestion de Tunis, Université de Tunis,
41, Avenue de La Liberté, Cité Bouchoucha, Bardo, Tunis, Tunisia
e-mail: houssemeddine.nouri@gmail.com

O.B. Driss
e-mail: olfa.belkahla@isg.rnu.tn

R. Silhavy et al. (eds.), *Software Engineering Perspectives and Application in Intelligent Systems*, Advances in Intelligent Systems and Computing 465,
DOI 10.1007/978-3-319-33622-0_2

Burke and his colleagues [5] note in this regard that this problem can be divided into two main categories: courses and exams. Different aspects separate these two categories. For example, we try to group the courses, but we prefer to move away exams from each other as possible. Or again, a course may take place at a given time in one classroom, while many exams may take place at the same time in the same classroom, or the same exam can be dispatched in many classrooms.

In this paper, we are interested to solve the university course timetabling problem. It can be defined as a set of university courses which take place throughout specific periods for five or six days in a week, directed by a limited number of teachers and classrooms requiring a better management in order to contain the large number of the registered students. Our aim is to get a best solution for this problem satisfying several hard and soft constraints while minimizing the temporal complexity.

This paper is organized as follows. In Sect. 2, we present how the university course timetabling problem is solved by previous works as well as its hard and soft constraints. We detail then in Sect. 3 our contribution based on multi-agent systems. Section 4 is devoted to the presentation of a real case study (instances of the Higher Business School of Tunis) in order to test our model as well as a set of scenarios evaluating its efficiency.

2 University Course Timetabling Problem

Many researchers are facing this problem from several points of view and with different approaches using different paradigms of resolution. The first attempts of resolution methods were based on the theory of graphs [6, 12], the integer linear programming [8] and the techniques of constraint satisfaction problem [1, 13, 15]. However, these methods have not given a solution dealing with all instances and constraints of this problem. That's why, they have given a way to other types of methods adapted to this type of problem, namely meta-heuristics such as the tabu search [16], the simulated annealing [7] and the genetic algorithms [2]. This family of approximate search has mechanisms that allow a good general investigation of the search space. But generally, it is nondeterministic and gives no guarantee of optimality. This has allowed the appearance of new approaches based on the multi-agent systems [3, 10–11, 14], but they did not succeed to well adapt this formalism to generate a solution satisfying all the problem constraints. That's why, and in this area we have proposed a new multi-agent model allowing to minimize the time complexity, to introduce new details that have not been taken into account by previous work and to attend a good satisfaction of the teachers preferences.

In order to get a best solution for this problem, we must take into account all the constraints of the problem that should be satisfied. These constraints are often classified into two categories, the first includes hard constraints and the second category includes constraints often called soft constraints:

Hard constraints: must be satisfied in any environment, because the violation of these constraints may cause the generation of an unsatisfiable solution:

- Two lectures cannot be programmed in the same classroom and at the same period of time,
- The lectures given by the same teacher cannot be programmed at the same period of time,
- A classroom can be assigned only to one lecture at the same period of time,
- A group lecture cannot takes place at the same period with another that is not a group lecture belonging to the same level of study,
- The number of students must be less than or equal to the capacity of the assigned classroom.

Soft constraints: the violation of these constraints has no effect on the generation of a satisfiable solution:

- The assignment of classrooms and periods of time must allow to satisfy at best the preferences of teachers,
- The assignment of classrooms to the different lectures must allow to satisfy at best some preferences.

In this work, we propose a multi-agent model based on cooperative agents, that we have named MATP a Multi-Agent model for university Timetabling Problem, enabling highly parallel and distributed processing of the problem. Our model incorporates several constraints that have not been taken into account by previous works.

Multi-Agent systems (M.A.S) are chosen because of their advantages in many different domains by means of the cooperation between a society of agents. In fact, each agent, concurrently and asynchronously, acquires information from its environment and from other agents to reason on and to act consequently, see the studies of [9].

3 Multi-agent Model for University Timetabling Problem

3.1 Agent Identification

We have equipped our multi-agent model MATP, see Fig. 1, with three classes of agents. The first class is composed of agents that we have named TA, "Teacher Agents", divided into three categories of teachers: *C1*: Professor, Associate-professor; *C2*: Assistant-professor, Assistant; *C3*: Contractual. The second class is composed of agents that we called CA, "Classroom Agents", divided into three types of class-rooms ("Course", "Tutorial Class", 'Practical Class",) related to the type of the lecture session. The third class contains three agents: two "Interface Agents" that we called IA1 and IA2 and one "History Agent" that we named HA.

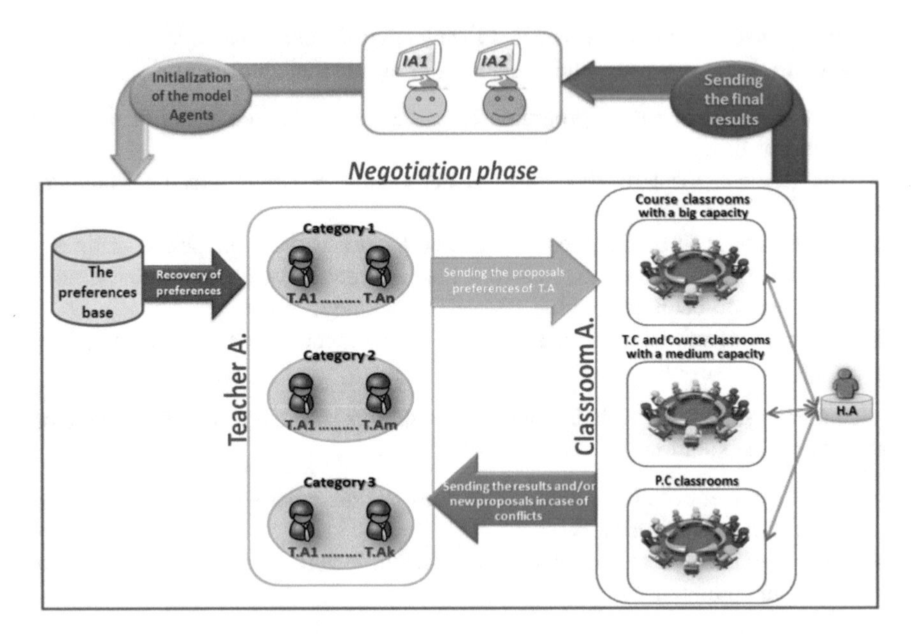

Fig. 1 Multi-agent model for university timetabling problem

3.2 Global Steps of MATP

The steps of our model MATP proceed in three phases detailed below: initialization, negotiation and transmission of final results.

Initialization phase. In this phase, we present the role of the agent IA1 which initializes the execution of the system agents. In fact, it allows the implementation of all agents based on the initial parameters fixed at the start by the user.

Negotiation phase. This phase is the kernel of our model. It is based on a messaging exchange system between the two agent's classes TA and CA in order to have in each case an agreement between them, respecting all the hard constraints of this problem. The first class of agents TA starts the negotiation process by sending all their allocation propositions (which were recovered from their preferences base) to the CA agents in order to get a better reservation of the most suitable classrooms and the most favorite time periods of the day.

The second class of agents CA will receive and analyze the TA agent's preferences. In fact, this class will ask the HA agent to verify the existence of duplication of time periods for a same TA agent in each reception of propositions. Thus, it allows either to validate, or to give a new proposition in the case of conflict. The CA may have 1 or n TA propositions asking the same period in the same day, and generating conflicts between them, see Fig. 2. That's why, we have added a

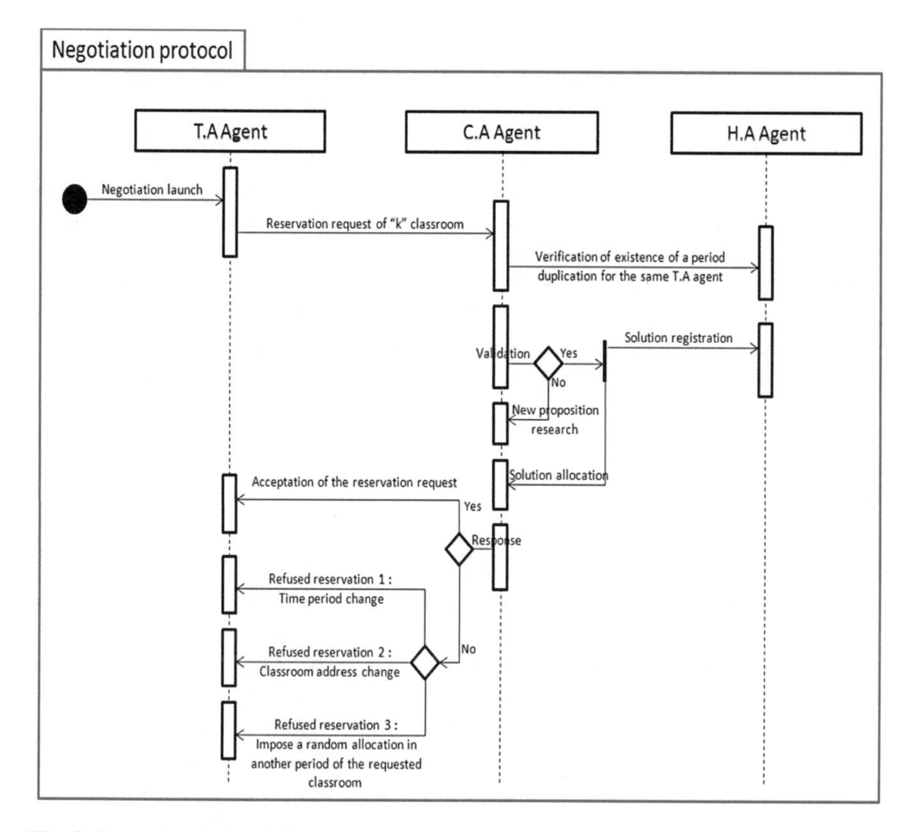

Fig. 2 Interaction protocol diagram

new hypothesis in which a classroom can be replaced by another one having the same characteristics, that we called the equivalence of classrooms (or vertical search) for the three categories of teacher agents. So we used a **V**ertical assignment priority **S**core ***VSi*** affected to each ***i*** category of teacher, where $i \in \{1, 2, 3\}$. This score ***VSi*** is incremental from zero to a maximum value ***VSimax***, where ***VSi*** $\in$ [0, ***VSimax***].

VSimax is the maximum value given by the user for this score that may have a TA agent with a category ***i*** where:

- Priority 1, ***VS1max*** (Rank of teacher): This score is given for each agent belonging to the first category of teacher agents TA having a rank of "Professor" or "Associate-professor".
- Priority 2, ***VS2max*** ("Course"): This score is given for each agent belonging to the second category of teacher agents TA and asking a lecture session with type "course".

- Priority 3, ***VS3max*** ("TC" or "PC"): This score is given for each agent belonging to the third category of teacher agents TA and asking a lecture session with type "TC" (Tutorial Class) or "PC" (Practical Class).

Also, by integrating many types of criteria for acceptance of a reservation (capacity of students for each lecture session, the teacher's category, type of classroom to be reserved and type of lecture session), we will have a decrease in the percentage of appearance of conflicts between TA agents.

Transmission of final results. Whenever a TA agent receives all solutions in response to its messages, it finishes its negotiation phase and transmits its final results to IA1 agent generating the form of teacher's timetable. Then the agent IA2 ends the process by generating the final timetable of the different classrooms.

3.3 Agent Behaviour

Interface agent behaviour. The behaviour of the IA1 is to initialize all the other agents of our model. Then, it moves to an inactive state pending the reception of the final TA agent messages to generate them in the form of a solution for the teacher timetabling problem. For the IA2 behaviour, this latter has to generate a solution for the classroom timetabling problem after the end of the negotiation process.

Teacher agent behaviour. A TA agent possesses a group of lectures (which can be a course, TD or TP) that it seeks to assign them to classrooms in the most favourite periods of the day. In fact, each TA begins its negotiation phase by sending its proposals to CA agents requesting the most preferred classrooms and teaching periods. Then, he receives a response message from CA:

- Verification of reservation request:

```
Begin
If Response = "Accepted"
  Then Display a message of acceptance.
Else-If A refusal reservation message 1, 2 or 3 has been
received.
  Then Send another message to C.A containing the new
  update of the assignment propositions.
End.
```

Classroom agents behaviour. A CA agent contains an array of periods to search solutions for the requested periods. Thus, this type of agent is composed of a set of rules for the negotiation management:

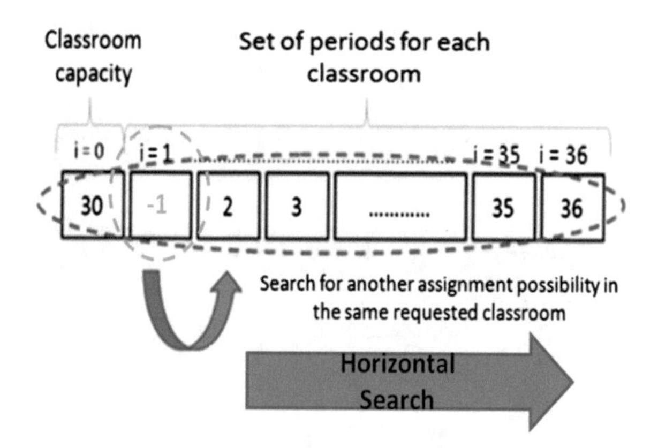

Fig. 3 Vertical assignment search

- Verification of duplication part:

```
Begin
If The requested period hasn't been duplicated for the
same TA agent.
   Then Go to the validation step.
   Else Change the requested period and send a message of
   a refused reservation 1 to TA agent.
End.
```

- Validation of proposition part (Figs 3 and 4):

```
Begin
If The requested period hasn't been reserved.
   Then Send an acceptance message of the requested period
   to the T.A agent and record the solution in the memory
   of the H.A agent.
Else-If The requested period hasn't been reserved and
VSi <= VSimax.
   Then Vertical assignment search: change the requested
   classroom address, increment the assignment priority
   score and send a message of a refused reservation 2 to
   TA agent, see Figure 3.
Else-If The requested period has been reserved and
VSi > VSimax.
   Then Horizontal assignment search: impose a random as-
   signment in another available period of the requested
   classroom and send a message of a refused reservation 3
   to TA agent, see Figure 4.
End.
```

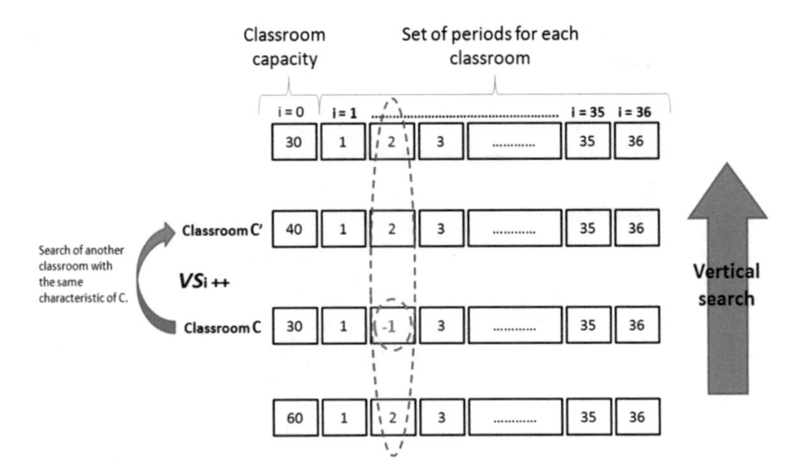

Fig. 4 Horizontal assignment search

4 Experimentation

4.1 A Case Study

To test our approach, we have chosen to conduct our study on a real case where we used data instances of the Higher Business School of Tunis.

Number and types of teachers.

- 5 Professors: 10 lecture sessions per week.
- 6 Associate-professors: 18 lecture sessions per week.
- 30 Assistant-professors: 120 lecture sessions per week.
- 40 Assistants: 320 lecture sessions per week.
- 50 Contractuals: 400 lecture sessions per week.

  - Total number of teachers = 131 teachers.
  - Total number of lecture sessions = 868 sessions.

Number of classrooms. For the teaching classrooms, we have 64 classrooms belonging to 5 blocks A, B, C, D, and I of building, that we have chosen to group them into three categories: Category 1: all course classrooms having a big capacity of students (A1, A2, A3, B5, B6). Category 2: all TC and course classrooms having a medium capacity of students (B1, B2, B3, B4, D1…D24, C2…C11, I8… I21). Category 3: all PC classrooms (I1, I2, I3, I4, I5, I6, I7).

Specialities and education levels. The school offers 25 specialities divided into 5 education levels: 1st year license, 2nd year license, 3rd year license, master M1, master M2.

4.2 Experimental Design

Furthermore, we have chosen to use the famous multi-agent platform Jade [4] to implement our model agents. Our choice was motivated by the benefits presenting this platform. For the development of our model, we have chosen to use the object-oriented programming language Java with the Eclipse Helios IDE. This choice was imposed because the different agents in our system are implemented on the Jade multi-agent platform and this latter has been entirely developed in Java.

4.3 Experimental Results

To evaluate the efficiency of our model, we realized a two test scenarios, where we have analyzed:

- The variation effect of the lecture and teacher numbers on the message number and the CPU execution time.
- The variation effect of the assignment priority score on the percentage of Teachers preferences satisfaction.

The variation effect of the lecture and teacher numbers on the message number and the CPU execution time. For this first test scenario, we analyzed the effect of varying the number of lectures and teachers on the messages number and the execution time. In fact, we realized 4 tests where we incremented in each case the lectures number as well as the teachers while fixing the assignment priority score (SV1 = 4, SV2 = 5, SV3 = 5: random choice) for the three TA agents categories until to the end of those tests.

According to the results, see Table 1, we can visualize the rapid increase of the message number and this is due to the increment of the lecture number for each test case, requiring more communication flows between agents to satisfy all the reservation requests in the negotiation phase. In other hand, the CPU time has known a small increase in each test, but its variation has always remained in second by 5.688 (for 540 lectures) to 8.531 (for 868 lectures) with an average of 6.976 s for this bound.

Table 1 Variation effect of the lecture and teacher number on the message number

Scenario 1	Lecture number	Teacher number	Assignment percentage (teacher—lecture—classroom) (%)	Message number	CPU time (s)
Test 1	540	85	100	2732	5.688
Test 2	660	100	100	3438	6.547
Test 3	728	111	100	3880	7.14
Test 4	868	131	100	4584	8.531

Table 2 Variation effect of the priority score VSi on the teachers preferences satisfaction

Scenario 2.1	VS1	VS2	VS3	Message number	CPU time (s)	Teachers preferences satisfaction in %
Test 1	2	3	3	4114	8.265	64.97
Test 2	4	5	5	4483	8.531	67.51
Test 3	4	7	7	4746	8.688	71.19
Test 4	4	10	10	5238	9.078	72.46

Table 3 Percentage of preferences satisfaction by category

Scenario 2.2	Preferences satisfaction in % (category 1)	Preferences satisfaction in % (category 2)	Preferences satisfaction in % (category 3)	Teachers preferences satisfaction in %
Test 1	100	77.27	49	64.97
Test 2	100	77.27	54.50	67.51
Test 3	100	77.80	62	71.19
Test 4	100	78.18	64.25	72.46

Moreover, the allocation percentage of the three resources "teacher—lecture—classroom" reached the 100 % for the 4 test cases. In fact, the execution process of our allocation algorithm between agents (TA, CA, HA) cannot be stopped only after a total assignment of all the "teacher—lecture" combinations to the different classrooms.

The variation effect of the assignment priority score on the percentage of Teachers preferences satisfaction. For this second test scenario, we chose to analyze the variation of the percentage of teachers preferences satisfaction by increasing in each time the assignment priority score for each category of Teacher Agents and fixing the lectures number to 868 and teachers to 131.

In view of these results, see Table 2, we distinguished that the first category of teachers has had the largest percentage of preferences satisfaction compared to the other categories, see Table 3. In fact, it knew a total satisfaction of all its allocation needs counting a percentage of 100 % from the first test case with a score of 2 assignment possibilities, this percentage remained unchanged until the last test.

The second category has had a percentage of 77.27 % with 3 assignment possibilities for the first test. This percentage has remained unchanged in the second test and then it knew a small improvement of 0.45 % by increasing its score to 7 assignment possibilities in the third test. Passing to the last test, it has been improved again by 0.46 % by increasing its score to 10 possibilities.

The third category has had a percentage of 49 % with 3 assignment possibilities for the first test and then it knew a small improvement of 5.5 % by increasing its score to 5 assignment possibilities. And this percentage was changed again in the third test by modifying the assignment score to 7 possibilities with an increase of 7.5 %. Passing to the last test, it knew another improvement of 2.25 % by increasing its score to 10 possibilities.

From these interpretations, we can conclude that the variation of the percentage of the preferences satisfaction for the three teacher's categories strongly depends on the assignment priority score SVi. In fact, see Table 3, the increase of the SVi will generate an improvement of the percentage of the teacher's preferences satisfaction by giving new assignment possibilities to other classrooms having the same characteristics of the requested one.

5 Conclusion and Perspectives

This paper proposed a multi-agent model based on cooperative agents, named MATP (Multi-Agent model for university Timetabling Problem), to solve the university course timetabling problem, enabling highly parallel and distributed processing of this problem and incorporating new details that have not been considered by previous work. The performances of the proposed model are exhibited through experimental results based on a real case study (instance of the higher business school of Tunis) by analyzing the variation effect of the lectures and teachers number on the messages number and the CPU execution time and the variation effect of the assignment priority score on the percentage of teachers preferences satisfaction.

Despite the encouraging results, some improvements are still possible. We can improve the intelligence level and the individual learning of our system agents, in such way that agents have a deliberative behavior such as B.D.I (Belief Desire Intention). Moreover, this model can be adapted to solve other forms of the personnel timetabling problem (hospitals, protection services and emergency, etc.) in the future works.

References

1. Abbas, A., Tsang, E.P.K.: Software engineering aspects of constraint based timetabling: a case study. Inform. Softw. Technol. J. **46**, 359–372 (2004)
2. Adewumi, A.O., Sawyerr, B.A., Ali, M.M.: A heuristic solution to the university timetabling problem. Eng. Comput.: Int. J. Comput.-Aided Eng. Softw. **26**(8), 972–984 (2008)
3. Babkin, E., Adbulrab, H., Babkina, T.: AgentTime: a distributed multi-agent software system for university's timetabling. In: Proceedings of the 4th European Conference on Complex Systems, a satellite Conference Emergent Properties in Natural and Artificial Complex Systems, pp. 10–22 (2007)
4. Bellifemine, F., Poggi, A., Rimassa, G.: JADE—A FIPA-compliant agent framework. In: Proceedings of the 4th International Conference and Exhibition on the Practical Application of Intelligent Agents and Multi-Agent Technology, pp. 97–108 (1999)
5. Burke, E.K., Petrovic, S.: Recent research directions in automated timetabling. Eur. J. Oper. Res. **140**(2), 266–280 (2002)
6. Burke, E.K., Marecek, J., Parkes, A.J., Rudová, H.: A supernodal formulation of vertex coloring with applications in course timetabling. Ann. Oper. Res. **179**, 105–130 (2010)

7. Ceschia, S., Di Gaspero, L., Schaerf, A.: Design, engineering, and experimental analysis of a simulated annealing approach to the post-enrolment course timetabling problem. Comput. Oper. Res. **39**, 1615–1624 (2011)
8. Daskalaki, S., Birbas, T., Housos, E.: An integer programming formulation for a case study in university timetabling. Eur. J. Oper. Res. **153**, 117–135 (2004)
9. Ferber, J.: Multi-agent Systems—An Introduction to Distributed Artificial Intelligence. Addison- Wesley (1999). ISBN: 0-201-36048-9
10. Henry-Obit, J., Landa-Silva, D., Ouelhadj, D., Khan-Vun, T., Rayner, A.: Designing a Multi-agent approach system for distributed course timetabling. In: Proceedings of the Hybrid Intelligent Systems Conference, pp. 103–108. IEEE Press (2011)
11. Oprea, M.: Multi-Agent system for university course timetable scheduling. In: The 1st International Conference on Virtual Learning, pp. 231–237 (2006)
12. Redl, T.A.: University timetabling via graph coloring: an alternative approach. Congr. Numer. **187**, 174–186 (2007)
13. Sheaufen, I.H., Safaai, D., Siti Zaiton, M.H.: Investigating constraint-based reasoning for university timetabling problem. In: Proceedings of the International Multi-Conference of Engineers and Computer Scientists vol. 1, pp. 139–143 (2009)
14. Xiang, Y., Zhang, W.: Distributed university timetabling with multiply sectioned constraint networks. In: the Twenty-First International FLAIRS Conference, pp. 567–571 (2008)
15. Zhang, L., Lau, S.: Constructing university timetable using constraint satisfaction programming approach. In: Proceedings of the International Conference on Computational Intelligence for Modelling, Control and Automation and International Conference on Intelligent Agents, Web Technologies and Internet Commerce, vol. 2, pp. 55–60. IEEE Press (2005)
16. Zhipeng, L., Hao, J.K.: Adaptive tabu search for course timetabling. Eur. J. Oper. Res. **200**, 235–244 (2008)

Optimized Clustering with Statistical-Based Local Model for Replica Management in DDM over Grid

M. Shahina Parveen and G. Narsimha

Abstract With the increasing complexity of the forms of the data (unstructured, massive, real-time, and heterogeneous), Distributed Data mining (DDM) approaches encounters a significant problem over grid infrastructure. The paper has identified a challenging problem i.e. an effective replica management which cost maximum resources to process the data from the warehouse in distributed data mining. The proposed system introduces a technique which presents a unique clustering mechanism of the data extracted from the replica (warehouse) and applies a novel statistical-based local model in order to extract the non-repetitive and unique data for accomplishing faster response time during distributed data mining. Powered by optimization using genetic algorithm, the proposed system offers better response time with increasing traffic load as compared to the similar existing technique of distributed data mining.

Keywords Clustering · Grid infrastructure · Distributed data mining · Response time · Genetic algorithm

1 Introduction

This advancement in communication technology has truly make the world a global village and has affected almost in any sector viz. education, healthcare, industries, enterprises, defense, meteorology, social network, entertainment etc. 30–40 years back, the data used to store in a standalone physical server, where a typical server-client relationship was used. However, with the advancement of database management system and ubiquitous computing, cloud computing has solved this

M. Shahina Parveen (✉)
JNTU, Hyderabad, India
e-mail: shahina1phd.jntu@gmail.com; shahinaparveenm@gmail.com

G. Narsimha
JNTUH, Kondagattu, Jagtial, Karimnagar, Telangana, India

R. Silhavy et al. (eds.), *Software Engineering Perspectives and Application in Intelligent Systems*, Advances in Intelligent Systems and Computing 465,
DOI 10.1007/978-3-319-33622-0_3

problem [1]. Using cloud, the data is made available to user at any point of time and place. Unfortunately, cloud offers a best and cost effective storage for any size of data, but question lies here—what to do with this data. The best answer to this question is data mining, which is a technique to extract a unique pattern of information hidden in the massive set of data [2]. Conventional data mining problem is also associated with some of the challenges e.g. clustering, classification, prediction, learning techniques etc. [3]. Such problems also lie in any sophisticated machine learning approach too. The process of conventional data mining technique initiates by exploring the source data and then extracting the data points, which are required to be evaluated and analyzed. After pulling out the pertinent data, the significant step in data mining is to find out the key value from the already pulled out set of data. The final step is to perform interpretation of the data [4]. But technology of the data management was subjected to certain amendments based on dynamic needs of our customers and users. The customer demands higher availability of data, virtual platforms, no downtime, no delay, better throughout etc. This could be only achieved if the data are stored and retrieved in highly distributed manner [5]. Now here comes the most confusive part to manage such forms of data i.e. distributed computing or grid computing. Interestingly distributed computing is all about managing more number of machines with lower computational capability. However, grid computing does the same thing with additional capabilities e.g. exploiting resource utilization of heterogeneous system, manages workloads, etc. The positive fact about grid infrastructure is its capability to execute on numerous domains of administration and is more inclined towards optimization technique that is missing is distributed computing [6]. Hence, there is always a difference between carrying out data mining operation using distributed computing or grid computing, whereas the suitability of grid infrastructure hold more appropriate for distributed data mining. The input to distributed data mining is various forms of warehouses that already have historical data. Now re-performing mining operation on them is equivalent to performing optimization of the existing data mining technique to make it represent more like the distributed. However, the process is not that easy as it seems like. At present the morphology of the data has entirely taken a shape of heterogeneity, unstructured, semi-structured, high-dimensional, etc. These all cases of complex and massive streams of data render conventional data mining technique ineffective. The problem thereby becomes stronger when it comes to grid infrastructure. The area of data mining has already gained enough paces in the area of research; however, distributed data mining over grid infrastructure is one of the less visited topics among the research communities globally. The proposed system introduces a technique which presents a unique clustering mechanism of the data extracted from the replica Sect. 2 discusses about the related work and problem identification has been discussed in Sect. 3. The discussion of proposed system has been discussed in Sect. 4. Section. Research Methodology has been presented in Sect. 5 and Implementation has been illustrated in Sect. 6. Discussion of research gap is made in Sects. 7 and 8 summarizes work.

2 Related Work

Lackovic et al. [7] have used Weka4WS architecture as well as service-oriented architecture for carrying out distributed data mining technique. Brescia et al. [8] have discussed about a project called as DAME or Data Mining Exploration which mainly targets to develop a distributed grid infrastructure. Hmida and Slimani [9] have presented as Weka4GML architecture for performing distributed data mining. Kantarcioglu and Nix [10] have applied game theory along with Vickrey-Clarke-Groves process for the purpose of validating the resultant data of distributed data mining.

Oyana [11] have presented a new technique of clustering for distributed data. Experimented over synthetic and real data, the presented technique shows an efficient query processing. The outcome of the study was evaluated with respect to mean squared error and response time on increasing percentage of data. Rao and Vidyavathi [12] have used multi-agent approach to carry out distributed data mining using game theory. The outcome of the study was evaluated using gain as performance parameter. Tlili and Slimani [13] have used association rule to perform distributed data mining as well as dynamic load balancing. However, the study is more inclined for load balancing. Prusiewicz and Zieba [14] have designed a data mining technique using service-oriented architecture and semantics. Santos et al. [15] have developed a distributed data mining considering a case study of healthcare sector. Mallik et al. [16] have presented an analysis for usage of asynchronous algorithm. Usage of multi-agent system was also found in the work of Pandey et al. [17], Prajapati and Menaria [18]. Zhang et al. [19] have presented an efficient learning process for enhancing distributed data mining. The authors have used Big Data approach for optimizing the accuracy of learning. Belbachir et al. [20] have presented a sequential algorithm for generating association rule. Vishvapathi et al. [21] have developed a distributed mining algorithm for grid system using supervised learning algorithm using Weka tool.

A unique form of study is carried out by Amini et al. [22] where the authors have used density-based clustering process over heterogeneous data of Internet-of-Things. The outcome is faster processing time with data quality. Maab et al. [23] have presented a study for processing data from smart grid using Big Data Analytics. Ogunde et al. [24] have applied association rule mining that suits better in distributed data storage. Rebbah et al. [25] have presented a technique for extracting association rule pertaining to grid computing. Srinivasan and Palanisamy [26] have used swarm intelligence-based concept to perform optimization of clustering process in high-dimensional data. The study outcomes were tested with respect to outliers. Zhou [27] has presented a data mining approach over cloud platform. The author has presented a simple mathematical modelling of establishing linear and non-linear relationship. The outcome of the study was shown to enhance the performance of distributed data mining with respect to running time, memory occupancy, and average clustering quality.

Hence, it can be seen that there are various approaches being presented by the researchers in the last decade out of which the multi-agent approach, game theory, service-oriented architecture, usage of open source machine learning and data mining e.g. Weka, Weka4WS, etc. are quite frequently and repetitively exercised. All the techniques have their own advantage as well as limitation.

3 Problem Description

The complications of the data mining process are directly proportional to dimensionality of the data. Higher the dimensionality of the data bigger is the complexity in analyzing data. One of the biggest challenges in implementing distributed data mining over grid is to design an integrated hypothesis of data mining algorithm. As the sources and types of data warehouse differs, so it is quite common the data mining algorithms will be majorly heterogeneous in nature over different data warehouse. This results in generations of knowledge to local model with difference in its values. Exploring the error in the knowledge generated from local data mining algorithm from aggregation view is one of the most challenging tasks over grid infrastructure. Another significant problem is the streaming of high speed data resulting to complex high dimensional data. Problems also lie in extracting knowledge from time-series data, sequenced data, unstructured data, and semi-structured data. As the data is generated from multiple data warehouses, the security protocols may be different on specific warehouses that significant result in delay in authentication and authorization problems for real-time data. Heterogeneous nature of data even from same source will pose a significant problem in implementing distributed data mining. Although distributed data mining process has the significant potential to perform knowledge discovery over grid, but it is not that easy to accomplish a precise amount of knowledge is a lesser duration of time with high data quality.

4 Proposed System

The prime purpose of the proposed system is to design a unique framework of clustering data from the warehouse in order to perform an efficient replica management in order to enhance the data quality. We define data quality as the amount of data which has less amount of repetitive information in order to save considerable computational capability. The schema of the proposed study is highlighted in Fig. 1.

Figure 1 shows that input to the proposed system is basically the historical data from the warehouses. The figure shows the extraction process of data from warehouse with presence of redundant data. Hence, if the quality of the data from warehouse is enhanced even before applying distributed data mining over the grid clusters than the post result of knowledge discovery will yield much superior outcome. Basically the proposed clustering mechanism significantly assists in identification of actual source of processed data that should be selected for distributed data mining process in grid infrastructure.

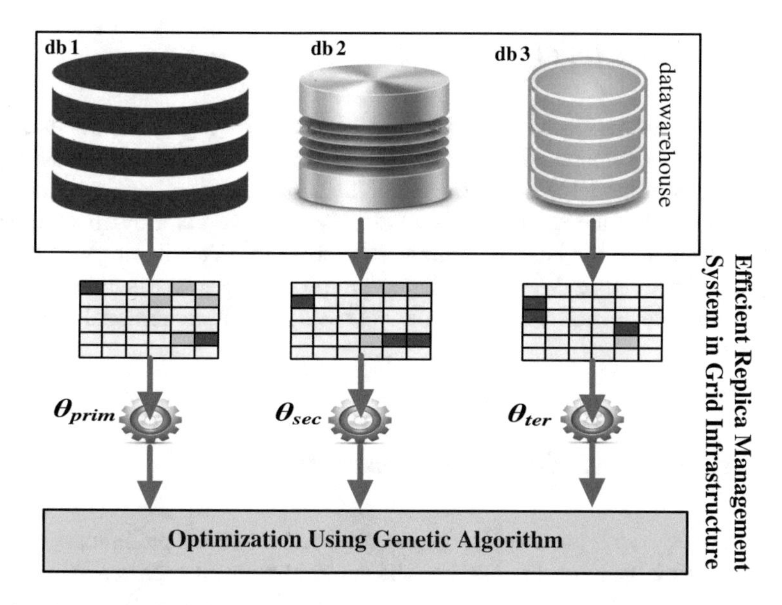

Fig. 1 Schema of proposed replica management system

The essential purpose is to enhance the data quality using novel replica management system. The contributions of the proposed system are as follows:

- To develop a novel clustering algorithm that can perform effective management of replica in distributed data storage system.
- To apply a statistical approach for identifying the extent of data replica using micro analysis (data vector) and select the data source with most unique data.
- To develop a cost-effective optimization using genetic algorithm for generating elite outcomes of clustering and best source of warehouse.

5 Research Methodology

The proposed system considers analytical research methodology for developing the proposed data mining algorithm. The proposed system mainly adopts statistical approach in a very unique fashion for introducing a framework that can perform an effective replica management for dynamic stream of high-dimensional data. The proposed system performs selection of clustered data from the replica and subjects it to three different empirical formulas for local model. The system also undergoes an optimization process using genetic algorithm for generating the best value of local model that indirectly states which replica corresponding to warehouse source should be accepted or rejected.

5.1 Processing Input

The input for the proposed system is considered to be highly unstructured and high-dimensional synthetic data. The proposed system considers that all the inputs towards the local model aggregation should be unique data and non-repeated. It is because forwarding the replicated data not only consume extra overhead towards network communication but also degrades the distributed data mining performance over grid infrastructure. We apply simple pre-processing to our synthetic data in order to eliminate noise and represent our data using a high-dimensional vector dv1, dv2… dvn.

5.2 Replica Clustering Using Local Mode

The system selects a replica randomly and considers it to be as preliminary centroid in the clustering plane of data points. The centroid point of each replica vector is essentially compared with the three local models using statistical theory. Adoption of probability theory is done for two reason (i) easier generation of inference of outcomes and (ii) faster computation. As we have developed recently three modes of local model (i.e. primary, secondary, and tertiary) for high-dimensional data, it is required that proposed comparative and search for non-redundant replica is computationally faster and leads to best and error free outcomes. In order to fasten the clustering process, we apply genetic algorithm. The outcomes will highly the best local model to recognize the best variant of replica. This replica can be further processed forward to be subjected to futuristic distributed data mining technique.

5.3 Replica Selection Process

The selection of the replica is strictly done based on the optimization done using genetic algorithm. The proposed system performs clustering of the data being retrieved from replica as a part of proposed replica management during data retrieval stage during analytics. However, even after clustering, it is not feasible to understand which the best data source is. It can be rather said that if there is a need of considering all the data sources or only few data sources which posse's non-redundant data. This is because presence of more number of replicated data will result in more data storage and more memory consumption. Hence, we apply three types of local models based on statistical approach, which can be extended to any number less than total number of data sources.

6 Algorithm Implementation

A laboratory prototype has been developed for the proposed system considering 12 machines of 32 bit and high configurations to formulate a grid infrastructure. We have also developed own synthetic dataset which is highly unstructured in size. As the data is synthetic, so it was essential to perform certain preprocessing. The core algorithm used for the proposed system is showcased below:-

---

Algorithm for Clustering Replica in Grid Infrastructure

Input: Data replica (d_1, d_2, ..d_n), n

Output: Best clustering value of replica

Start

1. Init D={d_1, d_2, d_3, ….d_n}
2. Apply $P_x(D)$
3. Apply local model

$$\sum_{i=1}^{m} L_{m_1} \Rightarrow \sum_{j=1}^{n} P_x(D_j) \forall m \subseteq n, m \leq n$$

4. $\sigma = 0$ // iteration
5. init pop
6. Evaluate *f*(pop) // fitness function
7. **While not** *termination condition* **do**
8. $\sigma = \sigma + 1$
9. Select p_1 and p_2 from pop
10. **begin** crossover operator
11. O^{best}=(O_1, O_2, …..O_n)
11. **end**
12. **begin** mutation operator with O^{best}
13. δ^{best}=(δ_1, δ_2, …..)
14. insert O^{best} in pop
15. **if** r< δ^{best}
16. $\arg_{max}$(δ_1, δ_2, ….)→best value
17. **else**
18. $\arg_{\max} = \max \sum_{k=1}^{m} f(L_m)$
19. **End**

The algorithm takes the input of multiple data sources d1, d2, etc. till dn, where n is the total number of the data warehouses. As the database for the proposed experiment is synthetic, we choose to consider some simple preprocessing techniques over our data set (Line-2), where Px(D) is a preprocessing function on the complete dataset.. In order to simplify the pre-processing, we consider the unstructured data in simple text form. The system than apply local model as shown in Line-3. Lm is a function of local model that are specific to each forms of data. For an example, if we apply Lm1 to d1, than we choose not to repeat Lm1 usage between d2 and dn. Therefore m represents number of local mode functions which should be always less than n. Although, we get the outcome for distributed data mining right after Line-3; however, we choose to perform more optimization on the top of it. We apply genetic algorithm. A population variable pop is initialized along with the fitness function f in Line-6. The iteration is incremented to highest level till the system arrives to optimal results. We consider the best value of population to be p1 (first best outcome) and p2 (second best outcome). The cross over operator is applied over the clustered data Obest followed by mutation operator to yield δbest. We insert Obest in the population (Line 14) and provide a condition if any arbitrary value is less than processed population (with Obest). In case of positive condition, it yields the mutation operator with highest value as best clustered value or else it applies specific local mode to find the highest value of cluster to be considered as best outcome. The algorithm takes the input of multiple data sources d1, d2 …etc. till dn, where n is the total number of the data warehouses. The design of the local mode clustering process is done by following manner. There is a potential possibility of resemblance between two set of replicated data from the warehouses that could possibly corresponds to empirical association between two set of data vector. The algorithm for Local Mode is as follows:-

---

Algorithm for Local Mode

Input: replica vectors dv_1 and dv_2.

Output: Value of primary, secondary, and tertiary local mode.

Start

1. init dv_1, dv_2

2. Apply Primary Local Mode

$$\theta_{pri} = \frac{dv_1.dv_2}{dv_1 x dv_2}$$

2. Apply Secondary Local mode

$$\theta_{\sec} = \frac{dv_1.dv_2}{dv_1^{\ 2} + dv_2^{\ 2} - dv_1.dv_2}$$

3. Apply Tertiary Local Mode

$$\theta_{ter} = \frac{E[dv_1.dv_2] - E[dv_1]E[dv_2]}{\sqrt{E[(dv_1)^2] - E[(dv_1)]^2}\sqrt{E[(dv_2)^2] - E[(dv_2)]^2}}$$ **End**

We consider dv1 and dv2 are a high-dimensional vectors of replica extracted from the warehouse. A closer look into the technique applied for all the local mode will yield a value between 0 and 1. Primary local mode allows the system to become independent of size of the replica. Therefore, if the primary local model between two vectors yields value equivalent to 1 than both the replicas doesn't have any discrete knowledge and it is continued for all the values of n.

7 Result Discussion

The outcome of the proposed system is compared with the work carried out by Dou et al. [28]. The prime reasons for selection of Dou et al. [28] are (i) Similar line of research problems of distributed data mining and (ii) usage of the genetic algorithm. The authors have developed a distributed mining technique for extracting knowledge from significant rules on numerous relational tables and integrated along with genetic algorithm for improving the knowledge discovery process. The authors have used number of rules as performance parameter to perform evaluation without comparative analysis. However, as proposed system doesn't implement any forms of rule, so it would be inappropriate to consider rules as performance parameters. Both the proposed system and the study presented by Dou et al. [28] is considered for comparative performance analysis using a common parameter of response time. The outcome of the comparative analysis is shown in Fig. 2.

Figure 2 shows the comparative analysis of response time for both proposed system as well as Dou et al. [28] approach. The outcome shows that implementation of genetic algorithm has resulted in almost similar increment pattern for both the approaches. However, proposed system has excelled better than Dou et al. [28] because of following reason: proposed system doesn't integrated genetic algorithm

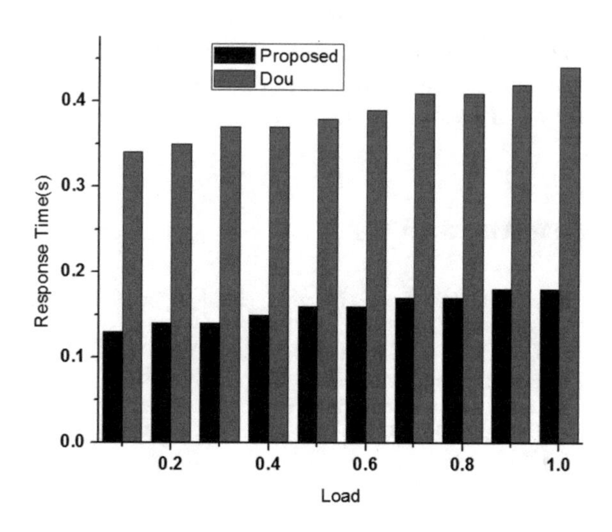

Fig. 2 Outcome of response time

on the complete clustering model. In fact it performs the clustering first, applies local model, and then applies genetic algorithm, which drastically minimize the computational as well as processing time. However, Dou et al. [28] have performed quite a complex schema using star model and finally the authors have reframed association rules to get better outcomes. Another difference is Dou et al. [28] approach cannot be recommended for managing replicas in the distributed data mining approach. It can also not be compatible for mining large streams of unstructured and massive data. It should be noted that proposed system is capable can equally process a standard dataset too as it has higher compatibility with synthetic dataset for carrying out clustering in grid infrastructure. Hence, proposed system can be claimed to be cost efficient model over grid.

8 Conclusion

There is already a list of unsolved problems in distributed data mining over grid interface. Such problems could be somewhat reduced if the grid posses a significant information about the replicas connected in its nodes. Replica management system is one of the significant operations in grid where replica acts both as asset and liability. In case of stream of non-static data, replica plays an asset and in case of static data management, replica plays somewhat liability if unused for long or found to be quite massive in size. This paper introduces a technique that takes the input from the data warehouse where the possibility of replicated data is too high. The proposed system initially extracts a data and applies simple preprocessing. It then performs clustering mechanism followed by extraction of values of local model towards the clustered data. The design of the local model is carried out using simple statistical approach which is further subjected to evolutionary technique like genetic algorithm. The outcome of the study shows much better response time in comparison to existing system. The presented model is meant to reduce the complexity of distributed data mining over grid interface. Therefore, our future work will be to further extend this model towards distributed data mining algorithm on the top of it for better knowledge discovery.

References

1. Sengall, R.S.: Research and Applications in Global Supercomputing, p. 672. IGI Global, Computers (2015)
2. Shaw, M.J., Subramaniam, C., Tan, G.W., Welge, M.E.: Knowledge management and data mining for marketing. Decis. Support Syst. **31**(1), 127–137 (2001)
3. Larose, D.T.:. Discovering Knowledge in Data: An Introduction to Data Mining. Wiley (2014)
4. Rollinson, H.R.: Using Geochemical Data: Evaluation, Presentation, Interpretation. Rutledge (2014)

5. Shi, G., Mortazavi, M., Chen, J., Kotha, V.G.R.: Method and apparatus for providing highly-scalable network storage for well-gridded objects. US Patent 8,996,803 (2015)
6. Weichhart, G., Molina, A., Chen. D., Whitman, L.E., Vernadat, F.: Challenges and current developments for sensing, smart and sustainable enterprise systems. Comput. Ind. (2015)
7. Lackovic, M., Talia, D., Trunfio, P.: A service-oriented framework for executing data mining workflows on grids. In: Grid and Pervasive Computing Conference, GPC'09, pp. 72–79. Workshops (2009)
8. Brescia, M., Cavuoti, S., Abrusco, R.D., Laurino, O., Longo, G.: DAME: a distributed data mining and exploration framework within the virtual observatory. In: Remote Instrumentation for eScience and Related Aspects, pp. 267–284. Springer (2012)
9. Hmida, M.B.H., Slimani, Y.: Meta-learning in grid-based data mining systems. Int. J. Commun. Netw. Distrib. Syst. **5**(3), 214–228 (2010)
10. Kantarcioglu, M., Nix, R.: Incentive compatible distributed data mining. In: 2010 IEEE Second International Conference Social Computing (SocialCom), pp. 735–742 (2010)
11. Oyana, T.J.: A new-fangled FES-k-means clustering algorithm for disease discovery and visual analytics. EURASIP J. Bioinf. Syst. Biol. **746021**(1) (2010)
12. Rao, V.S., Vidyavathi, S.: Distributed data mining and mining multi-agent data. (IJCSE) Int. J. Comput. Sci. Eng. **2**(04), 1237–1244 (2010)
13. Tlili, R., Slimani, Y.:. Executing association rule mining algorithms under a Grid computing environment. In: Proceedings of the Workshop on Parallel and Distributed Systems: Testing, Analysis, and Debugging, pp. 53–61 (2011)
14. Prusiewicz, A., Zieba, M.: The proposal of service oriented data mining system for solving real-life classification and regression problems. Technological Innovation for Sustainability, pp. 83–90. Springer, Heidelberg (2011)
15. Santos, M.F., Mathew, W., Portela, C.F.: Grid data mining for outcome prediction in intensive care medicine. Enterp. Inf. Syst. 244–253 (2011)
16. Mallik, R., Sarda, N., Kargupta, H., Bandyopadhyay, S.: Distributed data mining for sustainable smart grids. Proc. ACM SustKDD **11**, 1–6 (2011)
17. Pandey, T.N., Panda, N., Sahu, P.K.: Improving performance of distributed data mining (DDM) with multi-agent system. IJCSI Int. J. Comput. Sci. **2**(9) (2012)
18. Prajapati, R.B., Menaria, S.: Multi agent-based distributed data mining. Int. J. Adv. Res. Comput. Eng. Technol. (IJARCET) **1**(10), 76 (2012)
19. Zhang, Y., Sow, D., Turaga, D., Schaar, M.V.D.: A fast online learning algorithm for distributed mining of big data. ACM SIGMETRICS Perform. Eval. **41**(4), 90–93 (2014)
20. Belbachir, K., and Belbachir, H.: Parallel Mining Association Rules in Calculation Grids (2013)
21. Vishvapathi, P., Ramachandram, S., Govardhan, A.: GWSVM algorithm for a grid system. Int. J. Comput. Sci. Inf. Technol. **5**(5), 6871–6876 (2014)
22. Amini, A., Saboohi, H., Wah, T.Y., Herawan, T.: A fast density-based clustering algorithm for real-time internet of things stream. Sci. World J. (2014)
23. Maaß, H., Cakmak, H.K., Bach, F., Mikut, R., Harrabi, A., Süß, W., Jakob, W., Stucky, Kl-U, Kühnapfel, U.G., Hagenmeyer, V.: Data processing of high-rate low-voltage distribution grid recordings for smart grid monitoring and analysis. EURASIP J. Adv. Sig. Process. **1**, 1–21 (2015)
24. Ogunde, A.O., Folorunso, O., Sodiya, A.S.: The design of an adaptive incremental association rule mining system. In: Proceedings of the World Congress on Engineering, vol. 1 (2015)
25. Rebbah, M., Yemres, M.E.A., Khaldi, M., Debakla, M.: Hybrid Distribution for Association Rules Extraction on Grid Computing. Accessed 24 Nov 2015
26. Srinivasan, T., Palanisamy, B.: Scalable clustering of high dimensional data technique using SPCM with ANT colony optimization intelligence. Hindawi Sci. World J. 5 (2015)
27. Zhou, G.: Cloud platform based on mobile internet service opportunistic drive and application aware data mining. J. Electr. Comput. Eng. **50**, 357–378 (2015)
28. Dau, S.: The Book of Jonas: A Novel, p. 272. Penguin (2012)

The MOBIKEY Keystroke Dynamics Password Database: Benchmark Results

Margit Antal and Lehel Nemes

Abstract In this paper we study keystroke dynamics as an authentication mechanism for touchscreen based devices. A data collection application was designed and implemented for Android devices in order to collect several types of password. Besides easy and strong passwords we propose a new type of password—logical strong—which is a strong password, but easy to remember due to the logic behind the password's characters. Three main types of feature were used in the evaluation: time-based, touch-based and accelerometer-based. We propose a novel feature set—secondorder—which is independent of the length of the password. The preliminary results show that the lowest equal error rate (EER) is achieved by the logical strong password, followed by the strong password. The worst performance was achieved by the easy password; suggesting that the strong password is the best choice even in the case of keystroke dynamics based authentication systems.

Keywords Keystroke dynamics · Password difficulty · Mobile authentication · Performance evaluation · Sensors

1 Introduction

The pervasive presence of mobile devices equipped with many powerful sensors has led to new authentication mechanisms. One of them is user-authentication based on keystroke dynamics, an active research topic with remarkable results in the case of computers with hardware keyboards. Keystroke dynamics is a behavioural biometric which adds a second level security to alphanumerical passwords, by modelling the users' typing rhythms. Attempts to access the device by impostors, who have illegally

M. Antal (✉) · L. Nemes
Faculty of Technical and Human Sciences, Sapientia University,
Soseaua Sighisoarei 1C, 540485 Tirgu Mures/Corunca, Romania
e-mail: manyi@ms.sapientia.ro

L. Nemes
e-mail: nemes_lehel@yahoo.com

R. Silhavy et al. (eds.), *Software Engineering Perspectives and Application in Intelligent Systems*, Advances in Intelligent Systems and Computing 465,
DOI 10.1007/978-3-319-33622-0_4

obtained the user's password (through smudge-attack or shoulder surfing), can be detected based on the fact that they do not type the password in the same rhythm or that they handle the mobile device differently (device holding position, touchscreen usage).

In this paper we propose to investigate the influence of password difficulty on the authentication system's performance. The analysis is performed on our new dataset collected using mobile devices. This allows investigation not only of the effect of password difficulty, but also the influence of new features provided by the sensors of mobile devices.

Our work makes several contributions. One concerns the collected data, which contain the password typing patterns of three types of password i.e. easy, strong and logical strong. Data was collected using mobile devices therefore; besides time-based raw data we obtained additional data from sensors such as touchscreen and accelerometer. We have already made this data publicly available, hence it can be used by other researchers. Another contribution is the proposed secondorder feature set, independent of the length of the password and with equal error rates close to those obtained from the full feature set. The final contributions concern the evaluation results and the software used for the evaluation. Overall, we hope that our work will help focus attention on the opportunities provided by mobile device sensors in user identity verification.

The remainder of this paper is organised as follows. The next section (Sect. 2) presents related work with an emphasis on studies conducted on touchscreen-based mobile devices. Section 3 addresses research methods such as data collection, feature extraction and the different feature sets used in the evaluation. Section 4 offers evaluation results including two-class classifiers and anomaly detectors. The final section concludes our study and its findings.

2 Related Work

Keystroke dynamics is a well researched area. Several survey papers have been published to date [1, 4, 9, 17]. Most of this research has been carried out on computers or older mobile devices that utilise hardware keyboards. Less work has been carried out on touchscreen equipped mobile devices. However, the influence of key press pressure has been studied before the touchscreen smartphone era [8, 12, 14, 16]. In these studies special pressure-sensitive hardware keyboards were built. All these studies came to the conclusion that using key pressure as an addition feature increased the keystroke dynamic authentication system's performance.

In very recent years a few studies have been conducted on touchscreen-based mobile devices [2, 3, 6, 7, 10, 19, 21]. Except for Draffin et al.'s study [7], the other papers present results related to password-based authentication using keystroke dynamics. The most important aspects for the purpose of comparison are the datasets, the features, the methods and the results. Table 1 presents the characteristics of the datasets used in the aforementioned studies. It is important to note that

Table 1 Characteristics of keystroke datasets collected on touchscreen-based mobile devices

Study	# Users	Password	Raw data	Available	Best result(s) (%)
[19]	152	17-digit	Time	NO	FAR: 6.61
					FRR: 8.03
[21]	80	4–8 digit	Time	NO	EER: 3.75
			Touch		
			Accelerometer		
			Gyroscope		
[10]	20	7q56n5l144 phrase	Time	NO	EER: 13.6
			Space		
[3]	42	.tie5Roanl	Time	YES	EER: 12.9
			Space		
			Touch		
[6]	28	6–8 character	Time	NO	EER: 13.74
			Space		
			Touch		

not all studies saved the touch related raw data in the same way. Zheng et al. [21] and Buschek et al. [6] saved pressure and size (finger area) both at the moment of touch down and touch up. Conversely Antal et al. [2] saved this raw data only at the moment of key press. There are several differences between spatial raw data too. While Antal et al. saved the x, y coordinates only at the key press moment, Buschek et al. saved both the coordinates of the touch point at the moment of touch down and touch up. The differences between raw data imply different features for the analysed studies. Only Zheng et al. used raw data obtained from the accelerometer and the gyroscope sensors.

We have found only three papers which have studied the influence of password difficulty on the performance of keystroke dynamics system. Bartlow and Cukic [5] conducted the first study in this direction. Besides common short 8-lowercase letter passwords, such as `computer` and `swimming`, they used long 12-character length randomly generated passwords the typing of which required the usage of the Shift key. Example of such passwords include `+AL41fav8TB=` and `UC8gkum5WH`. In almost every EER performance measurement they observed a notable increase (at least 2 %) from short to long password, indicating that the usage of the shift key in a password plays a significant role. In feature ranking the shift key related features proved to be very discriminating.

Meng et al. [18] questioned the use of keystroke dynamics as biometrics. They built a training interface which allows intruders to train themselves in imitating another person's password typing rhythm. For this study they used two 8-character length passwords, an easy and a difficult one. They concluded that passwords that are easier to type are also easier to imitate.

Mondal et al. [15] introduced complexity measurement related to the typing of a password after which several performance measurements were conducted. In contrast to the previous two studies, they concluded that easier passwords are better choice for keystroke dynamics biometrics.

3 Methods

3.1 Data Collection

An Android application was designed and implemented with the aim of collecting typing data for different passwords. Users had to type in three different fixed passwords. The following passwords were used: easy—`kicsikutyatarka`; logical strong—`Kktsf2!2014`; strong—`.tie5Roanl`. The easy password contained only lowercase letters and was formed by the first three words of a Hungarian saying. Our proposal utilises the logical strong type and is based also on the same Hungarian saying, but in this case we took the first letters of the words and used `sf2!` for `sfsf` (two occurences of sf) followed by the year of data collection. The logic behind the logical strong password was explained to subjects before the data collection experiment. The strong password was used in the keystroke dataset collected by Killourhy [11].

54 volunteers took part in the experiment, 5 women and 49 male, with an average age of 20.61 years (range: 19–26). At the registration stage they stated their experience with touchscreen devices as follows: 2—inexperienced, 6—beginners, 17—intermediate and 29 advanced touchscreen users. Among them 4 users were left handed the others right handed. Data was collected in three sessions one week apart. In each session they typed at least 60 passwords, at least 20 passwords from each type. At the end of data collection each user had provided at least 60 samples from each type of password (easy: 3323 samples, strong: 3303, logical strong: 3308). The data was collected using 13 identical Nexus 7 tablets. Typos were not allowed, instead, the subjects had to retype the password. Each password had to be typed in the same way: the same keys had to be typed in the same order.

3.2 Feature Extraction

The application implemented a custom keyboard in order to store the time, touch and accelerometer related raw data during each user's typing. Raw data was saved at touch events initiated by the user for example, at the point of touch down and touch up. Touch down events were generated by the system when the user touched a key on the software keyboard, and touch up at the point of key release. Table 2 shows the raw data saved during the data collection process.

Table 2 The most important raw data saved during data collection

Data	Explanation
Key	The pressed key
Downtime	The timestamp at touch down event
Uptime	The timestamp at touch up event
Pressure	The pressure exerted on the screen at touch down event
Finger area	Touch area at touch down event
x, y	The x and y coordinate at touch down event
ax, ay, az	Acceleration measured along x, y, z axes

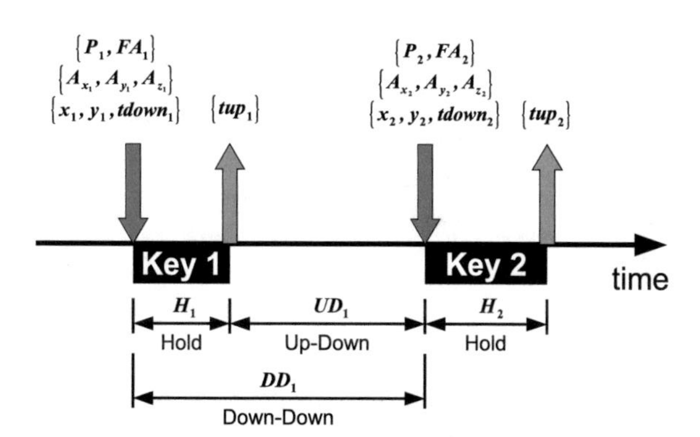

Fig. 1 Data collection. Raw data: *x*, *y*—coordinates; *tdown*, *tup*—timestamps; *Ax*, *Ay*, *Az*—directional accelerations; *P*—pressure; *FA*—finger area. Time-based features: *H*—hold time; *UD*—up-down time; *DD*—down-down time

Figure 1 shows the data saved at the moment of touch down and also the time-based features that can be extracted from these data such as hold time—the time between key press and release, down-down time—the time between consecutive key presses, and up-down time—the time between key release and next key press. The Nexus 7 tablet contains an embedded accelerometer with range $-2g$ and $+2g$ and measures the accelerations along three axes (the axes are device related). Its fastest sampling rate on sensor readings is about 50 Hz. During data collection these values were saved at the moment the user touched the screen. Using these directional accelerations we could characterise the device holding preferences of the users.

3.3 Feature Sets

Table 3 shows the full feature sets for each type of password. Because these feature sets contain features related to each key in a password, some feature types contain a different number of features for each password. Mean hold time (MHT) feature represents the average of key hold time values. The other mean values were computed similarly. The total distance feature (TD) was calculated as the sum of the distances (in pixels) between two consecutive buttons on the virtual keyboard. Total time (TT)

Table 3 Full feature sets for each type of password

Mnemonic	Feature type	Easy	Strong	Logical strong
HT	Hold time	15	13	13
DD	Down-down time	14	12	12
UD	Up-down time	14	12	12
P	Pressure	15	13	13
FA	Finger area	15	13	13
MHT	Mean hold time	1	1	1
MP	Mean pressure	1	1	1
MFA	Mean finger area	1	1	1
MAX	Mean X acceleration	1	1	1
MAY	Mean Y acceleration	1	1	1
MAZ	Mean Z acceleration	1	1	1
TD	Total distance	1	1	1
TT	Total time	1	1	1
V	Velocity	1	1	1
Total		82	72	72

represents the time needed to type in the password. Velocity (V) was computed as the quotient of the distance and the total time. Before evaluation data was normalized into the range [0, 1].

Besides the full feature sets presented in Table 3 some evaluations were performed on a so called—secondorder—feature set. This feature set contains 9 features: mean hold time, mean pressure, mean finger area, mean x acceleration, mean y acceleration, mean z acceleration, velocity, total time and total distance. The most important characteristic of this feature set is that the number of features is password-independent. All information related to this research is available at http://www.ms.sapientia.ro/~manyi/mobikey.html.

4 Evaluation and Results

Keystroke dynamics based authentication is a typical outlier detection problem. Given the keystroke data of a typed password the system has to decide whether the data belong to the genuine user. This problem can be formulated as a classification and as an anomaly detection problem. In the case of classification we typically employ a two-class classification algorithm, where the positive samples belong to the genuine user and negatives are selected from the others. Classifiers are more powerful since they yield information about the impostors (negative samples), whereas anomaly detectors can only check the deviation from the genuine user (positive samples). We should mention that in a real-world authentication system only the anomaly detection method is viable because of the lack of negative samples. However for comparison purposes, we present the evaluation of two-class classifiers too.

4.1 Two-Class Classification

In the case of two-class classification we call the data from the legitimate user positive samples and that from impostors we call negative samples. As our dataset contains data from several users and as each user typed the same password, one can easily select negative data for each user.

The general algorithm used for two-class classification measurements is depicted in Fig. 2. First we select positive and negative samples for a given user (*userData*). As negative samples we used two randomly selected samples from each other user. Then we repeat *nRuns* times the randomization of the data followed by n-fold cross-validation evaluation for the given user. The above two steps were repeated for each user.

Scores for positive and negative test samples were computed so as to form two sets, one for genuine users the other for impostors. Then a user-independent threshold was scanned through the two sets of scores and the False Negative (FN) and False Positive (FP) rates computed for each threshold. Plotted as error curves, these values show the system performance (see Fig. 3).

Besides Random Forests algorithm we chose to evaluate the k-nearest neighbours (kNN) and Bayes Net algorithms. All classification algorithms were used from the Weka Data Mining toolkit [20].

```
1: procedure MEASUREMENT(data, nFolds, nRuns)
2:     for user ← 1, numUsers do
3:         userData ← selectPositiveAndNegativeSamples(data, user)
4:         for run ← 1, nRuns do
5:             userData ← randomize(userData)
6:             for n ← 1, nFolds do
7:                 trainUserData ← trainCV(userData, n)
8:                 testUserData ← testCV(userData, n)
9:                 train two-class classifier for trainUserData
10:                evaluate the trained classifiers using testUserData
11:            end for
12:        end for
13:    end for
14: end procedure
```

Fig. 2 Two-class classification measurement algorithm using n-fold cross-validation

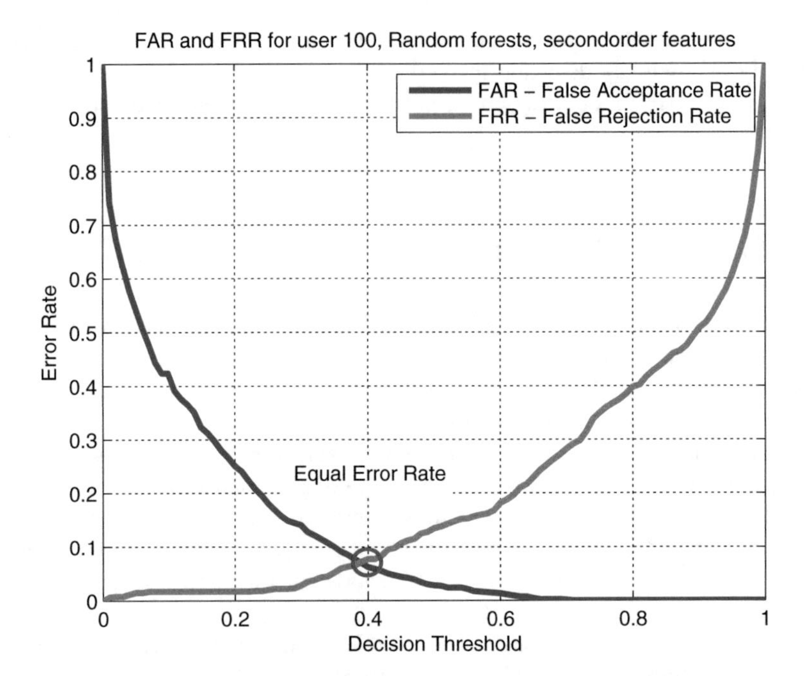

Fig. 3 EER computation for user 100 (Random forests classifier, secondorder features). EER for individual users were estimated as the intersection of FAR (False Acceptance Rate) and FRR (False Rejection Rate) curves

4.2 Anomaly Detection

In the case of anomaly detectors we used five detectors implemented in the R script provided by Killourhy and Maxion [11]. The detectors used were: Euclidean, Manhattan, Mahalanobis, Outlier count and Kmeans. This script works as follows: (i) it splits the data into three equal parts, each containing 20 samples from each user (in our case each part contained data from a single data-collection session) (ii) detectors are trained separately for each user using two-thirds of the data; evaluation was performed on the remaining third positive samples and two negative samples selected from each of the other users (20 positive + 53 * 2 negative); (iii) step (ii) is then repeated three times (threefold cross-validation), and the mean EER and its standard deviation computed.

4.3 Results

Results for classifiers and anomaly detectors are presented in Table 4. EER values were estimated for each user (see Fig. 3), then the mean and standard deviation were computed for each classifier or anomaly detector and each dataset.

Table 4 EER results for different methods and feature sets

Method	Features	Easy	Logical strong	Strong
Bayes net	Secondorder	0.074 (0.046)	0.058 (0.040)	0.067 (0.047)
kNN (k = 1)	Secondorder	0.056 (0.032)	0.048 (0.026)	0.054 (0.036)
Random forests (T = 100)	Secondorder	0.052 (0.029)	**0.045 (0.025)**	0.051 (0.032)
Bayes net	All	0.053 (0.039)	0.046 (0.037)	0.049 (0.038)
kNN (k = 1)	All	0.073 (0.036)	0.068 (0.033)	0.071 (0.043)
Random forests (T = 100)	All	0.032 (0.021)	**0.033 (0.025)**	0.033 (0.022)
Euclidean	Secondorder	0.208 (0.174)	0.149 (0.141)	0.181 (0.145)
Manhattan	Secondorder	0.202 (0.169)	0.144 (0.140)	0.169 (0.146)
Mahalanobis	Secondorder	0.191 (0.182)	0.154 (0.171)	0.159 (0.159)
Outlier count (th = 1.96)	Secondorder	0.208 (0.147)	0.164 (0.140)	0.178 (0.146)
Kmeans (k = 3)	Secondorder	0.177 (0.155)	**0.136 (0.132)**	0.143 (0.137)
Euclidean	All	0.238 (0.186)	0.183 (0.149)	0.195 (0.163)
Manhattan	All	0.203 (0.183)	0.154 (0.140)	0.167 (0.153)
Mahalanobis	All	0.256 (0.140)	0.193 (0.114)	0.210 (0.137)
Outlier count (th = 1.96)	All	0.160 (0.140)	**0.129 (0.126)**	0.143 (0.137)
Kmeans (k = 3)	All	0.173 (0.136)	0.128 (0.097)	0.131 (0.106)

The standard deviations are shown in parenthesis

We used 100 trees for the Random Forests classifier, $k = 1$ for the kNN classifier and the default Weka settings for the Bayes Net classifier. In the case of anomaly detectors the following settings were used: $k = 3$ clusters, at most 20 iterations for the kmeans detector; the *threshold* = 1.96 for the outlier count detector (used to count how many z-scores exceed a threshold) [11].

It can be seen that very low EER values were obtained by the classification algorithms, because these used the negative samples for building the user's model. However in real systems negative samples are not available (in the enrolment stage samples are collected only from the genuine user).

For the error curve we chose the DET error curve (Detection Error Tradeoff) [13], which is the most important error curve for biometric systems. Figure 4a, b show these error curves obtained for the Random Forests classifier (number of trees: 100) and Manhattan detector.

The best equal error rates were obtained by the Random Forests classifier, around 5 % for the secondorder feature set and around 3 % for the full feature set. We mention again that these classifiers use negative samples for building the user's typing model, which is not available in case of real systems. No significant differences were found in this evaluation between different types of password.

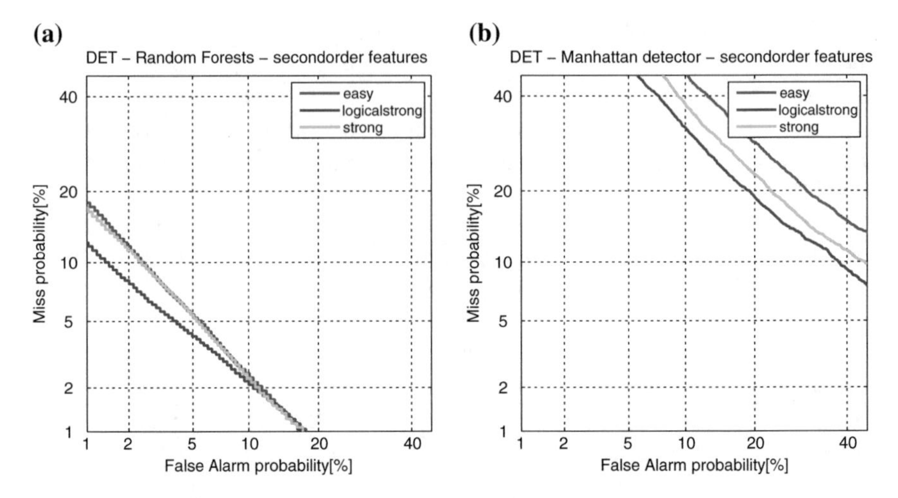

Fig. 4 DET curves—secondorder features. **a** Random Forests (T = 100). **b** Manhattan detector

In the case of anomaly detectors, where the user's model is based only on positive samples (the case of real systems), the equal error rates are always lower for logical strong and strong types of password.

5 Conclusions

Our objective in this work was to collect a dataset on mobile devices containing different types of password and to evaluate the influence of password difficulty on the performance of keystroke dynamics authentication. We provide both the datasets and evaluation methodology to the research community. The main contribution of this paper concerns the datasets, which not only contain three types of password, but contain raw data collected from mobile sensors too. Another contribution is the secondorder feature set which has the same number of features regardless of the password type. Measurements show the effectiveness of this novel feature set as very close to or sometimes better than the results obtained using the full feature set. Evaluations show that in the case of anomaly detectors the lowest equal error rates are obtained for the logical strong password, followed by the strong and the easy one. This is in concordance with the results obtained by Bartlow and Cukic [5] and Meng et al. [18].

Acknowledgments The research has been supported by the European Union and Hungary and co-financed by the European Social Fund through the project TAMOP-4.2.2.C–11/1/KONV–2012–0004–National Research Center for Development and Market Introduction of Advanced Information and Communication Technologies. The authors would like to thank all volunteers who participated in the data collection experiment.

References

1. Ahmad, N., Szymkowiak, A., Campbell, P.A.: Keystroke dynamics in the pre-touchscreen era. Front. Human Neurosci. **7** (2013)
2. Antal, M., Szabó, L.: An evaluation of one-class and two-class classification algorithms for keystroke dynamics authentication on mobile devices. In: 2015 20th International Conference on Control Systems and Computer Science (CSCS), pp. 343–350 (2015)
3. Antal, M., Szabó, L., Laszló, I.: Keystroke dynamics on android platform. Procedia Technol. **19**, 820–826 (2015). In: 8th International Conference Interdisciplinarity in Engineering, INTER-ENG 2014, 9–10 Oct 2014, Tirgu Mures, Romania
4. Banerjee, S.P., Woodard, D.L.: Biometric authentication and identification using keystroke dynamics: a survey. J. Pattern Recogn. Res. **7**(1), 116–139 (2012)
5. Bartlow, N., Cukic, B.: Evaluating the reliability of credential hardening through keystroke dynamics. In: 17th International Symposium on Software Reliability Engineering, 2006. ISSRE'06, pp. 117–126 (2006)
6. Buschek, D., De Luca, A., Alt, F.: Improving accuracy, applicability and usability of keystroke biometrics on mobile touchscreen devices. In: Proceedings of the 33rd Annual ACM Conference on Human Factors in Computing Systems. CHI'15, pp. 1393–1402. ACM (2015)
7. Draffin, B., Zhu, J., Zhang, J.: Keysens: Passive user authentication through micro-behavior modeling of soft keyboard interaction. In: Mobile Computing, Applications, and Services, Lecture Notes of the Institute for Computer Sciences, Social Informatics and Telecommunications Engineering, vol. 130, pp. 184–201. Springer (2014)
8. Eltahir, W., Salami, M.J.E., Ismail, A., Lai, W.: Design and evaluation of a pressure-based typing biometric authentication system. EURASIP J. Inf. Secur. **2008**(1) (2008)
9. Giot, R., El-Abed, M., Rosenberger, C., et al.: Keystroke dynamics authentication. Biometrics (2011)
10. Kambourakis, G., Damopoulos, D., Papamartzivanos, D., Pavlidakis, E.: Introducing touchstroke: keystroke-based authentication system for smartphones. Secur. Commun. Netw. (2014)
11. Killourhy, K., Maxion, R.: Comparing anomaly-detection algorithms for keystroke dynamics. In: IEEE/IFIP International Conference on Dependable Systems Networks, 2009. DSN'09, pp. 125–134 (2009)
12. Loy, C., Lai, W.K., Lim, C.: Keystroke patterns classification using the ARTMAP-FD neural network. In: Third International Conference on Intelligent Information Hiding and Multimedia Signal Processing, 2007. IIHMSP 2007, vol. 1, pp. 61–64 (2007)
13. Martin, A., Doddington, G., Kamm, T., Ordowski, M., Przybocki, M.: The DET curve in assessment of detection task performance. Technical report, DTIC Document (1997)
14. Martono, W., Ali, H., Salami, M.: Keystroke pressure-based typing biometrics authentication system using support vector machines. In: Computational Science and Its Applications ICCSA 2007, vol. 4706, pp. 85–93 (2007)
15. Mondal, S., Bours, P., Idrus, S.Z.S.: Complexity measurement of a password for keystroke dynamics: preliminary study. In: Proceedings of the 6th International Conference on Security of Information and Networks, pp. 301–305 (2013)
16. Nonaka, H., Kurihara, M.: Sensing pressure for authentication system using keystroke dynamics. In: International Computational Intelligence Society International Conference on Computational Intelligence. pp. 19–22 (2004)
17. Teh, P.S., Teoh, A.B.J., Yue, S.: A survey of keystroke dynamics biometrics. Sci. World J. **2013** (2013)
18. Tey, C.M., Gupta, P., Gao, D.: I can be you: Questioning the use of keystroke dynamics as biometrics. In: NDSS. The Internet Society (2013)
19. Trojahn, M., Arndt, F., Ortmeier, F.: Authentication with keystroke dynamics on touchscreen keypads-effect of different n-graph combinations. In: MOBILITY 2013, The Third International Conference on Mobile Services, Resources, and Users, pp. 114–119 (2013)

20. Witten, I.H., Frank, E., Hall, M.A.: Data Mining: Practical Machine Learning Tools and Techniques. Morgan Kaufmann, San Francisco (2011)
21. Zheng, N., Bai, K., Huang, H., Wang, H.: You are how you touch: user verification on smartphones via tapping behaviors. In: 2014 IEEE 22nd International Conference on Network Protocols (ICNP), pp. 221–232 (2014)

Joint Algorithm for Traffic Normalization and Energy-Efficiency in Cellular Network

P.T. Sowmya Naik and K.N. Narasimha Murthy

Abstract A base station is the core heart of the operation in cellular communication system. Various components e.g. base transceiver station, base station controller, mobile station controller are basically a sophisticated hardware device which require a massive amount of energy just to initiate, manage, and terminate an IP communication over mobile network. Inspite of presence of massive archives of literatures towards addressing energy consumption, there are very few evidence of standard prototype that can be actually adopted in real sense. Therefore, we present an analytical model for jointly addressing traffic normalization problem and energy efficiency problem using simple stochastic geometry. Supported by discussion of algorithms, the presented technique was also compared with one of the most recent study to find its superior performance with respect to energy consumption and processing time.

Keywords Cellular network · Mobile network · Energy consumption · Stochastic geometry

1 Introduction

The area of networking and telecommunication has undergone a significant revolution in past decades. The present scenario of telecommunication is completely different that what used to be there 20 years back. Usage of mobile devices started gaining pace 20 years back, where there are limited mobile phone manufacturer and obviously very limited service providers to obtain network connectivity. But with the increasing pace of technological advancement in semiconductors, the prices of

P.T. Sowmya Naik (✉)
Department of CSE, City Engineering College, Bengaluru, India
e-mail: sowmya.vturesearch@gmail.com

K.N. Narasimha Murthy
Department of CSE, Vemana Institute of Technology, Bengaluru, India

R. Silhavy et al. (eds.), *Software Engineering Perspectives and Application in Intelligent Systems*, Advances in Intelligent Systems and Computing 465,
DOI 10.1007/978-3-319-33622-0_5

mobile devices slashed down and more services evolved in Indian consumer market. The present base of users is found to be increasingly using the internet facilities on their mobile devices. Irrespective of economical status and ethnical background of user, mobile communication products and services are quickly establishing itself in every commercial market. Another interesting findings shows that 26 % as majority of user access internet all the 7 days a week. These statistics evidently shows that there are lot of increasing usage of cellular network in recent times which demands a closer look of its scale of effectiveness. Another biggest problem in cellular network is call drop, which even doesn't meet the standard of Telecom Regulatory Authority of India (TRAI). Till date, consumption of energy using services from cellular network is unsolved. A conventional research paper and theory speaks of cellular network in terms of hexagonal topology with essential elements like mobile station and base station. With a massive amount of research manuscript present in digital archives, still a better and benchmarked study towards energy efficiency has not been come across.

Therefore, this paper has introduced a simple framework that focuses jointly on traffic normalization and energy efficiency especially when the traffic is dynamic, uncertain, and unpredictable in nature. Section 2 discusses about prior studies towards energy efficiency followed problem discussion in Sect. 3. Section 4 presents the proposed study followed by research methodology in Sect. 5. Discussion of an algorithm implementation is carried out on Sect. 6 followed by result discussion on Sect. 7. Finally, Sect. 8 summarizes the contribution of the proposed system.

2 Related Work

Hu and Cao [1] have adopted A* search algorithm for minimizing the search space for scheduling offline and online for traffic aggregation for wireless network. Similar study has also been carried out by Gaikwad and Wagh [2] with a difference of focus in LTE network. Giri and Bodhe [3] have presented a technique by signifying importance of multi-agent system that can ensure better QoS in cellular network. The outcomes shows that better performance of call blocking probability can ensure better QoS which can directly contribute to energy conservation. Correa and Fernandes [4] have also emphasized about the QoS factor for better scalability, network efficiency, and Quality of Experience. Study towards QoS enhancement on heterogeneous network was carried out by Kaleem et al. [5], where the authors have aimed for enhancing the throughput for cell-edge user as well as to minimize the interference owing to allocation of dynamic bandwidth. Mtaho and Ishengoma [6] have investigated about various essential parameters that directly influence the QoS considering the case study of cellular data from Tanzania. Memis et al. [7] have carried out the study for allocating network resources for enhancing the QoS factor in cellular network. The outcome of the study shows decrease in power allocation for each link and increase in time allocation for each link. Lorincz and Matijevic [8] have emphasized on two parameters of energy (energy per unit area and energy per bit unit area) over heterogeneous network. Soh et al. [9] have implemented a probabilistic technique and stochastic approach to understand the energy efficiency in both homo/heterogeneous cellular network. Similar focus on coverage was also emphasized by the work of Zhang et al. [10]. The authors have studied small cells and macro cells for multiple deployment of spectrum. The study conducted by Guo et al. [11] has used involuntary forecasting method on low powered nodes in order to understand the states of power depletion in heaver network load. Wang et al. [12] have presented a study for confirming the energy efficiency towards multimedia delivery considering network interference. Study towards heterogeneous network was also carried out by Sambo et al. [13] where the authors have presented 2-tier deployment of transmission for minimizing the energy depletion. Panahi and Ohtsuki [14] have presented a scheme for cognitive network of heterogeneous type. The recent work carried out by Taranetz [15] has focused on analyzing interference. The authors have also used stochastic geometry for designing the heterogeneous cellular network. Studies conducted by Fan et al. [16] have presented a technique that ensures a sleep scheduling algorithm for heterogeneous cellular network. Tombaz et al. [17] have considered the idle energy as well as backhaul energy into consideration. Yu et al. [18] focused on the macro base station in order to enhance the probability of coverage. Similar problem has also been focused by Esmaeilifard and Rahbar [19], who laid emphasis on complex transmission over cellular traffic. The study has presented an analytical modelling for ensuring maximal energy conservation and efficient network capacity. Study towards energy conservation is also seen in the work of Huang et al. [20] considering the case study of railway communication system. Most recently, the work carried out by Xiang et al. [21] have introduced a topology-based mechanism to conserve energy in cellular network. The prime aim of the study was to minimize the switching frequency of the base station considering the Long-Term Evolution (LTE) network.

3 Problem Identification

Energy is one of the essential factors for successful operation of cellular network. The identified problems of the proposed study are energy consumption due to interface of antenna, energy consumption due to power amplifier, energy consumption die to transceiver, energy consumption due to baseband interface, and energy consumption due to cooling. The conventional study considers evaluating efficiency by dividing output power with input power. For traditional base station in cellular network, the amount of energy depletion is completely dependent on the amount of load due to dynamic traffic. However, due to energy being consumed by the power amplifier, the energy scales down as per the minimizing trend of traffic. The prime reason behind this is reduction of quantity of occupied subcarriers in passive mode of communication with presence of subcarriers without any data to carry. Therefore, such form of scaling over signal essentially depends on type of base station. The energy consumption by power amplifier for macro base station is 50–60 % while that of low-powered nodes it is around 30 %, which will mean that base station is the prominent victim of the event of energy drainage in cellular network. The operation of base station is also closely associated with other components e.g. BTS, BSC, MSC etc., which equally drags a large amount of power just to initiate, manage, and terminate one active call. Hence, we define out problem as—"It is a challenging task to design a framework that can offer an optimal traffic management along with energy efficient communication over cellular network". The next section discusses about proposed model.

4 Proposed System

The primary motive of the proposed study is to develop an intelligent framework that ensures an effective traffic management as well as cost effective energy conservation in cellular network. The proposed system offers a simple algorithm that is anticipated to be executed over the base station as it is the point of attention for complete traffic and thereby dissipates significant amount of energy. Energy efficiency of one base station will also conserve a significant amount of energy of other base stations too. The scenario of cellular network considered in the proposed system is quite different from existing approaches which uses normally a hexagonal figure. The proposed system considers asymmetric cell area and size with all the base station connected to each other. However, in order to map with real-world set up, we consider that there are base stations which are not connected directly. For an example in the Fig. 1, the base station B3 is not connected with B6 and B4.

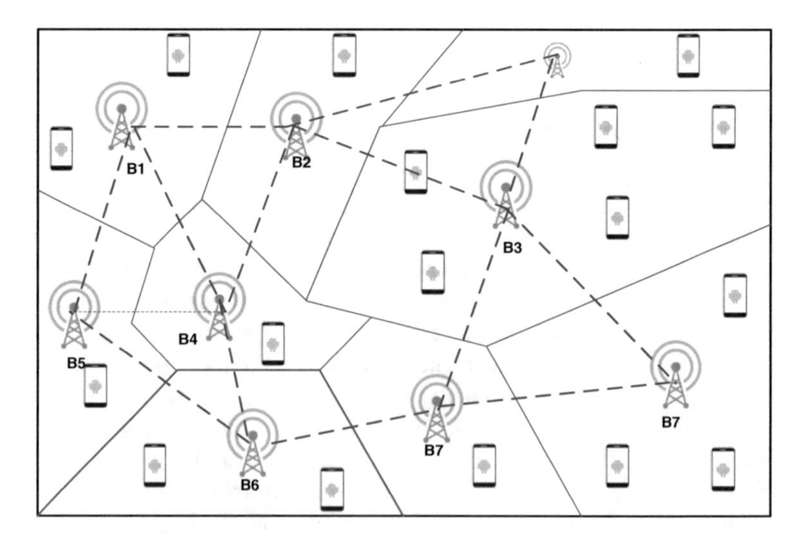

Fig. 1 Topology of cell considered in proposed system

The prime contributions of the proposed system are as follows:

- To gather intelligence by assessing the extent of energy dissipation using stochastic geometry approach of cellular network.
- To evaluate the traffic information of downlink transmission and find the best channel with less traffic.
- To compute the transmission power of the base station of each cells and determines its efficiency.

The proposed system is focused on the power consumption in base station considering both static and non-static power consumption. The static power consumption causes due to baseline power while non-static power consumption causes due to transmission circuits. The uniqueness of the proposed system is the adopted of stochastic geometry considering randomness of base station and mobile stations over the cellular area. The analytical modelling of the proposed system is carried out considering the cumulative interference estimated from interfering base station and mobile station as the core intelligence factor. The system also considers a scenario of natural problems on air e.g. scattering effect, fading effect, path loss, etc. Ultimately, the proposed system offers a joint algorithm that can effectively monitoring uncertain and unpredictable traffic states and offers better resiliency against faster energy drainage of base station.

5 Research Methodology

The proposed study considers the analytical methodology for carrying out the investigation towards energy efficiency in cellular network. The entire planning of the study was done in two phases (i) planning to mitigate traffic load and uncertain by modelling it and (ii) planning to ensure better energy efficiency using stochastic geometry.

The first phase of the planning is done by modelling Poisson's process over the cellular network considering random location of mobile station and base station. The system considers almost all forms of issues e.g. scattering, interference, fading, shadowing etc. The system than computes the cumulative load on traffic on particular cells considering that traffic is originated from the user (mobile station). It also computes the cumulative power on the base station during the transmission process which again depends on interference, rate of traffic, and channel conditions. The density of the traffic density is modelled using Eulerian function. The study in this phase also considers the weight factor for identification of scale of density in the simulation area. We also apply power law to control the traffic density considered in the system. Finally, the system estimates probability distribution of the traffic and after applying power law, we find the links with minimum traffic. Hence, an algorithm is also formulated that can do the similar task and return the best path of communication as output intelligence. The next phase of the study planning takes care of the energy dissipation in cellular network.

The study planning toward controlling energy dissipation initiates by evaluating the power required to perform downlink transmission. A downlink transmission is considered to be taking place between random users in the cells. The extent of interference between two same channels is assumed to be not more than 1. In order to discretely study the interference effect, we evaluate the receiving power for both the interfering mobile stations as well as base station. The study computes the SIR (ignoring noise) considering all the cumulative interference leading to the base station. A relationship between the interference and traffic density is established considering channel capacity assigned for mobile station difference between the channel capacity and the real-time coding as well as the module scheme for reducing rate of bit error. A relative distance between the interfering base station and mobile station is considered in this process.

Therefore, the proposed system finds the best channel and maintains optimal traffic to control unwanted energy consumption of the base station. It then applies energy efficient calculations to monitor any feasibility of outage. The simulation technique uses the concept of iterative random sampling process for developing an analytical model. The framework checks various scenarios of possible traffic density towards downlink transmission and evaluates the effect of weight factor on energy consumption. The outcome of the study is evaluated with respect to time complexity and energy consumption as the performance parameters. The proposed system is also compared with one of the recent on energy optimization, which has adopted optimization principle to enhance network lifetime of the associated base stations in cellular network.

6 Algorithm Implementation

The development of the proposed study was carried out in normal 32 bit machine using Matlab as the programming tool. The development of the proposed system is carried out by two core algorithm related to traffic management and energy efficiency.

6.1 Addressing Peak Traffic Condition

An algorithm is developed for addressing the uncertain and heavy traffic condition in cellular network. The algorithm takes the input of number of nodes to be distributed in random fashion in cellular area (Ca). We apply Eulerian integral (E(x)) to get the random distribution of the uncertain traffic condition in cellular network. The system assumes that traffic condition will be peak and uncertain with the number of mobile stations are more in the cellular network. For analysis of the traffic, we use the relative probability of the random variable using Line-3 of algorithm. The cells and the traffic of the mobile stations are evaluated considering the traffic density, which is evaluated with the total number of the mobile nodes present in one cell at one active session inside the cellular network. The equation shown in Line-3 consists of the variable a (inverse scale attribute of Eulerian integral function), b is considered as shape attributes, σ is considered as traffic density, and w is considered as weight factor. It is believed that less the weight factor the probability of the downlink traffic is expected to be quite bursty. Using probability logic, if the weight factor is between 0.8 and 1, the traffic density is considered to be quite dense.

Algorithm for traffic management
Input: cell area, nodes,
Output: intelligence of an efficient traffic with less density.
Start
1. Initialize C_a.
2. Define Eulerian integral

$$E(x) = \int_0^{\infty} \frac{t^{x-1}}{e^t} .dt$$

3. Evaluate relative probability of random variable

$$P_{c_a} = \frac{(a)^b}{E(x)} .x^{b-1} .\exp(-ax)$$

4. Evaluate traffic density
5. Evaluate relative probability of traffic density

$$P_\sigma(x) = w\sigma_{\min}^w \cdot \frac{1}{x^{w+1}} .$$

6. Apply power law of probability distribution

$$Traff = \arg_{\min}\left(\frac{w_{\sigma_{\min}}}{w-1}\right)$$

End

In order to extract the arbitrariness of the mobile station position as well as density of traffic, the cumulative load of traffic in a particular cell is evaluated. This evaluation is done by considering the specific cell to a point in Poisson distribution process. From the above lines involved in the algorithm, it is clear that the proposed system attempts to assess various active sessions on every cell and then it evaluates the exact size of traffic density. However, the filtering of the traffic is done by extracting only the links with less traffic originated from the base station. The next algorithm is responsible for evaluating and controlling the energy dissipation.

6.2 Addressing Energy Dissipation

The algorithm initially computes the Signal Interference Ratio (SIR) for the downlink transmission of the mobile station (Line-3) ignoring noise. The algorithm than draws a relationship between traffic density and SIR (Line-4) considering δ i.e. difference between bandwidth (C_{cap}) and real time-coding as well as the modulation technique for reducing bit error rate. It than starts evaluating the amount of power dissipation from each cell (Line-5) considering the variables x_i as position of mobile station and y_j as position of base station. It also considers variable R_{pow} as receiving power of the mobile station with n_bs as total number of base station (Line-5). Finally transmission power for all the associated base station is considered (Line-6) considering C_{gain} (channel gain). Finally, energy efficiency of the entire traffic is computed considering output power pout. $Traff_{real}$ is a variable considering the real-time traffic using first algorithm of traffic management.

Algorithm for Energy Efficiency
Input: Cell size, interference (I), γ (flag of interference)
Output: Energy efficiency
Start
1. Evaluate cumulative interference I_{cum} at mobile station (MS)
2. Initialize γ
3. Estimate SIR of MS

$$SIR = \frac{R_{Pow}}{I_{Cum}}$$

4. Construct density & SIR relationship

$$C_{cap}.\log_2(1+SIR/\delta)=\sigma(x)$$

5. Estimate power dissipation in each cell

$$T_{xpow}^{cell} = \sum \frac{\|x_i - y_j\|^{\alpha}}{n\_bs}.R_{pow}$$

6. Evaluate transmission power of base station

$$T_{xpow} = \left(\frac{R_{pow}}{C_{gain}(\|x_i - y_j)}\right)_{n\_bs}$$

7. Evaluate energy efficiency,

$$\eta = Traff_{real}(1-p_{out}) \text{ / mean total power}$$

End

The significant of both the algorithm is considering the entire traffic interference condition, location of base station and mobile station, it calculates the best traffic route (using first algorithm) and calculates energy efficiency (using second algorithm). The best part of the algorithm is its exploration for the best channel with reduced interference level and perform communication and thereby it also reduces the load of the traffic to and from base station in both uplink and downlink transmission. The next section discusses about the outcomes accomplished after implementing the algorithms.

7 Result Discussion

The outcome of the proposed system is compared with the recent work done by Ho et al. [22]. The reason behind selection of Ho et al. [22] work is similar aim i.e. energy enhancement. The authors have applied an optimization technique using Open StreetMap using non-linear load coupling equalization. The outcomes are studied with respect to energy consumption in Joule and Processing time in second over varying traffic load. For generalizing the outcomes, we consider traffic load in scale of probability 0–1.

Figure 2 shows that proposed system attempts to maintain better uniformity of the energy consumption with increasing load. Although, there is a slight increase in energy consumption but it is likely to happen in any real-time and dynamic traffic condition. On the other hand, Ho et al. [22] have applied a sophisticated optimization technique that is more focused on specific form of network (LTE), which results in increasing of energy consumption as compared to proposed system. Lack of randomness in the topology is another reason for poor outcome of Ho et al. [22] in energy consumption.

Figure 3 shows analysis of the processing time. We felt that processing time is an important performance parameter to judge network delay as well as algorithm complexity. We found that proposed system offers a stable processing time performance. The behavior of curve for proposed system is quite predictable in

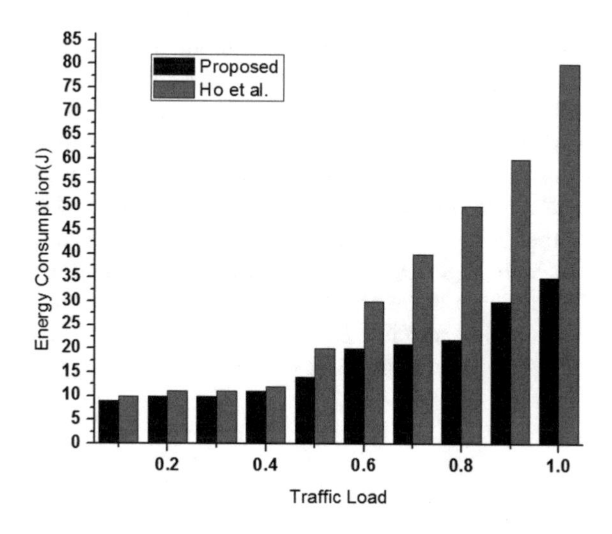

Fig. 2 Analysis of energy consumption

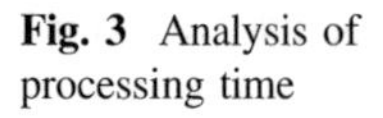

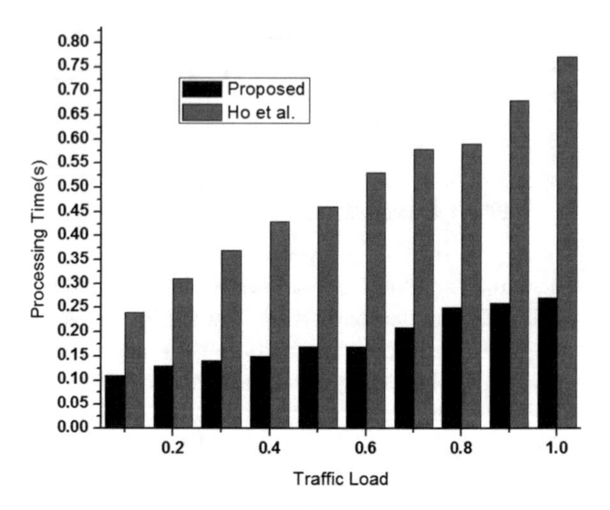

Fig. 3 Analysis of processing time

characteristics hence energy preservation algorithm will find appropriate condition of maintaining uniform performance of energy dissipation. Therefore, superior energy efficiency with lesser cost of resources can be claimed.

8 Conclusion

The presented paper attempts to implement a simple concept in different way. We have implemented a stochastic geometry approach which many researchers have used but implemented it to develop a random topology for positioning of both base station and mobile station in an intelligent manner. The study presents two contributions, where first we attempt to understand the dynamic traffic behavior in the form of preliminary intelligence and develop a simple algorithm to fine the traffic with least density as resultant intelligence. The second contribution of this paper is to develop a simple algorithm to achieve energy efficiency. The algorithm is tested considering the approximated numerical values considered in majority of the current research paper and benchmarking was carried out to find proposed algorithm outperforms existing technique with respect to reduction in energy consumption and well as processing time. Our future work will be to further enhance thus model to develop a novel architecture that can mitigate heavier traffic. Hence, our future work will be also in the direction of developing a further complicated traffic model and to perform more extensive analysis for energy efficiency.

References

1. Hu, W., Cao, G.: Energy optimization through traffic aggregation in wireless networks. IEEE Proc. INFOCOM 916–924 (2014)
2. Gaikwad, V., Wagh, T.: Overview of power optimization in LTE network. Int. J. Comput. Appl. (2015)
3. Giri, N., Bodhe, S.: Role of multi agent system For QoS guarantee in cellular networks. Int. J. Distrib. Parallel Syst. (IJDPS) **3**(5) (2012)
4. Correa, P., Fernandes, A.C.: End-to-end QoS in mobile networks. IEEE ComSoc (2013)
5. Kaleem, Z., Hui, B., Chang, K.: QoS priority-based dynamic frequency band allocation algorithm for load balancing and interference avoidance in 3GPP LTE HetNet. Springer-EURASIP J. Wirel. Commun. Netw. **185** (2014)
6. Mtaho, A.B., Ishengoma, F.R.: Factors Affecting QoS in Tanzania Cellular Networks. arXiv (2014)
7. Memis, M.O., Ercetin, O., Gurbuz, O., Azhari, S.V.: Resource allocation for statistical QoS guarantees in MIMO cellular networks. Springer-EURASIP J. Wirel. Commun. Netw. **217** (2015)
8. Lorincz, J., Matijevic, T.: Energy-efficiency analyses of heterogeneous macro and micro base station sites. Elsevier-Comput. Electr. Eng. **40**(2) (2014)
9. Soh, Y.S., Quek, T.Q.S., Kountouris, M., Shin, H.: Energy efficient heterogeneous cellular networks. IEEE J. Sel. Areas Commun. **31**(5) (2013)
10. Zhang, X., Su, Z., Yan, Z., Wang, W.: Energy-efficiency study for two-tier Heterogeneous Networks (HetNet) under coverage performance constraints. Springer J. Mobile Netw. Appl. (2013)
11. Guo, W., Wangy, S., Chux, X., Chen, J., Song, H., Zhang, J.: Automated small-cell deployment for heterogeneous cellular networks. IEEE Commun. Mag. (2013)
12. Wang, S., Guo, W., Khirallah, C., Vukobratovi, D., Thompson, J.: Interference allocation scheduler for green multimedia delivery. IEEE Trans. Veh. Technol. **63**(5), 2059–2070 (2014)
13. Sambo, Y.A., Shakir, M.Z., Qaraqe, K.A., Serpedin, E., Imran, M.A., Ahmed, B.: Energy efficiency improvements in HetNets by exploiting device-to-device communications. IEEE-Proccedings of 22nd Signal Processing Conference, pp. 151–155 (2014)
14. Panahi, F.H., Ohtsuki, T.: Stochastic geometry modeling and analysis of cognitive heterogeneous cellular networks. Springer-EURASIP J. Wirel. Commun. **141** (2015)
15. Taranetz, M.: System level modeling and evaluation of heterogeneous cellular networks. Dissertation of Fakultät für Elektrotechnik und Informationstechnik (2015)
16. Fan, S., Tian, H., Sengul, C.: Self-optimized heterogeneous networks for energy efficiency. Springer-EURASIP J. Wirel. Commun. Netw. **21** (2015)
17. Tombaz, S., Sung, K.W., Zander, J.: On metrics and models for energy efficient design of wireless access networks. IEEE Wirel. Commun. Lett. **3**(6), 649–652 (2014)
18. Yu, H., Li, Y., Kountouris, M., Xu, X., Wang, J.: Energy efficiency analysis of relay-assisted cellular networks. Springer-EURASIP J. Adv. Signal Process. **32** (2014)
19. Esmaeilifard, M., Rahbar, A.G.: A high capacity energy efficient approach for traffic transmission in cellular networks. J. Telecommun. Inf. Technol. (2015)
20. Huang, J., Zhong, Z., Huo, H.: A dynamic energy-saving strategy for green cellular railway communication network. Springer-EURASIP J. Wirel. Commun. Netw. **89** (2015)
21. Xiang, N., Li, W., Feng, L., Zhou, F., Yu, P.: Topology-aware based energy-saving mechanism in wireless cellular networks. In: IEEE International Symposium on Integrated Network Management, pp. 538–544 (2015)
22. Ho, C.K., Yuan, D., Lei, L., Sun, S.: Power and load coupling in cellular networks for energy optimization. IEEE Trans. Wirel. Commun. **14**(1), 509–519 (2015)

Architecture and Software Implementation of a Quantum Computer Model

Victor Potapov, Sergei Gushansky, Vyacheslav Guzik and Maxim Polenov

Abstract This paper considers the principles of architecture models of quantum calculators. It describes the existing problems of construction and implementation of their work, as well as ways to overcome these problems. The distinction model of the relevant modules in its composition is achieved. Withdrawal functionality number of three parts modules, their graphics (interface) components, which are produced as a result of differentiation of the model into separate modules included in its composition. We describe the interface of the model and place it in the auxiliary modules and libraries.

Keywords Qubit · Schrödinger equation · Decoherence · Quantum register · Entangled state · Modeling · Quantum computing · Model · Module · Modular structure · Calculator · Model of quantum calculator

1 Introduction

Currently in the world, including Russia, the company is actively working the study and physical implementation of quantum calculator. Prototypes of computers have already been built in various parts of the world at different times, but had not released yet been a full-fledged quantum computer that makes sense to perform

---

V. Potapov (✉) · S. Gushansky · V. Guzik · M. Polenov
Department of Computer Engineering, Southern Federal University, Taganrog, Russia
e-mail: vitya-potapov@rambler.ru

S. Gushansky
e-mail: smgushanskiy@sfedu.ru

V. Guzik
e-mail: vfguzik@sfedu.ru

M. Polenov
e-mail: mypolenov@sfedu.ru

R. Silhavy et al. (eds.), *Software Engineering Perspectives and Application in Intelligent Systems*, Advances in Intelligent Systems and Computing 465,
DOI 10.1007/978-3-319-33622-0_6

simulation of quantum computation on a computer with a classical architecture to explore and further construct quantum calculator. The goals of the construction of models are very different from modeling quantum channel data in cryptography to the simulation of quantum algorithms on the group qubit [1], so the models are constructed in completely different ways and approaches.

Model quantum computer is a kind of quantum calculator model interface, in which one party is represented by a person (user), and the other—by the model itself, which is a set of tools and methods of interaction with the real quantum device.

2 Architecture Calculator

Quantum methods for performing computing operations, as well as the transmission and processing of information, are already beginning to be translated into actual functioning experimental devices that stimulate efforts to implement quantum computers. A quantum computer consists of n qubits and allows one- and two-qubit operations on any of them (or any pair). These operations are performed under the influence of external field pulses controlled by classical computer.

Quantum register [2] is a collection of a number L qubits. Before entering information into the computer all of the qubits of quantum register should be listed in the base states |0>. This operation is called a preparation or initialization. Next, qubits certainly (not all) are subjected to selective external influence (for example, by external electromagnetic field pulses controlled by a classical computer), which changes the value of qubits, i.e., from the state |0> they pass to the state |1> (Fig. 1).

In this state the quantum register just goes into superposition of basic states, i.e. the state of the quantum register initially time will be determined by function:

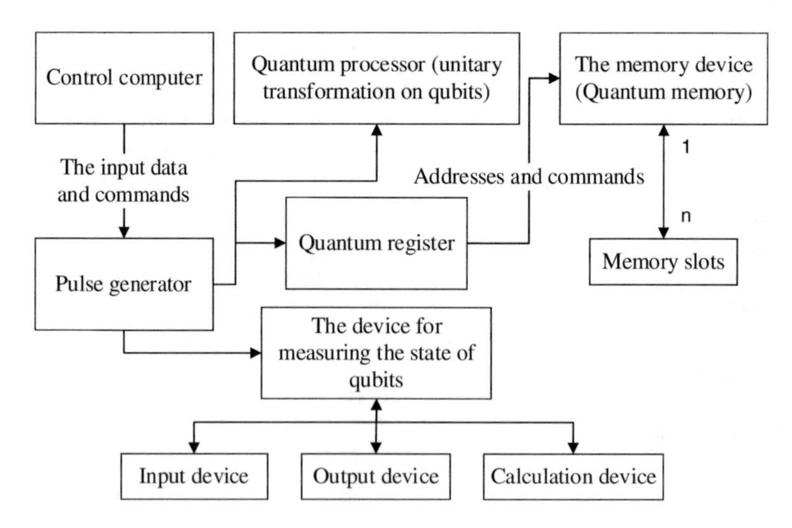

Fig. 1 Schematic structure of a quantum computer

$$|\Psi(0)> = \sum_{n=0}^{2^l-1} c_n|n> \tag{1}$$

In the quantum processor entries are subjected to a sequence of quantum logic operations. As a result, after a number of cycles of the quantum processor operation the initial quantum state of the system becomes a new kind of superposition:

$$|\Psi(1)> = \sum_{n,m=0}^{2^l-1} c_n U'_{mn}|n> \tag{2}$$

After the implementation of reforms in quantum computer, a new feature of superposition is the result of calculations in a quantum processor. We can only assume the values obtained, which are measured values of the quantum system. The final result, a sequence of zeros and ones, and, because of the probabilistic nature of the measurements, it can be any. Thus, quantum computer could have a chance to give any answer. In this scheme quantum computing is considered to be correct, if the correct answer is obtained with a probability sufficiently close to one. After repeating the calculation several times and selecting the answer that occurs most often, you can reduce the possibility of error to an arbitrarily small value.

One of the basic concepts of quantum, as well as classical information theory is the concept of entropy being a measure of the lack of (or uncertainty) information about the actual state of physical system. The description of a quantum system isolated from the environment needs to use the concept of net state, which is characterized by inexact values of coordinates and momenta, and a psi-function ψ [psi] (x, t) (x—a complete set of all continuous and discrete variables that determine the state of a quantum system, for example, it can be coordinate points, the polarization, the spin variables of all the particles, etc.). This complex wave function [3] allows to describe the properties of particles and determine the probability of certain events. The equation in the coordinate representation of Schroedinger [4] or the Heisenberg energy, which is subjected to this function, is a linear differential equation, and in this respect the behavior of the psi-function is perfectly computable and predictable, unlike the behavior described by its quantum objects:

$$|ih\frac{d}{dt}|\Psi(t)> = H'|\Psi(t)> \tag{3}$$

where ħ—Planck's constant, and Ĥ—some special self-adjoint operator, which is called the Hamiltonian operator or Hamiltonian of the system and is defined as the sum of the kinetic energy T and the potential function U:

$$H' = T^{\wedge} + U' \tag{4}$$

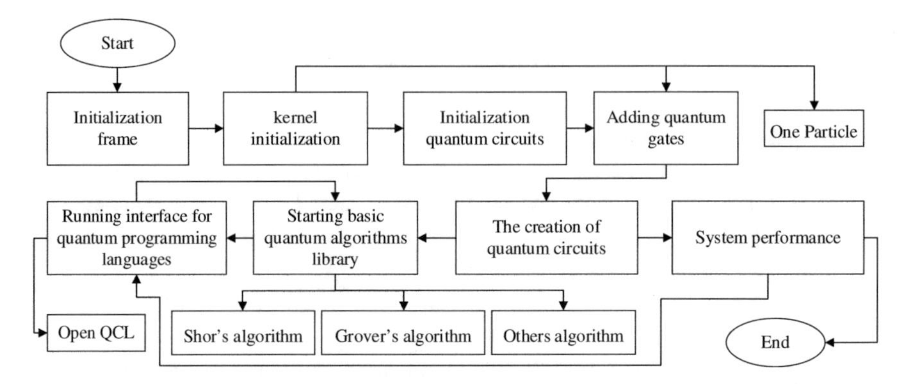

Fig. 2 The initialization of the system

3 The Interface of Model Quantum Calculator

Consider the algorithm of the calculator and the interaction of its various components. The dependence of the components from each other is very high. To begin with, consider the algorithm initialization components and analyze the relationship between the components as shown in Fig. 2.

After frame initialization you should first initialize the modeling kernel that has a default number of qubits in the system. Based on these data the quantum system is initialized, and the modeling of kernel displays the data on the qubits from the nucleus.

3.1 Common Interface Model

Common Interface of developed model [5] is shown in Fig. 3. In left-hand side are control buttons of quantum circuit. This area of the program allows automatic or incremental progress of model forward or backwards, you can also remove the last selected and entered into the scheme member or to completely clean the entire scheme. Above is a menu bar to control and configure the model, the top center is a set of quantum gates, below a state diagram of x-register and y-registers of entangled/not entangled qubits involved directly in the model. Quantum entanglement is a quantum mechanical phenomenon in which the quantum states of two or more qubits are interdependent. The entanglement between the qubits as a prerequisite for any quantum model of the calculator is a key factor responsible for determining the quantum parallelism and quantum advantage over conventional calculator. It is worth noting that the range of objects simulation of developed model is quite wide: gates, quantum algorithms and schemes that we ourselves create by adding the necessary gates, the behavior of particles, not to mention the intricacies of modeling implicitly.

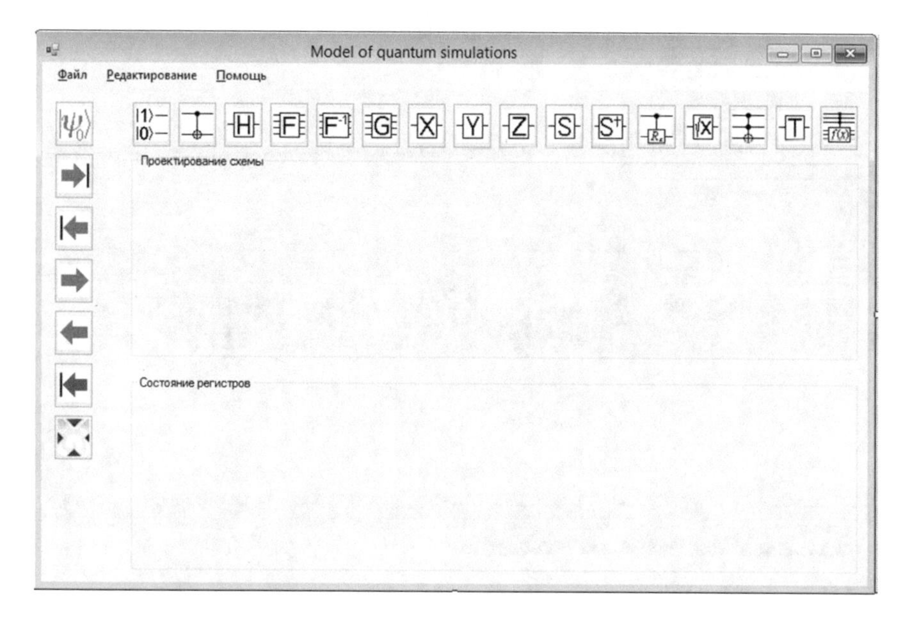

Fig. 3 The interface developed simulation environment

To get started with MQS you should initialize quantum circuit by pressing $|\psi_0\rangle$. This will lead to the conclusion of the window "Dimensions registers". It sets the number of branches of x-register and y-register. The field "design scheme" displays an empty quantum circuit in accordance with the entered values form "Dimensions registers".

Quantum scheme [6] is generated and updated automatically after adding a new gate in the scheme. Figure 4 shows the operation of model, in which intricate interconnected quantum states of the system are presented in different colors (one-color cell, color chart), indicating that the ability of entanglement of states of the developed model. Entangled states can be characterized by the magnitude (extent) of entanglement. The absence of confusion is indicated by y-register (one red cell), with is due to the lack of gates on the branches of the register, in other words, the model in this case has nothing to confuse.

After selecting the qubit, which will be applied to the gate, and agreements with the dialogue in the quantum circuit will add a new gate in the case of pressing the button , which is used to automatically perform quantum circuits, mathematical modeling of the core will perform the corresponding operation. Sometimes it is useful to consider the entire process step by step implementation of quantum circuits. For this purpose, there is a button . This is particularly useful for the study of the intermediate results of a quantum simulator. Each user who logs on to the simulator is given two quantum registers: x-y-register and register. You can use one of them.

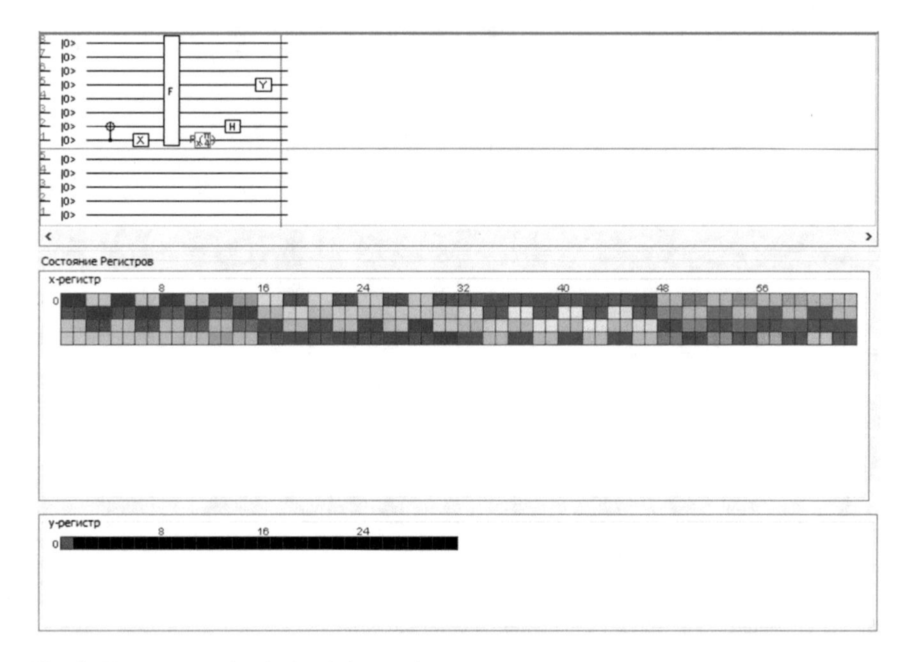

Fig. 4 The quantum circuit simulation environment

State of the quantum register at time t is not exactly the definition and is described by a linear combination with complex coefficients of m-bit states of the form

$$|\Psi> = c_1|0100101110 \quad 0011> + c_2|1100101110 \quad 0011> + \ldots \tag{5}$$

and the likelihood that the case is in the state $|01001011100011>$ is $|c_1|^2$, the probability of being in state $|11001011100011>$ is $|c_2|^2$, etc.

3.2 *Auxiliary Modules and Libraries in the Model of Quantum Calculator*

One of the most productive methods to increase the functionality of the model is to supplement the model with the subsidiary external libraries and modules. Developed and described earlier model [7] has been supplemented by external functionality, displayed in Fig. 1 in the lower left corner (highlighted in red), and called by pressing the corresponding button on the main form (Fig. 5).

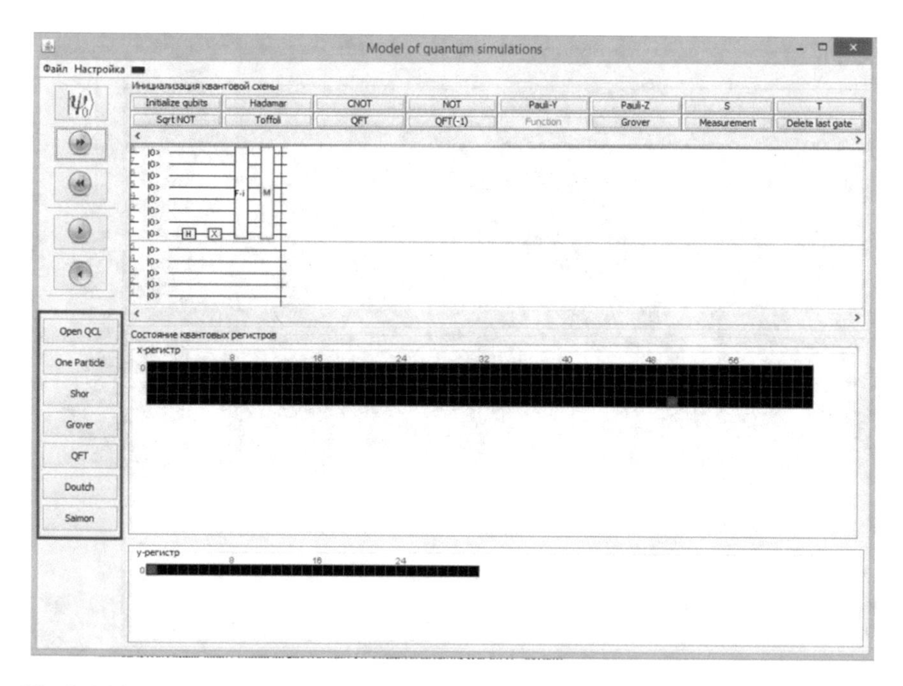

Fig. 5 Main window module (MQS) model of quantum calculator

The structure of plugins includes:

1. Open QCL. By clicking the button «Open QCL» started to write programs form a quantum programming language QCL (Fig. 2). QCL simulates a software environment, providing a classical program structure quantum data types and special functions to perform operations on them (Fig. 6).
2. One Particle. Java-applet simulating quantum mechanics, which describes the behavior of a single particle in bound states in one dimension. It solves the Schrödinger equation and allows visualization solutions.
3. Shor. If you press one of the keys located in the lower left corner, there will be an emulation of the corresponding quantum algorithm [8]: Shore (Fig. 3), Grover, Simon, Deutsch or the quantum Fourier transform (Fig. 7).

The result of the model should be considered:

- quantum circuits to store it in a separate file and load the parties;
- color chart model quantum computer;

The result of the quantum register in Fig. 4 can be interpreted and presented in mathematical (complex) with the help of a color table of probabilities/amplitude states of the qubit, which can then be used to study the degree of entanglement of certain pairs qubit model. Displays information about the state of the qubit system

```
C:\Windows\system32\cmd.exe
QCL Quantum Computation Language (32 qubits, seed 1428417371)
! Can't open default.qcl
[0/32] 1 |0>
qcl> qureg x1[2]
qcl> dump
: STATE: 2 / 32 qubits allocated, 30 / 32 qubits free
1 |0>
qcl>
```

Fig. 6 Interface QCL

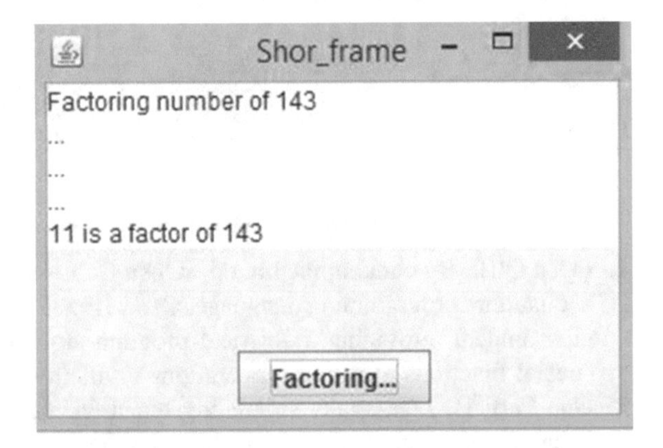

Fig. 7 Interface module factorization

using colors provides a more visual data that is easier to take than the numbers, but for the sake of completeness that it is sometimes not enough (Fig. 8).

- The result of the form to write programs on a quantum programming language QCL;

Naturally, that is QCL can be programmed only very small quantum computing. But this is enough to test the basic algorithms of quantum computing and work them before they will be able to use in full-scale quantum device, which is very useful.

- Visualization of the behavior of a single particle in bound states in one dimension;

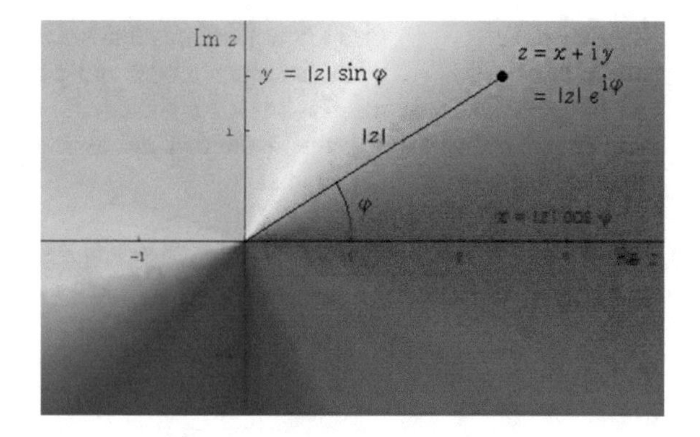

Fig. 8 Decomposition of flowers coloring cells

This external unit is able to visualize and simulate the behavior of a single particle in bound states in one dimension and, as a consequence, analyze and predict its future behavior by varying the available parameters. For example, changing the amplitude.

- The results of calculations of specific quantum algorithms.

The developed model is equipped with a set of specific simulation of quantum algorithms. The article is illustrated by simulation and made a conclusion the results of Shor's algorithm, the factorization of the number 143 (Fig. 7).

4 Conclusion

The interface of the program is no means the last aspect, which should be given to the development of quantum model of the calculator. It is intuitive and visually intuitive user interface simplifies the process of learning and working. The developed model stands out among its peers convenience, features and clarity. The main advantage of the developed modeling tools to the existing analogues is a modular architecture that allows use of mathematical models of several nuclei. Further studies will improve the GUI environment modeling and the possibility of its setting, as well as increase the functionality through the development of the following areas:

- Use other libraries, API, for comparative analysis of the performance and capabilities;
- Increasing the number of operators used;
- Update the graphics editor of quantum circuits new functionality that extends the current editing quantum circuits;

Analyzed and developed a set of features that will be implemented in a computer and described the core of the interface model and place it in the auxiliary modules and libraries. Was launched a series of third-party modules, the functionality of their graphics (interface) components that run as a result of the model in separate modules, included in its composition.

Acknowledgments This work was supported by RFBR, grant number 15-01-01270 NC\15.

References

1. Guzik, V.F., Gushansky, S.M., Potapov, V.S.: The operation algorithm of fundamental elements model of quantum calculator. In: Krukier, L.A., Muratov, G.V., Topolov, V.Y. (eds.) Modern Information Technology: Trends and Prospects: Materials of Conference, pp. 160–162. Southern Federal University Publishing House, South Federal University—Rostov-on-Don (2015)
2. Guzik, V., Gushanskiy, S., Polenov, M., Potapov, V.: Models of a quantum computer, their characteristics and analysis. In: 2015 9th International Conference on Application of Information and Communication Technologies (AICT), pp. 583–587. Institute of Electrical and Electronics Engineers (2015)
3. The wave function. https://ru.wikipedia.org/wiki/Wave_function. Accessed 27 Nov 2015
4. Schrödinger equation. https://ru.wikipedia.org/Schrödinger_equation. Accessed 28 Nov 2015
5. Guzik, V.F., Gushansky, S.M., Potapov, V.S.: Simulation models of quantum calculator. RF Patent number 2015662926, Russian Patent № 2015619794. Accessed 07 Dec 2015
6. Guzik, V.F., Gushansky, S.M., Polenov, M.Yu., Potapov, V.S.: Implementation of the components for the construction of an open modular models of quantum computing devices. Informatization Commun. **1**, 44–48 (2015)
7. Guzik, V.F., Gushansky, S.M., Potapov, V.S.: The model of quantum computing. Education and science: the current state and prospects of development. In: Collection of Articles of the International Scientific and Practical Conference, pp. 761–765. Tambov. Accessed 29 Sept 2015
8. Guzik, V.F., Gushansky, S.M., Potapov, V.S.: Difficulty rating quantum algorithms. Technology of development of information systems. In: Collection of Articles of the International Scientific and Practical Conference, pp. 81–85. Gelendzhik. Accessed 01 Sept 2014

MEEM: A Novel Middleware for Energy Efficiency in Mobile Adhoc Network

P.G. Sunitha Hiremath and C.V. Guru Rao

Abstract Achieving an enhanced network lifetime is still an open issue in mobile ad hoc network. After years of research and invention, a robust technique to ascertain energy efficiency is yet to be explored especially to overcome an adverse effect of dynamic topology and selection of highly stabilized links during routing. Therefore, we present a very simple and cost-efficient middleware called as MEEM i.e. Middleware for Energy Efficiency in Mobile ad hoc network that offers an effective solution towards this problem. MEEM offers cost effectiveness by enabling the middleware to increase its scope of selection of stabilized routes based on formulating new multiple decision-making parameters e.g. Time for Route Termination, Time for Stable Routing, Frequency of Route Error, Reduction in Signal Attenuation, and Remnant Network Lifetime. The design of MEEM is carried out using quadratic approach thereby retaining parallel processing of an algorithm to enhance the communication performance of the mobile nodes. The outcome of the study was found to excel better results in comparison to frequently used routing protocol e.g. Adhoc On-demand Distance Vector (AODV) and Optimized Link State Routing (OLSR) with respect to energy consumption and algorithm processing time. The algorithm is found to be compliant of time and space complexity thereby results in cost effective solution.

Keywords Energy efficiency · Middleware · Mobile adhoc network · Network lifetime · Route selection · Stabilized links

P.G.S. Hiremath (✉)
Department of Information Science & Engineering, BVBCET, Hubli, India
e-mail: sunitharesearch64@gmail.com; pgshiremath@bvb.edu

C.V.G. Rao
S R Engineering College, Warangal, India

R. Silhavy et al. (eds.), *Software Engineering Perspectives and Application in Intelligent Systems*, Advances in Intelligent Systems and Computing 465,
DOI 10.1007/978-3-319-33622-0_7

1 Introduction

A mobile ad hoc network is considered as one of the cost-effective communication technique for emergency [1]. However, owing to the dynamic topology in mobile ad hoc network, it is associated with multiple challenges [2]. A mobile node always dissipates energy whether it is active state, passive state or in sleep state [3]. Because of uncertainty in energy retention scheme among the nodes, there is always a risk of broken links [4], which is highly detrimental to the communication principle. For more than two decades, there has been enough evolution of problems towards energy efficiency in mobile ad hoc network. But till date, there is no evidence of 100 % solution towards energy efficiency owing to over-burdened responsibilities towards resource-constraint nodes [5]. Hence, introducing middleware system can significantly reduce the load of routing and multiple operations that are carried out by mobile nodes in ad hoc network and thereby introduce enough energy efficiency. However, retaining energy efficiency in mobile ad hoc network is not simpler and has multiple challenging factors as an impediment. Till date, there is no substantial study towards encouraging usage of middleware over ad hoc network to enhance network lifetime. The present paper discusses an efficient middleware design that targets to accomplish energy efficiency in the presence of dynamic topology of mobile ad hoc network. Section 2 discusses significant literature towards middleware design in present system followed by brief discussion of problem statement in Sect. 3. Section 4 discusses the proposed model followed by research methodology in Sect. 5. Algorithm implementation is discussed in Sect. 6 followed by a discussion of the accomplished result in Sect. 7. Concluding remarks are discussed in Sect. 8.

2 Related Work

This section discusses the significant work being carried out towards designing middleware applications in mobile ad hoc network most recently.

Most recently, Pasricha et al. [6] have developed a middleware for the purpose of optimizing energy for the mobile devices. The study has considered the case of Android-based smartphones, where the authors have developed a middleware that can optimize the processing and energy utilization to be controlled in the highest degree. The outcome of the study was found to conserve around 29 % of energy. However, the study considers ad hoc network for performing networking and investigation. Another author named as Silva [7] has also dedicated their research towards evolving up with a middleware design exclusively for the mobile system. The focus of the study was into attaining scalability by developing a UDP protocol executed over the mobile nodes for addressing the unstable or broken links. The outcome of the study has also assured enhanced packet delivery ratio for the presented middleware that supports various forms of data delivery too. However, just like the previous study, this study also doesn't emphasize on ad hoc network.

An interesting concept of middleware was formulated by Haschem et al. [8] who presented a system for processing information gathered from mobile internet-of-things. The authors have implemented service-oriented architecture to enhance the usability and scalability of the middleware framework using non-deterministic approach. The outcome of the study was assessed using response time of the system. Mehrotra et al. [9] have developed a middleware system focusing on social networking system. The system also uses contextual data from the social network applications from the information that is sensed from the mobile. It also makes use of a centralized server. The evaluation of the study was hypothetically carried out over multiple databases on server mainly focusing on context awareness. Akingbesote et al. [10] have developed a framework for middleware focusing on the healthcare sector. The authors have incorporated middleware layer in between the multimedia interface and grid infrastructure layer. The outcome of the study was evaluated with respect to waiting time on increasing server utilization. Gherari et al. [11] have introduced another middleware system that using profiling approach over cloud interface. The middleware is designed over a sophisticated architecture using contextual information of both cloud and mobile interface. However, the study doesn't emphasize much on data analysis.

Nikzad et al. [12] have presented a middleware that is responsible for maintaining energy efficiency on the mobile application considering Android operating system. The experimentation was carried out using sensed data from the mobile communication system where the outcome of the study shows 64 % of energy conservation. Similar sort of studies was also presented by Makki et al. [13] by presenting a middleware for android device focusing on the security aspect of mobile devices. Mohapatra et al. [14] have developed a middleware for enhancing the energy conservation for the mobile devices. Similar direction of the study was also carried out by Bajwa [15] by presenting a middleware for maintaining interoperability and integration in e-commerce. Lin et al. [16] have presented a schema of middleware for a mobile application using Bayesian approach using experimental approach. Zhuang et al. [17] have presented a scheme for middleware services for sensing processed information captured from the mobile device. Hence, it can be seen that that there are enough works being carried out in the direction of middleware design. The next section highlights the problem statement of the proposed study.

3 Problem Statement

The existing studies towards middleware are developed mainly focusing on the mobile devices. However, challenges involved in ad hoc network are quite higher in the form of computational complexity. Even if such middleware is design, the next challenge will relate to the selection process of stabilized links. At present, the frequently existing routing protocols e.g. AODV [18] and OLSR [19] are used for routing. Hence, if a middleware will be designed it will have to solely depend on the effectiveness of such routing protocols, which are associated with both advantages and disadvantages too. A successful and cost-effective design of middleware will call for incorporating a good interface between the middleware components and routing protocols along with energy efficiency. Unfortunately, enough studies have not been carried out in the direction of the mobile ad hoc network. This causes quite an uncertainty, in theory, formulation about how to define a novel middleware technique that can perform an efficient and faster selection of the routing in mobile ad hoc network. Besides, the present routing decision for selection of stabilized link is carried out by residual energy, which is not enough. Hence, there is a need for designing a novel middleware system that can take more information about the stabilized and energy-efficient links considering the problem of spontaneous dissipation of energy among mobile nodes owing to issues of dynamic topology. The problem statement of the study can be stated as —*It is a computationally challenging task to develop an efficient middleware for restoring significant network lifetime of mobile nodes in adhoc network.* The next section presents a discussion about the proposed system that addresses the problems of developing a middleware system for enhancing energy and network lifetime in mobile adhoc network.

4 Proposed System

The prime purpose of the proposed system is to design a novel middleware system that can perform significant conservation of energy among the nodes in mobile ad hoc network. The study introduces a technique called as MEEM (Middleware for Energy Efficiency in Mobile ad hoc network). The present study is an extension of our prior design of middleware system called as MERAM (Message Exchange with Resilient and Adaptive Middleware system) [20]. MERAM was designed for mitigating the replication issues of a message in order to facilitate quick exchange of message for time-critical applications in mobile ad hoc network. The advantages of MERAM are (i) higher resilience to link failures on dynamic topology, (ii) applicable on delay tolerant protocol, (iii) outcomes witnessed with increased delivery probability and reduced message exchanging time. However, MERAM didn't focus on energy efficiency, which may result in minimization of network lifetime. Therefore, in order to accomplish the objective of energy-efficiency over MERAM, the present study performs following contribution:

- To design an ad hoc-based connectivity for wireless access technology that can track and surveil the emergency situation.
- To develop a middleware system that can minimize or control energy drainage from the mobile nodes for data dissemination process in mobile ad hoc network.

5 Research Methodology

The proposed system MEEM considers analytical methodology and intends to design an extended version of the proposed middleware system for incorporating further resiliency against drainage of battery and thereby enhancing the lifetime of hybrid mobile ad hoc network. The proposed system will design a framework on mobile application that is purely on the basis of the MERAM system. The nodes in mobile adhoc network (e.g. Smartphone, laptops, tablets etc.) can always have multiple radio interfaces, although the conventional research-based study only consider single radio interface. Therefore, the proposed system considers multiple radio interfaces for connecting various mobile nodes and to increase data rates. In order to minimize any forms of radio frequency interference, the proposed MEEM considers smaller transmission range. The design of this part of the system will be based on message-oriented middleware system as well as to provide an asynchronous communication system between the communication mobile nodes. The prime focus of this part of the study is to develop an analytical modeling of energy dissipation during the communication between the two mobile nodes. The intention is being to understand the cost involved in the transmission in ad hoc nature and give a proper solution to it and thereby present a robust middleware system. The proposed MEEM is a technique that will run over the mobile nodes to perform a certain operation which is mainly associated with communication control for ensuring energy efficiency. MEEM will be responsible for reviewing the amount of various resources being used during routing and it will offer an empirical means of evaluating the following:

- **Time for Route Termination (TRT)**: This is the time witnessed after an established route expires.
- **Time for Stable Routing (TSR)**: This factor calculates the time between two nodes with sufficient residual energy. We say stable routing to be only formulated between two nodes of sufficient level of remnant energy.
- **Frequency of Route Error (FRE)**: This variable calculates the total number of the occurrences of error witnessed in a particular route.
- **Reduction in Signal Attenuation (RSA)**: This parameter checks for the level of drop of signal attenuation as a quality signal to be evaluated.
- **Remnant Network Lifetime (RNL)**: It is the approximated residual power of all the nodes in the simulation network.

The system architecture used in MEEM is highlighted in Fig. 1, which shows that MEEM performs aggregation of the information for TRT, TSR, FRE, RSA, and RNL on every routing cycle. All these metrics are used for finalization of one probable route to be established, hence, MEEM is an algorithm of the first kind in MANET where the establishment of stabilized link depends on multiple parameters. All these parameters can be taken from the novel design of a control message

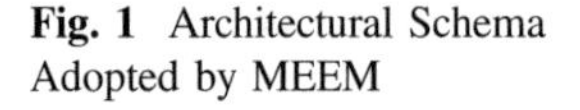

Fig. 1 Architectural Schema Adopted by MEEM

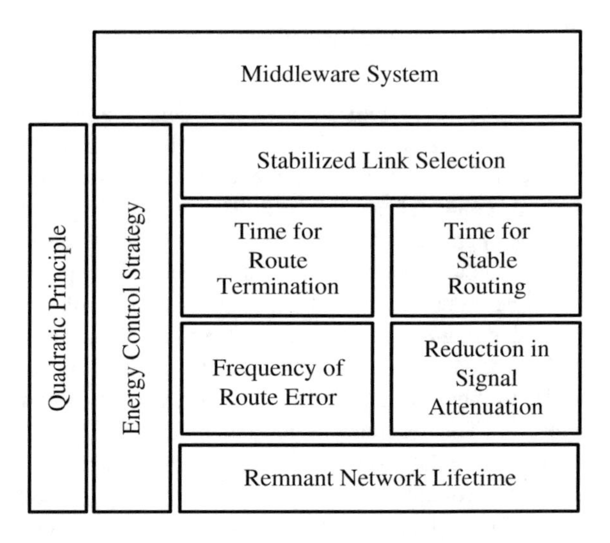

shown in Fig. 1. Therefore, as a consequence MEEM minimizes the probability of broken links as now the routing technique have more decision parameters, which are not found in conventional energy efficient routing protocols in MANET [21]. The mobile nodes are considered to follow the multi-path propagation of the message for faster message delivery in proposed schema.

The next section discusses the algorithms that are designed for proposed middleware for energy efficiency in mobile ad hoc network.

6 Algorithm Implementation

This section discusses the implementation of the proposed MEEM. The proposed system considers *n* number of mobile nodes that are positioned in simulation area using random mobility model. The design of the algorithm mainly emphasized on accomplishing following novel components of middleware system in ad hoc network as Time for Route Termination (TRT), Time for Stable Routing (TSR), Frequency of Route Error (FRE), Reduction in Signal Attenuation (RSA), Remnant Network Lifetime (RNL). The brief discussion of the algorithms is as follows:

Algorithm for Time for Route Termination (TRT)
Input: *n* (number of nodes), *r* (transmission range), u_1/u_2 (node speed), φ_1/φ_2 (orientation of two nodes), α β γ, δ (orientation-based route termination parameters),
Output: Time for Route Termination.
Start

1. Init *n*, *r*
2. Init node speed u_1 and u_2.

3. Init direction φ_1 and φ_2.
4. Apply Random Waypoint
5. Applying quadratic principle
6. $E_1 = (\alpha\beta + \beta\gamma)$
7. $E_2 = |\sqrt{(\alpha^2+\gamma^2)\cdot \arg_{\max} r^2 - (\alpha\delta - \gamma\beta)^2}|$
8. $E_3 = 1/(\alpha^2 + \gamma^2)$
9. $E = E_1 \cdot E_2 \cdot E_3$
10. $TRT = \arg_{\min}|E|$

End

Normally, the termination of the route takes place owing to two reasons e.g. (i) dynamic topology, (ii) sudden node death. Hence, MEEM addresses both the problems by TRT algorithm. After suitable initialization of nodes (*n*), transmission range (*r*), speed and orientation of two communicating nodes (u_1/u_2 and φ_1/φ_1), we apply random waypoint as the mobility model and quadratic approach is considered for algorithm formulations. The prime reason behind the adoption of quadratic approach is to incorporate optimization in the information for better middleware design. The system formulates three entities E_1, E_2, and E_3 corresponding with the standard variables (*a*, *b*, *c*) in any quadratic approach. The parameter α computes the cosine difference of first node speed ($u_1 \cdot \cos \varphi_1$) with second node speed ($u_2 \cdot \cos \varphi_2$). Similarly, The parameter δ computes the sinusoidal difference of first node speed ($u_1 \cdot \sin \varphi_1$) with second node speed ($u_2 \cdot \sin \varphi_2$).The parameter β and δ are the positional difference between two nodes and corresponds to $x_1 - x_2$ and $y_1 - y_2$ respectively. Finally, the time is computed that assist proposed middleware to decide about the selection process of a route during the route discovery process in mobile ad hoc network.

Algorithm for Time for Stable Routing (TSR)
Input: C_E (Cut-off energy of a node), n_s/n_d (nodes), E_{sd} (remnant energy of source-destination)
Output: Time for Stable Routing
Start

1. init C_E
2. If $E_{sd} > C_E$
3. $n_s \rightarrow n_d$
4. or else,
5. reject (n_d)
6. search_next(n_d);
7. Evaluated time ($n_s \rightarrow n_d$)

End

The TSR algorithm is responsible for exploring the total time found with the higher probability of stabilized link. An algorithm is designed with a cut-off energy level of a node, which is a permissible battery level till which the node functions properly. It can be different for different applications. If the sender node finds the remnant energy of its neighbor node to be less than cut-off energy, it is believed that such link formation will not be reliable to carry forward the data and hence rejected. Hence, link formation is only supported by proposed middleware of the remnant energy is more than cut-off.

Algorithm for Frequency of Route Error (FRE)
Input: *rec_msg* (RREQ), C_{err} (Cut-Off Error)
Output:
Start

1. Capture *rec_msg* (n_d) & Calculate BER
2. if (BER>C_{err})
3. remove those nodes
4. or else
5. Consider those nodes for communication

End

The above algorithm is responsible for the selection of route based on the error rate. In order to do so, our middleware system captures the control message from the communicating nodes and keeps on computing bit error rates. A cut-off error rate can be defined based on different applications which will be compared with accomplished BER. In the case of permissible limit, proposed middleware will give positive feedback for selection of such communicating nodes to establish a link.

The next part of the implementation will focus on capturing the signal attenuation (in dB) (RSA) while the MEEM can keep on collecting the remnant energy information of the entire node in the simulation for analysis purpose (RNL). The prime performance will be observed from the extent of energy consumption in mobile ad hoc network.

7 Results and Discussion

The outcome of the proposed system is evaluated with respect to energy consumption per node in Joule and algorithm processing time in second. The proposed MEEM is compared with the frequently used AODV and OLSR routing algorithm by using uniform simulation parameters. The simulation study considers 500–1000 mobile nodes with a variation of 100–550 m of transmission range. With omni-antenna considered on each node, the initialized power is 0.5 J considering MAC protocol for IEEE 802.11 standard.

Figure 2 shows the comparative analysis fo proposed middleware-based communication system on mobile ad hoc network with AODV and OLSR. The sequence number of AODV is higher but definitely not the latest, for which purpose each intermediate node has to allocate an extra energy in case of broken links. Owing to a periodic delivery of control message, OLSR keeps more updated route entries compared to AODV, but it requires an extra processing power to do so. However, proposed MEEM undertakes this decision of routing based on Time for Route Termination (TRT), Time for Stable Routing (TSR), and Frequency of Route Error (FRE). Hence, the amount of information required to avoid re-routing as well as reliable communication is quite high enough compared to AODV and OLSR and hence can take decision faster thereby requiring less processing power. In increasing iterations, the middleware becomes less dependent on routing information and hence processing power increases in a very slower pace thereby restoring sufficient amount of energy (Fig. 3).

Algorithm processing time is one of the significant parameter for scaling the effectiveness of the proposed middleware system. The outcome shows that processing time of MEEM is considerably less than existing routing mechanism e.g. AODV and OLSR. Although delay for establishing the connection is less in AODV but it requires a massive channel capacity to process the message for 500-1000

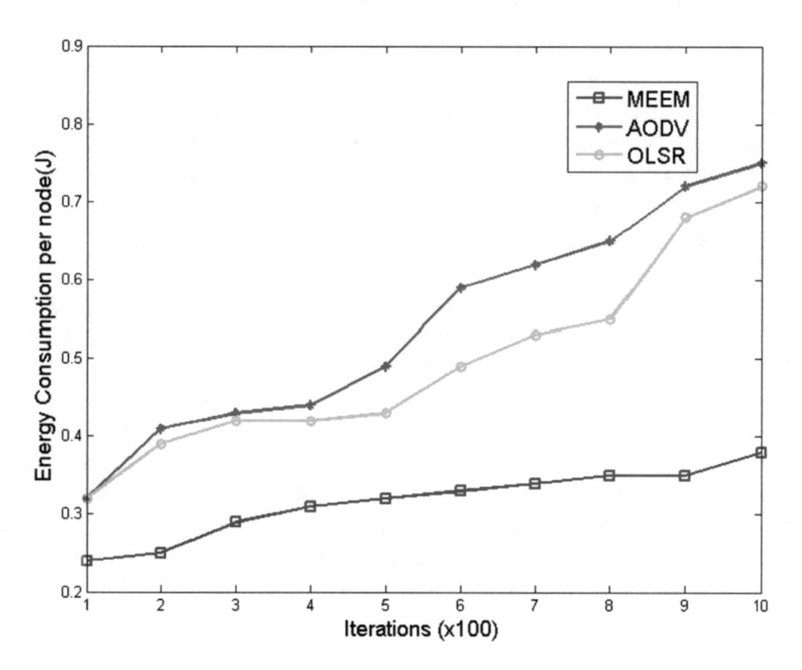

Fig. 2 Comparative Analysis of MEEM, AODV and OLSR (Energy)

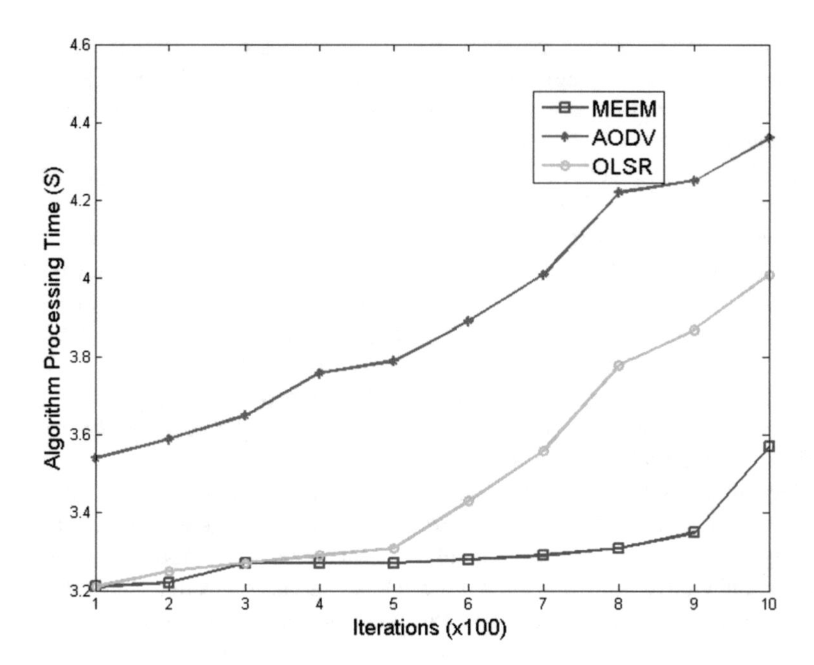

Fig. 3 Comparative Analysis of MEEM, AODV and OLSR (Algorithm Processing Time)

mobile nodes. This increases the time of construction of routes. On the other hand, OLSR is not at all dependent on such control message to ascertain the link stability. Unfortunately, OLSR consumes more processing time to find out the unstable routes as well as any broken links. We address this problem in MEEM by using cut-off level of remnant energy which avoids the routing decision to be made on less stabilized nodes. Similarly, the system performs continuous monitoring of Time for Route Termination (TRT) and Time for Stable Routing (TSR) that significantly accumulates up with relevant routing data thereby requiring fewer algorithms processing time. Moreover, the cut-off values are completely dependent on the application which means MEEM can be customized for any futuristic applications of MANET.

From storage complexity viewpoint, the algorithm doesn't require more than 20–30 Kb of space for 5000 iterations. This fact will mean that proposed MEEM can be used for any low-powered embedded mobile device in ad hoc network. Therefore, the proposed middleware is found with higher energy preservation, lower algorithm processing time and it is highly compliant of time and storage complexity from the performance of the MEEM viewpoint.

8 Conclusion

There are various adverse effect of dynamic topology where the prominent one is on energy efficiency and routing. For more than a decade there has been an enough evolution of various energy efficient technique as well as the routing protocol for delivering better communication standards. However, till date, there is no such existence of energy efficient routing that can overcome the adverse effect of dynamic topology in mobile ad hoc network. This paper appraises that there is an adoption of middleware and has been quite a common in mobile devices; however, there was a less attempt towards investigation of middleware towards energy efficiency. This paper has introduced a technique called as MEEM that formulates multiple parameters e.g. Time for Route Termination (TRT), Time for Stable Routing (TSR), Frequency of Route Error (FRE), Reduction in Signal Attenuation (RSA), and Remnant Network Lifetime (RNL) for selection of stable routes. The outcome of the simulation study was compared with frequently used AODV and OLSR routing protocols to see superior performance in energy conservation and lowered processing time.

References

1. Basagni, S., Conti, M., Giordano, S., Stojmenovic, I.: Mobile Ad Hoc Networking: The Cutting Edge Directions, John Wiley & Sons Technology & Engineering, pp. 888 (2013)
2. Gour, K.: Mobile Multimedia Communications: Concepts, Applications, and Challenges: Concepts, Applications, and Challenges, IGI Global Computers, pp. 420 (2007)
3. Tan, K.-L., Franklin, M.J., Lui, J.C.-S.: Mobile Data Management: Second International Conference, MDM 2001 Hong Kong, China, 2001 Proceedings, Springer Computers, pp. 290, 8–10 Jan 2001
4. Fischhoff, B.: Communicating Risks and Benefits: An Evidence-Based User's Guide, Government Printing Office, Health & Fitness, pp. 240 (2012)
5. Ahsan Rajon, S.A.: Energy Efficient Data Communication for Resource Constrained Systems, pp. 85. LAMBERT, Academic Publishing (2014)
6. Pasricha, S., Donohoo, B.K., Ohlsen, C.: A middleware framework for application-aware and user-specific energy optimization in smart mobile devices. Pervasive Mob. Comput. (2015)
7. Silva, L.D.N: A Scalable Middleware for Context Information Provision and Dissemination in Distributed Mobile Systems, Depart of De Informatics (2014)
8. Hachem, S., Pathak, A., Issarny, V.: Service-Oriented Middleware for the Mobile Internet of Things: A Scalable Solution, HAL Achieves Overrates (2014)
9. Mehrotra, A., Pejovic, V., Musolesi, M.: SenSocial: a middleware for integrating online social networks and mobile sensing data streams. In: Proceedings of the 15th International Middleware Conference, pp. 205–216 (2014)
10. Akingbesote, A.O., Adigun, M.O., Xulu, S., Jembere, E.: Performance modeling of proposed GUISET middleware for mobile healthcare services in E-marketplaces. J. Appl. Math. (2014)
11. Gherari, M., Amirat, A., Oussalah, M.: Towards smart cloud gate middleware: an approach based on profiling technique. In: Conference francophone sur-les Architectures Logicielles (CAL) (2014)

12. Nikzad, N., Chipara, O., Griswold, W.G.: APE: an annotation language and middleware for energy-efficient mobile application development. In: Proceedings of the 36th International Conference on Software Engineering, pp. 515–526 (2014)
13. Makki, S.K., Abdelrazek, K.: An active middleware for secure automatic reconfiguration of applications for android devices. Int. J. Digit. Inf. Wirel. Commun. (IJDIWC), **4**(3), 284–291 (2014)
14. Mohapatra, S., Rahimi, R., Venkatasubranian, N.: Power-aware middleware for mobile applications (2011)
15. Bajwa, I.S.: Middleware design framework for mobile computing. Int. J. Emerg. Sci. **1**(1), 31–37 (2011)
16. Lin, K., Kansal, A., Lymberopoulos, D., Zhao, F.: Energy-accuracy aware localization for mobile devices. In: Proceedings of 8th International Conference on Mobile Systems, Applications, and Services (MobiSys'10) (2010)
17. Zhuang, Z., Kim, K-H., Singh, J.P.: Improving energy efficiency of location sensing on smartphones. In: Proceedings of the 8th international conference on Mobile systems, applications, and services, pp. 315–330 (2010)
18. Perkins, C., B-Roye, E., Das, S.: Ad hoc On-Demand Distance Vector (AODV) Routing, The Internet Society, RFC. 3561 (2003)
19. Clausen, T., Jacquet, P.: Optimized Link State Routing Protocol (OLSR), The Internet Society, Network Working Group, RFC. 3626 (2003)
20. Hiremath, P.G.S., Dr. Rao, C.V.G.: MERAM: message exchange with resilient and adaptive middleware system in MANET. In: IEEE International Conference on Computational Intelligence and Computing Research (ICCIC) (2015)
21. Chawda, K., Gorana, D.: A survey of energy efficient routing protocol in MANET. In: 2nd International Conference on Electronics and Communication Systems (ICECS), pp. 953–957, (2015)

A 3D Visualization Design and Realization of Otrokovice in the Nineteen-thirties

Pavel Pokorný and Markéta Mazáčová

Abstract This paper briefly describes a visualization method of the Otrokovice Municipality in the Nineteen-thirties. The rapid growth of this town began at the beginning of the 20th century, and in 1938, Otrokovice had around 8000 inhabitants. Many building plans, cadastral maps and historical photographs are preserved from this period. This is the reason why a rendering set in these years was created. All of the collected information was chronologically sorted, and on this basis, a 3D visualization of the Otrokovice town center was created. To begin with, a terrain model of Otrokovice was created, based on the altitude values published on the Internet. All buildings and accessories were separately modeled (the standard polygonal representation was used) and textured by the UV mapping technique. Then, a more complex 3D scene from the individual models and accessories was created. The visualization output is performed by rendered images and animations in these years.

Keywords Computer graphics · 3D visualization · Modeling · Texturing · Animation · Rendering

1 Introduction

Data visualization is a hot topic. A simple definition of data visualization says: It's the study of how to represent data by using a visual or artistic approach rather than the traditional reporting method [1]. Represented data are most often displayed through texts, images, diagrams or animations in order to communicate a message.

P. Pokorný (✉) · M. Mazáčová
Faculty of Applied Informatics, Department of Computer and Communication Systems, Tomas Bata University in Zlín, Nad Stráněmi 4511, 760 05 Zlín, Czech Republic
e-mail: pokorny@fai.utb.cz

M. Mazáčová
e-mail: mazacova.marketa@gmail.com

R. Silhavy et al. (eds.), *Software Engineering Perspectives and Application in Intelligent Systems*, Advances in Intelligent Systems and Computing 465,
DOI 10.1007/978-3-319-33622-0_8

Visualization today has ever-expanding applications in education, science, engineering (e.g. product visualization), interactive multimedia, medicine, etc. Typical example of a visualization application is the field of computer graphics. The invention of computer graphics may be the most important development in visualization since the invention of central perspective in the Renaissance period. The development of animation also helped the advance of visualization [2].

With the development and performance of computer technology, the possibilities and limits of computer graphics are still increasing. The consequences of this are 3D visualizations, which are used more frequently, image outputs get better quality [3] and the number of configurable parameters is rising [4]. As mentioned above, these visualizations are used in many scientific and other areas of human interest [5].

One of the fields is visualizations of history. Based on historical documents, drawings, maps, plans and photographs, it is possible to create 3D models of objects that exist no more—things, products, buildings or extinct animals. If we assign suitable materials and the corresponding textures to these models, we can get a very credible appearance of these historic objects. From these individual objects, we can create very large and complex scenes that can be very beneficial tool for all people interested in history.

This paper describes a 3D visualization method of Otrkovice in the Nineteen-thirties.

1.1 The History of Otrokovice

Otrokovice is a town in the Zlín Region, Czech Republic. It is located in a hilly area centrally located in the region called Moravia and is located on the Morava River.

The first written record of Otrokovice's existence dates back to 1141, when the Bishop, Henry Zídka, mentioned this village in his documents—where the ownership of the Olomouc Diocese was calculated. In the first half of the 14th Century, Otrokovice still remained in the Church's ownership and it subsequently passed into secular hands and formed the joint property of the Tečovice Estate. This estate was purchased by William Tetour of Tetovo in 1492, who was the owner of the nearby Zlín Estate and a Black Company commander in the service of Matthias Corvinus (King of Hungary) [6]. During this time, the whole estate developed into a successful economy [7].

In the early 18th Century, a total of 305 people lived in Otrokovice, 198 of them were peasants. The village recorded growth over the next few years, and in 1805, its own school was built. The rural character of the village began to diminish in the early 20th Century. In 1929, the Bata Company purchased a large parcel of ground in an area on the right bank of the Morava River, which was used as a suitable place for building houses and a new factory area. Construction took place along the main road and, among many other buildings, the Sokol Building was built in 1925

Fig. 1 The Sokol Building after 1925 [7]

Table 1 Number of inhabitants and houses in Otrokovice in 1930–1938 [7]

Year	1930	1932	1933	1934	1935	1937	1938
Inhabitants	2009	3300	3654	4788	5567	6571	8002
Houses	339	406	585	655	761	821	901

(Fig. 1). These various constructions significantly contributed to the expansion of the city boundaries. A few years later, Otrokovice was fully electrified [8].

Otrokovice became the most important offshoot of the Bata Company's Zlín factories during the Nineteens-thirties and grew into the second most important place in the Zlín District. The significant growth of the population was hardly experienced or equaled by any other community throughout Czechoslovakia at this time. Its growth during the 1930–1938 period, is described in Table 1 [7].

The 3D visualization, designed and created for this purpose, shows the town of Otrokovice at this time.

2 Resources and Software

The first phases that needed to be done were to collect all available historical materials relating to Otrokovice and to select appropriate programs for visualization creation purposes.

2.1 Acquiring Resources

The overall progress of this work was initiated by the collation of available historic materials and information about the town of Otrokovice in the Nineteen-thirties. The main resources were the Moravian Land Archive in Brno [9], the State District Archive in Zlín – Klečůvka [10], books and the Internet. Attention was mainly focused on building plans, cadastral maps and historical photos.

The cadastral maps of that period are a very important part of the visualization process. In the Moravian Land Archive in Brno, there are two of these maps of Otrokovice available—from 1877, and 1926. Both maps were acquired and, because they were divided into several parts, they were loaded and combined in the GIMP software [11] in order to obtain the entire maps. These maps are colorized and in them, it is easy to distinguish wooden or brick buildings, gardens, roads, rivers, fields, etc. A part of the cadastral map of the center of Otrokovice, dating from 1877, is shown in Fig. 2. We used these maps for the creation of the model's terrain (e.g. the proper placement of edges and faces), terrain texture drawings, ground plans of buildings and the correct placements and orientations of the 3D models of buildings and accessories in the final scene.

Period photographs are the next important resources. We used two historical books [8, 12] that were published by the town of Otrokovice. In addition, in these books, the period photographs are supplemented by textual information, which complements the former conditions. The other suitable historical resources were

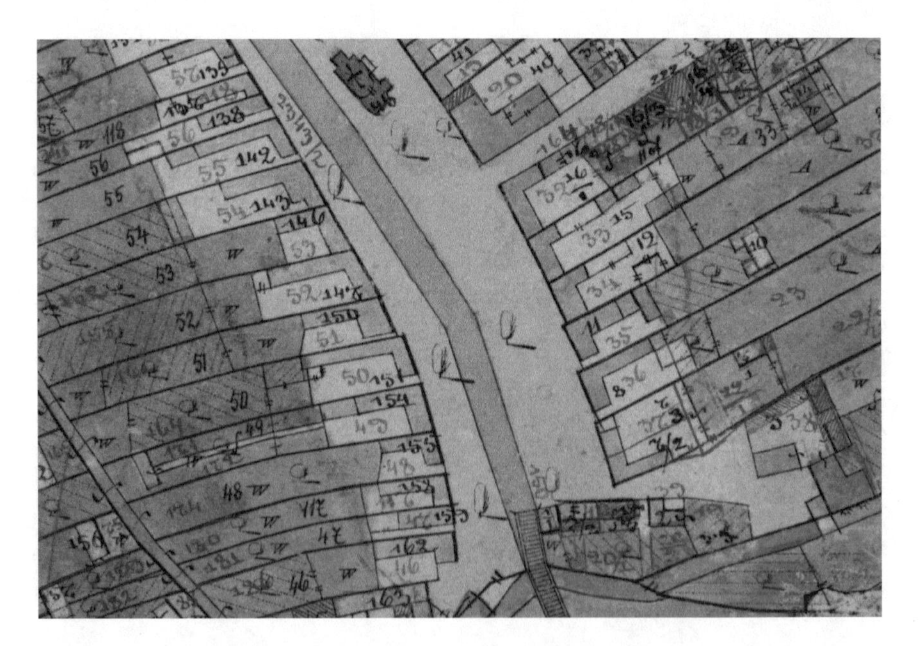

Fig. 2 A part of the cadastral map of the Otrokovice Center from 1877 [9]

found in the Zlín – Klečůvka Archive, where many historical archival materials are stored. The Internet was the last resource used. Some historical photographs were found on various www pages, mainly of the Otrokovice town center.

Approximately 150 photographs were obtained from the above-mentioned resources, which captured the appearance of the town between the years 1830–1960.

2.2 *Used Programs*

Where possible, "free to use" software was preferred. For 3D modeling, texturing and rendering, the Blender software suite was therefore used [13]. Textures were drawn in GIMP [11]. Microdem [14] was the last software package that was used.

Blender is a fully integrated creation suite, offering a broad range of essential tools for the creation of 3D content, including modeling, uv mapping, texturing, rigging, skinning, animation, particle and other simulation, scripting, rendering, compositing, post-production, and game creation [15, 16]. Blender is cross-platform, based on the OpenGL technology, and is available under GNU GPL license.

GIMP is an acronym for GNU Image Manipulation Program. It is a freely distributed program under the GNU General Public License. It is mainly a digital image editor and a drawing tool. It allows one to retouch photos by fixing problems affecting the whole image—or parts of the image, adjust colors in photographs to bring back the natural look, image compositing or image authoring [17].

Microdem is a freeware microcomputer mapping program, designed for displaying and merging digital elevation models, satellite imagery, scanned maps, vector map data or GIS databases [14]. This software was used in order to convert landscape elevation map data into a bitmap image (i.e. a heightmap).

3 A Landscape Model of Otrokovice

Data files which contained text information about the earth elevations were used in order to create the landscape model of Otrokovice and its vicinity.

The Digital Elevation Model data (i.e. DEM) was used, which was provided by the NASA Shuttle Radar Topographic Mission (SRTM) in 2007. The data for over 80 % of the globe is stored on [18] and can be freely downloaded for noncommercial use.

So the data of the Otrokovice region was downloaded, and then opened with the Microdem freeware program, which was described above. This software is able to convert the obtained data into a bitmap image. Microdem can clip and convert these images into grayscale (e.g. into a heightmap) that was applied to the Otrokovice region. Specifically, a heightmap area of 17.5 km^2 (a rectangle with 3.5 km width and 5 km height, centered on the center of Otrokovice—Fig. 3) was created.

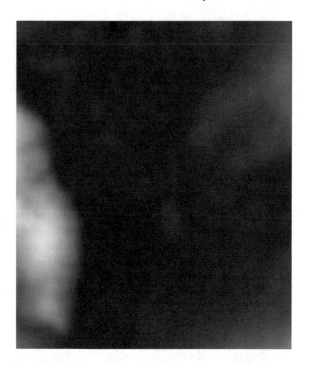

Fig. 3 The heightmap of the Otrokovice's Region

This heightmap in PNG format was saved (it is very important to use lossless compression). In Blender, a square was inserted (the Plane object) in the new scene and its edges were scaled to the ratio 3.5:5. After that, it was divided several times with the Subdivide tool in order to get a grid with a density of several thousand vertices. Then, the Displace modifier was used, to which a texture was assigned (for the obtained heightmap). The Displace modifier deforms an object, based on the texture and setting parameters (Fig. 4). The model of Otrokovice's landscape was

Fig. 4 Settings for the displace modifier in the Bender environment

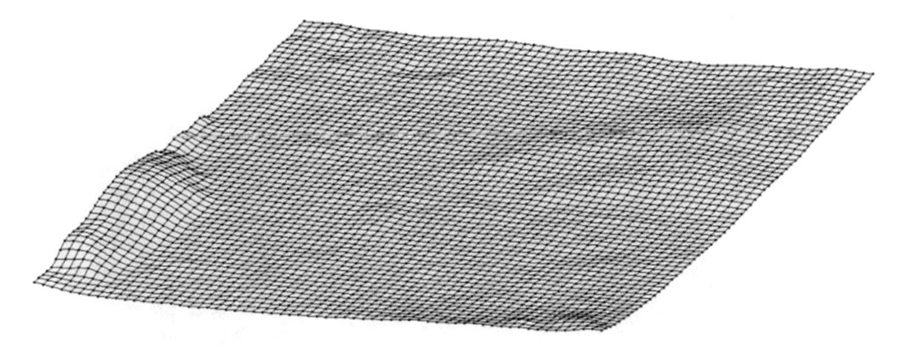

Fig. 5 The final mesh model of the Otrokovice Region's landscape

obtained by using this method in this way. This is shown in Fig. 5. Although this model is not entirely accurate—but given the scale, the whole scene, and the quality of the Otrokovice model buildings, it is sufficient.

4 Modeling Buildings and Accessories

The modeling of the buildings was always performed in the Blender program, and according to the same scenario. Before the modeling phase, the appropriate part of the cadastral map with the selected building described in Sect. 2.1 was incorporated into the background screen in Blender. After that—the Plane object was inserted and then its profile (i.e. the size and shape) was modified according to the floor plan of the building in the 3D scene. Mainly, two tools were used to shape this object. The Extrude tool allows one to alter the selected face in a specifically chosen direction; and then the Subdivide tool—which breaks down the selected face into an even greater number of smaller parts [16].

When the floor shape was finished, this profile was extruded to a height corresponding to the existing photographs. In cases where the photographs were missing, the building height and its accessories were improvised according to the neighboring buildings—whose photographs did still exist.

The next step was the necessity to model the building's roof. Buildings in Otrokovice at that time had only one type of roof—a pitched roof, whose shape was always adapted to the building's floor plan. The pitched roof was created by dividing the top face into two parts, and the newly created edge was subsequently moved to the required height.

To make the windows and doors embedded in the buildings, the Subdivide tool was used again on the relevant faces. After this, these new sub-faces were deleted to model holes. Additionally, the border edges of these holes were extruded to the depth of embedment of windows and doors. Chimneys were created from the Cube objects with changed scales in selected axes.

Fig. 6 A 3D model of the Otrokovice Church

An example of one modelled building is shown in Fig. 6. Here is for example, the Otrokovice church model from the Nineteen-thirties in the Blender environment.

3D models of trees and shrubs were created by the help of Blender's internal script called Sapling. This script allows one to create different trees, based on set initial parameters. The tree objects used in this project were simplified as much as possible so that the scene was not too complicated for the rendering process.

In order to make the whole rendered scene look more realistic, models of grasses were added. These models were based on Plane objects, on which the grass textures with transparency were mapped later. Groups of these objects created tufts of grasses that have good visibility from different viewpoints. Later, these groups were inserted as multiples into the scene and thereby large lawns were generated.

5 Texturing

The UV mapping technique was used for texturing objects. This process starts by the decomposition of each object into 2D sub-surfaces (a UV map). At the beginning of the decomposition process, it is necessary to mark the edges, which should be ripped from one another. In Blender, this process is performed by the Mark Seam command. After that, it is possible to finish decomposing by using the Unwrap tool. The UV map created in this way is saved into the .png raster graphic format (it is also possible to save it into another raster graphic format, but it is necessary to use a lossless compression algorithm). The resolutions 512 × 512 or 1024 × 1024 pixels for the UV maps were used, because the rendering process usually needs the power of two texture resolution.

All textures were drawn in the GIMP software environment. Therefore, all of the UV maps created in this way were also opened in GIMP. In these pictures, the location of each part of the 3D object is visible. With this information, a user can fill each individual sub-surface as necessary. Most of these textures were drawn by hand, and in some cases, pre-created textures from the CGTextures website were used [19]—these textures were edited and modified in order to use them on the models so created. For texture creation and editing purposes, standard GIMP

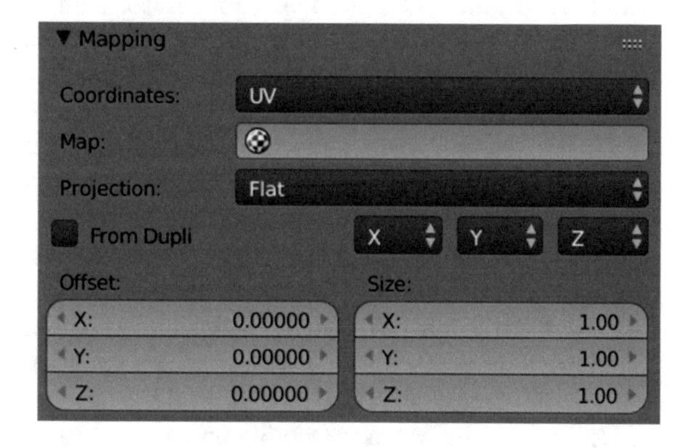

Fig. 7 The user environment for the UV mapping settings in Blender

Fig. 8 Textured 3D model of the Otrokovice Church

drawing, coloring and transforming tools were used [17]. Once this process was finished, they were saved in .jpg format. In this case, the graphic format with a lossy compression algorithm can be used in order to save computer memory—and the .jpg format is ideal for that. The next step was to re-load these created textures in Blender and to correctly map them into 3D objects. This process is performed by correcting the set parameters in the Texture Mapping panel (Fig. 7).

All 3D sub-models of Otrokovice center in the Nineteen-thirties were also created and textured in this way. An example of the Otrokovice Church model with textures is shown in Fig. 8.

6 Rendering and Animation

After the completion of the modeling phase of separate buildings and accessories with textures, they were inserted into a single complex 3D scene of the landscape model. In addition, here, it was necessary to set the other suitable parameters—like the surroundings, lighting and cameras. The surrounding area setting parameters are performed by the World window in Blender. It is possible to set simple colors for the horizon and zenith and to blend them, or to use the internal (i.e. procedural) or external texture (any bitmap file). In this project, the Clouds procedural texture was used, which looks close to reality in the scene.

Lighting of the scenes can be realized in several ways in Blender. It is possible to use light objects (called Lamps in Blender) for local lighting or global influences (i.e. Ambient Light, Ambient Occlusion, Environment Lighting and Indirect Lighting). The Environment Lighting technique combined with the Emit material settings of selected objects was used in this 3D scene.

The last step before the rendering process was to select a suitable position for the cameras. In order to capture the scene from different positions and prepare for the virtual tour output in the future, 8 camera objects were inserted in the scene and placed in suitable positions. After that, these cameras were correctly oriented in order to capture the most graphic images of the whole scene.

The Render command performs the rendering calculation process in the Blender environment. Additionally, the user can set many of the accompanying parameters. The basic parameters are the choice of a rendering algorithm, image or animation resolution, type of output file format, antialiasing, motion blur, enable/disable ray-tracing and shadows. In this project, the decision was made to use Blender's internal renderer with image resolutions of 1920×1080 pixels, 25 frames per second and the MPEG-2 output format to render animations. Figure 9 shows one rendered image of the Otrokovice Center in the Nineteen-thirties.

Fig. 9 The rendered image of the Otrokovice Center model in the Nineteen-thirties

7 Conclusion

This paper has presented a visualization method for the Otrokovice town in the Nineteen-thirties. Based on the historical materials, a more complex 3D scene was created that contains 3D models of terrains, houses and buildings, trees, shrubs, grasses and even more accessories in order to achieve the best possible authenticity. The output of this work includes two different ways of doing so. The rendered images are the first output. They captured the scene in the places where the most photographs of that time were created in order to provide a comparison. The second output consists of an animation. This animation is performed by the Curve object, which links through the whole scene and the camera is connected to it and follows the curve's shape during the defined time.

The next goal, in the future, is to expand and improve the whole scene in order to make it even more realistic. This process will include the creation of more models of houses and buildings—mainly on the edge of town, and the modeling of more types of accessories, like other constructions, fences, benches, etc. The authenticity mainly depends on obtaining other historical materials.

A further extension could be the rendering of a virtual tour. This output requires panoramic rendered images, which are possible to create in the Blender environment by the appropriate camera parameter settings. After that, the panoramic images have to be joined, corrected and exported to another application that can process them and generate the virtual tour from them.

References

1. Brand, W.: Data Visualization for Dummies. Wiley, New Jersey (2014)
2. Visualization (computer graphics): http://en.wikipedia.org/w/index.php?title=visualization_(computer_graphics)
3. Qiu, H., Chen, L., Qiu, G., Yang, H.: Realistic simulation of 3D Cloud. WSEAS Trans. Comput. **12**(8), 331–340 (2013)
4. Congote, J., Novo, E., Kabongo, L., Ginsburg, D., Gerhard, S., Piennaar, R., Ruiz, O.E.: Real-time volume rendering and tractography visualization on the web. J. WSCG **20**(2), 81–88 (2012)
5. Qiu, H., Chen, L., Qiu, G., Yang, H.: An effective visualization method for large-scale terrain dataset. WSEAS Trans. Inf. Sci. Appl. **10**(5), 149–158 (2013)
6. Historie města Otrokovice: http://historie-otrokovice.czweb.org/
7. Gazda, Z., et al.: Otrokovice – dějiny a současnost 1131–1981. Otrokovice, Městský národní výbor (1981)
8. Klokočka, P., et al.: Otrokovice objektivem času. Otrokovice, Město Otrokovice (2009)
9. Moravský zemský archív v Brně: http://www.mza.cz/
10. Státní okresní archív Zlín: http://www.mza.cz/zlin/
11. GIMP—The GNU Image Manipulation Program: http://www.gimp.org
12. Tichý, J.: Toulky minulostí Otrokovic. Otrokovice, Město Otrokovice (2000)
13. Blender.org—Home: http://www.blender.org
14. Microdem Homepage: http://www.usna.edu/Users/oceano/pguth/website/microdem/microdem.htm
15. Skala, V.: Holography, stereoscopy and blender 3D. In: 16th WSEAS International Conference on Computers ICCOMP12, pp. 166–171, Kos, Greece (2012)
16. Hess, R.: Blender Foundations—The essential Guide to Learning Blender 2.6. Focal Press (2010)
17. Peck, A.: Beginning GIMP: From Novice to Professional. Apress (2008)
18. CG Textures—Textures for 3D, graphic design and Photoshop!. http://www.cgtextures.com/
19. Jarvis, A., Reuter, H.I., Nelson, A., Guevara, E.: Hole-filled seamless SRTM data V4, International Centre for Tropical Agriculture (CIAT). http://srtm.csi.cgiar.org (2008)

Cost Effective Framework for Complex and Heterogeneous Data Integration in Warehouse

P. Amuthabala and M. Mohanapriya

Abstract Ever increasing customer requirements and highly competitive business atmosphere as forced the enterprise to enhance the quality of service by understanding the customer's needs. Understanding customer needs through data is accomplished using data warehouse, which allows the enterprises to analyze customer's data, which includes various queries, solutions as well expectations based on their various actions and requirements. Though the customer data are specific to the certain domain and fail to provide a complete understanding of the customer requirement. Hence, it calls for the need of integrating multiple platforms to arrive at a clear conclusion on customer requirement. Considering the complexity involved data integrity over multiple domains, this paper suggests an algorithm to achieve data integration and data synthesis to enhance data quality. The evaluation of the framework is carried through application run on at test bed comprising 100's of the system and is found to be efficient and effective in performing the data integration and data synthesis.

Keywords Application · Data ware house · Data integration · Data synthesis

1 Introduction

In this digital era, information plays a vital role in our everyday life, especially for the enterprises wherein every bit of information as a significant value. Due to its significance this information in any form must be properly stored so that they are

P. Amuthabala (✉)
CS&E Department, Karpagam Academy of Higher Education, Coimbatore, Tamil Nadu, India
e-mail: amuthabala79@gmail.com

M. Mohanapriya
Department of Computer Science & Engineering, Karpagam Academy of Higher Education, Coimbatore, Tamilnadu, India

R. Silhavy et al. (eds.), *Software Engineering Perspectives and Application in Intelligent Systems*, Advances in Intelligent Systems and Computing 465,
DOI 10.1007/978-3-319-33622-0_9

easily and readily available at the time of need. However, redundancy of the data also makes the process of data analytic as well as knowledge extraction tedious especially in the case of commercial enterprises [1]. Data warehousing is a mechanism wherein a large amount of electronically generated data over a period is stored, so that it can be used at a time of urgent need to achieve the goals that are beyond the routine tasks connected to regular processing [2]. For instance, we can consider a large corporation that consists of different department and modules. To evaluate the contribution of each and every department the corporation may store the data of every department based on the task performed by that department in a database so as to be used for evaluation when needed. To assess the performance, tailor-made queries are made and by these queries information is retrieved from the database [3]. To accomplish this task, the desired query is formulated based on the database catalogs. Later these queries are processed [4]. The processing may require huge time depending on various factors like amount of the data, the complexity of the data and so on. Based on these factors a final report was generated which was usually in the form of a spreadsheet and was used for analysis [5]. Over period database designers started realizing this mechanism of analysis is not feasible as it demands an enormous amount of time and resources and also fails to deliver the expected results. Furthermore, the system will slow down when processing a mixture of analytical data as well as routine transactional queries [6]. Also fails to meet the requirement of users of either type. Current advanced data warehousing mechanisms separately process the Online Analytical Processing (OLAP) from Online Transactional Processing (OLTP) [7]. By generating new information database that is integrated with different sources, aligns data formats properly and ensures the availability of data for evaluation and analysis, which is aimed at planning and decision-making method. The main constituent of the data warehousing is the Decision Support System (DSS) [8]. DSS is a class of expandable as well as interactive IT mechanisms and tools that are designed to process and analyze data to support decision making or policy making. This is achieved by matching the individual resource with the computer resources to enhance the quality of policy making [9]. Data warehousing as a wide range of application and at present it is widely applicable in almost all domain, few examples are in trade and commerce to analyze sales and claims, customer care and public relation, in financial service for risk analysis and fraud detection, in transportation service to perform vehicle management, in telecommunication domain to analyze call flow and customer profile, in health care sector to maintain the patient profile and diagnosis results and history. Data warehouse is applicable in the domain such as education, e-commerce, transportation, social media telecommunication, and public service as well as government service was storage of huge data for latter analysis is required [10]. In this paper, a novel algorithm to integrate complex data and knowledge synthesis is presented. The paper mainly aims at integrating data's from the different heterogeneous database and performs knowledge synthesis to enhance customer service provider relationship. The paper is organized as follows, Sect. 2 illustrates prior research work, Sect. 3 briefs about problem description,

Sect. 4 discusses proposed model, Sect. 5 illustrates research methodology and implementation and Sect. 6 provides results and outcome of the proposed system and Sect. 7 gives conclusion and future work.

2 Prior Research Works

This section discusses the prior research work carried by various researchers in the data warehouse and highlights the significance and drawbacks of these works. In any research work the initial step is performing the extensive survey on the domain, so as to have a clear knowledge of the pros and cons of the technology or mechanism. So that it will help in developing a better system for future work. Such similar review on data warehouse was carried by Abai et al. [11]. Another similar review was performed by Gosain and Heena [12]. George et al. [13] highlighted the need for the data warehouse in health care. The authors also distinguished the difference between the traditional business Intelligence and the one required in the healthcare system. Jiang et al. [14] proposed a Retrospective Audit method to integrate heterogeneous data and improve the data quality using associative rules. Authors suggest that the work can be extended to study optimization of parameters by a traceable audit.

Another similar work on integrating data was proposed by Boumakoul et al. [15] in which authors has integrated data from moving an object that are related to urban transportation. Authors have utilized distributed cloud infrastructure along with big data NoSQL database. The author also suggests in future a software solution will be developed to accomplish prime field texts. In any enterprise customer relationship is a crucial factor in determining the success of the enterprise. Hence, enterprises make use of data warehouse to enhance Customer Relationship Management (CRM). One such research work is carried by Khan et al. [16] wherein the author as the analyzed the impact on the enterprises that have switched to data warehousing for CRM application especially in the Pakistan. The author also highlighted the benefits achieved from proposed system reduced ETL processing, alignment with enterprise goal, minimizing operational cost, advancement in CRM, and customer retention.

Another similar work is proposed by Zhao et al. [17] where the authors have suggested Case-Based Reasoning (CBR) and Multiplicative Analytic Hierarchy process) which will assist the enterprises in meeting the customer requirement and improving profit margin. The author suggests that the effectiveness of the work was illustrated through simulation experiment. Another work in related to enhancing customer satisfaction is carried by Yaakub et al. [18] wherein the authors have developed a multi-dimensional model to perform opinion mining that will capture the customers review by subjective expression. Authors have used techniques like OLAP and Datacube. The authors suggest that future work will focus on the evaluation of multiple products.

To improve the response time, business intelligence tools have started utilizing approximate query answering system so as to provide useful data for policy making. One such research on approximate query answering system is proposed by Tria et al. [19] where the author as suggested the use of metadata for approximate query answering systems. The authors suggest that from the result the proposed system successfully identified which data can be used for analytical processing by approximate methodologies. Through the user can formulate queries and system to provide the automatically solution to access minimized data based on user defined query. Aydin et al. [20] suggested a scalable and distributed architecture that gathers sensor data, stores it and performs an analysis. The authors have developed the system using different open source technologies and executed the system on a cluster of virtual servers. The authors suggest that through test result it is seen that system can successfully use for executing computationally complex data analysis algorithms and has performed better with huge sensor data. A work on multimedia data warehouse was carried by Vora et al. [21] where the authors suggested multimedia warehouse to enhance performance. The authors incorporated compression technique and partitioning technique to improve storage capacity. Authors enhanced the access and analysis efficiency through representing multimedia data by multilevel features and application of indexing technique. Ghani et al. [22] suggested an integrated telemedicine framework that is assisted by the data warehouse. Authors have evaluated the framework using test case mechanism and from the result it is seen that the framework provides useful information for telemedicine that is helpful especially during consultations.

Data warehouse are extensively being used in the healthcare sector, one such work of data warehouse in healthcare was suggested by Giradi et al. [23] has proposed an oncology based clinical data warehouse for scientific research. Authors suggest that their proposed system will assist domain expert throughout the complete mechanism of knowledge extraction from data unification to utilization. Abello et al. [24] suggested a framework to assist self-servicing business intelligence and associated challenges. The authors have utilized notion of fusion cube as the core idea in the sense multi-dimensional cubes that are extended both in their schema and well as instances. Also, the situational data and metadata are also associated with quality as well as provenance annotations. Andreescu et al. [25] have suggested the use of two tools to highlight scope and relevance of data quality evaluation in analytical projects through the tools such as Oracle Warehouse Builder and SAS 9.3. Faridi et al. [26] suggested the use of data warehouse and data mining in making an interactive decision especially in textile sector to improve the productivity and manage the resources. Authors have conducted the experiments in the textile industry and focused on providing the top management inputs in improving profits. Kumar [27] suggested integrating bitmap indexing, iceberg querying, and uncertain data processing to gather with data processing in the data warehouse to achieve efficient query processing speed. The authors also suggested in future generating time factor to the three techniques mentioned above. A work on Data warehouse being used in e-shopping is proposed by Suchanek et al. [28] wherein the authors a new strategy for creating e-commerce system methods as well as to manage e-commerce with the help of modern software tools in assisting the decision support. The outcome is analyzed through simulation. In the following section, we will discuss the various issues, challenges in using data warehouse in a different domain.

3 Problem Identification

The increasing use of Data warehouse in enterprises and organization as also possess certain limitation and challenges. Few drawbacks or imitations identified in our work is as follows initially before storing any data in a data warehouse it needs to be cleaned and extracted. A major issue observed in this process is it takes huge time with increasing amount of time. Another frequently observed issue is the compatibility issues; this is frequently seen in many enterprises especially when a new transaction takes place. The system does not respond as it is not compatible with the new data. This kind of problem will have severe implication with workers working with data warehouse unless they are trained properly. Another severe problem associated with this issue is, people accessing the data warehouse through the internet may face huge security implication. One common problem with all enterprise related to the data warehouse is the maintenance, organization or enterprise need to be aware of the benefits over the short comes of the data warehouse. So that it can opt the data warehouse service, it should be able to bear the maintenance cost also.

Another problem seen is the storage technique; usually there are two forms of storage technique which are used in data ware house. On form is dimensional technique suitable for less amount of data and is easily understandable and also it is fast in operation. Current day data warehouse face serious challenges of adaptability that is crucial in the overall progress of the enterprises. But the biggest problem with such storage is if the business mode of the organization changes it's difficult to accommodate new data's. Other form is the data normalization that provides an easy way to add data's but is tedious to produce reports. Another major problem that is seen in the data warehouse it the problem of data quality several factors affect the quality of the data such as the data source the various operation carried on the data in before storing it in the data warehouse. The lack of quality will result in the poor decision making or policy making, herby making the business less productive. The biggest problem in the data warehousing is integrating the data from a different source, especially heterogeneous data from a different source. In an organization it may happen different department use different forms of data and these different forms of data are saved in different data warehouse, for instance, one, a department may use SQL database to store the data's concerning that department. Likewise, different departments may choose different data warehouse of a different platform suitable to store their respective data's. To make policy or decision in the organizational view all these data's needed to be integrated. So that these data's are extracted to gain sufficient information related to the decision making and improving the performance as well as profit of the organization or enterprise. But, the process of integrating heterogeneous data warehouse is complicated and challenging. To overcome this problem of integrating heterogeneous data ware house, the paper presents a proposed system in which an application interface is developed to perform the integration of the heterogeneous data ware house. The proposed system is illustrated in the next section.

4 Proposed System

The following section provides the description about the proposed system. The architecture of the proposed system is shown in the Fig. 1.

The aim of the proposed system is to develop an application interface in order to achieve the following objectives.

- The prime aim of the proposed system is to develop a multiple enterprise application which is connected to the multiple heterogeneous databases. The raw operational data's are derived from these databases.
- Designing a common application interface to integrate heterogeneous data from different data warehouse.
- Permitting multiple user enrollments through the common application interface.
- Performing cleansing of the heterogeneous data.
- Allowing the user query through common interface application based on the user profiling.
- Providing a collaborative platform that will assist in interaction between client and service provider.
- Performing the analysis of the heterogeneous data.
- Enhancing the adaptability of the data warehouse system by assisting in processing heterogeneous data.

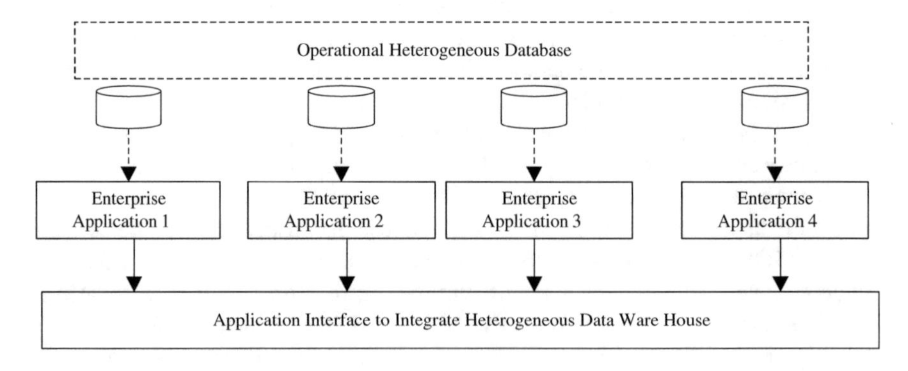

Fig. 1 Architecture of the proposed system

- Refining the decision making process by providing valuable information through heterogeneous data process which is limited in existing data warehouse techniques.
- Evaluation of thc common interface application on the basis of the response time.
- Providing a cost efficient solution for data quality assessment.
  The proposed system consists of three major modules and several constituent modules. The three major modules are as follows.

1. Enterprise Applications
2. Heterogeneous databases
3. Common interface application

The above modules along with the constituent modules are illustrated in the research methodology.

5 Research Methodology

This section illustrates the research methodology of the proposed system. As mentioned earlier the proposed system is made up of three major modules and several constituent models. All the necessary modules are briefly explained in this section.

5.1 Enterprise Application

Is a java based application developed for specific business purpose. It is capable of supporting multi-user and can be installed in multiple systems and is also a multi-component application. This application can perform manipulation of huge data and is designed to make efficient use of parallel processing and distributed network resources. It is also platform independent and is also capable of operating along with other applications. Enterprise application acts an interface between the customer and retailer. Application also incorporates certain security features like authentication for both the user and retailer. Authentication is performed on the basis of credentials provided by the user and retail and is similar to standard authentication. Overall application work is monitored and controlled by the admin. The application consists of its own database, which is used to store all the information related to business involving both user and retailer.

5.2 *Heterogeneous Database*

In any electronic application, there needs to be transaction recording facility. In this enterprise application all, the business transactions are recorded and stored in the database. The database may be of any type for instance relational database such as Microsoft SQL server, My SQL, Oracle, IBM DB2 and so on. The type of the database selection is dependent on the application characteristic. Here different types of database are used in different machine to gather heterogeneous data.

5.3 *Application Interface*

Application interface is prime contribution of our work. This application interface is developed using java. The main objective of this application interface is to facilitate the integration of heterogeneous database consists of variety of unrelated data's. The main modules within the application interface are explained below.

- Customer Module: Through this module, the interface receives all the information and data related to the customer present in the databases. The data may be of different formats, size or nature depending on the database they are retrieved from. Customer data contains all necessary information of the customer right from his credentials to his shopping history and also the transactions along with his opinion or review put up in the dashboard of the shopping portals.
- Retailer Module: In this module, all information related to the retailer as well as product is maintained. It contains variety of data such as text or images of different size and formats. The information includes the details of retailer that contains his credential, transaction detail, his shop detail and also the products available in the shop, description as well as details of the product along with opinion on the retailer as well as the product.
- Purchase Module: This module constituent the details related shopping, it also include details such as the product viewed, added to the cart, Products deleted from the cart, financial transaction of purchase between the user and retailer and so on. The interface extracts the necessary information from all above module through the help of constituent modules. These extracted information's have a significant role in policy making related to boosting the overall growth of the enterprises.
- Constituent Models: The constituent modules are the various supporting modules that are part of application directly or indirectly. The different constituent modules are as follows.
  - Structuring Module: This module is responsible for structuring the data in the database, i.e. it aligns the different data's obtained from different database in accordance to their format. The structuring is performed so as to make the process of knowledge synthesis easier.

- Data Management: The raw data as well as the processed data are needed to be managed in order to have efficient performance of the system. Data management module is responsible for performing this task. Data management involves various tasks like performing the data backup, storage of data, and retrieval of data and so on. The data management involves both structured as well as unstructured data. An efficient data management system contributes to increase the efficiency and performance of the system.
- Common Interface and Analytics: The common interface is associated with enterprise application. Through this common interface, the user and retailer are performing the business via enterprise application. This is accomplished by using the integration of the web services through which it is possible to integrate different web portals. This common interface also allows the user to generate some queries. These queries are answered through the process of data synthesis. The process of data synthesis involves different operation of performing data extraction in order to perform the analysis of the data. The data extraction is crucial since the knowledge synthesis result is dependent on the quality of the result obtained by the data extraction. Analytics module is responsible for performing the analysis of the data, the analysis is carried on all form of data irrespective of the format and size of the data. The data analytics is crucial in understanding the needs of the customer, and provides the retailer with information in order to provide better information to maximize his businesses and improve his customer relation. These processes are also collaborated with the data mining. The mining is performed through a conventional algorithm. In the following section we will discuss about the result obtained through the proposed system and provide the comparison of the result with respect to the existing system highlight the contribution of the proposed in enhancing the performance compared to existing system.

6 Prototype Dissection

This section discusses about the proposed prototype and its advantages over the existing system. Figure 2 shows the schematic of the existing system and the proposed system. The prototype is designed on java platform with following system configuration, 64 bit windows operating system, 2 GB of RAM, I3 Intel processor. The prototype is simulated on 15 different system of different configuration. In compare to the existing system, the advantage of our proposed system is that unlike existing system it does not just provide a single shopping portal. In existing system user are only able to view or shop in a single portal which is similar to single window system. In which the users are limited to buy from a products sold by that portal only, and are limited with purchase option. Proposed system is successfully integrated in number of different shopping portal as shown in the Fig. 2.

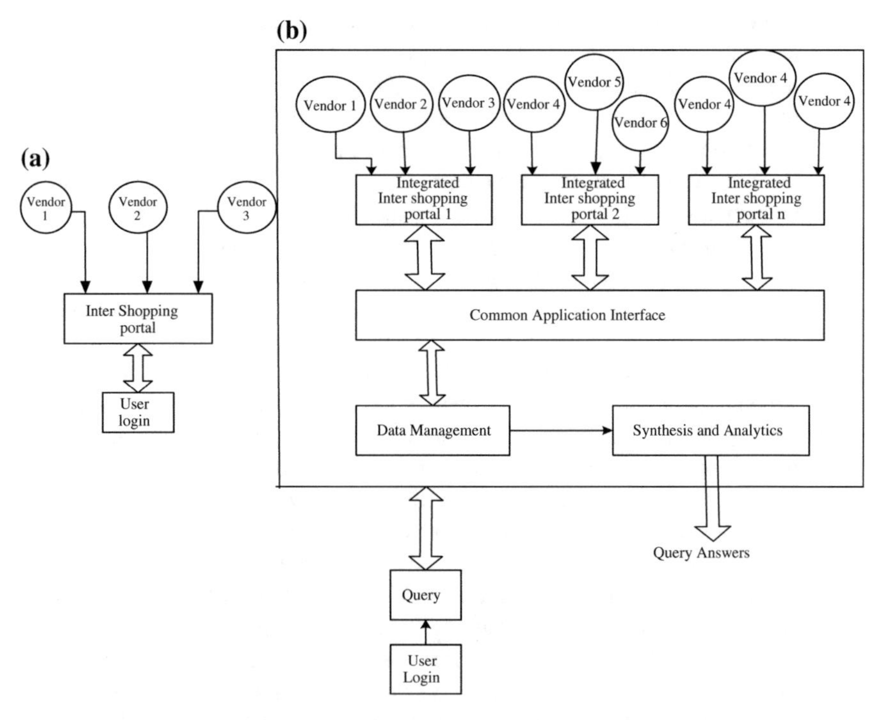

Fig. 2 Schematic of existing system. **a** Existing system. **b** Proposed system

The proposed system provides several benefits to the shoppers as well as vendors. As the shoppers are allowed to compare the products with different vendors and can select the product they wish to buy on basis of their budget. Wherein the retailers also benefits by getting to know, how much the product is priced by other vendors. All these features are achieved by web service integration. Through this it is possible integrate all web portal that are associated with shopping application. Initially in the proposed system multiple web portal corresponding to different shopping portal or service providers are integrate through web service integration service. Since these webs portal services have their databases to reposit their data's, the proposed system uses a proprietary database that are capable of combining all the data's present in these databases into this proprietary database. On achieving that, all the data are subjected to different data management mechanism as illustrated in Fig. 2b. The data management unit classifies the data on the basis of the different database on the basis of data attributes like size, format, and content and so on. On completion of this classification, the crucial phase of knowledge synthesis is started; here all the data are subjected to extraction kind of process to retrieve needful information. The analytics is performed on the basis of the query generated, and the query answer is generated through an algorithm designed to perform a semantic analysis of the user produced by the user that will help in identifying various requirements as well as user emotion. The proposed system as also aimed at providing a responsive query analytic system in future, by incorporating advanced features like data synthesis as well as data analytics.

However, accomplishing an effective task of data integration is a complex process which includes structureness of the data. Normally performing structuring of the data in our proposed data management will make use of open source software framework for conversion of unstructured/semi-structured data to highly structure data, which can be stored in warehouse. We will also develop a novel mathematical model that can assess the conversion rate and can perform it in highly cost efficient manner using resource sharing process. Interestingly, data conversion process using resource sharing in data warehouses is quite novel approach and that is where our proposed system will have a unique contribution in processing heterogeneous data. Hence, the system can easily identify the issues and perform cost-efficient transformation of the historical data before repositing in warehouse.

7 Conclusion

In general data warehouse is used to incorporate different forms of data, as well as different queries like relational as well as other form of database and collect huge amount of data. So as to serve the customer by proper understanding of the queries and build a better customer-client relationship. This paper presents a novel approach in understanding the customer requirement and emotions, thereby assisting the enterprises to improve their business and increase their profit. This paper as designed a common application interface to integrate different web portal by using a simple web service integrating mechanism. There by providing a cost-effective solution. In future, the proposed system will be included with different advanced features like data management, and data synthesis and analytic which also include designing an algorithm to perform efficient as well as lightweight semantic analysis of the data.

References

1. Gelbstein, E., Kamal, A.: Information Insecurity: A Survival Guide to the Uncharted Territories of Cyber-Threats and Cyber-Security, vol. 198. United Nations Publications (2002)
2. Hoffer: Modern Database Management Systems, 9th edn. Pearson Education India (2009)
3. Golferalli: Data Warehouse Design. Tata McGraw-Hill Education (2009)
4. Despande, A., Ives, Z., Raman, V.: Adaptive Query Processing. Now Publishers (2007)
5. Manning, C.D., Raghavan, P., Schutze, H.: Introduction to information Retrieval. Cambridge University (2008)
6. Manyika, J., Chui, M., Brown, B., Bughin, J., Dobbs, R., Roxburgh, C., Byers, A.H.: Big Data: The Next Frontier for innovation, Competition, and Productivity. McKinsley Global Institute (2011)
7. Han, J., Kamber, M., Pei, J.: Data Mining, Concepts and Techniques, Southeast Asia Edition. Morgan Kaufmann (2006)

8. Power, D.J.: Decision Support Systems: Concepts and Resources for Managers. Greenwood Publishing Group (2002)
9. Forgionne, A., Guisseppi: Decision-Making Support Systems: Achievements and Challenges for the New Decade. Idea Group Inc. (2002)
10. Golfarelli, M., Rizzi, S.: Data Warehouse Design: Modern Principles and Methodologies. McGraw-Hill Inc. (2009)
11. Abai, Z., Yahaya, H., Deraman, A.: User requirement analysis in data warehouse design: a review. Procedia Technol. **11** (2013)
12. Gosain, A., Heena: Literature review of data model quality metrics of data warehouse. Procedia Comput. Sci. **48** (2015)
13. George, J., Kumar, V., Kumar, S.: Data Warehouse Design Considerations for a Healthcare Business Intelligence System. In: World Congress on Engineering (2015)
14. Jiang, L., Chen, H., Ouyang, Y., Li, C.: A multisource retrospective audit method for data quality optimization and evaluation, Hindawi Publishing Corporation. Int. J. Distrib. Sensor Netw. Article ID 195015 (2015)
15. Boulmakoul, A., Karim, L., Mandar, M., Idri, A., Daissaoui, A.: Towards scalable distributed framework for urban congestion traffic patterns warehousing, Hindawi Publishing Corporation. Appl. Comput. Intell. Soft Comput. Article ID 578601(2015)
16. Khan, A., Ehsan, N., Mirza, E., Sarwar, S.Z.: Integration between Customer Relationship Management (CRM) and data warehousing, Elsevier. Procedia Technol. (2012)
17. Zhao, Y.J., Luom, X.X., Deng, L.A.: CBR-based and MAHP-based customer value prediction model for new product development, Hindawi Publishing Corporation. Sci. World J. Article ID 459765 (2014)
18. Yaakub, M.R., Li, Y., Zhang, J.: Integration of sentiment analysis into customer relational model: the importance of feature ontology and synonym. Procedia Technol. (2013)
19. Tria, F.D., Lefons, E., Tangorra, F.: Metadata for approximate query answering systems, Hindawi Publishing Corporation. Adv. Softw. Eng. Article ID 247592 (2012)
20. Aydin, G., Hallac, I.R., Karakus, B.: Architecture and implementation of a scalable sensor data storage and analysis system using cloud computing and big data technologies, Hindawi Publishing Corporation. J. Sensors Article ID 834217 (2015)
21. Vora, M., Vora, J., Jani, N.: Modelling architecture for multimedia data warehouse. IJIRSET **4** (1) (2015)
22. Ghani, M.K.A., Jaber, M.M., Suryana, N.: Telemedicine supported by data warehouse architecture. ARPN J. Eng. Appl. Sci. **10**(2) (2015)
23. Giradi, D., Dirnberger, J., Giretzlehner, M.: An ontology-based clinical data warehouse for scientific research. Saf. Health (2015)
24. Abell: Fusion cubes: towards self-service business intelligence. Int. J. Data Warehouse. Min. **9** (2) (2013)
25. Andreescu: Measuring data quality in analytical projects. Database Syst. J. (1) (2014)
26. Faridi, M.S.: Usability of data warehousing and data mining for interactive decision making in textile sector. Glob. J. Comput. Sci. Technol. 12(7) (2012)
27. Kumar: Improvement of query processing speed in data warehousing with the usage of components—bitmap indexing, iceberg and uncertain data. Int. J. Comput. Appl. (2015)
28. Suchánek, P., Perka, R.S., Dolák, R., Kus, M.M.: Intelligence decision support systems in e-commerce. In: Jao, C. (ed.) Efficient Decision Support Systems—Practice and Challenges in Multidisciplinary Domains (2011)

Uplink Channel Performance and Implementation of Software for Image Communication in 4G Network

N.R. Deepak and S. Balaji

Abstract Most contemporary Internet applications and services require high speed data transfer with high levels of quality of service. Hence, the next generation mobile communication with the above requirement is expected to provide better link quality at higher data rate as compared to the existing systems. Single Carrier Frequency Division Multiple Access (SC-FDMA) and Orthogonal Frequency Division Multiple Access (OFDMA) are considered as uplink, multiple and strong access point for International Mobile Telecommunications-Advanced (IMT-Advanced). This paper discusses the SC-FDMA and OFDMA implementation and analysis in uplink of 4G networks. The advantages of SC-FDMA and OFDMA in terms of good link performance (Symbol Error Rate) and low Peak to Average Power Ratio (PAPR) in transmission scenario of relay. The simulations carried out for hybrid techniques to achieve the end to end link performance compared with the pure SC-FDMA technique and same value of PAPR over access link. With this, low PAPR value is achieved as compared to OFDMA case, that is much required in uplink transmission because of limited power (battery) of user equipment constraints.

Keywords 4G network · IMT-advanced · OFDMA · PAPR · Quality of service · SC-FDMA · Uplink

---

N.R. Deepak (✉)
Department of CS&E, Research Scholar-VTU, City Engineering College,
Bangalore 560061, India
e-mail: deepaknrgowda@gmail.com

S. Balaji
Center for Emerging Technologies, Jain Global Campus, Jain University,
Jakkasandra Post, Kanakapura Taluk, Ramanagara Dist 562112, Karnataka, India

R. Silhavy et al. (eds.), *Software Engineering Perspectives and Application in Intelligent Systems*, Advances in Intelligent Systems and Computing 465,
DOI 10.1007/978-3-319-33622-0_10

1 Introduction

Since the decades ago, the wireless technology is growing in such way that it is ubiquitous now. Recently, the broadband standards of wireless in Long Term Evolution (LTE) has obtained the much attention from communication industries of mobile due to its good capability towards providing faster broadband services with simple quality of service (QoS) management in mobile communication [1]. Orthogonal frequency domain multiple access (OFDMA) will used for LTE downlink while single-carrier frequency division multiple access (SC-FDMA) will used for LTE uplink. In multipath fading OFDMA is more powerful and it has high scalability and the frequency efficiency. But in mobile devices the OFDMA will cause the power problem due to high peak to average power ratio (PAPR); in some cases of OFDM symbols, the sudden changes in the RF will cause the increase in the PAPR. Because of single carrier characteristic property and multiple access features the SC-FDMA is used for LTE uplink [2, 3]. The progress of optical fibers with limitless bandwidth assurance and predictions of high speed wireless access over the internet in entire world has given future thrives for both popular technical and press journals [1].

The wireless network communication has fastest growth in the history because of its unprecedented evolution in communication field. The wireless communication kids are experiencing the many wireless standards like GSM, Wi-Fi, LTE and Wimax. These wireless standards will operate within the lower range of microwave of about 2-4GHZ. Because of propagation losses in this low range frequency and multipath fading problem will require the robustness solution against narrowband interferences, in multipath environment and in efficiency. OFDM has given promise in all these aspects by providing high capacity, wireless broadband network communication for multimedia at high speed but coexists with systems of present and future. Orthogonal frequency-division multiplexing (OFDM) is a digital modulation method where the signal will be split into many narrowband channel signals at many different frequencies. OFDM has adopted by many technologies like IEEE 802.16a, IEEE 802.11a/g, A symmetric Digital Subscriber Line (ADSL) services, Digital Audio Broadcast (DAB), and digital terrestrial television broadcast such as DVD (in Europe), ISBD (in Japan) 4G,IEEE 802.20,IEEE 802.11n and IEEE 802.16. The OFDM will convert the selective channel frequency into parallel frequency collection of sub channels [2]. OFDM is derived from the frequency division multiplexing (FDM) and it provides the many more advantages over the conventional technique. The OFDM will choose the subcarrier frequencies by which the signals can be mathematically orthogonal over OFDM, one symbol period. Both multiplexing and modulation can be obtained digitally by using the inverse fast Fourier transform (IFFT) and hence, the required accurate orthogonal signal is generated. The most significant challenge of OFDM is that it is based on high Peak to Average Power Ratio (PAPR) in transmitter that needs use of high linear amplifier that will cause lower power efficiency [4, 5]. When the OFDM is operated over amplifier non linear area, it will go via degrade in the performance error and nonlinear distortions.

The remainder of the paper is organized as follows: Sect. 1 gives an overview of LTE system model. Section 2 describes SCFDMA subcarrier mapping. Section 3 addresses pulse shaping method. Section 4 presents the results obtained through computer simulation, Sect. 5 gives the conclusions and lists the references considered for this paper.

2 Related Work

Adachi [6] has discussed the revaluation of wireless network in mobile communication. Wireless is considered as the core for advancement in the mobile communication. The author has given the necessary requirement for advancement in future technology of mobile communication. Deb et al. [7] have proposed a paradigm of machine learning for uplink interference management by power control over LTE commonly known as Leap algorithm. The authors have used radio network plans to perform the extensive evaluations from LTE network. The study outcome gives the high gain. Oularbi et al. [8] have discussed the vertical handover towards the network based on OFDM by physical layer matrices. The study focuses on present and next generation networks for multiple access of technique of Physical layers of OFDM with OFDMA or CSMA/CA. The several signal synthesis feature is presented with complementary matrices set such as channel rate occupancy, collision rate and SNR as inputs to the vertical handover decision algorithm. The authors concluded different mechanisms to estimate the link quality among the mechanisms. Maeder et al. [9] have presented the present and future challenges in OFDMA. The study has focused on the operation and deployment challenges of OFDMA network. The paper has given the outlook for future deployment of 4G and beyond the 4G network. Myung [10] has given the overview of Single carrier frequency division multiple accesses (SC-FDMA). SC-FDMA will utilize frequency domains and the single carrier modulation in transmitter. The most useful advantage is that it has low peak-to-average power ratio (PAPR) value which does not exist in orthogonal frequency division multiple accesses (OFDMA). The author has given an in-depth SC-FDMA overview by focusing on the aspects of resource management and physical layer.

The study also illustrates the PAPR characteristics and the SCFDMA channel dependency on resource scheduling. karakaya et al. [11] have presented the channel interpolator for LTE uplink to adopt in environments with high Doppler based on Kalman filter. The authors have given a method and are evaluated in four different kinds of scenarios of different settings to reflect many performance many performance parameters like speed, size and propagation environment of the resource block. Kedia et al. [12] have given performance analysis techniques for 4G

communications. The SC-FDMA is the most popular and useful technique in the multiuser communication as it has low PAPR value. Still there is an requirement of new technologies which are useful in further reduction of PAPR without the degradation in the BER of the system. The study gives the comparison between the PAPR and BER performance of various techniques. The simulation results of the paper give that the better technique of SC-FDMA and DSCDMA performance over OFDMA-
CDMA. Wu et al. [13] gives the comparison between the SC-FDMA and OFDMA, in which linear frequency domain equalizer is been assumed for selected with the combat frequency channels for both the systems. By decoding the both SC-FDMA and OFDMA independently for selective frequency channels among the data (received) blocks, the authors have analytically proved that SC-FDMA will have the rate loss as compared with the OFDMA. Ciochina et al. [14] have given the history of two major techniques in wireless communication such as OFDMA and SC-FDMA. The study says that both the techniques have the better virtues than other. The OFDMA has the better performance of higher modulation that is used for the useful propagation condition. The OFDMA also has the low SNR threshold above that the high modulation and code rates are used. The study concludes that the OFDA will offer higher capacity of cell and SC-FDMA will lead for cell range extension.

The authors have concluded that these losses can be resolved by spatial diversity and linear frequency domain equalizer over multiuser systems with the help of multiple receiving antennas. Berardinelli et al. [15] have presented turbo equalization for SC-FDMA performance improvement in LTE uplink. The authors have considered a new solution of adoptive coefficient for frequency domain equalization. The study concludes with the simulation results as the performance can be improved up to 1 dB by doing few iterations and SC-FDMA performance improvement over OFDMA with high coding rate. Lin et al. [16] have derived an improved mathematical expression for algorithm of the frequency domain receiver for LTE uplink using SC-FDMA. The analytical and the simulation results show superior SINE and BER performance for LTE uplink system. Islam et al. [17] have given the study of signal transmission in multi-view video by using simulation. The system presented uses many signal detecting and processing schemes. The study concludes that the systems can be used for different data transmission in channels of hostile fading, in which the noise induced should be comparable with the transmitted signal power. Thomas et al. [18] have given the different modulation schemes in SC-FDMA operation. The authors have focused on the high PAPR value problem in OFDMA and are overcome by using the SC-FDMA. They used the modulation techniques such as QAM, CPM and QPSK and interleaved the mapping of localized (LFDMA) and interleaved (IFDMA). The study concludes with the better BER performance in LFDMA than IFDMA and the better PAPR reduction in IFDMA as compared with LFDMA. Xiao et al. [19] have presented a multiuser system frequency domain equalization (FDE) based on OFDMA. An improved algorithm of FDE for multiuser MIMO is derived with improper signals.

The study with the simulation results states that the system proposed has good BER performance than the conventional FDE and also results the novel iteration for FDE. Gupta et al. [20] have discussed analysis of the PAPR and spectral density of FDMA and SC-FDMA. They focused on the robustness against the channels for frequency of selective fading will cope the issues of limiting bandwidth and transmission of power at higher data rate over wireless system. For analysis purpose authors have used simulation with lower modulation scheme like QPSK, 16-QAM and BPSK to get lesser PAPR. The outcome concludes that the SC-FDMA is highly power efficient.

3 Proposed System

PSK is the digital modulation techniques of different forms such as BPSK (2-PSK), QPSK (4-PSK), 8-PSK and 16-PSK are evaluated to select the perfect maculation technique for the available technique. The communication system transmission quality will be quantified by Packet Error Rate (PER) or Bit Error Rate (BER), in which the packets contains number of bits. The communication design must have least errors and effective channel bandwidth utilization.

The SCFDMA transmit signals were characterized by fluctuations of low signals; the degradation of performance because of amplification of non linearty will affect the system link performance. The Peak to Average Power Ratio (PAPR) of discrete time signal is the ratio of maximum peak power and the signal average power and is mathematically expressed as below:

$$PAPR = \frac{\max|x(t)|^2}{E\left\{|x(t)|^2\right\}} \tag{1}$$

The high PAPR will be the major limitation in uplink communication as it will degrade the system BER performance. To attain the large PAPR of a signal the input power need to be reduced. The no reduction in the input power will lead the signal distortion that will result the band spectral out and signal re-growth as signal can be amplified at non-linear range. The signal distortion can be avoided if the front end must posess linear wide ranges have transmitted waveforms with the peaks. The problems can be solved with the help of SC-FDMA with various mapping schemes.

3.1 Pseudo Code for Browse Image

The Pseudo code for browsing an image is stated below with its input, variable initializations, functions and computational statements. In this code, the input image is get digitized and stored into array of [I]. Then this input image is compared to check whether this image is higher dimension or 2D image. If the image is higher dimension than do the dimension reduction to the 2D image, otherwise use this image for next steps.

The Pseudo code of the browsing image is described below with its input, variable initializations, computational statements and functions.

1. Start
2. Input I;
3. Array [f, p] ← get file name and path of input image;
4. Path1 ← array [p f];
5. Array [I] ← get digitized image signal;
6. Display (image);
7. End;

3.2 Pseudo Code for Simulate System

The Pseudo code of the simulate OFDMA is described below with its input, variable initializations, computational statements and functions.

1. Start
2. Input high_SNR, low_SNR, M_array, I;
3. Perform QAM modulation;
4. Perform OFDMA modulation;
5. Calculate no. of generated bits;
6. Calculate all transmitted symbols;
7. Divide signal frame wise;
8. Calculate data per frame for 64 and 128 samples;
9. Initialize length of guard interval;
10. Extract data samples from first frame and add zero padding to frames;
11. Add cyclic prefix;
12. Perform OFDMA demodulation;
13. Recover the data from received signal;
14. Extract data samples from frame;
15. Perform IFFT operation and recover the original data;
16. Remove guard interval;
17. Demodulate signal using 16-QAM;
18. Calculate spectrum OFDMA;
19. Calculate PAPR, CCDF, BER;
20. End;

4 Results and Discussion

This paper uses matlab simulation to get the performance on uplink 4G physical layer under different conditions. The system transmitter and receiver are implemented in order perform the evaluation of simulated system, so the various conditions bit rate error is calculated. SDFMA promises low PAPR, which is helpful in power efficiency for mobile communication at transmitter of mobile terminal. In resource allocation certain flexibility and scheduling can be achieved by using DFT-spread OFDM in uplink. The 4G uplink frequency 20 MHz is maximum and this is considered as the sampling frequency. The Fig. 1. Shows the input image considered in simulation of this paper and the array is selected and the SNZ value is chosen between the range of –30 and +30.

Figure 2. shows the noise image for the selected array size as an input. Figure 3. Shows the resulted received image after the demodulation.

Figure 4 shows the transmitted spectrum of OFDM signal and is represented in the spectrum of bands.

The Fig. 5 shows the plot of low PAPR Vs Complementary Cumulative Distribution Function (CCDF) SC-FDMA array number. In plot low PAPR

Fig. 1 Shows input image

Fig. 2 Shows noise image

Fig. 3 Shows received image

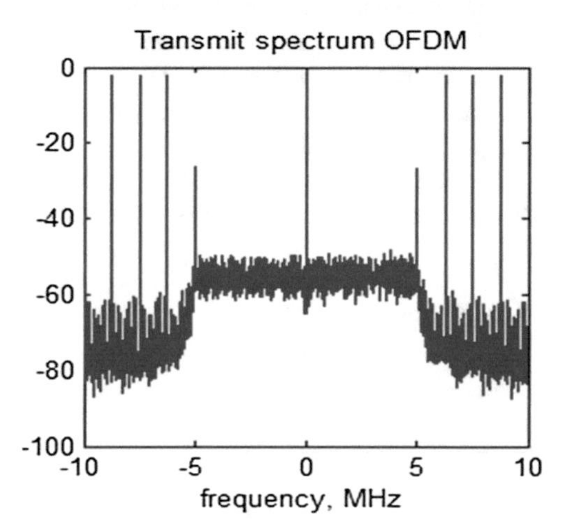

Fig. 4 Shows OFDMA spectrum

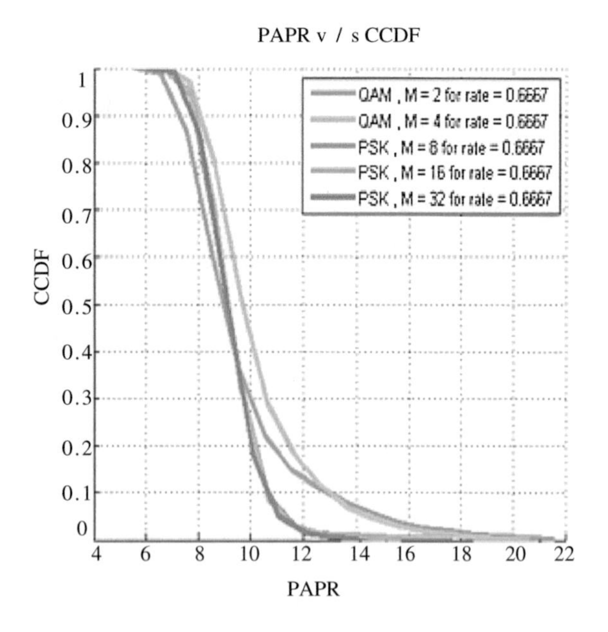

Fig. 5 Shows PAPR for SC-FDMA system

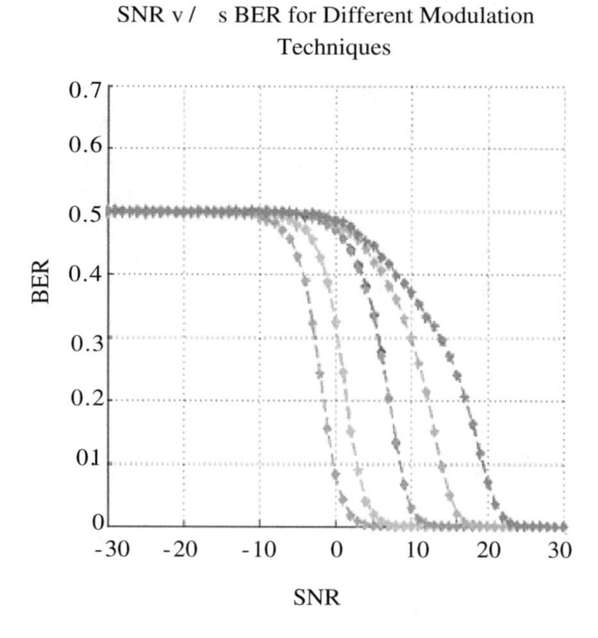

Fig. 6 Shows SC-FDMA BER using different modulation techniques

indicates the SC-FDMA will be used for uplink of 4G for optimizing the frequency range and uplink power consumption. The sufficiently transmitted low PAPR is helpful to avoid size, excessive cost and the power consumption.

Figure 6. Plots the bit error rate of different signal values to the noise ratio. It is seen that BER posses minimum values while using BPSK and at starting increases with different modulation techniques. QPSK and BPSK have shown good results with least errors compared with other techniques. Single-Carrier Frequency-Division Multiple Access (SC-FDMA) scheme is used as the alternative scheme that gives the reduced PAPR value in compression with the high PAPR of OFDMA. A lower PAPR signal will be desired as the non-linear amplifiers power efficiency can be improved. At transmitter SC-FDMA uses single carrier modulation and equalization of frequency domain at receiver side and this achieves the low PAPR.

The SC-FDMA is shown good attention towards alternative of OFDMA, especially for uplink communication in which the low PAPR is more useful in mobile communication with power efficiency transmission. SC-FDMA is presently working on the uplink multiple access scheme of Long term Enhancement (LTE). In both the cases the bit error rate is inversely proportional to the modulation index, rather than FFT size. Finally, because of low PAPR in SC-FDMA can be used in current and upcoming generations in mobile communication.

5 Conclusion

This paper uses mat lab simulation to get the performance on uplink 4G physical layer under different conditions. The SC-FDMA is shown good attention towards alternative of OFDMA, especially for uplink communication in which the low PAPR is more useful in mobile communication with power efficiency transmission. SC-FDMA is presently working on the uplink multiple access scheme of Long term Enhancement (LTE). In both the cases the bit error rate (BER) is inversely proportional to the modulation index, rather than FFT size. Finally, because of low PAPR in SC-FDMA can be used in current and upcoming generations in mobile communication.

References

1. Mullett, G.J.: Wireless Telecommunications Systems and Networks, Thomson Delman Learning (2011)
2. Stüber, G.L., Barry, J.R., Mclaughlin, S.W., Li, Y., Ingram, M.A., Pratt, T.G.: Broadband MIMO-OFDM Wireless Communications. Proc. IEEE **92**(2), 271–294 (2004)
3. Armstrong, J.: OFDM for optical communications. J. Light Wave Technol. **27**(3), 189–204 (2009)
4. Sari, H., Svensson, A., Vandendorpe, L.: Multicarrier systems. EURASIP J. Wirel. Commun. Netw. Article ID 598270 (2008)
5. Melood, M., Abdased, A., Ismail, M., Nordin, R.D.: PAPR Performance Comparison between Localized and Distributed-Based SC-FDMA Techniques, ACEEE DOI 02.WCMCS (2013)
6. Adachi, F.: Wireless past and future-evolving mobile communications systems. IEICE Trans. Fundam. Electron. Commun. Comput. Sci. **84**(1), 55–60 (2001)
7. Deb, S., Monogioudis, P.: Learning-based uplink interference management in 4G LTE cellular systems. IEEE/ACM Trans. Netw. **23**(2), 398–411 (2015)
8. Oularbi, M.R.: Physical layer metrics for vertical handover toward OFDM-based networks. EURASIP J. Wirel. Commun. Netw. **1**, 1–25 (2011)
9. Maeder, A., Zein, N.: OFDMA in the field: current and future challenges. ACM SIGCOMM Comput. Commun. Rev. **40**(5), 71–76 (2010)
10. Myung, H.G.: Introduction to single carrier FDMA. In: Proceedings of the 15th European Signal Processing Conference (EUSIPCO'07) (2007)
11. Karakaya, B., Arslan, H., Cırpan, H.A.: An adaptive channel interpolator based on Kalman filter for LTE uplink in high Doppler spread environments. EURASIP J. Wirel. Commun. Netw. p. 7 (2009)
12. Kedia, D., Modi, A.: Performance analysis of a modified SC-FDMA-DSCDMA technique for 4G wireless communication. J. Comput. Netw. Commun. (2014)
13. Wu, H., Haustein, T., Hoeher, P.A.: On the information rate of single-carrier FDMA using linear frequency domain equalization and its application for 3GPP-LTE uplink. EURASIP J. Wirel. Commun. Netw. p. 9 (2009)
14. Ciochina, C., Sari, H.: A review of OFDMA and single-carrier FDMA and some recent results. Adv. Electron. Telecommun. **1**(1), 35–40 (2010)
15. Berardinelli, G.: Improving SC-FDMA performance by turbo equalization in UTRA LTE uplink. In: Vehicular Technology Conference, VTC Spring (2008)

16. Lin, Z.: Analysis of receiver algorithms for LTE SC-FDMA based uplink MIMO systems. IEEE Trans. Wirel. Commun. **9**(1), 60–65 (2010)
17. Islam, S.F., Kafi, M.H., Islam S., Ullah, S.E.: Multiview video signal transmission in a dual polarized DWT aided MIMO SC-FDMA wireless communication system. Int. J. Sci. Eng. Rcs. **6**(7) (2015)
18. Thomas, P.A., Mathurakani, M.: Effects of Different modulation schemes in PAPR reduction of SC-FDMA System for Uplink Communication
19. Xiao, P.: Frequency-domain equalization for OFDMA-based multiuser MIMO systems with improper modulation schemes. EURASIP J. Adv. Signal Process. **1**, 1–8 (2011)
20. Gupta, P., Raina, J.P.S: To analyze the power spectral density and PAPR of FDMA and SC-FDM. Int. J. Appl. Sci. Eng. Res. **3**(3) (2014)

A Brief Analysis of Reported Problems in the Use of Function Points

Andreia Silva, Plácido Pinheiro and Adriano Albuquerque

Abstract Know the software size is a key issue to guide the planning and management of a software project. In this context, Function Point Analysis (FPA) has been consolidated as a strategic tool for measuring the functional size of software. The function point metric is the most widespread in the world, but despite its growth has received several criticisms from its users. This paper presents an investigation of the problems and difficulties on the application of FPA. As a result, the reported problems were analyzed and proposed solutions to these problems were presented.

Keywords Reported problems · Function Point Analysis · Software project

1 Introduction

The Function Point Analysis consists of a functional size metrics software that has been consolidated as an important tool to support the management of software development and maintenance contracts. Created by Allan Albrecht of IBM, the metric allows measuring the functional size of a project or application software, considering the features requested and received by the user, regardless of the technology and the development process used.

The use of Function Point Analysis has grown over the years, and this has become the most used technique in the world. However, many issues still need to be

A. Silva (✉) · P. Pinheiro · A. Albuquerque
University of Fortaleza, Av. Washinton Soares, 1321, BL J,
SL 30, Fortaleza, Ceará 60833-155, Brazil
e-mail: andrearsp@gmail.com

P. Pinheiro
e-mail: placido@unifor.br

A. Albuquerque
e-mail: adrianoba@unifor.br

R. Silhavy et al. (eds.), *Software Engineering Perspectives and Application in Intelligent Systems*, Advances in Intelligent Systems and Computing 465,
DOI 10.1007/978-3-319-33622-0_11

dealt with and this may be reflected in several published critiques through scientific articles related to the technique application.

Thus, a study based on systematic review was conducted in order to identify the main problems related to the use of Function Point Analysis technique. Systematic review is a means to identify, evaluate and interpret impartially and fairly relevant searches to a particular topic [1, 2]. According [3], systematic reviews also represent an important tool to support new research activities. Furthermore, it is believed that, because a rigorous research method, the use of systematic review can provide more reliable results, as compared to an activity informal review, since systematic reviews can be audited and repeated [1].

This paper is organized as follows: Sect. 2 briefly discusses the Function Point Analysis; Sect. 3 presents the research protocol used to guide the study; Sect. 4 describes the obtained results; Sect. 5 discusses the results; and Sect. 6 presents our final considerations.

2 Function Point Analysis—ISO/IEC 20926

The Function Point Analysis measures the size of the functional requirements of a software from the user's point of view. The functional size is composed of requirements of data and transaction requested by user. Thus, for each requirement identified as function of data or transaction is assigned a degree of complexity (low, medium or high), which according to the type of function takes a value in function points. The total function points corresponds to the sum of the contribution according to points all the features involved in the count.

However, this number represents the unadjusted function points, in other words, includes the features that the system will provide the user, disregarding the technical requirements of the system. To cover these technical requirements, the metric has an adjustment factor obtained through the determination the level of influence of a set formed by fourteen (14) general characteristics of the system. The application of the adjustment factor on the unadjusted function points can produce a variation of ±35 %.

However, to meet the requirements of ISO/IEC 14143, which consists of a pattern of functional measurement, the use of the adjustment factor is no longer required since 2002 because it treats some characteristics considered non-functional requirements.

3 The Research Protocol

Seeking to identify the difficulties and problems in the use of the technique, a study was conducted based on a systematic review of literature. This study aims to examine experience reports and scientific publications in the context of Function

Point Analysis, in order to characterize reported problems and difficulties as well as proposed approaches, methods, processes and tools to address these issues.

For the desired information, identified a main research question (*MQ*) and some secondary questions (*SQ*):

- MQ—What difficulties or problems have been reported on the application of the technique of Function Point Analysis in software projects?
- SQ1—Solutions were proposed to solve the reported problems?
- SQ2—Some tooling support was suggested?
- SQ3—The approached context is academic or industrial?
- SQ4—How the solution was validated or evaluated?
- SQ5—There were compared to approaches presented in other works?

After defining the research questions were defined some criteria to ensure the scope and feasibility of the study. That said, the selected digital libraries were: *ACM Digital Library*; *Compendex*; *IEEExplore (IEEE)*; *Scopus*; and *Web of Science*. These libraries were selected because provide easy access and allow recovery of the full text of articles, in addition were cited by [2] as important sources for software engineers.

Thus, based on the research questions, was set the following search string to run on selected digital libraries: *((“software development” OR “software project” OR “software projects” OR “software process” OR “software processes” OR “software engineering” OR “system development”) AND (“fpa” OR “function point” OR “function points”) AND (“problem” OR “problems” OR “trouble” OR “troubles” OR “question” OR “questions” OR “difficulty” OR “difficulties” OR “difficult” OR “weakness” OR “weaknesses”))*.

Knowing that the search string only checks the syntactic aspect of the text of the publications, it was necessary to establish criteria that would ensure that only the useful publications to the context of this study were selected. Thus, after being previously cataloged, the abstract of all articles returned by the search string passed through an analysis. All who attended at least one of the criteria were excluded:

- EC1—Publications that presented no problems or difficulties resulting from the application of Function Point Analysis in software projects.
- EC2—Written publications with language other than english or portuguese.
- EC3—Publications classified as norms, models or national or international standards.
- EC4—Publications concerning the presentation of keynote speeches, tutorials, courses, workshops and the like.
- EC5—No publications available for download via digital libraries or by any other means at no cost to the researcher.

Thus, all the publications that did not fit any of these criteria were selected. Is worth mentioning that, when there were doubts about the classification of the publication in one of the criteria, it was not deleted. Finally, the selected publications in the previous phase were read in this third phase and analyzed with respect to the same defined exclusion criteria, but the analysis was performed considering all the text of the publication.

4 Results

The conduct of the study returned 176 different publications distributed through informed libraries. In the second stage of selection of publications, the abstract of each work was read and 43 publications were selected from the defined exclusion criteria. Among these articles, three works could not be considered because they are not available for free by the electronic libraries.

So, all 40 publications obtained were analyzed, this time considering the full text, and classified in relation to the defined exclusion criteria. At this stage, 29 publications were selected. For these publications, a summary was made containing the answers to the survey questions defined in this study.

Regarding the first secondary question of the study about which solutions have been proposed to solve the reported problems, the results were grouped according to the cited solutions by the authors of publications. Most of the works proposed a new way to count to solve the problems reported in the use of Function Point Analysis.

About provide tooling support for the proposed solution to the problems reported, only a small part of the works, corresponding to 10.3 %, presented some type of tool support.

About the context in which the work was developed, was can be seen that most studies were performed in industrial context, while another significant portion was held at the academy. Only a small percentage did not report or did not make clear the application context of the proposed work.

With respect to the validation of the proposed approach, 58 % of the studies were evaluated through their use in industry, while 17 % had their assessment conducted at the academy. There was also an evaluation report in both contexts: industrial and academic. Other proposals, about 13 %, showed only an example of how the solution could be used; the other did not mention any form of evaluation of the proposal.

Concerning the comparative analysis with other approaches, 41.4 % of the articles presented information about other papers related to the study context.

So as you can see below, the problems and difficulties reported in the use of Function Point Analysis were grouped according to a set of criteria, in order to facilitate the achievement of the analysis of the main question of the study.

- *AS FOR SCOPE*
  - Does not allow measuring reused components [4, 5].
  - Does not take into consideration programming languages, hardware, or any other data processing technology [4–10].
  - Not suitable types of applications generally dealing with complex algorithms [6, 11].
  - The functional measurement process does not measure the data elaboration effort, but only the movement of those data [12].

  - Calculation of function point counting tends to make the system view as a black box [13].

- *PRESENT*
  - The weights used in the Function Point count have not changed since they were defined [14–16].
  - The value assigned to the function point weights were determined based on a limited amount of project types from a single organization and are applied universally without limitation type or organization [5, 17].
  - The types of user-defined function cannot be fully adequate to current technologies [13].

- *REABILITY*
  - The counting usually requires much more detail than has been available in the early stages of development [9, 18–20].
  - The classification of types of functions in simple, average and complex looks very simplified [11, 13].
  - Measures functionality may be incorrect because the analyst which defines the requirements and which measures the software can be different people [21].
  - The accuracy of the estimate cannot be improved if the system is further detailed [22].
  - The function point technique does not deal the complexity of software problem adequately, which can result in disproportionate measures software size for different system types [23, 24].
  - There is a strong relationship between the components of the Function Point Analysis and this may result in unreliable data [18].

- *AUTOMATION*
  - The function point count is a hardly automatable process [14, 18, 22, 25].

- *TIME/EFFORT*
  - The count is usually a slow process [14, 21, 25–27].
  - It makes possible a low level of abstraction, requiring a detailed analysis [5].

- *COST*
  - The counting process usually has a high cost [14, 25, 27].

- *KNOWLEDGE*
  - It requires a lot of knowledge and experience [7, 14, 18, 26].

- *SUBJECTIVITY*
  - The determination of complexity [7] and the assignment of weights is subjective [26, 28].

  - The counting process is described using natural language [14, 29].
  - Different people may estimate a same software differently [6, 21, 30, 31].

- *GENERAL SYSTEM CHARACTERISTICS*

  - The degree of influence of each factor varies from 0 to 5, which is a very simple measure [11].
  - The weight of general characteristics is the same for different types of systems [32].
  - The application of the adjustment factor contributes little to the cost estimating process [13].

5 Discussion

Regarding criterion of coverage, it is clear that there is an understanding that the technique is not appropriate for measuring certain types of systems or software features. However, it is important to note that Function Point Analysis is a technique for measuring the functional size of software [33] and criticism, in general, are associated with non-functional requirements. In this case, the Function Point Analysis probably well fulfills his role when it proposes to measure the functional size. One indication of this is that, even with the creation of new techniques that, according to the authors, work the weaknesses of the technique originally proposed by Allan Albrecht [34], the Function Point Analysis is still the most technical widespread in the world [35]. Still, it should be noted that the software size is composed of two perspectives: functional and non-functional. The latter should be properly treated and measured, using appropriate methods. An alternative would be to use together with the Function Point Analysis, the Software Non-functional Assessment Process (SNAP), an initiative of IFPUG to establish a link between the non-functional size and the effort to provide non-functional requirements of software.

In relation to technique upgrading, the fact is that the values applied as weights to measure software size were attributed in technique creation in the 70 s and defined based on a specific set of software projects related to the study context the author. Since then, the IFPUG, responsible for maintaining the technique, published some updates Counting Manual, but they treat, in most cases, improvements focused on the interpretation of the original rules in new environments [36].

On the issue of reliability, some points should be noted. First, as defined in the Counting Practices Manual [36], the Function Point Analysis technique can be used in different phases of the software lifecycle. At times, application of the technique will result in a size estimate, such as their use in the proposed time, i.e. when the users are only expressing the needs of the software. It is important to note that in the early stages of the lifecycle, the initial requirements may be the only information available, yet the size obtained with the technical application can be very useful to produce an early estimate [36]. On the other hand, from the following stage, the technique can already be applied to obtain a proper size measurement.

Also with respect to reliability, in addition to being guided by counting rules, according to [35], the Function Point Analysis is also a review process of the requirements. Given that, if performed by different people may have even greater benefits.

Another point mentioned in the study is related to the classification of function types in simple, average and complex that, according to some authors, seems to be greatly simplified. However, increasing the complexity of the counting process could be impractical because, in addition to hinder the application of the technique, might require more time for analysis of the options. Moreover, one of the premises of Function Point Analysis technique is to be simple enough to minimize the cost inherent to the measurement process [36].

Finally, the authors cite the Function Point Analysis do not treat the software problem complexity and may result in disproportionate measures to the size of different types of software.

As previously mentioned, the Function Point Analysis was created with the purpose of measuring the functionality delivered to the user, so the technical issues related to the software need to be treated separately, with appropriate techniques. However, because there are no appropriate techniques for measuring non-functional requirements for a long time, may have generated in the community the desire that the Function Point Analysis assume this role. This can be confirmed by the amount of criticism concerning the inadequacy of Function Point Analysis to non-functional context of software presented in this work.

Although there are already tools that automate part of the process, there is still the need to evaluate every feature and rank them according to existing rules.

With regard to the time required for technique application, largely depends on the rigor with which it is applied. In one estimate, when you do not have a lot of information and it is possible to make assumptions about certain situations, the count can be relatively simple and fast. Moreover, large and complex software will require a more detailed analysis and consequently longer.

Likewise, the cost of applying the technique to estimate or measure a software depends in part from accuracy and the type count desired. The more detailed the count is greater time and cost. However, it is noteworthy that the cost of not planning can bring higher losses and, therefore, applying a technique that provides subsidies to make better decisions in a project should not be seen only from the perspective of cost, but also of investment.

Regarding the knowledge to application of the technique, really the understanding the rules contained in Counting Practices Manual and its correct application influence the counting result. The IFPUG, concerned about promoting adequate training to an increasing satisfactory use of the technique, provides training in partnership with companies and also has a certification program to ensure that professionals are effectively prepared to work with the Function Point Analysis.

About subjectivity, in fact the application of the technique can be considered, to a certain degree, subjective, given that the rules to be evaluated, at times, be a result of human opinions. Furthermore, the document describing the counting rules determines not all the possibilities of software scenarios, then some scenarios will be new and need to be mapped into some form of counting. However, this is not a characteristic only of Function Point Analysis. All functional measurement techniques require some interpretation for their application and, in this context, it is important to note that in any search attempt for objectivity should not forget that interpretation is, above all, a human activity and therefore subjectivity is omnipresent [37]. Seek mechanisms to reduce the degree of subjectivity is what should be done.

Finally, the definition of an adjustment factor, based on fourteen (14) general system characteristics, was an attempt to include in the Function Point Analysis the treatment of technical aspects of software. However, to meet the requirements of ISO/IEC 14143 standard, which consists of a pattern of functional measurement, the use of the adjustment factor is no longer required in 2002. According to [35], when using this adjustment was still required, a survey conducted with the support of IFPUG showed that many users have used the metric without the adjustment provided by the factor. In addition, the IFPUG already recognized that the use of the factor was not approved by the user community, who considered obsolete and subjective characteristics. As [38], the fact that the variation obtained with the application of the factor, corresponding to ±35 %, not be enough to represent all the general characteristics of application and there is no evidence that the application of factor presents benefits for software effort estimation are the main criticisms related to the adjustment factor and general system characteristics.

6 Final Considerations

As can be seen, despite the many criticisms of Function Point Analysis, to date, no approach could provide better results in order to replace the technique satisfactorily.

However, we must remember that some aspects, especially with regard to non-functional software requirements, deserve to be treated. On the other hand, maybe the problem is not related only to measure the non-functional size, but the appropriate treatment of these requirements throughout the lifecycle of software development. According to [39], the non-functional requirements are often specified improperly and the reasons for this treatment are probably related to the high level of abstraction and lack of understanding of these requirements.

In this sense, there needs to be a broader initiative to provide, initially, support mechanisms for the identification of non-functional features of the software to allow organizations to acquire the work culture with these requirements and may, therefore, consider them in an estimated size.

In this context, we are developing a guide to support the elicitation of non-functional requirements based on ISO/IEC 25000. The aim of this study is to provide subsidies for organizations to define their requirements and, consequently, use a metric to measure the size of these requirements.

Acknowledgments The second author is thankful to National Counsel of Technological and Scientific Development (CNPq) via Grants #475239/2012-1.

References

1. Kitchenham, B.: Procedures for performing systematic reviews. Keele, UK, Keele University 33. 1–26 (2004)
2. Kitchenham, B., Dyba, T., Jorgensen, M.: Evidence-based software engineering. In: 26th International Conference on Software Engineering (2004)
3. Kitchenham, B.A., Charters, S.: Guidelines for performing systematic literature reviews in software engineering. In: Technical report. Ver. 2.3 EBSE Technical Report. EBSE (2007)
4. Gao, X., Lo, B.: A modified function point method for CAL systems with respect to software cost estimation. In: International Conference Software Engineering: Education and Practice (1996)
5. Moser, S., Nierstrasz, O.: The effect of object-oriented frameworks on developer productivity. Computer **29**, 45–51 (1996)
6. Matson, J., Barrett, B., Mellichamp, J.: Software development cost estimation using function points. IEEE Trans. Softw. Eng. **20**, 275–287 (1994)
7. Gao, X., Lo, B.: An integrated software cost model based on COCOMO and function point approaches. In: Software Education Conference (SRIG-ET'94) (1995)
8. Zheng, Y., Wang, B., Zheng, Y., Shi, L.: Estimation of software projects effort based on function point. In: 4th International Conference on Computer Science & Education (2009)
9. Živkovič, A., Rozman, I., Heričko, M.: Automated software size estimation based on function points using UML models. Inf. Softw. Technol. **47**, 881–890 (2005)
10. Abdullah, N.A.S., Abdullah, R., Selamat, M.H., Jaafar, A.: Software security characteristics for function point analysis. In: IEEE International Conference on Industrial Engineering and Engineering Management (2009)
11. Lokan, C.: An empirical analysis of function point adjustment factors. Inf. Softw. Technol. **42**, 649–659 (2000)
12. Lavazza, L., Robiolo, G. The role of the measure of functional complexity in effort estimation. In: 6th International Conference on Predictive Models in Software Engineering- PROMISE'10 (2010)
13. Jeffery, D., Low, G., Barnes, M.: A comparison of function point counting techniques. IIEEE Trans. Softw. Eng. **19**, 529–532(1993)
14. Dai, Y.B., Ren, X.L.: Size measurement in cost estimation. In: Fourth International Symposium on Information Science and Engineering (2012)
15. Ahmed, F., Bouktif, S., Serhani, A., Khalil, I.: Integrating function point project information for improving the accuracy of effort estimation. In: The Second International Conference on Advanced Engineering Computing and Applications in Sciences (2008)
16. Xia, W., Capretz, L.F., Ho, D., Ahmed, F.: A new calibration for Function Point complexity weights. Inf. Softw. Technol. **50**, 670–683 (2008)
17. Lavazza, L., Garavaglia, C.: Using function points to measure and estimate real-time and embedded software: Experiences and guidelines. In: 3rd International Symposium on Empirical Software Engineering and Measurement (2009)
18. Macdonell, S., Shepperd, M., Sallis, P.: Metrics for database systems: an empirical study. In: Fourth International Software Metrics Symposium (1997)

19. Sheetz, S.D., Henderson, D., Wallace, L.: Understanding developer and manager perceptions of function points and source lines of code. J. Syst. Softw. **82**, 1540–1549 (2009)
20. Kaur, M., Sehra, S.K.: Particle swarm optimization based effort estimation using Function Point analysis. In: International Conference on Issues and Challenges in Intelligent Computing Techniques (ICICT) (2014)
21. Jeffery, R., Stathis, J.: Function point sizing: structure, validity and applicability. Empir. Softw. Eng. **1**, 11–30 (1996)
22. Heričko, M., Živkovič, A.: The size and effort estimates in iterative development. Inf. Softw. Technol. **50**, 772–781 (2008)
23. Horgan, G., Khaddaj, S., Forte, P.: Construction of an FPA-type metric for early lifecycle estimation. Inf. Softw. Technol. **40**, 409–415 (1998)
24. Rabbi, M.F., Natraj, S., Kazeem, O.B.: Evaluation of convertibility issues between IFPUG and COSMIC function points. In: Fourth International Conference on Software Engineering Advances (2009)
25. Quesada-López, C., Jenkins, M.: Function point structure and applicability validation using the ISBSG dataset. In: 8th ACM/IEEE International Symposium on Empirical Software Engineering and Measurement- ESEM'14 (2014)
26. Wu, S.I.K.: The quality of design team factors on software effort estimation. In: IEEE International Conference on Service Operations and Logistics, and Informatics (2006)
27. Jones, C.: Function points as a universal software metric. In: ACM SIGSOFT Softw. Eng. Notes **38**(1) (2013)
28. Al-Hajri, M.A., Ghani, A.A.A., Sulaiman, M.N., Selamat, M.H.: Modification of standard Function Point complexity weights system. J. Syst. Softw. **74**, 195–206 (2005)
29. Rao, K.K., Nagaraj, S., Ahuja, J., Apparao, G., Kumar, J.R., Raju, G. Measuring the Function Points from the Points of Relationships of UML. In: International Conference on Computer and Electrical Engineering (2008)
30. Lavazza, L., Morasca, S., Robiolo, G.: Towards a simplified definition of Function Points. Inf. Softw. Technol. **55**, 1796–1809 (2013)
31. Turetken, O., Top, O.O., Ozkan, B., Demirors, O.: The impact of individual assumptions on functional size measurement. In: Software Process and Product Measurement Lecture Notes in Computer Science. pp. 155–169 (2008)
32. Peng, H., Yang, G.X., Cai, L.: Research on VAF of IFPUG method based on fuzzy analytic hierarchy process. In: IEEE/ACIS 11th International Conference on Computer and Information Science (2012)
33. Implementation note for IEEE adoption of ISO/IEC 14143-1:1998: Information technology—software measurement—functional size measurement. Part 1: definition of concepts, in IEEE Std 14143.1-2000 (2000)
34. Symons, C.: Function point analysis: difficulties and improvements. IIEEE Trans. Softw. Eng. **14**, 2–11 (1988)
35. Vazquez, C.E., Simoes, G.S., Albert, R.M.: Análise de Pontos de Função: Medição, Estimativas e Gerenciamento de Projetos de Software. 13a. Edição. Érica. São Paulo. (2013)
36. IFPUG: Counting Practices Manual. Version 4.3. January 2010. http://www.ifpug.org/
37. Bana, E., Costa, C.A.: Structuration, Construction et Exploitation Dún Modèle Multicritère D'aide à la Décision. Thèse de doctorat pour l'obtention du titre de Docteur em Ingénierie de Systèmes– Instituto Técnico Superior, Universidade Técnica de Lisboa (1992)
38. Lokan, C., Abran, A.: Multiple viewpoints in functional size measurement. In: International Workshop on Software measurement-IWSM'99. Canada. 121–132 (1999)
39. Hasan, M.M., Loucopoulos, P., Nikolaidou, M. Classification and qualitative analysis of non-functional requirements approaches. In: Enterprise, Business-Process and Information Systems Modeling Lecture Notes in Business Information Processing 348–362 (2014)

Task Allocation in Distributed Software Development Aided by Verbal Decision Analysis

Marum Simão Filho, Plácido Rogério Pinheiro and Adriano Bessa Albuquerque

Abstract One of the most critical activities in distributed project management is the allocation of tasks among remote teams. In distributed software development projects, to allocate a task for a team in any of the locations, the project manager needs consider several factors such as team maturity and time zone difference among the sites. Deciding which task to allocate for each team constitutes a decision-making task. This decision is usually made subjectively. The verbal decision analysis is an approach based on solving problems through multi-criteria qualitative analysis, which means it considers the analysis of subjective criteria. This paper describes the application of a verbal decision analysis methodology called ORCLASS to classify the most relevant factors that the project managers should take into account when allocating tasks in distributed software development projects.

Keywords Distributed software development · Task allocation · Verbal decision analysis · ORCLASS · ORCLASSWEB

1 Introduction

Distributed Software Development (DSD) is an increasing approach adopted by software development companies. The expansion of business into new markets, the need to expand the capacity of the workforce and cost reduction perspective are

M.S. Filho (✉) · P.R. Pinheiro · A.B. Albuquerque
University of Fortaleza, Fortaleza, Brazil
e-mail: marum@unifor.br; marum@fa7.edu.br

P.R. Pinheiro
e-mail: placido@unifor.br

A.B. Albuquerque
e-mail: adrianoba@unifor.br

M.S. Filho
7 de Setembro College, Fortaleza, Brazil

R. Silhavy et al. (eds.), *Software Engineering Perspectives and Application in Intelligent Systems*, Advances in Intelligent Systems and Computing 465,
DOI 10.1007/978-3-319-33622-0_12

some of the causes of the increasing number of distributed software development projects [17]. On the other hand, the distribution brings many challenges, such as language differences, time zone differences and increased complexity of coordinating and controlling the project [15]. The allocation of tasks is a critical activity for project planning. This allocation becomes even more complex in a distributed context where we must consider factors inherent to distribution [14]. The allocation of tasks to remote teams can be seen as a fundamental activity for the success of a distributed project. However, this activity is still a major challenge in global software development due to limited understanding of the factors that influence task allocation decisions [3].

We are facing a decision-making problem. Deciding which task we should allocate for each team configures itself as one of the most critical activity for the project planning, specially whether the project is distributed. Typically, the project manager makes this decision based on their experience and knowledge about the project and the teams involved. We mean that a high degree of subjectivity is present in the decision-making process. Verbal decision analysis (VDA) is an approach based on multicriteria problem solving through its qualitative analysis [6], which means the VDA methods take into consideration the criteria's subjectivity.

This paper describes the application of a VDA methodology to select the most relevant factors to be considered by project managers when allocating tasks in distributed software development projects. Expert interviews were conducted to identify the criteria and the criteria values. A questionnaire was applied to a group of project managers to characterize each factor through the criteria and criteria values. The ORCLASS method was then applied to divide the factors into preference groups.

The rest of the paper is organized as follows: Sect. 2 deals with issues involving task allocation in distributed software development. Section 3 provides a brief description of the verbal decision analysis framework. Section 4 describes the application of ORCLASS to aid the selection of the most important factors on task allocation in DSD. Section 5 presents the results of our work. Finally, in Sect. 6, we provide the conclusions and suggestions for further work.

2 Task Allocation in Distributed Software Development

Distributed tasks within the global software development context bring both many risks and many opportunities. Nowadays, distributed development is often driven by a few factors, or even by just a single factor, such as labor costs. Risks and other relevant factors such as the workforce skills, innovation potential of different regions, or cultural factors are often insufficiently recognized [4].

Many studies about the allocation of tasks in DSD have been carried out along the years. Lamersdorf et al. [5] conducted a survey on the state of practice in DSD in which they investigated the criteria that influence task allocation decisions.

Lamersdorf and Münch [3] presented TAMRI (Task Allocation based on Multiple cRIteria), a model based on multiple criteria and influencing factors to support the systematic decision of task allocation in distributed development projects. They enumerated some criteria for allocation of tasks in the development of standard and customized software.

Ruano-Mayoral et al. [19] presented a methodological framework to allocate work packages among participants in global software development projects. They claim that the allocation of work packages among the participants of the project is not a simple task. Traditionally, this allocation has been made based on availability and competence, but the distribution introduces more complexity in a process that is already complex. Besides, they pointed out several factors, with their respective metrics, that influence the allocation of tasks in DSD environments.

Simão Filho et al. [20] conducted a quasi-systematic review of studies of task allocation in DSD projects that incorporate agile practices. The study brought together a number of other works, allowing the establishment of the many factors that influence the allocation of tasks in DSD, which we can highlight: technical expertise, expertise in business, project manager maturity, proximity to client, low turnover rate of remote teams, availability, site maturity, personal trust, time zone, cultural similarities, and willingness at site.

3 Verbal Decision Analysis

Decision-making is an activity that is part of people's and organizations' lives. A wrong decision can lead to undesirable situations, such as financial losses and waste of time. Therefore, it is an activity of great importance.

In most problems, to make a decision, a situation is assessed against a set of characteristics or attributes, also called criteria. Decision making based on various criteria is supported by multi-criteria methodologies [2]. The multi-criteria methodologies support the decision-making process involving the analysis of objects from different points of view. These methodologies favor the generation of knowledge about the decision context, which helps raise the confidence of the decision maker [1, 27].

The verbal decision analysis is an approach to solve multicriteria problems through qualitative analysis [6]. That is, the VDA methodologies consider that most decision-making problems can be qualitatively described. In other words, the VDA supports the decision-making process through the verbal representation of problems [10, 16, 22–26].

The VDA methodologies can be used for ordering or sorting the alternatives. According to [7], the VDA methodologies for classification can be used to sort the possible alternatives of a decision problem in ordered or unordered groups. Among them, we can mention ORCLASS, SAC, DIFCLASS, and CYCLE methodologies [18]. Figure 1 shows the VDA classification methods.

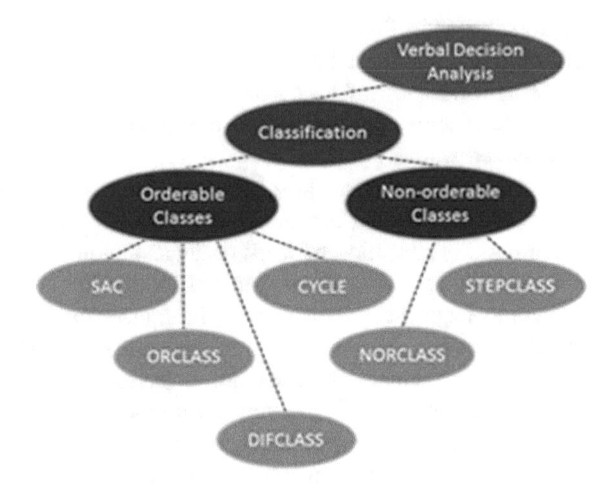

Fig. 1 Verbal decision analysis methods for classification (Adapted from [21])

3.1 The ORCLASS Method for Classification

The ORCLASS method [7] aims at classifying the alternatives into ordered decision groups, which are predetermined on the problem formulation stage.

In the first stage of problem's formulation, the set of criteria, criteria values, and the decision groups are defined. The criteria values must be sorted in an ascending order of preference (from most to least preferable). Then, the construction of the classification rule will be carried out based on the decision maker's preferences. We use the same concepts presented in [7], based on which a classification task is presented as a set of boards. Each cell is composed of a combination of values for each criterion defined for the problem, which represents a possible alternative to the problem [13]. More information about the ORCLASS method is available in [7].

In order to facilitate the decision-making process, a tool was developed and made available for use over the Internet [11]. The ORCLASSWEB tool (http://www2.unifor.br/OrclassWeb) was proposed to automate the comparison process of alternatives and to provide the decision maker a concrete result for the problem, according to ORCLASS definition [12]. Other applications of the ORCLASS method can be found in [2, 9, 13].

4 The Case Study

4.1 Bibliography Research to Find Out the Influencing Factors

First, we conducted a literature research in order to identify the main influencing factors that should be considered when allocating tasks in distributed software development projects. Table 1 shows the factors found as result of this research and which worked as the alternatives to our decision problem.

4.2 Interview with the Experts to Define Criteria and Criteria Values

Next, we interviewed a group of 4 project management experts to define the criteria and the criteria values. This is the definition stage of the criteria. For each criterion, we established a scale of values associated with it [8, 9, 25]. The criteria values were ordered from the most preferable value to the least preferable one. The list of criteria and criteria values for the problem of selecting the most important factors to be considered on task allocation in DSD projects is as follows:

1. Criterion A: Facility for carrying out the task remotely, i.e., how much easier it becomes to implement the remote task if the factor is present.
    - A1. It facilitates much: The implementation of the remote task is much easier if the factor is present.
    - A2. It facilitates: The implementation of the remote task is easier if the factor is present.
    - A3. Indifferent: The presence of the factor is indifferent to the implementation of the remote task.
2. Criterion B: Time for the project.
    - B1. High gain: The presence of the factor can cause much reduction of the period referred to perform the task.
    - B2. Moderate gain: The presence of the factor may cause some reduction of the time limit for performing the task.
    - B3. No gain: The presence of the factor does not cause changes to the deadline to execute the task.

Table 1 Influencing factors on task allocation in DSD projects

ID	Alternatives
Factor1	Technical expertise
Factor2	Expertise in business
Factor3	Project manager maturity
Factor4	Proximity to client
Factor5	Low turnover rate
Factor6	Availability
Factor7	Site maturity
Factor8	Personal trust
Factor9	Time zone
Factor10	Cultural similarities
Factor11	Willingness at site

3. Criterion C: Cost for the project.
   - C1. High gain: The presence of the factor can cause a lot of cost reduction expected to perform the task.
   - C2. Moderate gain: The presence of the factor may cause some reduction of the time limit for performing the task.
   - C3. No gain: The presence of factor induces no change compared to the estimated cost to perform the task.

4.3 Survey with the Project Managers to Define Alternatives, Decision Groups, and Alternatives' Characterization

A questionnaire was created in order to gather information and opinions about the factors that influence the allocation of tasks in DSD projects. The questionnaire was applied over the Web to a group of 20 project managers and consisted of two parts. The first part aimed to trace the respondents profile about his/her professional experience and education.

The second part of the questionnaire inquired the views of experts on the factors that influence the allocation of tasks in DSD projects. For our problem, such influencing factors were described as alternatives. Thus, in every question, the professional analyzed the influencing factors in relation to a set of criteria and criteria values, and selected what criterion value that best fitted the factor analyzed. An example of question is as follows:

1. Factor: Technical expertise—knowledge of the techniques, languages, frameworks, tools, APIs, etc. needed by the team to accomplish the task.
   (a) Criterion A: Facility for carrying out the task remotely
   ( ) A1. It facilitates much. ( ) A2. It facilitates. ( ) A3. Indifferent.
   (b) Criterion B: Time for the project
   ( ) B1. High gain. ( ) B2. Moderate gain. ( ) B3. No gain.
   (c) Criterion C: Cost for the project
   ( ) B1. High gain. ( ) B2. Moderate gain. ( ) B3. No gain.

We did the same for the other ten factors. The responses were analyzed to determine the criteria values representing the alternatives. For each influencing factor, the final table was filled based on the responses of the majority of professionals. We then selected the value of the criterion that had the greatest number of choices to represent the alternative. Table 2 summarizes the responses to the questionnaire, showing the sum of responses and characterization of alternative according to the values of each criterion (represented in the "Final Vector" column). The bold numbers in gray cells in the table indicate the criteria values selected by most of the interviewed professionals to represent a certain factor.

Table 2 Characterization of alternatives according to answers collected in the questionnaire

Criteria/ alternatives	Facility for carrying out the task remotely			Time for the project			Cost for the project			Final vector
	A1	A2	A3	B1	B2	B3	C1	C2	C3	
Factor1	**11**	7	2	**13**	6	1	**11**	7	2	A1B1C1
Factor2	**15**	3	2	**13**	7	0	**10**	8	2	A1B1C1
Factor3	8	**11**	1	5	**14**	1	7	**10**	3	A2B2C2
Factor4	**13**	4	3	8	**10**	2	8	**10**	2	A1B2C2
Factor5	**14**	6	0	**15**	4	1	**12**	7	1	A1B1C1
Factor6	**10**	8	2	**13**	5	2	**9**	6	5	A1B1C1
Factor7	**16**	3	1	**11**	9	0	9	**11**	0	A1B1C2
Factor8	8	**10**	2	6	**11**	3	3	**13**	4	A2B2C2
Factor9	3	**12**	5	3	8	**9**	3	6	**11**	A2B3C3
Factor10	4	**13**	3	3	**10**	7	3	8	**9**	A2B2C3
Factor11	**10**	8	2	**10**	8	2	**9**	6	5	A1B1C1

We emphasize that the various answers given by professionals, considering they are experienced project managers, were related to the fact that they have different professional backgrounds. Thereby, the characterization of a particular factor was based on answers given by most professionals.

Thus, the decision groups were defined as follows:

- Group I: The first group was chosen to classify the influencing factors that will be selected after the application of ORCLASS as the most important factors that project managers should take into account when allocating tasks to remote teams (preferable factors);
- Group II: The second group will contain the set of influencing factors that should be less considered by project managers when they need to allocate tasks to remote teams (not preferable factors).

4.4 The ORCLASS Method Application

The ORCLASS method application was aided by ORCLASSWEB tool, which was divided into four stages:

1. Criteria and criteria values definition.
2. Alternatives definition.
3. Construction of the classification rule.
4. Presentation of results obtained.

OrclassWeb
Criteria Alternatives Restart
Criteria Definition
Criterion Name: C: Cost for the project.
Criterion Value: C3. No gain: The presence of factor induces no change compared
+ Add
Values
C1. High gain: The presence of the factor can cause a lot of cost reduction expected to perform the task.
C2. Moderate gain: The presence of the factor may cause some reduction of the time limit for performing the task.
Save
Next
Name
Values
A: Facility for carrying out the task remotely.
A1. It facilitates much: The implementation of remote task is much easier if the factor is present.
A2. It facilitates: The implementation of remote task is easier if the factor is present.
A3. Indifferent: The presence of the factor is indifferent to the implementation of the remote task.
B: Time for the project.
B1. High gain: The presence of the factor can cause much reduction of the period referred to perform the task.
B2. Moderate gain: The presence of the factor may cause some reduction of the time limit for performing the task.
B3. No gain: The presence of the factor does not cause changes to the deadline to execute the task.

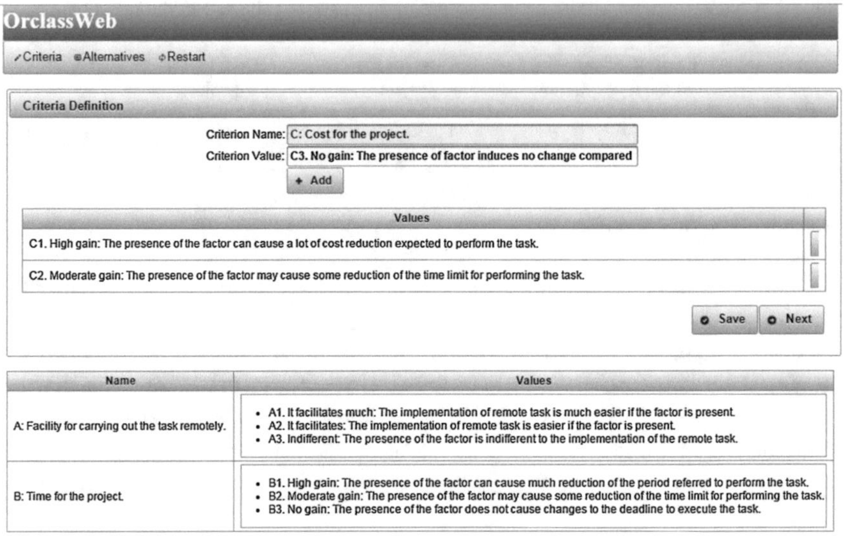

Fig. 2 Introducing criteria into the ORCLASSWEB tool

We introduced the problem's criteria into the ORCLASSWEB tool. In this step, we specified the criteria's names and their possible values. The tool allowed us to insert all the necessary criteria, as shown in Fig. 2. Next, we introduced the problem's alternatives into the ORCLASSWEB tool. The tool allowed us to inform the alternatives' names, and their representations in criteria values, according to the criteria defined in the previous step (and in the column "final vector" in Table 2).

The ORCLASSWEB tool also supported the construction of the classification rule. The tool calculates which question would be the next one that the decision maker is supposed to answer according to the ORCLASS method's rules for the selection of the most informative alternative. In this step, we had the support of an experienced project manager to answer the questions to classify the alternatives. The classification rule was completed based on the decision-maker choices. In the end, the ORCLASSWEB tool processed the complete classification of the alternatives.

5 Results

The ORCLASSWEB tool selected the following factors as the most important ones that project managers should consider when allocating tasks in distributed software development projects, by analyzing the alternatives classified by the ORCLASS method (Group I—the preferable factors): Factor 1—Technical expertise, Factor 2—Expertise in business, Factor 3—Project manager maturity, Factor 4—Proximity to the client, Factor 5—Low turnover rate, Factor 6—Availability, Factor 7—Site maturity, Factor 8—Personal trust, and Factor 11—Willingness at site. The Group II, with the not preferable factors, consisted of following factors: Factor 9—Time zone and Factor 10—Cultural Similarities.

6 Conclusion and Future Works

For large software development projects, working with distributed teams has been an alternative increasingly present in large companies. However, the distribution brings many challenges, particularly concerning the allocation of tasks among remote teams, since there are many factors that project managers should take into consideration. Typically, this multi-criteria decision-making problem involves subjective aspects. The verbal decision analysis methods support decision-making through multi-criteria qualitative analysis.

The main contribution of this work was to apply the VDA ORCLASS method to select the most important factors that project managers should consider when allocating tasks among distributed teams. A tool called ORCLASSWEB supported this work allowing to develop the case study in a fast and practical way. Previously, we conducted interviews and applied questionnaires to a group of project management experts so that we could identify the factors, criteria and criteria values to use in the ORCLASS method.

As future work, we intend to evolve this research by working on a hybrid model joining ORCLASS to ZAPROS method in order to provide an ordering rank of the preferable alternatives (factors). The application of the factors resulting of this research on actual projects would be useful to validate the model, and this could be an important future work. In addition, we intend to apply VDA methods to help choose the team that should be assigned a specific task, based on the task characteristics and teams profiles.

Acknowledgments The first author is thankful for the support given by the "Coordination for the Improvement of Higher Level-or Education- Personnel" (CAPES) and 7 de Setembro College during this project. The second author is grateful to National Counsel of Technological and Scientific Development (CNPq) via Grants #305844/2011-3. The authors would like to thank The Edson Queiroz Foundation/University of Fortaleza for all the support.

References

1. Evangelou, C., Karacapilidis, N., Khaled, O.A.: Interweaving knowledge management, argumentation and decision making in a collaborative setting: the KAD ontology model. Int. J. Knowl. Learn. **1**(1/2), 130–145 (2005)
2. Gomes, L.F.A.M., Moshkovich, H., Torres, A.: Marketing decisions in small businesses: how verbal decision analysis can help. Int. J. Manage. Decis. Making **11**(1), 19–36 (2010)
3. Lamersdorf, A., Münch, J.: A multi-criteria distribution model for global software development projects. Braz. Comput. Soc. (2010)

4. Lamersdorf, A., Münch, J., Rombach, D.: Towards a multi-criteria development distribution model: an analysis of existing task distribution approaches. In: IEEE International Conference on Global Software Engineering, ICGSE 2008 (2008)
5. Lamersdorf, A., Münch, J., Rombach, D.: A survey on the state of the practice in distributed software development: criteria for task allocation, In: Fourth IEEE International Conference on Global Software Engineering, ICGSE 2009 (2009)
6. Larichev, O.I., Brown, R.: Numerical and verbal decision analysis: comparison on practical cases. J. Multicriteria Decis. Anal. **9**(6), 263–273 (2000)
7. Larichev, O.I., Moshkovich, H.M.: Verbal Decision Analysis for Unstructured Problems. Kluwer Academic, Dordrecht (1997)
8. Machado, T.C.S., Menezes, A.C., Pinheiro, L.F.R., Tamanini, I., Pinheiro, P.R.: The selection of prototypes for educational tools: an applicability in verbal decision analysis. IEEE Int. Joint Conf. Comput. Inf. Syst. Sci. Eng. (2010)
9. Machado, T.C.S., Menezes, A.C., Pinheiro, L.F.R., Tamanini, I., Pinheiro, P.R.: Applying verbal decision analysis in selecting prototypes for educational tools. In: IEEE International Conference on Intelligent Computing and Intelligent Systems, pp. 531–535. Xiamen, China (2010)
10. Machado, T.C.S., Pinheiro, P.R., Albuquerque, A.B., de Lima, M.M.L.: Applying verbal decision analysis in selecting specific practices of CMMI. Lecture Notes in Computer Science vol. 7414, pp. 215–221 (2012)
11. Machado, T.C.S.: Towards Aided by Multicriteria Support Methods and Software Development: A Hybrid Model of Verbal Decision Analysis for Selecting Approaches of Project Management. Master thesis, Master Program in Applied Computer Sciences, University of Fortaleza (2012)
12. Machado, T.C.S., Pinheiro, P.R., Tamanini, I.: OrclassWeb: a tool based on the classification methodology orclass from verbal decision analysis framework. Math. Probl. Eng **2014**(Article ID 238168), 11 (2014). doi: 10.1155/2014/238168
13. Machado, T.C.S., Pinheiro, P.R., Tamanini, I.: Project management aided by verbal decision analysis approaches: a case study for the selection of the best SCRUM practices. Int. Trans. Oper. Res. **22**(2), 287–312 (2014). doi:10.1111/itor.12078
14. Marques, A.B., Rodrigues, R., Conte, T.: Systematic literature reviews in distributed software development: a tertiary study. IEEE Int. Conf. Global Softw. Eng. ICGSE **2012**, 134–143 (2012)
15. Marques, A.B., Carvalho, J.R., Rodrigues, R., Conte, T., Prikladnicki, R., Marczak, S.: An ontology for task allocation to teams in distributed software development. In: IEEE 8th International Conference on Global Software Engineering, ICGSE 2013, pp. 21–30 (2013)
16. Mendes, M.S., Carvalho, A.L., Furtado, E., Pinheiro, P.R.: A co-evolutionary interaction design of digital TV applications based on verbal decision analysis of user experiences. Int. J. Digit. Cult. Electron. Tourism **1**, 312–324 (2009)
17. Miller, A.: Distributed Agile Development at Microsoft patterns & practices, Microsoft patterns & practices (2008)
18. Pinheiro, P.R., Machado, T.C.S., Tamanini, I.: Verbal decision analysis applied on the choice of educational tools prototypes: a study case aiming at making computer engineering education broadly accessible. Int. J. Eng. Educ. **30**, 585–595 (2014)
19. Ruano-Mayoral, M., Casado-Lumbreras, C., Garbarino-Alberti, H., Misra, S.: Methodological framework for the allocation of work packages in global software development. J. Softw. Evol. Proc. (2013)
20. Simão Filho, M., Pinheiro, P.R., Albuquerque, A.B.: Task allocation approaches in distributed agile software development: a quasi-systematic review. In: 4th Computer Science On-line Conference 2015. Proceedings of the 4th Computer Science On-line Conference 2015 (CSOC2015). Software Engineering in Intelligent Systems, 2015, vol. 3, pp. 243–252. Zlín (2015)
21. Tamanini, I.: Hybrid Approaches of Verbal Decision Analysis Methods. Doctor thesis, Graduate Program in Applied Informatics, University of Fortaleza (2014)

22. Tamanini, I., Carvalho, A.L., Castro, A.K.A., Pinheiro, P.R.: A novel multicriteria model applied to cashew chestnut industrialization process. Adv. Soft Comput. **58**(1), 243–252 (2009)
23. Tamanini, I., de Castro, A.K.A., Pinheiro, P.R., Pinheiro, M.C.D.: Towards an applied multicriteria model to the diagnosis of Alzheimer's disease: a neuroimaging study case. IEEE Int. Conf. Intell. Comput. Intell. Syst. **3**, 652–656 (2009)
24. Tamanini, I., de Castro, A.K.A., Pinheiro, P.R., Pinheiro, M.C.D.: Verbal decision analysis applied on the optimization of Alzheimer's disease diagnosis: a study case based on neuroimaging. Adv. Exp. Med. Biol. **696**, 555–564 (2011)
25. Tamanini, I., Machado, T.C.S., Mendes, M.S., Carvalho, A.L., Furtado, M.E.S., Pinheiro, P.R.: A model for mobile television applications based on verbal decision analysis. Adv. Comput. Innovations Inf. Sci. Eng. **1**(1), 399–404 (2008)
26. Tamanini, I., Pinheiro, P.R.: Challenging the incomparability problem: an approach methodology based on ZAPROS. Model. Comput. Optim. Inf. Syst. Manage. Sci. Commun. Comput. Inf. Sci. **14**, 338–347 (2008)
27. Tamanini, I., Pinheiro, P.R.: Reducing incomparability in multicriteria decision analysis: an extension of the ZAPROS method. Pesquisa Operacional **31**(2), 251–270 (2011)

Application of Evolutionary Algorithm in Supply Chain Management for Internet Marketing

Sayed Sayeed Ahmad, Manuj Darbari and Harsh Purohit

Abstract The paper highlights the issue of development of a framework for internet based supply chain management system using the concept of Evolutionary Algorithm. Authors have developed objective function of supply chain and optimize it using EMO (Evolutionary Multi Objective Optimization). The paper discusses how a mobile company "XEMO" selling its mobile phones online was able to optimize between the demand and supply using EMO.

Keywords Urban traffic system · Cognitive radio · EMO

1 Introduction

With the development of new technologies and product Market demand for Internet based product has increased management has increased by many folds. Internet based goods supply has certain advantages like shortening of the product cycles and proliferation of product variety.

S.S. Ahmad
College of Engineering and Computing, Al Ghurair University, Dubai, United Arab Emirates
e-mail: saeed.ks@gmail.com

M. Darbari (✉)
Department of Computer Science & Engineering, Babu Banarasi Das University, Lucknow, Uttar Pradesh, India
e-mail: manuj_darbari@acm.org

H. Purohit
Faculty of Management Studies, WISDOM, Banasthali Vidyapith, P.O. Banasthali Vidyapith, Vanasthali 304022, Rajasthan, India
e-mail: deanwisdom@banasthali.in

R. Silhavy et al. (eds.), *Software Engineering Perspectives and Application in Intelligent Systems*, Advances in Intelligent Systems and Computing 465,
DOI 10.1007/978-3-319-33622-0_13

There are many issues on which supply chain management is based like Dis-intermediation, improving on the customer value, process innovation and data management [1–4].

The use of internet purchases has increases by many folds. Starting from the basics of permanent of goods from the vendors to final delivery to the customers and moreover taking the return back of the goods are all part of internet based system of marketing. Generally the use of internet takes place during transportation process [5], where marketers' have to coordinate between frequency of arrival of goods [6], order delivery areas and total count of orders received by the firm.

Faisal et al. [7] reduces the supply chain risk by managing the flow. Cucchiella and Gataldi [8] dealt with uncertainty aspect of the supply chain management and suggests some methods of deducing it. Tang [9] reviews quantitative models for management of risk in supply chain. Gohet [10] also presents a stochastic model providing a cascading model for global supply chain network. Ritchie and Brindley [11] tried to correlate risk management and supply chain management by developing a model. Khan and Burnes [12] also identifies a competitive analysis of qualitative and quantitative approaches to mitigate the risk in supply chain management. Li and Chandra [13] linked knowledge management and supply chain management.

Lastly Rao and Goldsby [14] correlated SCRM and typology of risk in supply chain.

2 A Brief Overview of Supply Chain Management

A supply network ranges from procurement of raw materials to reaching out the finished product to customers. Supply chain exists in both the sectors and a manufacturing unit. There are various factors [15–17] which can enhance the performance of the supply chain like efficient stock management, focusing on customer related demand, production planning etc.

The aim of the supply chain can be categorized into two broad categories to observe the movements in supply chain and decide the policy accordingly. Several approaches for modeling and optimisation [18] of a supply chain can be classified into five classes:

- project of the supply chain
- integer-mixed programming optimisation
- stochastic programming
- heuristic methods
- Simulation-based methods

The main aim of SCM is to get the required material at the right moment of time.

3 Internet Marketing

Internet marketing by Chaffey can be simply defined as:

> Achieving marketing objectives through applying digital technologies.

It uses the concept of extensive use of digital technologies like websites of the particular company providing online promotions scheme thorough it thereby acquiring large number of new customer base and building a constant repo with the older customers i.e. CRM.

The basic idea of internet marketing is using a PULL strategy as customer's will be interacting to the business website of single entity with full attention. Secondly exploited thoroughly using web semantic where the machine is made intelligent to provide personalization service based on his/her liking.

4 Evolutionary Algorithms

Evolutionary Algorithms [3, 5, 14, 19–23] provide robust optimizer for engineering design problems. In evolutionary algorithm we manage the tradeoff between the objective values. In evolutionary algorithm a new set of candidate solutions are generated depending on the cost assigned to it.

Evolutionary algorithm has certain rules about the fitness function area they are specified. The traditional systems involve the following necessary steps:

- Ranking of the entire population.
- Marking the best individuals to the worst individuals.
- Assign the fitness value to the individuals with same rank so that all of them are sampled at the same rate.

5 Our Proposed Model

The focus of our research will be to apply Evolutionary Optimisation Algorithm [24] like NSGA II which mainly summarizes on a trade-off [25] of one complete solution over another in multiobjective space.

The focus of the problem is to analyse the Internet User's Profile and their likes and dislikes about the particular product and based on their product liking the product is customized accordingly, which finally rippled to Supply Chain.

As an outcome of our study we would be applying the features like Attributes, Criteria, Objectives, and Goals and develop a Parerto Front based on the characteristics.

```
Step 1: MC:= Initialize (MC)
Step 2: while termination condition is not satisfied, do
Step 3: MC′:= Selection (MC)
Step 4: MC″ := Genetic Operations(MC′)
Step 5: MC := Replace (MC U MC″)
Step 6: end while
Step 7: return (non dominated solutions (MC))
```

The algorithm for Internet Marketing based Supply Chain.

```
{
// Initialize all the order points of the products
(01,02,…,0n);
// Evaluate the fitness of each pair as per the function
F_min  (α LLP,α Mean LLP);
// Set the termination criteria
  F_min  (α LLP,α Mean LLP)>α LLP;
Do
{
//Select the most optimized delivery plan
Crossover (Various combinations of delivery plan);
  Mutation(Change the delivery plan according to LLP;)
  Evaluate the new population;
}while terminating criteria is satisfied;
}
```

$$Demand\ Management(DM) = \frac{Online\ Order\ fullfilment(OF)}{Liking\ level\ of\ the\ product(LLP)}$$

$$DM = \frac{OF}{LLP_{\alpha+1} - LLP_{\alpha}}$$

We have to maximize the demand to increase the reach of particular product.

$$Stock\ Density(SD) = \frac{Online\ Order\ Fullfillment(OOF)}{Order\ Frequency(OF)} * Number\ of\ times\ it\ is\ viewed(V_{\alpha})$$

$$SD = \frac{OOF}{OF_{\alpha+1} - OF_{\alpha}} * V_{\alpha}$$

In order to manage the smooth flow of stock density we have to maximize it.

Experimental Setup:

Population Type	Bit String
Number of Population	0 to 400
Selection Function	Tournament
Tournament Size	3
Crossover Fraction	0.8
Mutation	Uniform
Crossover function	scattered

Solutions in the current population are sorted to assign a rank to each solution. The outcome is formulated using crowding distance which is the sum of the calculated distance over all the objectives. On plotting this for two objective(variables) we got large number of non-dominated solution [26] for multi objective optimization at the value of 67,136 runs (Figs. 1 and 2).

6 Analysis

The Fig. 3 shows the Demand Management Score Histogram and Stock Density and Order frequency histogram. From the above plots we conclude that in case of Online Demand Management has direct impact on Online order fulfillment and Liking of the Individuals. Similarly, Stock density dependent upon Order fulfillment and Order frequency.

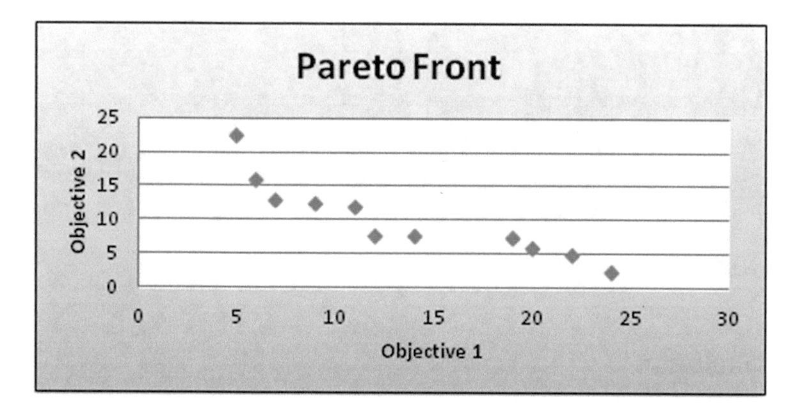

Fig. 1 Pareto front 1

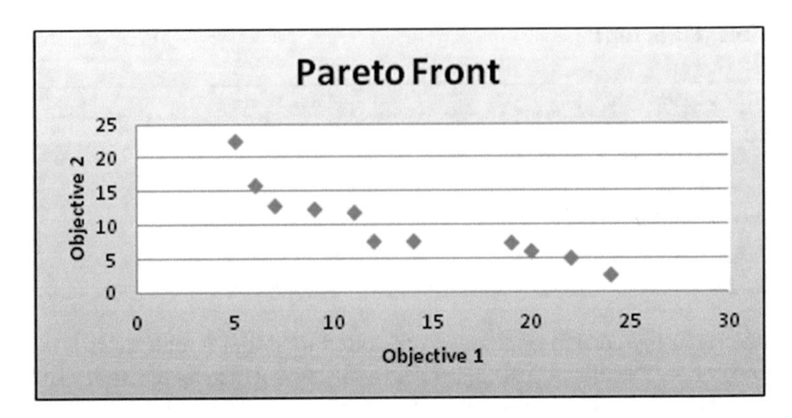

Fig. 2 Pareto front 2

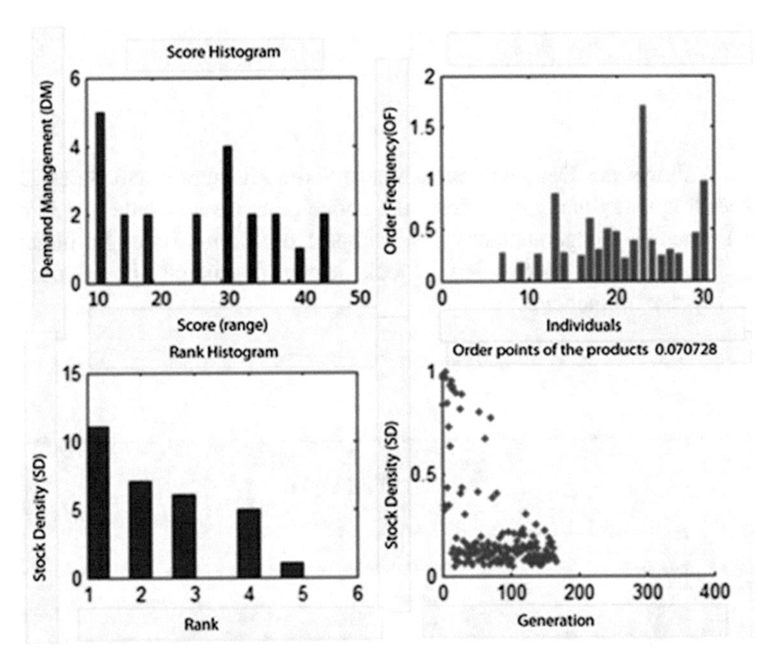

Fig. 3 Analysis of the setup

7 Conclusion

We will try to integrate the following areas like: Demand management, Manufacturing flow management, Order fulfillment, Product development and Commercialization, Customer service Management, Customer relationship management by developing a framework for internet based supply chain management. Using

Pareto front on different runs we were able to find out the optimization level between Demand Management and Stock Density which helps the mobile company "XEMO" in identifying the tradeoff between the two variables. The main advantages of the proposed integration architecture are:

- It is independent of any agent-oriented methodology;
- The pattern matching process gives simple explanations about the correspondences applied in the translation;
- The addition of new methodologies just requires the specification of their mappings;
- Mappings with evolutionary computing provide a suitable resource to discover and analyze missing features in agent-oriented languages;
- The architecture provides necessary automation tools to support Unified framework.

References

1. Lavania, S., Darbari, M., Ahuja, N.J., Siddqui, I.A.: Application of computational intelligence in measuring the elasticity between software complexity and deliverability. In: 2014 IEEE International Advance Computing Conference (IACC). IEEE (2014)
2. Bansal, S., Darbari, M.: Designing knowledge based expert system for handling business dynamics. Int. J. Sci. Eng. Res. **2**(11) (2011)
3. Bansal, S., Darbari, M.: Multi-objective intelligent manufacturing system for multi machine scheduling. Int. J. Adv. Comput. Sci. Appl. **3**(3) (2012)
4. Darbari, M., Sahai, P.: Adaptive e-learning using granulerised agent framework. Int. J. Sci. Eng. Res. **5**(3), 167–171 (2014)
5. Asthana, R., Ahuja, N.J., Darbari, M., Shukla, P.K.: A critical review on development of urban traffic models and control systems. Int. J. Sci. Eng. Res. **3**(1) (2012)
6. Yagyasen, D., Darbari, M.: Quantification of business dynamics using multi-agent and Petri-Sim. Int. J. Sci. Eng. Res. (2015). ISSN: 2229-5518
7. Nishat Faisal, M., Banwet, D.K., Shankar, R.: Mapping supply chains on risk and customer sensitivity dimensions. Ind. Manag. Data Syst. **106**(6), 878–895 (2006)
8. Cucchiella, F., Gastaldi, M.: Risk management in supply chain: a real option approach. J. Manuf. Technol. Manag. **17**(6), 700–720 (2006)
9. Tang, C.S.: Perspectives in supply chain risk management. Int. J. Prod. Econ. **103**(2), 451–488 (2006)
10. Ritchie, B., Brindley, C.: Disintermediation, disintegration and risk in the SME global supply chain. Manag. Decis. **38**(8), 575–583 (2000)
11. Khan, O., Burnes, B.: Risk and supply chain management: creating a research agenda. Int. J. Logistics Manag. **18**(2), 197–216 (2007)
12. Li, X., Chandra, C.: A knowledge integration framework for complex network management. Ind. Manag. Data Syst. **107**(8), 1089–1109 (2007)
13. Rao, S., Goldsby, T.J.: Supply chain risks: a review and typology. Int. J. Logistics Manag. **20**(1), 97–123 (2009)
14. Lavania, S., Darbari, M., Ahuja, N.J., Shukla, P.K.: Application of evolutionary algorithm in managing the trade-off between complexity of software and its deliverables. Int. Rev. Comput. Softw. **7**(6), 2899–2903 (2012)

15. Lavania, S., Darbari, M., Ahuja, N., Siddqui, I.A.: Application of computational intelligence in measuring the elasticity between complexity and deliverability. In: 4th IEEE International Advanced Computing Conference—IACC (2014)
16. Yagyasen, D., Darbari, M.: Application of semantic web and petri calculus in changing business scenario. In: Advances in Intelligent Systems and Computing. Springer (2014). ISSN: 2194-5357
17. Ahmad, S.S., Darbari, M., Purohit, H.: Handling web dynamics of internet marketing supply chain using evolutionary algorithms and semantic breakdown strategy. In: International Business Information Management Conference (25th IBIMA). Netherlands (2015)
18. Shukla, P.K., Darbari, M., Singh, V.K., Tripathi, S.P.: A survey of fuzzy techniques in object oriented databases. Int. J. Sci. Eng. Res. **2**(11) (2011)
19. Yagyasen, D., Darbari, M., Ahmed, H.: Transforming non-living to living: a case on changing business environment. IERI Procedia **5**, 87–94 (2013)
20. Asthana, R., Ahuja, N.J., Darbari, M.: Model proving of urban traffic control using neuro petrinets and fuzzy systems. Int. Rev. Comput. Softw. **6**(6) (2011)
21. Tahilyani, S., Darbari, M., Shukla, P.K.: A new multi agent cognitive network model for lane-by-pass approach in urban traffic control system. Int. Rev. Comput. Softw. **7**(5), 2179–2182 (2012)
22. Lavania, S., Darbari, M., Ahuja, N.J., Shukla, P.K.: Application of evolutionary algorithm in managing the trade-off between complexity of software and its deliverables. Int. Rev. Comput. Softw. **7**(6) (2012)
23. Tahilyani, S., Darbari, M.: Cognitive framework for intelligent traffic routing in a multiagent environment. In: Advances in Intelligent Systems and Computing. Springer (2015). ISSN: 2194-5357
24. Darbari, M., Sahai, P.: Adaptive e-learning multi-agent systems with swarm intelligence. Int. J. Appl. Inf. Syst. **7**(3), 16–20 (2014)
25. Lavania, S., Darbari, M., Ahuja, N., Siddqui, I.A.: Genetic algorithms-fuzzy based trade-off adjustment between software complexity and deliverability. In: 9th Annual International Joint Conferences on Computer, Information, Systems Sciences, and Engineering. Springer (2013)
26. Ahmad, S.S., Purohit, H., Alshaikhly, F., Darbari, M.: Information granules for medical infonomics. Int. J. Inf. Oper. Manag. Educ. **5**(3) (2013)

Intelligent Mechanism for Cloud Federation and Requirements Changes Management

Nassima Bouchareb, Nacer Eddine Zarour and Samir Aknine

Abstract Cloud Computing is a new paradigm which provides ubiquitous access to shared computing resources for organizations and end users. In this paper, we propose an agent-based mechanism to automatically manage Cloud federations and requirements change to accept the maximum of requests with minimum cost and energy consumption. First, we present the mechanism's strategies: gain strategy, acceptance strategy and offer strategy. During the implementation of the mechanism (or after a certain time) the user's needs may change (some are modified, others are added or deleted). So, we present how to manage the requirements change of the previous strategies. Finally, simulation results indicate that these policies enhance the provider's profit.

Keywords Cloud Computing · Resource management · Federation · Requirements change · Green computing · Multi-agent systems

N. Bouchareb (✉) · N.E. Zarour
LIRE Laboratory, Faculty of New Technologies of Information and Communication, Department of Software Technologies and Information Systems, University Constantine, 2 – Abdelhamid, Mehri, Algeria
e-mail: nassima.bouchareb@univ-constantine2.dz

N.E. Zarour
e-mail: nasro.zarour@univ-constantine2.dz

S. Aknine
LIRIS Laboratory, Department of Computer Science, University Lyon 1, Villeurbanne, France
e-mail: samir.aknine@univ-lyon1.fr

R. Silhavy et al. (eds.), *Software Engineering Perspectives and Application in Intelligent Systems*, Advances in Intelligent Systems and Computing 465,
DOI 10.1007/978-3-319-33622-0_14

1 Introduction

Cloud Computing (CC) provides a pool of computing resources. One of its essential characteristics is the availability and sharing of resources. The resource management in CC encounters difficulties, especially when there are a big number of requests. The selection of suitable resources and provider cost are among these problems. It has been argued that energy costs are among the most important factors impacting on provider total cost [1, 2]. To increase the provider's gain, they must be able to dynamically increase their available resources to accept more requests by establishing *Cloud federations*. So, the aim of the suggested mechanism is to select resources when the provider cannot satisfy the new request (insufficient resources), by forming federations. But how does the Cloud decide which federations to form to get more gain? How does it compute the Cloud trust and the federation utility? As Cloud federations are distributed, scalable, and every agent belonging to the federation is a selfish member, their needs may be influenced by other goals (like the respect of Service-Level Agreement) that can change through time, which leads to an evolution of the Cloud system requirements. Therefore, the adaptation of the system to these changes must be treated early in the development process because the resolution of any problem must be managed from the root. Also, it is well known that the requirements changes at a later stage of the software development is an important source of software defects and higher costs "more an error is introduced early and later detected, more expensive to correct it". That is why we propose in this paper a strategy that allows our mechanism to adapt to some requirements changes that can arise.

The CC is a distributed and complex system, when it comes to design this type of systems; the agent technology proves suitable, because agents do not only allow the sharing or distribution of knowledge, but also the fulfillment of a common goal. The agent-specific aspects here are in their autonomy, communication, cooperation, negotiation, and the key feature is their intelligence which results from these properties, and engenders collective behaviors [3]. Also, multi-agent systems have shown their effectiveness in the treatment of the concept of adapting systems to changes of their environments. Consequently, we propose an agent-based mechanism that aims to maximize provider's gains, and minimize the energy consumption. We present strategies that help the provider to form or leave a federation, accept or not an offer, compute the Cloud trust, and especially adapt to some requirements changes that can arise.

The remainder of this paper is organized as follows. In the next section, we discuss the similar works. After, we present the proposed Cloud architecture in Sect. 3. In Sect. 4, we detail the federation management mechanism and in Sect. 5, we present the requirements change management strategy. Finally, we present the experimental results in Sect. 6, before concluding and giving some future directions in Sect. 7.

2 Related Work

Among works that have examined the resource management in Cloud federations: [4], authors just compare between liberating the occupied resources and outsourcing the new request. However leaving a federation may be the best solution. This case is treated in the proposed mechanism (The difference between these two works is detailed in Sect. 6). In [5], authors allow the provider to contact only its Cloud neighbors, while a non neighbor Cloud can be more beneficial than another neighbor Cloud. So in the proposed mechanism we just give priority to Clouds according to their neighborhood when calculating their trusts. Among works that treat Green Computing in CC: [6] where authors propose efficient green enhancements in CC, using power-aware scheduling techniques, and live migration of VMs, but they don't propose any mechanism for resource management. Our contribution is based on the results of this framework. A large majority of studies have examined the issue of requirements change [7, 8] but in a general way, not in CC. In [9], authors have developed an agent-oriented approach, which is effective in the management of systems to be adapted to changes in their environments starting with the first step of development. In our work, we have used the same steps of their process with some modification (the process steps are detailed in Sect. 5.1). In [10], authors have adopted CC to solve the problem of requirement change management in global software development environment. Authors in [11] motivate the need for a new requirements engineering methodology for systematically helping businesses and users to adopt cloud services and for mitigating risks in such transition. Among works that treat the concept of requirements engineering in the CC [12], it is intended to set the foundation for a reference model for requirements engineering for cloud-based solutions. According to these insufficiencies, we treat in this paper the requirements change management in Cloud resources and especially in Cloud federations.

3 An Agent-Based Cloud Architecture

The proposed Cloud architecture contains four cognitive agents, which reason before making decisions (see Fig. 1). *Cloud agent* (*CA*): Is the first agent that receives the request; it detects the quantity of VMs (Q) needed for the request, the duration of use (D), price (P) and the customer's country. We suppose that resources are grouped according to their geographical locations, so depending on the origin of the request, *CA* selects the corresponding *AA* to minimize time, cost of transfer and energy consumption. *Allocator agent* (*AA*): When *AA* receives Q and D, it selects available resources according to Green Computing. *Coalition agent* (*CoA*): When *AA* does not find sufficient resources; *CoA* finds the best solution with external resources (by forming federations). It computes the gain of the proposed solution, the Cloud trust and the federation utility. *Supervisor agent* (*SuA*): oversees the anomalies in the federations' requirements and captures the dynamic changes.

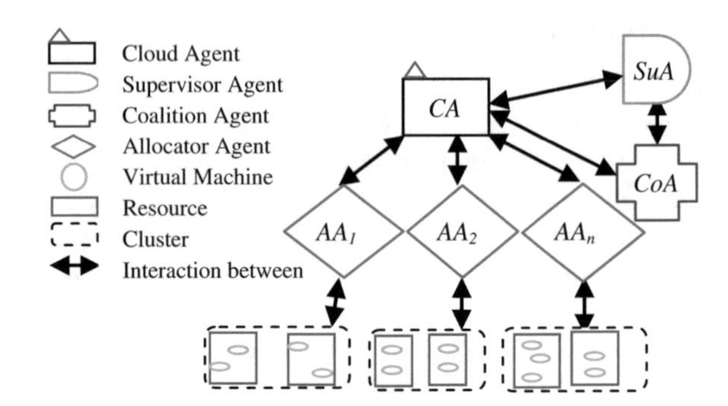

Fig. 1 Agent-based Cloud architecture

4 Federation Management Strategies

We present the compute of the gain, the Cloud trust and the federation utility.

4.1 Gain Strategy

The decision between forming and leaving a federation is made according to the gain (G) of each proposal. To compute it, the following parameters are considered "R_{cust}, R_{coal}, P_{cust}, P_{coal}, C_{cust}, C_{coal}, C_{out}, N_{cust}, and N_{coal}". Where R_{cust}, R_{coal}, P_{cust}, and P_{coal} are revenues of customers requests, federations offers, customers penalties and federations members penalties. P_{cust} are penalties paid by customers when they cancel their requests and P_{coal} are penalties paid by Cloud providers of the formed federations when they leave the federations. C_{cust} and C_{coal} are the operational costs of satisfaction of customers and federations requests respectively. C_{out} is the cost of the outsourced VMs that a provider pays to federation members hosting its requests. N_{cust} and N_{coal} are penalties paid by the Cloud provider to customers and federations members respectively, when he cancels the requests. So:

$$G = R_{cust} + R_{coal} + P_{cust} + P_{coal} - C_{cust} - C_{coal} - C_{out} - N_{cust} - N_{coal} \quad (1)$$

From this equation, we compute the gain when the Cloud receives a federation offer or customer request: (1) In the case of forming a new federation with other clouds to use their resources. (2) In the case of leaving a federation to free up its VMs.

4.2 Offer Strategy

When the Cloud decides to form a federation, its *CoA* selects the best according to their trusts. The provider computes the $Cloud_j$ trust ($trst_j$); it consults the $Cloud_j$

Table 1 Denotations 1

Symbol	Signification	Symbol	Signification
$trst_j$	$Cloud_j$ trust	F_j^{ref}	Refusing frequency of $Cloud_j$
tr_j	$Cloud_j$ behavior	F_j^{ign}	Ignorance frequency of $Cloud_j$
$dist_j$	Geographical distance between the two Clouds	F_j^h	Honesty frequency of $Cloud_j$ in gains sharing
$decis_j$	$Cloud_j$ decision	F_j^q	Honesty frequency of $Cloud_j$ in respecting the QoS
$reslt_j$	Results of old federations with $Cloud_j$	F_j^d	Honesty frequency of $Cloud_j$ in respecting delay
F_j^{coa}	Federation formation frequency with $Cloud_j$	F_j^r	Frequency of revenues type paid by $Cloud_j$
F_j^{can}	Cancellation frequency of $Cloud_j$		

behavior (tr_j). The geographical distance between the two Clouds ($dist_j$) is another parameter in this strategy. The $trst_j$ is computed as follows (see Table 1):

$$trst_j = w_1 tr_j + w_2 dist_j \tag{2}$$

For each parameter in the following equations, there is a corresponding weight w which quantifies its importance for the Cloud provider. In each equation, w sum is equal to 1. The provider gives priority to neighbors Clouds to gain more time in data transfer and minimize the cost of communication and energy consumption, but with a very low weight comparing to w_1. tr_j is calculated as follows:

$$tr_j = w_3 decis_j + w_4 reslt_j \tag{3}$$

When the provider solicits $Cloud_j$, counter θ_j is incremented. If $Cloud_j$ accepts, accepts then cancels, refuses, or ignores the offer, the counter coa_j, can_j, ref_j, or ign_j is incremented respectively. Then, we divide the counter values by θ_j to obtain the different frequencies values of acceptation, cancelation, refusal and ignorance (F_j^{coa}, F_j^{can}, F_j^{ref}, and F_j^{ign} respectively). So:

$$decis_j = w_5 F_j^{coa} + w_6 F_j^{can} + w_7 F_j^{ref} + w_8 F_j^{ign} \tag{4}$$

If $Cloud_j$ is honest in sharing gains, it respects the QoS, the delay and offers important revenues, counters h, q, d, and r are incremented respectively, else they are decremented respectively. F_j^h, F_j^q, F_j^d, and F_j^r are computed as in ($decis_j$). So:

$$reslt_j = w_9 F_j^h + w_{10} F_j^q + w_{11} F_j^d + w_{12} F^r \tag{5}$$

Table 2 Denotations 2

Symbol	Signification	Explanation
U_{rev}	Offered revenues	It prefers offers with highest revenues to maximize its gain
U_{all}	Resources allocation	It prefers requests that use activated resources. After, those which activate resources and finally requests that oblige the provider to form/leave federations
U_{trst}	Cloud trust	It prefers providers with highest trusts

4.3 Acceptance Strategy

Each Cloud computes the federation utility to accept or reject it. It is computed as follows (see Table 2 for denotations):

$$Utility = w_{13} * U_{rev} + w_{14} * U_{all} + w_{15} * U_{trst} \tag{6}$$

5 Requirements Change Management Strategy

We propose now a new strategy that helps the system to take automatically the appropriate decisions when it encounters some federations' requirements changes. To ensure these changes, we have used the *SuA*. This agent oversees the anomalies in the federations' requirements and executes the changes according to the following process.

5.1 The Process Steps

Bendakir and Zarour have developed in [9] an effective approach for management of systems to be adapted to changes in their environments. We are inspired from their process, because it is an agent-oriented approach and they treat the change management at the first step of development and from the perception of the change to its validation. So, we project this process in our context. Any request of change must have a life cycle of five phases (see Fig. 2): perception, evaluation, request submission, achieving change, verification and validation.

Perception. When the supervisor agent perceives a change, it identifies the source of change (internal or external), it indicates the reason of this change, it identifies stakeholder's requirements that must be changed (the concerned strategy of the *CoA*) and indicates the type of change (modification/addition/deletion).

Table 3 presents the changes that our system can encounter (see Sect. 5.2).

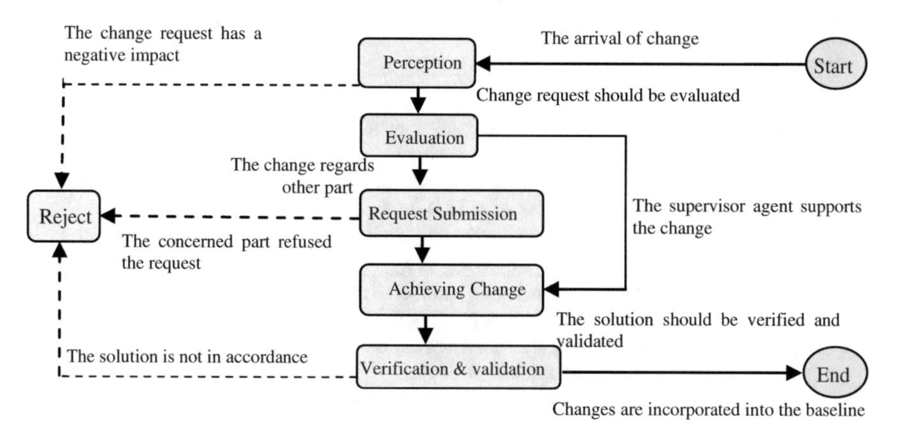

Fig. 2 The process steps [9]

Table 3 The different changes of the system

The change	The concerned strategy	The type of change
Service-level agreement violation	Gain strategy	Addition (Pn_{cust}, Pn_{coal})
Non-neighbors Clouds more efficient than neighbors Clouds	Offer strategy	Deletion ($dist_j$)
Federations with the same utility	Acceptance strategy	Modification (w_{13}, w_{14}, w_{15})

Evaluation. Any change request must take into account an impact study (cost, security, time ...), by analyzing its effect on the system. In our case, the supervisor agent decides whether to perform the change or not.

Request Submission. The request submission is done after the identification of the concerned strategy of the *CoA* (in the first step). The request of change accompanied by a report containing the arguments "the *SuA* evaluation" is submitted to the *CA* which first contacts the administrator. This later evaluates the request in order to decide whether to accept the change or not. In the case where the administrator accepts the request, the *CA* sends the request to the *CoA* to achieve it.

Achieving Change. This phase involves the realization of change. After making the change, the *CoA* returns the result to the *SuA*.

Verification and Validation. This phase aims to validate the successful implementation of the request. The final result must be screened or verified by the *SuA* in order to avoid any type of conflict to validate the change.

The process steps are inspired from the process of [9], but there are some differences. Table 4 resumes these differences.

Table 4 The differences between the two processes

Differences		Arguments
Process of [9]	Our process	
The supervisor agent just perceives the change and verifies it at the end, and the analyst agent accomplishes the other steps (evaluation and request submission)	There is no analyst agent; it is the supervisor which accomplishes all the steps (except of course the achieving change which is done by the *CoA*)	Because the agent which verifies at the end must be the same which evaluates at the beginning
The agent concerned by the change is identified at the 3rd step (request submission)	There is one agent concerned by the change, it is the *CoA*, but with 3 strategies, the identification of the concerned strategy is done at the 1st step when identifying the type of change	Because it is necessary for the evaluation (2nd step)
The concerned agent is directly contacted to achieve the request of change	The *CA* is first contacted to inform the administrator	Because the administrator must be aware of this change
The concerned agent evaluates also the request of change and it can refuse it	The *CoA* does not evaluate the request of change, it automatically achieves it	Because the *SuA* and also the administrator have evaluated it
If the concerned agent refuses the request, the analyst agent tries to negotiate a solution	There is no negotiation in the process	Because the *CoA* does not take decisions

5.2 Requirements Changes in Different Strategies

We present in this section the three cases of the requirements change.

Requirements Change in Gain Strategy. It was assumed in the proposed mechanism that the Cloud providers must meet the requirements of customers and respect the SLA (Service-Level Agreement), i.e. if the Cloud discovers that it cannot satisfy the received request, it tries to contact other Clouds to form a federation and obtain external resources or leaves an old federation to release its own resources. This may change over time; the system may violate the SLA. In this case, the *SuA* perceives this change; it follows the precedent process. The change which must be happen concerns the gain strategy. New parameters must be added to the gain equation: penalties paid to customers (Pn_{cust}) or to federation members (Pn_{coal}), it depends of the request origin. So Eq. (1) becomes (7):

$$G = R_{cust} + R_{coal} + P_{cust} + P_{coal} - C_{cust} - C_{coal} - C_{out} - N_{cust} - N_{coal} - Pn_{cust} - Pn_{coal} \quad (7)$$

These parameters are not inserted from the beginning in the gain formula (1) because the system is not meant to violate the request SLA, and especially to not weigh down the system with unused parameters.

Requirements Change in Offer Strategy. The change which can be done at the offer strategy concerns the geographical distance parameter. This parameter can be completely removed from the trust function, if the *SuA* perceives that the non-neighbors Clouds are more efficient than the neighbors Clouds.

Requirements Change in Acceptance Strategy. The *SuA* can also perceive that the Cloud provider has received two federations with the same utility, but it cannot accept the two offers at the same time (lack for resources). In this case, the *CoA* eliminates the parameter joined by the lowest weight in the Eq. (6), and it accepts the offer which has the greatest utility value. If the new utilities are still the same, it removes again the parameter with the lowest weight. If with a single parameter, the utilities are equal, the provider accepts the offer which occupies the minimum resources; otherwise it checks the reputation of the soliciting Clouds.

6 Simulation and Experimentation Results

We have compared the performance of the proposed Agent Based Resource Management (*ABRM*) mechanism with three policies proposed in [4], which differ in how Cloud handles requests when they can't be served by internal resources:

- *Non-Federated Totally In-house* (*NFTI*): provider terminates VMs. If this action does not release enough resources for the request, it is rejected.
- *Federation-Aware Outsourcing Oriented* (*FAOO*): provider outsources the request to the provider that offers the cheapest price (form a federation). If outsourcing is not possible, VMs are terminated as a last resort.
- *Federation-Aware Profit Oriented* (*FAPO*): provider compares the profit of outsourcing with VMs termination.
- *The proposed Mechanism* (*ABRM*): provider compares the profit of leaving and forming a federation before deciding.

Figure 3 shows the impact of varying the number of requests on the gain of the provider in the different policies. *FAPO*, *FAOO*, and *ABRM*, which support outsourcing, has higher profit by increasing load. They outperform the *NFTI* policy. The difference between *FAPO*, *FAOO*, and our *ABRM* is more observable at higher loads. Our strategy gives the best results. However, we note that sometimes *FAPO*

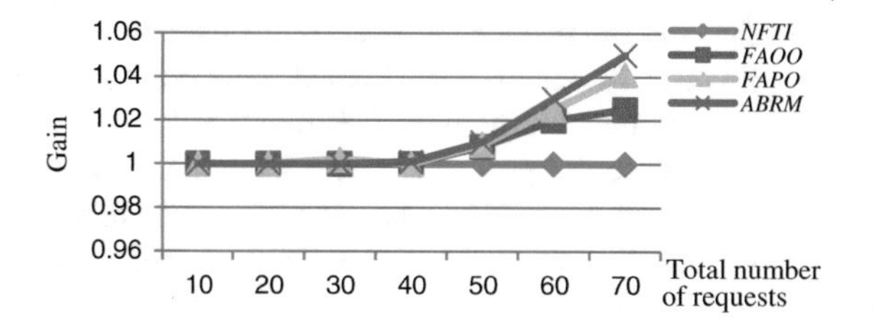

Fig. 3 Impact of load on gain for a provider with different mechanisms

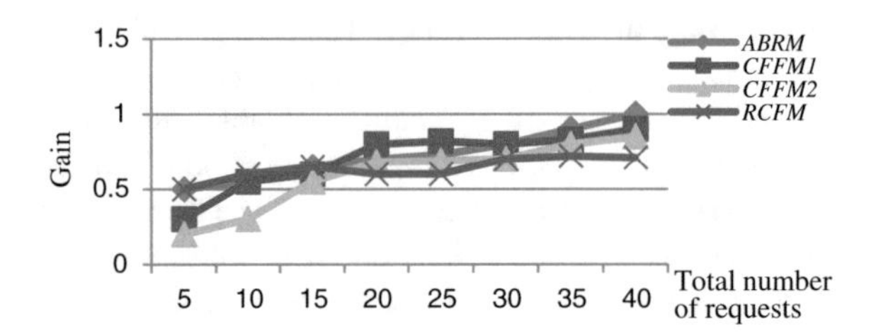

Fig. 4 Impact of load on the gain of the solicited Cloud with different mechanisms

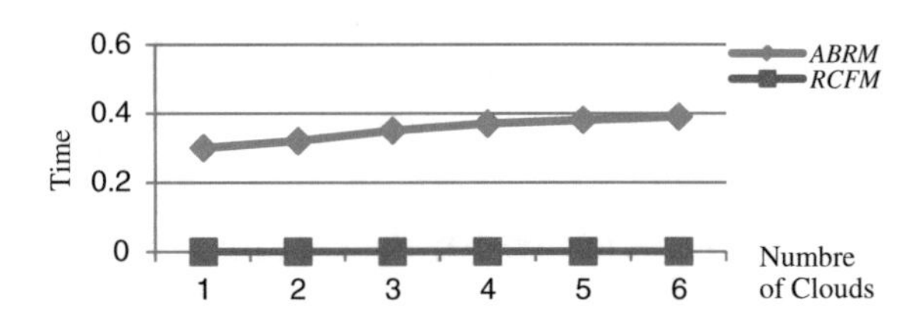

Fig. 5 Impact of the number of Clouds on time with *ABRM* and *RCFM*

is more efficient than *ABRM* as in Fig. 2—30 requests (paying penalties involves more costs than terminating VMs). We have also compared the acceptance strategy with the federation utility used in the Cloud Federation Formation Mechanism (*CFFM*) proposed in [13]. The federation utility of *CFFM* is only based on revenues offered by this federation. Mashayekhy and Grosu [13] do not detail in their work if the costs of energy consumption are taken into account when calculating revenues or not. So, we have compared our acceptance strategy with *CFFM1* which takes into account the energy consumption costs, but doesn't interest to the Cloud trust, and *CFFM2* which only interests to the offer revenues.

We have also *RCFM* (Random Cloud Federation mechanism) which accepts federation offers randomly. Figure 4 shows that the acceptance strategy presented in the proposed *ABRM* mechanism typically offers the best gains for the solicited provider. Sometimes *CFFM1* and *RCFM* can give the best results. For example, when the requesting is a new Cloud, the solicited provider does not have its trust degree to calculate the federation utility, so he avoids it and coincidentally *CFFM1* or *RCFM* accepts it and gives the best results. Figure 5 summarizes the execution time when calculating various Cloud trusts. We note that the calculation time increases by increasing the number of the discovered Clouds i.e. the trust degree when there is just one discover Cloud is calculated more quickly than several trust degrees, but it is not a significant difference. It is true that *RCFM* execution time is always close to zero, because the provider selects Clouds randomly, but the good result of the federation is not guaranteed. According to these results, the proposed strategies of the *ABRM* mechanism are usually the most beneficial compared to the other strategies.

7 Conclusion

We have presented in this paper strategies that increase the provider's profit: gain strategy, offer strategy, acceptance strategy and requirements change management strategy. This work opens many perspectives: We hope first testing the scalability of the mechanism. Then, we will interest to data integrity and security aspects.

References

1. Guazzone, M., Anglano, C., Canonico, M.: Energy-efficient resource management for Cloud Computing infrastructures. In: The 3rd International Conference on Cloud Computing Technology and Science, pp. 424–431, Athens, Greece (2011)
2. Li, H., Jeng, J.: CCMarketplace: a marketplace model for a hybrid Cloud. In: The Conference of the Center for Advanced Studies on Collaborative Research, pp. 174–183, Toronto, Ontario, Canada (2010)
3. Talia, D.: Cloud computing and software agents: towards Cloud intelligent services. In: The 12th Workshop on Objects and Agent, pp. 02–06, Rende, CS, Italy (2011)
4. Toosi, A., Calheiros, R., Thulasiram, R.K., Buyya, R.: Resource provisioning policies to increase IaaS provider's profit in a federated Cloud environment. In: The International Conference on High Performance Computing and Communications, pp. 279–287, Banff, Canada (2011)
5. Ye, D., Zhang, M. and Sutanto, D.: Integrating self-organization into dynamic coalition formation. In: The International Foundation for Autonomous Agents and Multi-agent Systems, pp. 1253–1254, Valencia, Spain (2012)
6. Younge, J., Laszewski, G. Wang, L. Lopez-Alarcon, S., Carithers, W.: Efficient resource management for Cloud Computing environments. In: The International Conference on Green Computing, pp. 357–364, Chicago, IL (2010)
7. Dam, K.H., Winikoff, M.: Evaluating an agent-oriented approach for change propagation. In: Luck, M., Gomez-Sanz, J.J. (eds.) Agent-Oriented Software Engineering IX, pp. 159–172. Springer, Berlin, Heidelberg (2009)
8. Lim, S.L., Finkelstein, A.: Anticipating change in requirements engineering. In: Avgeriou, P., et al. (eds.) Relating Software Requirements and Architectures, pp. 17–34. Springer, Berlin, Heidelberg (2011)
9. Bendakir, S., Zarour N.E.: The agent-oriented approach to support requirements changes. In: The 3rd International Conference on Information Systems Post-implementation and Change Management, pp. 199–206, Lisbon, Portugal (2014)
10. Bibi, S., Hafeez, Y., Hassan, M.S., Gul, Z., Pervez, H., Ahmed, I., Mazhar, S.: Requirement change management in global software environment using Cloud Computing. J. Softw. Eng. Appl. **7**, 694–699 (2014)
11. Zardari, S., Bahsoon, R.: Cloud adoption: a goal-oriented requirements engineering approach. In: The 2nd International Workshop on Software Engineering for Cloud Computing, pp. 29–35, Honolulu, HI, USA (2011)
12. Schrödl, H., Wind, S.: Requirements engineering for Cloud Computing. J. Commun. Comput. **8**, 707–715 (2011)
13. Mashayekhy, L., Grosu, D.: A coalitional game-based mechanism for forming Cloud Federations. In: The 5th International Conference on Utility and Cloud Computing, pp. 223–227, Chicago, Illinois, USA (2012)

Comparison of the Intrusion Detection System Rules in Relation with the SCADA Systems

Jan Vávra and Martin Hromada

Abstract Increased interconnectivity, interoperability and complexity of communication in Supervisory Control and Data Acquisition (further only SCADA) systems, resulted in increasing efficiency of industrial processes. However, the recently isolated SCADA systems are considered as the targets of considerable number of cyber-attacks. Because of this, the SCADA cyber-security is under constant pressure. In this article we examine suitability of current state signature based Intrusion Detection System (further only IDS) in SCADA systems. Therefore, we deeply evaluate the Snort and the Quickdraw rules based on signatures in order to specify their relations to SCADA cyber security. We report the results of the study comprising more than two hundred rules.

Keywords Cyber security · Intrusion detection system · Industrial control system · Signature

1 Introduction

An increasing number of the cyber-attacks relating to the Supervisory Control and Data Acquisition (further only SCADA) systems have the eminent influence on the SCADA cybersecurity. Accordingly, there is necessity to implement an Intrusion Detection System (further only IDS) in the SCADA systems, because IDS will become an essential part of the SCADA system. Moreover, there is a prediction of

J. Vávra (✉) · M. Hromada
Tomas Bata University in Zlin, Zlin, Czech Republic
e-mail: jvavra@fai.utb.cz

M. Hromada
e-mail: hromada@fai.utb.cz

R. Silhavy et al. (eds.), *Software Engineering Perspectives and Application in Intelligent Systems*, Advances in Intelligent Systems and Computing 465,
DOI 10.1007/978-3-319-33622-0_15

increasing dependency of the SCADA systems on IT; therefore, the percentage of industrial companies utilizing the IDS will rapidly grow. The implementation of the IDS in the SCADA systems has been widely investigated. Furthermore, Verba and Milvich [1] suggest that the current state of signature or anomaly based IDS are not suited to be widely deployed in the SCADA systems; accordingly, future research is needed. Moreover, there is a little research dealing with the current state of IDS rules related to the SCADA. The previous research has not fully addressed the type, priorities and number of SCADA rules. This study was designed to investigate the IDS SCADA rules. In this paper, we present the comparison and evaluation between two groups of SCADA-based rules (the Snort and the Quickdraw rules).

2 Supervisory Control and Data Acquisition

The SCADA systems are developed for monitoring, management and control of industrial systems. Moreover, the SCADA is an internal part of the Critical Information Infrastructure (further only CII). Nowadays, the CII is an essential part almost every sector of the critical infrastructure (transportation systems, power plants, dams, water treatment, oil production, chemicals, gas distribution, etc.). Therefore, every cyber-attack on the CII systems can be considered as a lethal attack. It can result in fatal damage to the environment, population or a country.

The SCADA have a positive influence on contemporary society; nevertheless, these systems are under increasing pressure to improve efficiency and interoperability. Thus, the recently isolated systems are becoming more dependent on interconnection with external technologies [2]. This trend resulted in the emergence of new vulnerabilities and, accordingly, the protected system becomes more vulnerable to new cyber-attacks.

3 Evaluation of ICT and SCADA Cyber Security

The Evaluation of the main differences between ICT and SCADA cyber security is crucial for this research and especially for the IDS. Accordingly, we used three security criteria (availability, confidentiality and integrity) to describe differences between ICT and SCADA cyber security. Their relationship can be seen in Fig. 1. As a result, confidentiality is the most important security element for ICT. On the other hand, availability is the most important for the SCADA. That is why every threat to the continuity of the SCADA processes is considered as critical.

Fig. 1 The comparison of ICT and SCADA cyber security [3]

4 Intrusion Detection System

An Intrusion Detection System is characterized as a system for detection of an intruder. It is a reactive tool used for monitoring, detecting and recording dangerous behavior in a computer network or a computer system. The IDS inspecting each packet within protected computer network and looking for malicious content [4]. Thus, there are two main detect methodologies (signature and anomaly based). However, we deal only with signature based methodology. Every registered intrusion is reported to the operator who will make appropriate decision to eliminate this threat.

4.1 IDS Architecture

The IDS architecture is composed of four main parts. All parts are interconnected and disruption of one of them has serious consequences to the whole system. The results, given in Fig. 2, show entire common IDS.

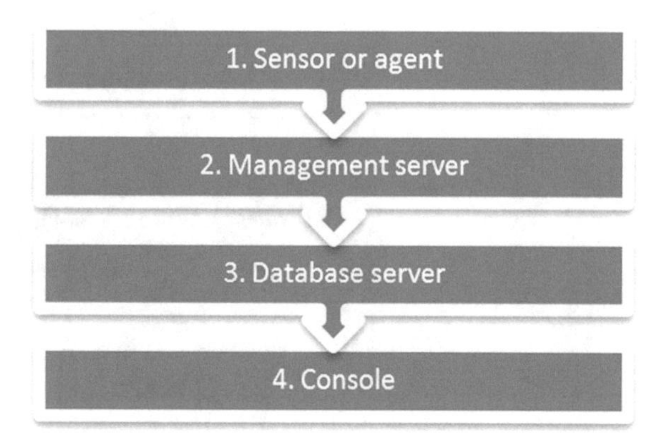

Fig. 2 Common architecture of the IDS [5]

There is a specification of each IDS component:

- The first and essential element in Fig. 2 is a sensor, which is also known as a agent. It is used for monitoring, recording and evaluating network traffic. The sensor is being used in conjunction with the network intrusion detection system (further only the NIDS) while the agent is used in conjunction with the host intrusion detection system (further only the HIDS) [5].
- The management server is a centralized device that accepts information from the agents or the sensors. These data can be further processed to a qualitatively higher level. The management server can correlate inputs from multiple sensors or agents in order to increase detection capabilities. However, the outputs from the management server can be used in the control and management improvement of particular sensors or agents.
- The database server is designed to store data communication from the sensors or the agents. It can also store knowledge database based on signatures and other detection methodologies for detecting an intrusion.

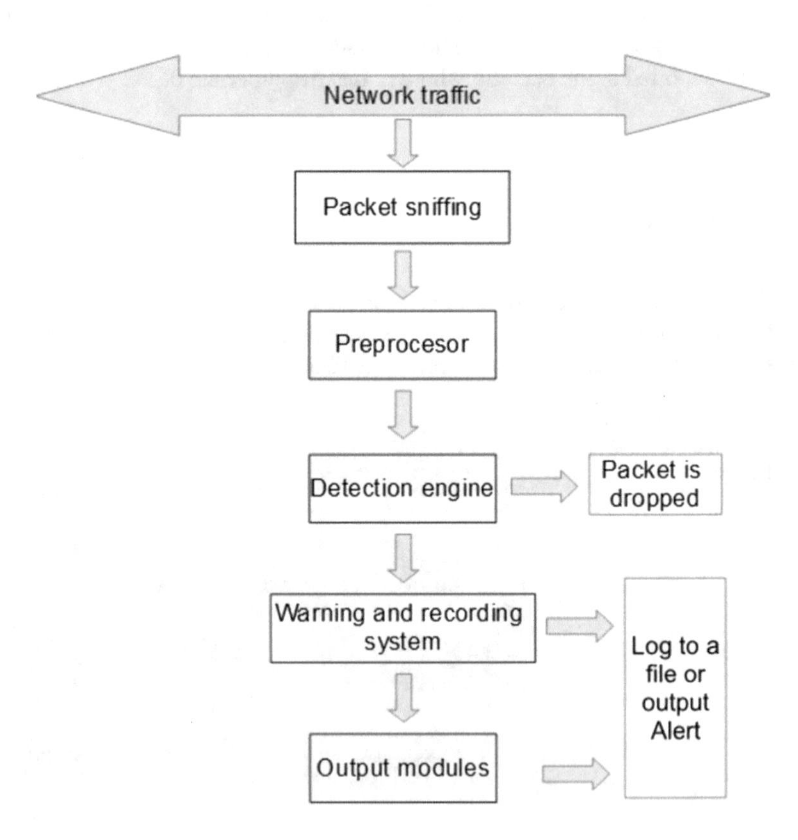

Fig. 3 Snort IDS components [6]

- The Console is a software interface between the user and the IDS. The console is commonly used for administration or configuration of the sensors or the agents [5].

4.2 Snort IDS Components

Snort is an open source IDS. This system is divided into five main components which are hierarchically organized. The whole components are shown in Fig. 3.

The diagram in Fig. 3 determines the basic architecture of famous Snort IDS. The first fundamental operation of every IDS is to capture particular packets of network traffic. The captured data are modified for the preprocessor. The preprocessor is a component of IDS which is responsible for modifying packets into a standardized form for the detection engine. The detection engine is the most important part of the IDS. Its main purpose is to detect suspicious activity in network traffic. Snort can use more than one detection methodologies. However, if the evaluated packet is harmless, then it is discarded. On the other hand, if the packet is detected as an intrusion attempt, then the record and warning are generated. Output modules finally generate a final message to the user [6].

5 Signature-Based IDS Detection

The signature-based detection is one of the most basic methods for malware detection. This methodology is based on the comparison. Therefore, the IDS generally use rules to compare real traffic patterns with cyber security signatures. They matches to particular cyber-attacks, moreover each of them is stored in the signature database. This methodology is very effective against known cyber threats and has a low rate of false positive detection. On the other hand, it is often ineffective against unknown threats like zero-day attacks or the modifications of already known cyber-attacks. Furthermore, due to the comparison of every packet in network traffic with all signatures; there is a momentous time-consuming process, which is responsible for slowing down the system operation. This is the serious problem for the SCADA systems.

5.1 Definition of IDS Rule Based on Signature

Each rule-based signature starts with a header. Moreover, within the header are particular criteria, which allow the comparison between the rule and the network traffic. The header also specifies an action for a case when a match is found. The architecture of a Snort header shows Fig. 4.

Action	Protocol	Address	Port	Directional operator	Address	Port

Fig. 4 Components of the Snort header [7]

Each segment in Fig. 4 is subsequently discussed:

- **Action**—this is an action that will take place when the rule exactly matches the packet. This segment is determined by actions such as alerting, logging, ignoring, blocking packet and many others.
- **Protocol**—this segment determines a communication protocol for which the rule is proposed.
- **Address**—the IP address used as source or destination depending on the directional operator.
- **Port**—it can be described as a source or a destination port used in a network communication.
- **Directional operator**—determines, if the address and the port are considered as the source or the destination port.

The remaining part of the rule contains additional criteria for detection of cyber-attacks and the background information. An example of the whole rule can be seen in Fig. 5.

Figure 5 shows the rule for the DNP3 SCADA communication protocol. The IDS is designed to generate the alert message when a rule matches with the data. Moreover, the rule is dedicated for TCP protocol. "$EXTERNAL_NET" and "20000" were set as source IP address and source port. On the contrary, destination IP address and destination port were set as "$HOME_NET" and "any".

The remaining part of the rule is known as a rule options is responsible for providing additional information. The content of this segment is not static, furthermore, it may change. However, the Fig. 5 shows common rule composition:

- **msg**—it is used for quotations of the rule. In this case, the msg describes communication protocol and background information.
- **flow**—the flow is used for TCP sessions. Moreover, it describes packet direction for which the rule is made. Figure 5 shows the rule that is applicable only on TCP session.
- **dnp3_ind**—the dnp3_ind is a particular rule option for the DNP3 protocol. It is used to match against flag bits in a DNP3 packet header.
- **reference**—this segment is responsible for accompanying information about the rule. It is not necessary for detection; therefore, it can be ignored.

```
alert tcp $EXTERNAL_NET 20000 -> $HOME_NET any (msg:"PROTOCOL-
SCADA DNP3 unsupported function code error";
flow:established,to_client; dnp3_ind:no_func_code_support;
reference:url,www.dnp.org/About/Default.aspx; classtype:protocol-
command-decode; sid:15718; rev:5;)
```

Fig. 5 Architecture of the Snort rule [8]

- **classtype**—the classtype provides a classification of the alert and its priority for the rule. The sample of the classtype is shown in Fig. 5. However, there is a classification of alert without the priority; accordingly, the priority is set on default value in the classification.config file. Furthermore, low priority value represents a high threat.
- **sid**—the sid can be described as a rule ID. Moreover, the sid number up to one million is reserved for Snort rules. The sid number over one million is dedicated for local rules.
- **rev**—the rev determines how many times the rule was modified. It can be also determined as a version of the rule.

6 Methods

The Snort and the Quickdraw SCADA rules were collected due to the evaluation of possibilities of deployment the IDS in the SCADA system. The Snort and the Quickdraw provide databases of SCADA rules used by a considerable number of organizations. This article is dealing with the comparison between the Snort and the Quickdraw SCADA rules. The research is examined according to the following criteria: the number of SCADA rules, the type of cyber-attack alerts and priority of the rule. Thus, the rules were collected from the Snort and the Quickdraw databases. The collected rules are usually used for the Modbus and the DNP3 communication protocols. The Modbus and the DNP3 rules were selected based on their ports; whereas, the Modbus communication protocol uses port 502 and the DNP3 communication protocol uses port 20000. The total sample consist more than 100 rules. In the follow-up phase of the study, we evaluated collected data in order to obtain crucial information for the purpose of the article. In the interest of determining the relationship between rules, a quantitative data analysis was used. Each rule was evaluated and classified.

7 Results

The aim of the article is the evaluation of the cyber threats in relation to the SCADA systems. In order to evaluate the SCADA rules, we determined three objectives. The first objective of the research is to evaluate the data in term of number. The comparison of the SCADA rules is shown in Fig. 6.

Figure 6 shows the distribution of rules per SCADA communication protocols and provides comparison between the Snort and the Quickdraw rules. As can be seen, Snort provides the highest amount of Modbus rules (58 rules), compared to Quickdraw with 14 rules. On the other hand, Quickdraw has the highest amount of the DNP3 rules (14 rules), compared to Snort with 10 rules.

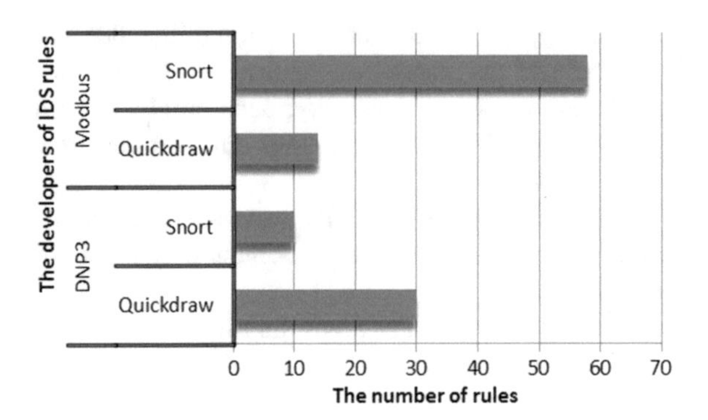

Fig. 6 The comparison of IDS SCADA rules

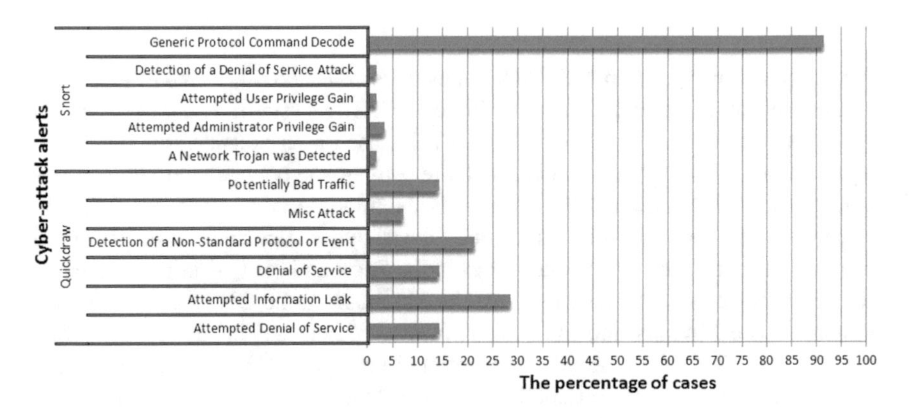

Fig. 7 The comparison of the cyber-attack alerts in relation with Modbus

The second objective of the research is to evaluate cyber-attack alerts. The Fig. 7 shows a representation of the alerts generated by the Snort and the Quickdraw rules in relation with Modbus protocol. Each alert is represented by the percentage of cases. There is an eminent difference between Snort and Quickdraw. The distribution of the rule is divergent. Almost 92 % of all Snort rules are focused on protection against Generic Protocol Command Decode. On the other hand, the Quickdraw rules are mainly focused on the defense against Attempted Information Leak with 28 % and the Detection of Non-Standard Protocol or Event.

In order to meet the second objective, we need to determine the Snort and the Quickdraw cyber-attack alerts in relation with the DNP3 communication protocol. As can be seen in Fig. 8, there is only one alert type of the Snort DNP3 rule. All rules are focused on the cyber-attack based on Generic Protocol Command Decode (100 %); whereas, the Quickdraw rules are much more diverse than Snort rules. The Figure shows the distribution of all Quickdraw rules. The rules are mostly responsible for the protection against Attempted Denial of Service (27 %) and Potentially Bad Traffic (23 %).

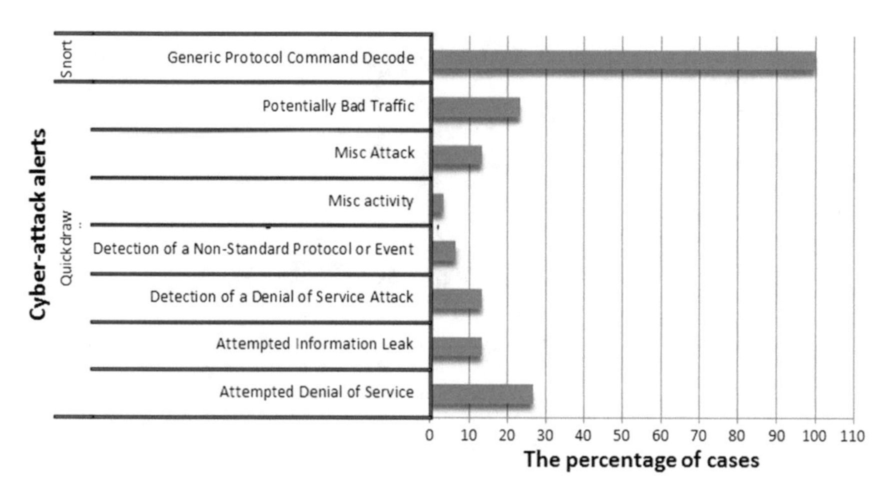

Fig. 8 The comparison of the cyber-attack alerts in relation with DNP3

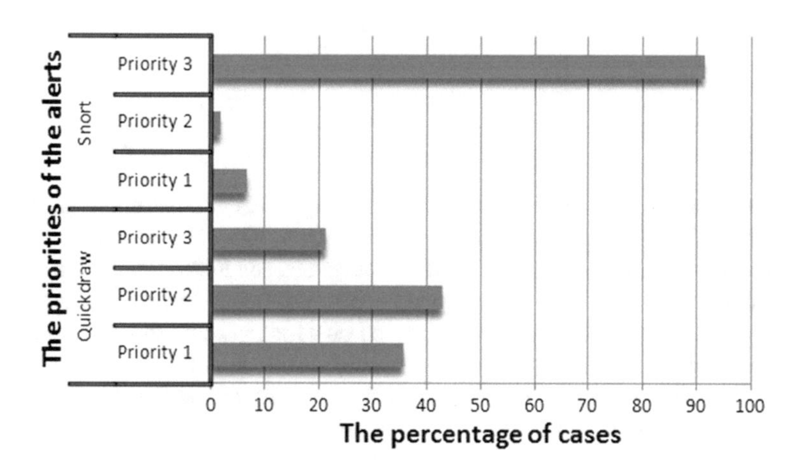

Fig. 9 The distribution of the Modbus priorities

The third objective is focused on the specification of the Snort and the Quickdraw rules priorities in relation with the Modbus and the DNP3 communication protocols. The rule with priority 1 is the most crucial for the SCADA system while the rule with priority 3 is the least important for the SCADA system. As can be seen in Fig. 9, the most of the Snort Modbus rules are classified by priority 3 (91 %). Moreover, the Quickdraw Modbus rules can be mostly characterized by priority 2 (43 %) and priority 1 (36 %).

The second part of the third objective is dealing with the specification of the Snort and the Quickdraw rules priority in relation with the DNP3 communication

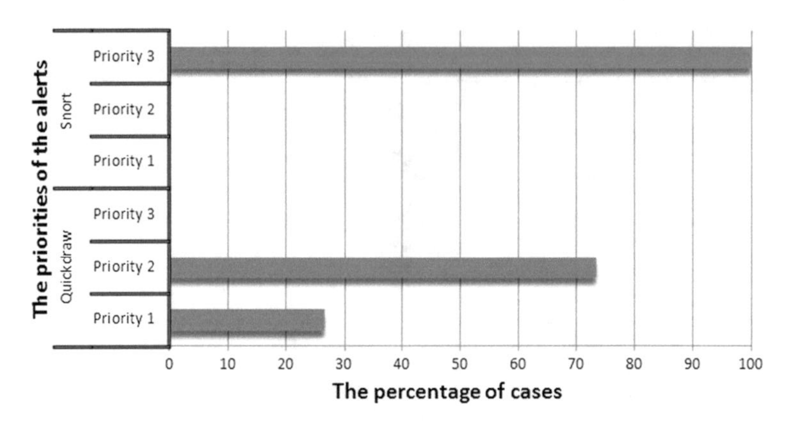

Fig. 10 The distribution of the DNP3 priorities

protocol. The Fig. 10 shows that 100 % of the Snort rules are determined by priority 3. In the case of Quickdraw, there is a large group of rules with priority 2 with 73 %.

8 Discussions

The objective of the article was to evaluate the SCADA cyber security. Therefore, this case study was based on Snort and Quickdraw IDS rules in relation to the SCADA. The results are consistent with earlier studies conducted with IDS rules [7].

The assessment of the IDS rules is performed according to different perspectives. In general, the Snort rules are mostly focused on Modbus communication protocol compared to Quickdraw. Moreover, the overall results indicate the dominance of Snort rules in terms of number. However, the diversity of the Snort rules is low; furthermore, the Snort rules are mainly focused on Generic Protocol Command Decode alert. This type of alert has a low priority; accordingly, the Snort rules are not essential for the SCADA cyber security. On the other hand, the Quickdraw rules provide the appropriate diversity. Furthermore, they are mainly aimed at dangerous cyber-attacks. An interesting fact is that Quickdraw rules protecting the SCADA systems against Denial of Service attack. This type of attack represents an eminent danger to the availability of the system; therefore, it is the most critical threat for the SCADA. The overall results indicate that the difference between Snort and Quickdraw rules is noticeable. Every SCADA system relying only on IDS with SNORT rules is vulnerable by a considerable number of cyber-attacks. Nonetheless, it is notable that the Snort provides another layer of cyber security, even though it is not an effective solution.

It should be noted that this study has been primarily concerned with the SCADA-particular rules. However, the considerable number of cyber-attacks is focused on the IT systems interconnected with the SCADA. Marginalization of the IT cyber security can be critical for SCADA. Nonetheless, more research is required in this area in order to determine the reliable cyber defense of the critical information infrastructure.

Acknowledgments This work was founded by the Internal Grant Agency (IGA/FAI/2016/014) and supported by the project ev. no. VI20152019049 "RESILIENCE 2015:RESILIENCE 2015: Dynamic Resilience Evaluation of Interrelated Critical Infrastructure Subsystems" supported by the Ministry of the Interior Security Research Programme of the Czech Republic in the years 2015–2020. Moreover, this work was supported by the Ministry of Education, Youth and Sports of the Czech Republic within the National Sustainability Programme project No. LO1303 (MSMT-7778/2014) and also by the European Regional Development Fund under the project CEBIA-Tech No. CZ.1.05/2.1.00/03.0089.

References

1. Verba, J., Milvich, M.: Idaho National Laboratory Supervisory Control and Data Acquisition Intrusion Detection System (SCADA IDS). In: IEEE Conference on Technologies for Homeland Security, pp. 469–473 (2008)
2. Stouffer, K., Falco, J., Scarfone, K.: Guide to Industrial Control Systems (ICS) Security. National Institute of Standards and Technology. http://csrc.nist.gov/publications/nistpubs/800-82/SP800-82-final.pdf (2011) [cit. 2015-12-01]
3. Macaulay, T., Singer, B.: Cybersecurity for industrial control systems: SCADA, DCS, PLC, HMI, and SIS, 193 pp. CRC Press, Boca Raton, FL (2012). ISBN: 14-398-0196-7
4. Fisk, M., Varghese, G.: Fast content-based packet handling for intrusion detection. Los Alamos National Lab, NM Computing Communications and Networking Div (2001)
5. Scarfone, K., Mell, P.: Guide to Intrusion Detection and Prevention Systems (IDPS): Recommendations of the National Institute of Standards and Technology (2007) [cit. 2015-12-03]
6. Pritika, M.E.H.R.A.: A brief study and comparison of snort and bro open source network intrusion detection systems. Int. J. Adv. Res. Comput. Commun. Eng. **1**(6), 383–386 (2012)
7. Shah, S.N., Singh, Ms.P.: Signature-based network intrusion detection system using SNORT and WINPCAP. Int. J. Eng. Res. Technol. (ESRSA Publications) (2012)
8. The Snort Intrusion Detection System: https://www.snort.org/ (2015)

An Optimization Scheduler in the Intranet Grid

Petr Lukasik and Martin Sysel

Abstract Scheduling of processes is the basic task for a grid computing. This role is responsible for the allocation of time for computational agents. The calculation agent can include a wide range of devices, based on various types of computer systems. Is it possible to efficiently build a grid infrastructure in the company environment. The grid can be used in scientific and technical computing, as well as better load distribution of the individual computing systems and services. The scheduler is a major component of grid computing. The main task is to effectively distribute the load of the system and allocate tasks to places that are not sufficiently utilized at a given moment. The article also focuses on the relation between conflicting parameters, which relate to the quality of the planning process. Time calculation of the optimization algorithm affects the quality of the draft plan. It has a direct impact on the total period of the job processing. In the strategy of the scheduling there is a point where extensions of time have no effect on quality of the draft of the plan but getting worse the overall runtime of the job. The aim was to compare the common metaheuristic algorithms. From the measured values to propose a methodology for determining the optimum time for planning process.

Keywords Grid · JSDL · POSIX · Precedence · Optimization scheduler

1 Introduction

Distribution of tasks and scheduling are essential elements of a grid services. A tool that allows easy definition of the role and its distribution in the environment is a prerequisite for high-quality and user-acceptable Grid Services. The user should have a

P. Lukasik (✉) · M. Sysel
Faculty of Applied Informatics, Department of Computer and Communication Systems, Tomas Bata University in Zlin, nam. T. G. Masaryka 5555, 760 01 Zlin, Czech Republic
e-mail: plukasik@tajmac-zps.cz
URL: http://www.fai.utb.cz

M. Sysel
e-mail: sysel@fai.utb.cz

R. Silhavy et al. (eds.), *Software Engineering Perspectives and Application in Intelligent Systems*, Advances in Intelligent Systems and Computing 465,
DOI 10.1007/978-3-319-33622-0_16

freedom as well as resources to easily tracking of their own processing. An important feature is that the Grid service has the least restrictive conditions for a successful job execution. (Type or version of software, operating system and hardware features). The user of the grid should have a certain freedom. Not to be tied up of restrictive rules, except the rules relating to information security and data processing [2, 4].

The aim of this work is the description of the scheduler. The basic functionality is to create a schedule of tasks that will meet the priority, precedence and optimization requirements for the distribution and processing of batch jobs in a grid computing environment. The result is the design of optimal scheduling time in relation to the number of jobs that are released into the processing and comparison of various scheduling strategies.

The Task is defined by using the standard JSDL (Job Submission Definition Language). JSDL is an XML-based computational specification for the management and distribution of batch jobs in a Grid environment, developed by OGF-JSDL-WG [1, 3, 9]. A current version 1.0 has also the definition of the POSIX support. JSDL includes support to define a range of computing resources (the length of processing number of threads, disk space). The parameters to be defined by the user (type of system, type of processor, memory, network bandwidth) are basic inputs for the scheduler [6].

2 Long-Term Job Scheduler

The scheduler is a major component of the Intranet Grid. Logical and temporal sequence of tasks is the primary activity for him. Is designed as a set of three main modules.

The priority scheduler is responsible for optimizing the solution of tasks based on priority. The priority of the process is an optional input parameter that must be entered by the user when the task starts. Precedence scheduler is responsible for the order of the processed parts. The optimization solver searches the best possible use of computing resources. Thereby contributes to optimizing the job time makespan. Scheduler is subdivided into three independent components (Fig. 1).

2.1 A Priority Scheduler

This scheduler uses a modified cyclic Round-Robin queue. The priority of jobs is managed by a time-slice t_q, which is assigned to a particular job.

The t_q determines the activity for a specific task. After the expiration of this period, the current job is paused and control is passed to the next job. A Round-Robin mechanism of working time is not prone to neglect the tasks with lower priority (starvation process).

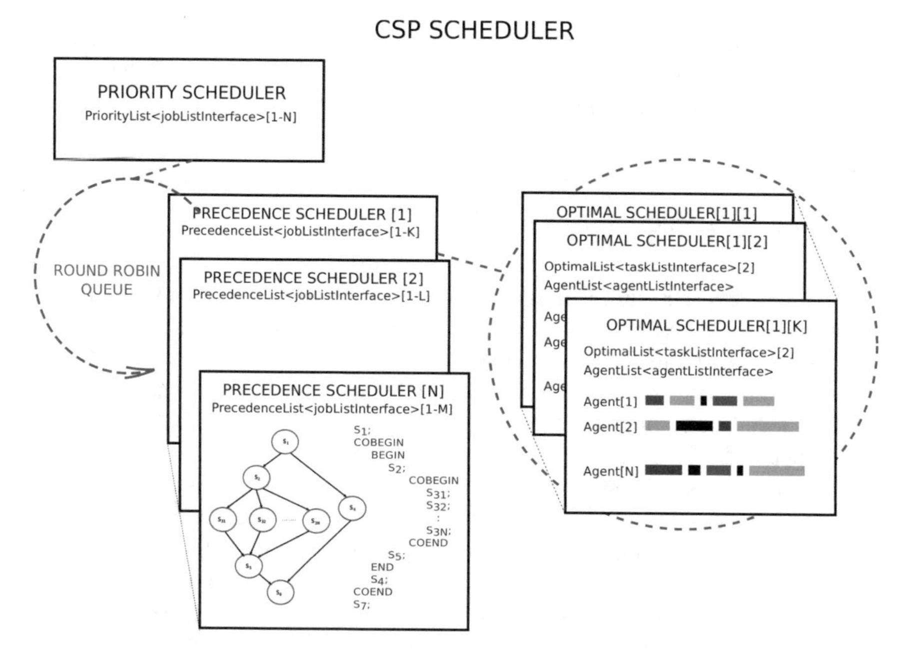

Fig. 1 A block diagram of the optimization scheduler

Each job has a guaranteed periodical running time. The quality of proposed scheduling strategy is depending on the determination of the optimal time period t_q for the task context switching.

A short period of time switching represents a significant increase of load on the system. A long interval of time t_q degrades Round-Robin scheduler to the FCFS queue. The timetable of FCFS scheduler is inefficient in this case, see Sect. (4.1—Priority Scheduler).

2.2 *Role of the Precedence Scheduler*

The precedence scheduler is responsible for managing the data flow in the individual tasks. Performs the role of Mid-Term scheduler for a current task.

This scheduler is designed as a priority FIFO queue with feedback. The scheduler monitors the progress of the running tasks in the current job which provides the results for the next tasks. Solves a generic acyclic graph (acyclic Directed Graph), which describes the workflow of the solved task. Scheme of the precedence graph is defined by the user mandatory.

The precedence scheduler also solves check-pointing of the system. When an exception occurs, the system can restore the status before failure. Subsequently this defective part is sent to the processing again. Is able to solve only a transient type of errors. These errors are defined as temporary and correctable errors (recoverable errors). Can not solve permanent types of errors. These errors mean failure of the entire system, including the schedulers and also the server.

2.3 The Optimization Scheduler

The optimization scheduler is designed for the finding the optimum distribution of computational load for each computing agents connected to the grid.

The aim is to minimize the computational time—makespan. Makespan is defined as the time difference between the start and end of a sequence of jobs in an environment of the independent parallel machines. The makespan is a good indicator for the throughput of the Grid Services.

The optimization scheduler together with a priority scheduler solves the problem of dynamic scheduling. This is applied during adding another task to an already running process. It also applies in a case of failure of one of computing resources. The first input parameter for optimizing scheduler is the capacity of computing resources. The capacity is evaluated by the system for all computational agents, see Sect. (3—Evaluation of the Capacity of Computing Resources). A second input value is the duration of the task. This value is entered by the user when the job is started. It is a mandatory input parameter.

3 Evaluation of the Capacity of the Computing Resources

The input variables for the evaluation of the computational capacity:

- *Memory size*: Capacity of the memory.
- *network throughput*: Rate the data transmission speed in the network.
- *Properties of the computing units*: The CPU and GPU parameters.

The variables described above are the basic parameters for the test and determine the performance capacity of the computational resources. For the benchmark test is designed a simple transcendental equation $cos(x) - x = 0$, which can be solved by iteration.

In the first phase, the performance of CPUs is measured. The algorithm starts with the number of parallel threads based on the number of CPU cores. The presence of the GPU is also included in the determination of capacity the computing resources.

The algorithm for evaluation of performance computing resource was chosen according to the following criteria. The number of tasks running in parallel must not exceed the number of processor cores. The CPU computational capacity is an indicator of the performance characteristics of computing resources. The GPU computational capacity gives the possibility of using the graphics card for a special parallel tasks.

Table 1 Capacity of the computing resources

Evaluation of the computational capacity								
	GPU is detected				GPU not detected			
CPU time [ms]	1 000	500	250	125	1 000	500	250	125
C_{CPU}	10	12	13	14	10	12	13	14
GPU time [ms]	100	50	25	15				
C_{GPU}	2	4	6	8				
$C = C_{CPU} + C_{GPU}$	**12**	**16**	**22**	**24**	**10**	**12**	**13**	**14**

For evaluating and comparing the performance of sources requires a certain standard. Based on this standard is evaluated the measured values of any other sources. For this standard was chosen normal office computer with a standard power (CPU) + memory without a graphics processing unit (GPU). The performance was measured by the algorithms described above. This standard was evaluated by the lowest scores from the set $C_{cpu} = (10, \ldots, 24), C \in N$.

The capacity of the GPU was empirically described by the set $C_{gpu} = (2, \ldots, 8)$, $C_{gpu} \in N$. Calculations on the GPU requires a different approach (technology CUDA, OpenCL, OpenGL) (Table 1).

The measurement shows that a significant impact on the capacity of the system are the I/O operations. It is evident in Fig. 3 and the Hill Climbing algorithm sorted section. It was suggested the optimal distribution of computing capacity. Sorting by the length of time duration was assigned to some of the computational agents, a large number of the tasks with short run-time.

This has a negative impact on processing time. In practice, it has proven advantageous the random distribution of tasks with different length processing. Load based on the I/O operations is better distributed.

4 Definition of the Time Interval of the Process Scheduling

4.1 Priority Scheduler

Priority scheduling is designed as cyclic (Round Robin) queue.

Input parameters:

N	The number of tasks running concurrently
t_n	The predicted time of the job run—specified by the user
$P \in (1, \ldots, 5)$	Priority of the task—specified by the user
t_q	Time interval for the job switching

Output Parameters:

t_J	The time allocated for the task

The time interval that is allocated for the running job affects the behavior of the priority scheduler [5]. A selection of short time interval significantly increases the system load due to the context switching and also increases the risk of starvation processes. Starvation may occur so that the following process has no computing resources at time t. They may be busy with other processes. A long interval limit degrades cyclic queue to the FCFS. It will handle tasks sequentially. This condition has a negative impact on the optimization of processing time.

Time interval for the job switching

$$t_q = \frac{\sum_{n=1}^{N} t_n}{N} \tag{1}$$

The time allocated for the task

$$t_j = t_q \left(\frac{1}{P_{max} - P + 1} \right) \tag{2}$$

The Exception—starvation process is solved by increasing the priority tasks on priority $P + 1$. A subsequent operation returns the priority tasks to its original value.

5 The Optimization Scheduler

The optimization scheduler together with precedence scheduler solves the distribution of parallel and sequential of tasks on each computer's agent for the current job. The aim is to minimize the makespan processing tasks in an identical computational machine.

Input parameters:

$m = (1, \ldots, M)$	The number of the computational agents
$j = (1, \ldots, N)$	Set of the tasks running concurrently

Output parameters:

$t = t_{scd} + l_i$	Run time of the job = time scheduling + makespan

The batch job i, consumes t_{ij} units of time. Load of the computing agent is

$$l_i = \sum_{(j \in J_i)} t_{i,j} \tag{3}$$

and

$$l_{max} = \max_{(i \in m)} l_i \tag{4}$$

is the maximum load.

The value l_{max} is called a makespan of the job. In this case, can be said that grid computing agents belong to a set of identical machines. Identical in the sense that the job can be started on any of them. In this case is $t_{i,j} = t_j$ for $i \in M$ and $j \in N$ [7].

The principal task for this scheduler is a minimize of the makespan l_{max}. The criterion for optimizing is job time in the process and the best distribution of load on all computing agents who is available. Total time is the sum of the run time of the tasks in the current job (makespan) and the time needed to create a scheduled task. This fact must be included in the design of optimization criteria.

$$t = t_{scd} + l_i \tag{5}$$

The proposal of the parameters of the scheduler is dependent on two conflicting values. It is necessary to find their optimal size. The quality of the scheduling process is depending on the algorithm and time during which is running.

For the design of appropriate parameters were studied some metaheuristic algorithms. The main task was to find the most appropriate algorithm and determining a reasonable time interval t_{scd} for the scheduling process. The time period for finding the optimal solution has a significant impact on the overall processing time. It is necessary to compromise between quality the proposed plan and time to create a suboptimal schedule. Run time of the schedule was determined from the total task time measurement. It was measured the total time for 1 000, 3 000 and 10 000 tasks. The time interval for scheduling was 0, 10, 60 and 300 s.

Five measurements were performed for each type of algorithm and the time interval. Figure 2 shows the average values of these measurements. Time for scheduling in the length of 0 s downgraded optimization schedule to the FCFS.

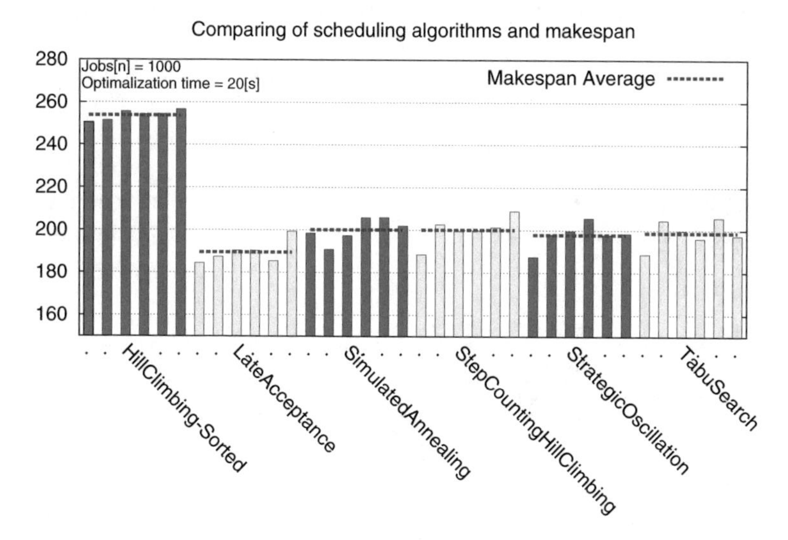

Fig. 2 Comparing of scheduling algorithms and makespan

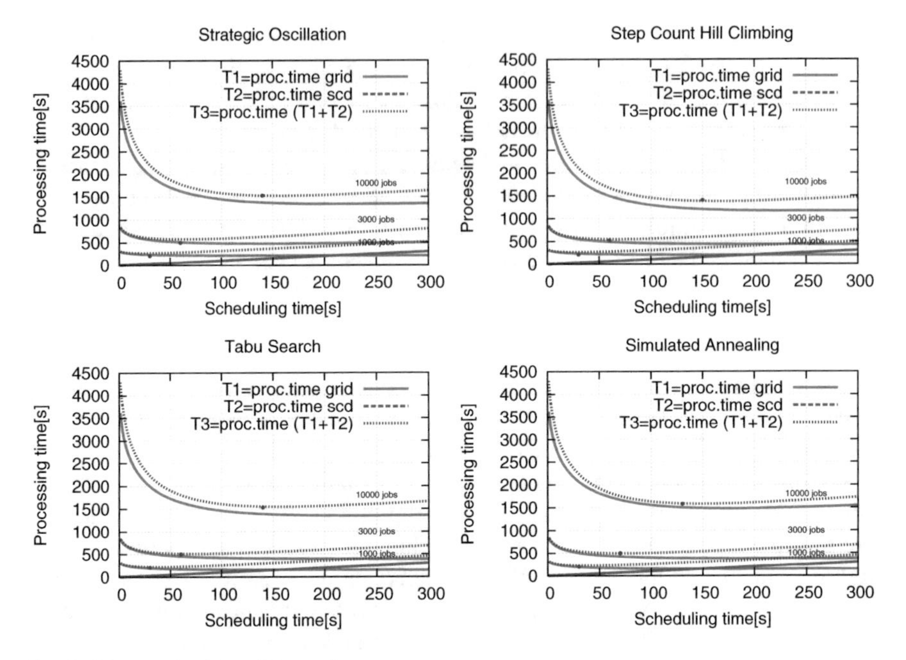

Fig. 3 Measured values of the makespan

The optimal time is derived from the measured values. The optimum time is deducted in Fig. 3. The values t_{scd} for the individual algorithms were not much different. Using linear regression has been found the dependency—a straight line which describes the relationship between a number of the scheduled tasks and time t_{scd} for the scheduling with sufficient accuracy (Table 2).

Relation between the number of task in the current job and time for the scheduling

$$t_{scd} = 0.0122J + 20.69 \tag{6}$$

During the measurement was evaluated run time of elaboration for different optimization algorithms. Measurements were carried out so that each optimization algo-

Table 2 Run time of scheduling algorithm t_{scd} [s]

Run time of scheduling algorithm

The number of tasks in the current job	1 000	3 000	10 000
	t_{scd} [s]		
Strategic oscillation	30	60	140
Step count hill climbing	30	60	150
Tabu search	25	60	140
Simulated annealing	30	70	135

rithm had the task to propose a timetable for 1 000 tasks. For each algorithm, was performed five measurements. The diameter measurements were evaluated. The best results were achieved with the Late Acceptance algorithm. Poor results were measured with the hill climbing algorithm. The result was that for some agents were assigned to a large number of short tasks. Thee computing nodes have been burdened with a large overhead I/O operations. Algorithms with random task run length showed much better results.

6 Conclusion

The aim of this work was the description and design of appropriate strategies and parameters in the grid computing environments to optimize performance, response time and throughput of the service. It was shown that for solving job shop Scheduling type which belong to a category of NP—complete problems it is necessary to select some compromises between quality results and a total length of treatment. The dependence on the quality of the final plan and the time required for calculation defines the point that determines the threshold at which further improve the quality of the final timetable of the plan is ineffective. The proposed approximation dependence of scheduled tasks and time of the runtime scheduling algorithm t_{scd} see (6), describes this dependence well. Approximation proposal does not address the limitations resulting from Amdahl's law [8].

Segmentation into three separate objects priority precedence and optimization has led to simplify the design of the scheduler.

Measurements showed that the effect of time for task scheduling has similar results for all investigated optimization algorithms. For all types of algorithms that were examined, it is possible to use the same methodology to determine the length of time (t_{scd}) for the scheduling algorithm.

The aim of future work is to verify the properties of other stochastic optimization algorithms based on the principle of evolutionary strategies. Particularly Ant Colony Optimization, which observes the behavior of ants in search of food. Ant Colony Optimization well simulates the concept of finding the shortest path. The advantage of this algorithm is less sensitive to premature convergence to the insignificant local extremes.

References

1. Anjomshoaa, A. (EPCC, Fred Brisard, CA), Drescher, M. (Fujitsu), Fellows, D. (UoM), Ly, A. (CA), McGough, S. (LeSC), Pulsipher, D. (Ovoca LLC), Savva, A. (Fujitsu): GFD-R.136 Job Submission Description Language (JSDL) Specification. http://forge.gridforum.org/projects/jsdl-wg. Accessed 28 July 2008. Copyright (C) Open Grid Forum (2003–2005, 2007–2008). All Rights Reserved

2. Ezugwu, A.E., Frincu, M.E., Junaidu, S.B.: A multiagent-based approach to scheduling of multi-component applications in distributed systems. In: Advances in Intelligent Systems and Computing, Artificial Intelligence Perspectives and Applications, vol. 347, pp. 1–12. Springer International Publishing (2015)
3. Humphrey, M. (UVA), Smith, C. (Platform Computing), Theimer, M. (Microsoft), Wasson, G. (UVA): JSDL HPC Profile Application Extension, Version 1.0. 14 July 2006. Updated: 2 Oct 2006. Copyright Open Grid Forum (2006–2007). All Rights Reserved
4. Lukasik, P., Sysel, M.: A task management in the intranet grid, modern trends and techniques in computer science. In: Advances in Intelligent Systems and Computing, vol. 349, pp. 77–85. Springer International Publishing (2015). doi:10.1007/978-3-319-18473-9. ISBN: 978-3-319-18472-2 (Print)
5. Noon, A., Kalakech, A., Kadry, S.: A new round robin based scheduling algorithm for operating systems: dynamic quantum using the mean average. IJCSI Int. J. Comput. Sci. Issues **8**, 224–229 (2011)
6. Rodero, I., Guim, F., Corbalan, J., Labarta, J.: How the JSDL can exploit the parallelism? In: Sixth IEEE International Symposium on Cluster Computing and the Grid, 2006. CCGRID'06, vol. 1, pp. 8, 282, 16–19 May 2006. doi:10.1109/CCGRID.2006.55. http://ieeexplore.ieee.org/stamp/stamp.jsp?tp=&arnumber=1630829&isnumber=34197
7. Souza, A.: Combinatorial Algorithms Lecture Notes, Winter Term 10/11. Humboldt University, Berlin
8. Sun, X.-H., Chen, Y.: Reevaluating Amdahl's law in the multicore era. J. Parallel Distrib. Comput. **70**, 183–188 (2010)
9. Theimer, M. (Microsoft Corporation), Smith, C. (Platform Computing Corporation): An Extensible Job Submission Design, 5 May 2006. Copyright (C) Global Grid Forum (2006), All rights reserved. Copyright (C) 2006 by Microsoft Corporation and Platform Computing Corporation, All rights reserved

A Hybrid Clustering Metric-Based Algorithm for Wireless Sensor Networks

Bo Dong and Xue Wang

Abstract In wireless sensor network (WSN), clustering of nodes is an effective way to organize the network structure. The cluster head (CH) is responsible for the creation of the network dominating set, the maintenance of the network topology structure and the data collection. Each cluster head is responsible for managing all nodes belonging to its subset. In this paper, we propose a new distributed clustering algorithm called HCMA based on a previous algorithm. The HCMA algorithm calculates the link state between nodes, node congestion state and energy consumption of nodes as the mixed measure in order to elect the cluster head. This solution is primarily intended to provide better performance, such as the maximum survival time, and reduce the packet loss rate. The simulation results show that HCMA can reduce the packet loss quantity and improve the network lifetime compared with other similar methods.

Keywords Hybrid algorithm · Cluster head · Wireless sensor network · Network lifetime · Subnetwork

1 Introduction

Wireless Sensor Networks have started to receive growing attention from the research and engineering communities in recent years. Potential applications of WSN include detecting and countering pollution in coastal areas, performing oceanic studies of bird/fish migration and weather phenomena, detection and prevention of forest fires, deterring terrorist threats to ships in ports, destruction of mines in different environments facilitating/conducting urban search and rescue (USAR), detecting

B. Dong
Computing Center of Liaoning University, Shenyang, Liaoning, China

X. Wang (✉)
Information Technology Center of Liaoning University,
Shenyang, Liaoning, China
e-mail: wangxue@lnu.edu.cn

R. Silhavy et al. (eds.), *Software Engineering Perspectives and Application in Intelligent Systems*, Advances in Intelligent Systems and Computing 465,
DOI 10.1007/978-3-319-33622-0_17

suspiciously active chemical/biological agents, etc. [1]. Each of these micro-sensor is basically equipped with a sensing device to collect data from the environment, a processing unit to do some operations on data, a transceiver to send and receive collected, and an energy source to provide the required energy to operate (usually a battery) [2]. The data collected is transmitted over multiple hops to a sink also called controller or base station (BS) [3].

The concept of grouping a large area into a number of small regions based on geographic location has been presented in literature called clustering [4]. The clustering technique takes part in partitioning network nodes into groups called clusters, giving to the network a hierarchical organization. It has been widely pursued by researcher that the grouping of sensor nodes into clusters in order to accomplish the objective of the network scalability.

The novelty and contribution of our work is proposing a distributed algorithm to select these cluster head nodes. The head election is based on a hybrid metric taking into account the four parameters: link metric, congestion metric, the energy consumption and the distance metric from nodes to the base station. Performance of the scheme is evaluated and compared through qualitative and quantitative analyses; results show the present scheme's dominance over the competing schemes.

The rest of this paper is organized as follows: Sect. 2 describes relative works for clustering on WSN. In Sect. 3, we propose Hybrid Clustering Metric-based Algorithm (HCMA) Simulation results are presented in Sect. 4 while conclusions are offered in Sect. 5.

2 Related Research Works

Many algorithms for cluster formation and cluster head selection have been proposed to choose cluster heads in ad hoc networks, and wireless sensor networks. In [5] paper, distance based Cluster head selection algorithm has been proposed for improving the sensor network life time. In [6] paper, Cluster Head selection protocol using Fuzzy Logic (CHUFL) has been presented. Cluster Head Election mechanism using Fuzzy Logic protocol (CHEF) is discussed in [7]. The evolution of these algorithms has proven that the combination of several metric is more efficient for better performance.

2.1 Single Metric for Cluster Head Election Algorithms

In [8], the highest-connectivity (degree) algorithm is proposed. The highest degree node in a neighborhood becomes the clusterhead. The algorithms are described below: The node with the highest degree in its one-hop neighborhood becomes cluster head. Nevertheless, due to the criterion of degree, the highest-connectivity (degree) algorithm will build dense cluster in the situation that there are few nodes.

In fact that clusters are very sensitive to the mobility of nodes. So, in the WSN, this results lead to reconstructions of cluster structure frequently. The highest-connectivity (degree) algorithm operates in synchronous mode and uses TDMA access method in order to avoid collisions.

The most important challenge in WSN is the limitations in communication bandwidth and the energy [9]. The authors have proposed a protocol called LEACH. LEACH enables point-to-point connectivity and does not use low-energy routing or MAC. It builds on this work by creating a new ad-hoc cluster formation algorithm that better suits micro-sensor network applications. In LEACH, the nodes organize themselves into local clusters, with one node acting as the cluster head. All non-cluster head nodes transmit their data to the cluster head, while the cluster head node receives data from all the cluster members, performs signal processing functions on the data (e.g., data aggregation), and transmits data to the remote BS. The cluster head election which based on generation of a random number assigns this role to different nodes according to a Round-Robin policy to ensure fair energy dissipation between nodes. This rounds have about the same time interval which previously determined [4].

LEACH is a distributed cluster formation algorithm that offers no guarantee about the placement and/or number of cluster head nodes. Since the clusters are adaptive, obtaining a poor clustering set-up during a given round will not greatly affect overall performance. However, using a central control algorithm form the clusters may produce better clusters by dispersing the cluster head nodes throughout the network.

LEACH-C [4] is the centralized version of LEACH in which the BS coordinates the clustering process. The BS is assumed to have location and residual energy data of every sensor node. Based on gathered data, the BS selects CHs and generates data gathering schedules, then broadcasts schedules back to sensor nodes. Since LEACH-C is a centralized approach, it provides better formation of clusters than LEACH in terms of energy balancing. However, a centralized approach is not appropriate in many practical environments.

2.2 *Multiple Metric for Cluster Head Election Algorithms*

Sometimes, it is no wise to use of a single metric to elect cluster head to generate stability on cluster head formation. Other approaches that used multiple attribute/criteria decision-making cluster head selection have been proposed.

2.2.1 Hybrid Energy-Efficient Distributed Clustering (HEED)

HEED [10] is a distributed clustering scheme in which CH nodes are picked from the deployed sensors. HEED considers a hybrid of energy and communication cost when selecting CHs. It improves the LEACH according to use residual energy as primary parameter and network topology features as secondary parameters to break connection between candidate cluster head. Unlike LEACH, it does not select cell-head nodes randomly. Only sensors that have a high residual energy can become cell-head nodes.

The probability that two nodes within each others communication range becoming CHs is low. Unlike LEACH, this means that CHs are well distributed in the network. Energy consumption is not assumed to be uniform for all the nodes. For a given sensors transmission range, the probability of CH selection can be adjusted to ensure the connectivity of inter-CH.

2.2.2 Weighted Clustering Algorithm (WCA)

The authors of WCA [11] take into account its degree, transmission power, mobility and battery power. So, there is high overhead induced by WCA. The disadvantage of WCA is, if a node moves into an area that is not covered by any clusterhead then the cluster set-up procedure is invoked again which causes reaffirmations. Its combination of four metrics of a node: each node's degree Δ_n, which the sum of the distances between node n and each of its neighbors, the mobility metric M_n, the sum of the distances with all its neighbor D_n, the cumulative time P_n. The weight of a node is obtained by the following formula:

$$W_n = w_1\Delta_n + w_2D_n + w_3M_n + w_4P_n \tag{1}$$

For each w_1, w_2, w_3, w_4 in the Eq. 1 is weight fixed in the system. And $w_1 + w_2 + w_3 + w_4 = 1$. The objective of WCA is to minimize the number of changes in each of cluster set update. All the four parameters are normalized, that means that the values of parameters are made to a pre-defined region. The related weights w_1, w_2, w_3 and $w4$ are fixed for the network. Note also that the weighting factors give the flexibility of modifying the effective contribution of each of the parameters in calculating the combined weight W_n. As an issue that, in a network where the battery power of each node is adequate while the mobility is important, the $w4$ associated with P_n can be set larger.

2.2.3 A Distributed Weighted Clustering Algorithm for Mobile Ad Hoc Networks (DWCA)

The goals of DWCA [12] are maintaining stable clustering structure, minimizing the overhead during the clustering set up and maintenance procedure. DWCA works like WCA except that the process of power management and distributed cluster setting up is done by subset configuration and reconfiguration of clusters. The consumption is a better than the cumulative time during which the node acts as a clusterhead that is used in WCA because it reflects the actual amount of power usage. If in a situation where there is not enough battery power then lifetime of network topology can be increase by switching the role of the clusterhead node to an ordinary one.

DWCA consists of the clustering set up and clustering maintenance phases. In clustering setting up phase, election of the head of clusters is based on the weight values of the neighbor nodes. Each node calculates its multi-weight value based on the four following factors: Degree difference, Sum of distances, Running average of speed, Consumed battery power. All the factors are based on factors used in WCA [11], except the consumed battery power. Computing consumed battery power is a better measure than the cumulative time during which used in WCA. In clustering maintenance phase, two situations may invoke the clustering maintenance: the time when the node move to the outside of its cluster boundary, the other is the time when excessive battery consumption at a clusterhead.

When an ordinary node moves at the edge of its cluster boundary, it is required to find a new clusterhead to belong to. If it finds a new clusterhead, it will hand over to the new one cluster. Otherwise, it declares itself as a new clusterhead. Each clusterhead updates the amount of consumed battery power during the time when it sends and receives packets. If the consumed battery power becomes more than a reasonable value that pre-defined threshold then the clusterhead becomes an ordinary node.

3 HCMA: A Hybrid Clustering Metric-Based Algorithm

In this section, we present the proposed HCMA: A Hybrid Clustering Metric-based Algorithm. HCMA is a distributed clustering algorithm which takes into account several parameters like link state between nodes, node congestion state and energy consumption. Nodes changes the role over time based on the value of the hybrid metric to accommodate a particular situation that the changing network structure and to be reliable to small topology changes. HCMA is also a algorithm provided energy balance. Nodes changes role between cluster head and ordinary node by the value of metric.

3.1 Calculation of Metric

Initially, each node broadcasts a *Hello* message to notify its status to the neighbors. A *Hello* message includes ID and position value of the node. Each node builds its neighbor table based on the *Hello* messages received from its neighbors. For a node n, the hybrid metric $M(n)$ that depend on four parameters: link state between two nodes (LS_n), congestion metric (CM_n), contributing towards $M(n)$ in efficient neighborhood functioning (N_n), efficient energy consumption (E_n).

3.1.1 Measuring the Link State Metric

The received signal strength (RSS_{ix}), from a distance x, to a neighbor node i can be expressed as [11],

$$RSS_{ix} = \frac{G_r * G_t * S_t}{(4\pi * x/\lambda)^2} \tag{2}$$

where λ is wavelength in WSN, G_r is the receiving antenna gain, G_t is the transmitting antenna gain, and S_t is the maximum transmitting power of transmitting antenna.

While we set the max threshold value (T_{maxi}) of the received signal strength of the ith neighbor node as,

$$T_{maxi} = \frac{G_r * G_t * S_t}{(4\pi * 0.9284R/\lambda)^2} \tag{3}$$

where R denotes the range of the antennas.

According to this threshold value (T_{maxi}), a node calculates each Link State for each of its Links. Link State (LS_i) for the ith neighbor node as,

$$LS_i = \begin{cases} 0, & \text{if } RSS_{ix} < T_{maxi} \\ (1 - \frac{T_{maxi}}{RSS_{ix}}), & \text{if } RSS_{ix} \geq T_{maxi} \end{cases} \tag{4}$$

So, a node n gets all the Link State (LS_i), can calculates Link State Metric (LS_n) for all the links as,

$$LS_n = \sum_{i=1}^{n} LS_i \tag{5}$$

3.1.2 Measuring the Congestion Level

We use the scheme as discussed in [13]. Every node calculates a Node Congestion Metric (CM_n) depending upon three parameters: buffer occupancy, channel load measurement and packet drop rate. The congestion metric between 1 and 0.

3.1.3 Measuring the Neighbors Level

- $N_{connected}$ the numbers of nodes in the neighborhood of node n.
- N_{max} the maximum number of nodes that a cluster head can processing. This parameter is to ensure that the nodes election to the cluster head are not overloading.

Therefore we can calculate the neighbors level N_n as,

$$N_n = \frac{N_{connected}}{N_{max}} \tag{6}$$

3.1.4 Measuring the Energy Consumption

- E_{cn} the consumption of a node n
- E_{init} the initial energy of a node n

Then we can calculate the energy consumption E_n as,

$$E_n = \frac{E_{cn}}{E_{init}} \tag{7}$$

Finally, we defined the hybrid metric for each node n as:

$$M_n = \theta_1 LS_n + \theta_2 CM_n + \theta_3 N_n + \theta_4 E_n \tag{8}$$

For each θ_1, θ_2, θ_3, θ_4, in the Eq. 8 is weight fixed in the system. And $\theta_1 + \theta_2 + \theta_3 + \theta_4 = 1$. The related metric θ_1, θ_2, θ_3, θ_4, are fixed for the network. Note also that the four metric give the flexibility of alterable the effective contribution of each of the parameters in calculating hybrid metric M_n. For an example that in a network where the battery power of each node is infinite while the mobility is important, the θ_4 associated with E_n can be set lower.

3.2 *Election of Cluster Head Election Algorithm*

Based on the calculation of hybrid metric M_n, we put forward to a new distributed algorithm that decides to which node is to be the cluster head. We suppose that all the nodes in WSN are statics even if the algorithm can be used in mobility scene. The reason is that nodes mobility makes clustering very challenging sometimes it's no necessary to be nodes into clusters. If the mobility of nodes is not very frequency we can use HCMA again to elect the cluster head.

We model a WSN as a graph $G = (V, E)$, where V is the set of sensor nodes while E is the set of wireless links *nm* between each pair of sensors n and m which are in antenna range of each other. The notations we used below:

- AD_n the MAC address or the IP address of a node n.
- H_{ADn} the address or the cluster head elect by a node n.
- NE_n the neighborhood of a node n.
- M_n the metric of node n computed in Eq. 8.
- H_n the cluster head node of a node n.

- MH_n the metric of the cluster head node of a node n.
- L_n the list of children node that chose n as cluster head.

Algorithm 1 HCMA distributed algorithm runs at node i

```
1: initialize node N_i;
2: H_i = {φ};
3: MH_i=maxvalue; //the max metric of the cluster head
4: L_i = {φ};
5: if NE_i == {φ} then
6:     H_i ← i;
7:     MH_i = M_i;
8:     H_AD = AD_i;
9: end if
10: for each j ∈ NE_n do
11:     if M_i ≥ M_j and MH_i ≥ M_j and L_i = {φ} then
12:         H_AD = AD_j
13:         MH_i = M_j
14:     end if
15: end for
16: if H_AD ≠ AD_i then
17:     for each j ∈ {NE_n − {H_i}} do
18:         if H_j == i or H_ADj == AD_i then
19:             add j into L_i
20:         end if
21:     end for
22: end if
```

The function of Algorithm 1 is to select to its cluster head for each node i. The node i has n neighbors. When the metric of each node has been computed, HCMA runs at each node. Firstly, each node i initialize, and make the cluster head metric a value that big enough (*instruction* 1–4). Then, if node i has no neighbor, it will be the cluster head of itself (*instruction* 5–9). For each node j in the neighbors of i, if the three condition that the metric of i and the metric of cluster head of node i are both higher than the metric, the third that node i is not the cluster of another node (*instruction* 10–11) j be satisfied at the same time, the node j will be the cluster head of node i. If the node i is the cluster head of another node, it will be add the nodes that belong to itself into the list of neighbors L_i (*instruction* 16–22).

4 Performance Analysis

In this section, we perform simulations by using the NS-3 simulator. We simulate a system of N nodes on an 1000 m × 1000 m area. The value of N that the numbers of nodes is chosen between 20 and 70. Channel capacity is set to 2 Mbps. The range radio of each node is set to 80 m. Initial energy for each node is set to 120 J. MAC layer protocol is IEEE802.11. The nodes move randomly to all directions with max

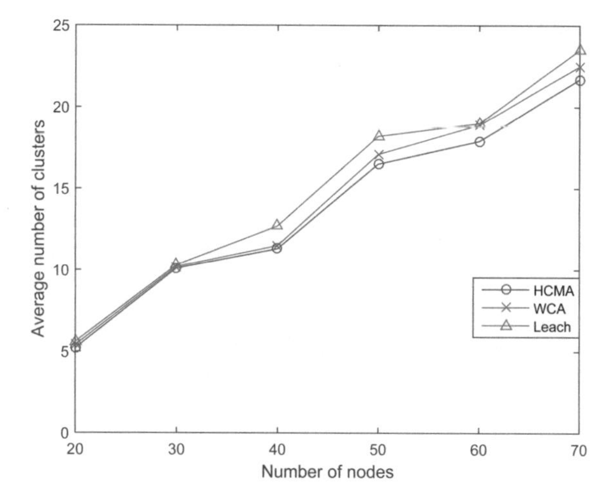

Fig. 1 Average number of clusters

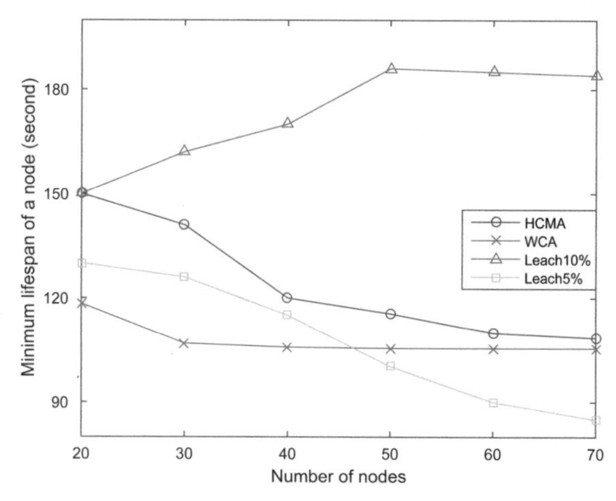

Fig. 2 Average number of clusters, minimum life span of a node

speed of 18 m/s. The values of metric factors used in calculation W_n as: $\theta_1 = 0.7$, $\theta_2 = 0.2$, $\theta_3 = 0.05$, $\theta_4 = 0.05$. While, the ideal degree of each node is set to 3. Figure 1 shows the average number of clusters related to the total number of nodes in the WSN. We can see in Fig. 1 the HCMA produced less clusters than WCA and Leach. When nodes in the spares network, HCMA nearly the same as WCA and Leach. While in dense network, our algorithm produced about 3.5 % less clusters than WCA and 7.6 % less clusters than Leach.

Figure 2 shows the minimum lifespan of nodes. The reason for a node become invalid is that due to battery exhaustion. Nodes consume more battery power as the mobility. With the increase of the number of nodes, the minimum lifespan in three algorithm (WCA, Leach with $p = 5\,\%$ and HCMA) are decreased. The Leach with $p = 10\,\%$ don't decreased due to the parameter p. It means that the more node to be the cluster head, then more nodes retransmission data thus less minimum lifespan in

Fig. 3 Network life time

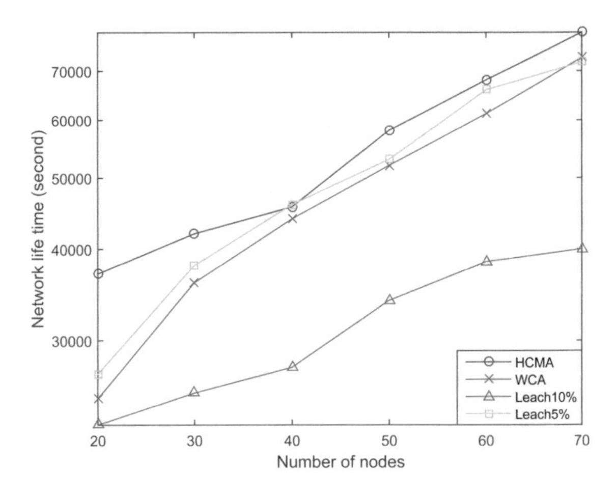

Fig. 4 Delivery ratio for different nodes

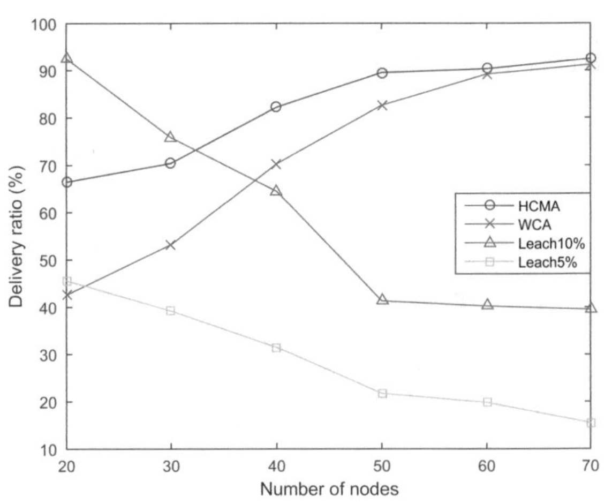

Leach with 10 %. But we can see from Fig. 3 the network life time that the key parameter representing the performance of the algorithm is much lower than the others. Results show that with our algorithm HCMA, the network lifetime is prolonged by respectively 48 %, 60 % and 37 % compared to WCA, Leach with $p = 5\,\%$ and Leach with $p = 10\,\%$ for 20 nodes, and by respectively 8.2 %, 9.74 % and 9.72 % compared to WCA, Leach with $p = 5\,\%$ and Leach with $p = 10\,\%$ for 50 nodes as shown on Fig. 3. This is because, by the fact that in HCMA, when cluster head are shutdown or the hybrid metric have increased, other nodes in the network take replace to cluster head immediately until dying as well, and the process will be carried out in the loop until all the nodes died. This method takes an efficient energy balance over WSN nodes, therefore enhanced the network lifetime.

Figure 4 shows the delivery ratio of three algorithm for the scenarios from 20 to 70 nodes. The delivery ratio is calculated by dividing the amount of data received by the base station divided by the amount of data received. Figure 4 shows that the LEACH with different parameters both lose more data than HCMA, nevertheless the amount of cluster head increased, while the more nodes added into the network. In LEACH, nodes transmit data the only way is to sending the data to their cluster head. If node can not find cluster head their data will be lost. Although there are a lot of cluster heads, while the numbers of cluster increased, some packages with LEACH are still lost due to the reason that the cluster head is overload that may not have time to transmit all data.

5 Conclusions

In this paper, we proposed a new clustering algorithm which is based on hybrid metric values considering link state between nodes, node congestion state and energy consumption. It has also the flexibility of using different weights in order to adapt to different situations.

Results obtained from simulations show that HCMA improves minimum life span, network lifetime and average delivery ratio. Though our algorithm performs better than LEACH and WCA in particular situation, it considers more realistic parameters. In the future, we will compare our algorithm with more algorithms and apply it to network which the nodes moving rate more faster.

References

1. Akyildiz, I., Kasimoglu, I.: Wireless sensor and actor networks: research challenges. Ad Hoc Netw. **2**(4), 351–367 (2004)
2. Chen, F., Chandrakasan, A.P., Vladimir, M.S.: Design and analysis of a hardware-efficient compressed sensing architecture for data compression in wireless sensors. IEEE J. Solid-State Circuits **47**(3), 744–756 (2012)
3. Kour, H., Sharma, A.K.: Hybrid energy efficient distributed protocol for heterogeneous wireless sensor network. Int. J. Comput. Appl. **4**(6), 1–5 (2010)
4. Heinzelman, W.R., Chandrakasan, A., Balakrishnan, H.: Energy-efficient communication protocol for wireless microsensor networks. In: Proceedings of the 33rd Hawaii International Conference on System Sciences, vol. 8, p. 8020. Washington, DC, USA (2000)
5. Kumar, B., Sharma, V.K.: Distance based cluster head selection algorithm for wireless sensor network. Int. J. Comput. Appl. **57**(9), 41–45 (2012)
6. Gajjar, S., Sarkar, M., Dasgupta, K.: Cluster head selection protocol using fuzzy logic for wireless sensor networks. Int. J. Comput. Appl. **97**(7), 38–43 (2014)
7. Kim, J.M., Park, S.H., Han, Y.J., Chung, T.M.: CHEF: cluster head election mechanism using fuzzy logic in wireless sensor networks. In: 10th International Conference on Advanced Communication Technology (ICACT 2008), pp. 654–659. Gangwon-Do (2008)
8. Gerla, M., Tsai, J.T.C.: Multicluster, mobile, multimedia radio network. Wireless Netw. **1**(3), 255–265 (1995)
9. Heinzelman, W.B., Chandrakasan, A.P., Balakrishnan, H.: An application-specific protocol architecture for wireless microsensor networks. IEEE Trans. Wireless Commun. **1**(4), 660–670 (2002)

10. Younis, O., Fahmy, S.: HEED: a hybrid, energy-efficient, distributed clustering approach for ad hoc sensor networks. IEEE Trans. Mob. Comput. **3**(4), 366–379 (2004)
11. Chatterjee, M., Das, S.K., Turgut, D.: WCA: a weighted clustering algorithm for mobile ad hoc networks. Cluster Comput. **5**(2), 193–204 (2002)
12. Choi, W., Woo, M.: A distributed weighted clustering algorithm for mobile ad hoc networks. In: Advanced International Conference on Telecommunications, 2006. AICT-ICIW'06. International Conference on Internet and Web Applications and Services, p. 73. IEEE (2006)
13. Kang, J., Zhang, Y., Nath, B.: Accurate and energy-efficient congestion level measurement in ad hoc networks. In: Wireless Communications and Networking Conference, vol. 4, pp. 2258–2263. IEEE (2005)

On Robust Computation of Tensor Classifiers Based on the Higher-Order Singular Value Decomposition

Bogusław Cyganek and Michał Woźniak

Abstract In this paper a method of faster training of the ensembles of the tensor classifiers based on the Higher-Order Singular Value Decomposition is presented. The method relies on the fixed-point method of eigenvector computation which is employed at the stage of subspace construction of the flattened versions of the input tensor patterns. As verified experimentally, the proposed method allows up to five times speed-up factor at no significant difference in accuracy.

Keywords Tensor classifiers · Subspace classification · Higher-Order singular value decomposition

1 Introduction

Tensor based classifiers showed high accuracy and fast response time. Their good properties can be appreciated especially for multidimensional data, such as images, videos, medical records, etc. since tensors directly account for multi factors of such data. One of the interesting solutions in this domain are ensembles consisted of the classifiers operating in the orthogonal tensor subspaces computed with the Higher-Order Singular Value Decomposition (HOSVD) of the input multidimensional patterns [2–4, 16]. However, tensor classifiers exhibit large memory consumption as well as high computational costs required for training. In this paper we propose a method to overcome the latter problem. Since training of the HOSVD

B. Cyganek (✉)
AGH University of Science and Technology, Al. Mickiewicza 30,
30-059 Kraków, Poland
e-mail: cyganek@agh.edu.pl

B. Cyganek · M. Woźniak
Wroclaw University of Technology, Wybrzeże Wyspiańskiego 27,
50-370 Wrocław, Poland
e-mail: Michal.Wozniak@pwr.wroc.pl

R. Silhavy et al. (eds.), *Software Engineering Perspectives and Application in Intelligent Systems*, Advances in Intelligent Systems and Computing 465,
DOI 10.1007/978-3-319-33622-0_18

classifier requires the SVD decomposition of the flattened version of the input tensor, our idea is to substitute the expensive SVD procedure with a much faster eigenvector computation method. Although it is an iterative procedure, it allows much faster training and the same accuracy of the resulting classifiers as compared to the full SVD decomposition.

The rest of the paper is organized as follows. Section 2 provides an introduction to the classification method of the multidimensional data with HOSVD based classifiers. Experimental results are included in Sect. 4. The fast HOSVD decomposition method is described in Sect. 3. The paper ends with experiments, discussed in Sect. 4, as well as conclusions in Sect. 5.

2 Introduction to Classification of Multidimensional Data with HOSVD Based Classifiers

Many machine learning applications deal with multidimensional data, such as images, video, sell values, etc. When training with such data, it was shown that tensor based classifiers usually provide better results compared to classifiers with vectorized input patterns [5, 11]. In other words, in a tensor approach each degree of freedom is represented by a separate index of a tensor [1]. Details of tensor processing can be found in literature [4, 10].

In the HOSVD based classifiers, the training patterns are contained in a prototype tensor [4, 6]. This is then HOSVD decomposed which leads to the orthogonal tensor subspaces of some dimensions [1, 10, 12]. In other words, the bases of these spaces are spanned by orthogonal tensors, rather than simple vectors as. In the following we briefly outline principles behind the HOSVD based classifiers. The HOSVD decomposition allows any P-dimensional tensor $\mathcal{T} \in \Re^{N_1 \times N_2 \times \ldots N_m \times \ldots N_n \times \ldots N_P}$ to be equivalently represented by the following product [12, 13]

$$\mathcal{T} = \mathcal{Z} \times_1 \mathbf{S}_1 \times_2 \mathbf{S}_2 \ldots \times_P \mathbf{S}_P, \tag{1}$$

where $\times_k$ stands for a k-mode product of a tensor and a matrix, $\mathbf{S}_k$ are $N_k \times N_k$ unitary matrices, and $\mathcal{Z} \in \Re^{N_1 \times N_2 \times \ldots N_m \times \ldots N_n \times \ldots N_P}$ is a core tensor [12, 13]. For 3D tensors, the above formula can be equivalently represented in the following form

$$\mathcal{T} = (\mathcal{Z} \times_1 \mathbf{S}_1 \times_2 \mathbf{S}_2) \times_3 \mathbf{S}_3. \tag{2}$$

The factor in parenthesis in formula (2) forms orthogonal tensors [4]. Thus, (2) can be rewritten as

$$\mathcal{T} = \sum_{h=1}^{N_3} \mathcal{T}_h \times_3 \mathbf{s}_3^h. \tag{3}$$

where $\mathbf{s}_3^h$ are columns of the unitary matrix $\mathbf{S}_3$. Since each $\mathcal{T}_\mathrm{h}$ is of dimension two, then $\times_3$ means the outer product of a 2D tensor and 1D vectors $\mathbf{s}_3^h$. Computation of $\mathcal{T}_\mathrm{h}$ can be done after simple rearrangement of (2)

$$\mathcal{T} \times_3 \mathbf{S}_3^T = (\mathcal{Z} \times_1 \mathbf{S}_1 \times_2 \mathbf{S}_2). \tag{4}$$

Matrix $\mathbf{S}_3$ can be computed from the SVD decomposition of the third flattening mode $\mathbf{T}_{(3)}$ of the tensor $\mathcal{T}$. This can be written as follows

$$\mathbf{T}_{(3)} = \mathbf{S}_3 \mathbf{V}_3 \mathbf{D}_3^T. \tag{5}$$

The orthogonal property of $\mathcal{T}_\mathrm{h}$ in (3) are used for construction of the orthogonal tensor subspaces which server for pattern recognition. One set of tensor bases is used to represent one training class. For multi-class problems, a set of $\mathcal{T}_h^i$ is built. Given a test pattern $\mathbf{X}$, its projections into each of the spaces spanned by the set of the bases $\mathcal{T}_h^i$ in are computed. Finally, an output class c is the one which fulfills the following optimization problem [4, 16]

$$c = \arg\min_i \left\{ \sum_{h=1}^{N^i \leq N_3^i} \langle \mathcal{T}_h^i, \mathbf{X} \rangle^2 \right\}. \tag{6}$$

In practice, also important is the choice of the number of important components N^i in the summation in (6), as will be discussed.

3 Fast HOSVD Decomposition

The main idea presented in this paper is that training of the HOSVD based classifier can be greatly speeded up substituting computation of the full SVD decomposition of the flattened matrix of the input tensor with computation of only a limited number of leading eigenvectors. Following the works by Bingham and Hyvärinen, for this purpose we propose to use the fixed-point algorithm, as will be discussed [1]. A similar approach was then used in tensor based signal filtering method proposed by Marot et al. [15].

Let us now observe that Eq. (5) can be rewritten as follows

$$\mathbf{T}_{(3)} \mathbf{T}_3^T = \mathbf{S}_3 \mathbf{V}_3^2 \mathbf{S}_3^T. \tag{7}$$

Thanks to this simple operation we gain at least two goals. The first is that the resulting product $\mathbf{T}_{(3)} \mathbf{T}_3^T$ (usually) has lower dimensions than a single flattened matrix $\mathbf{T}_{(3)}$, since a number of its columns is a product of P-1 dimensions of a tensor $\mathcal{T}$. This comes at a cost of a matrix multiplication, though. However, the second goal is that $\mathbf{T}_{(3)}\mathbf{T}_3^T$ becomes a symmetric matrix, and SVD computations can be reduced to eigenvectors problem. This means that its eigenvectors are real and orthonormal and for their computation an algorithm faster than full Jacobi based SVD can be used. Moreover, as will be shown, only a number of leading eigenvectors need to be determined which greatly speeds up the whole procedure.

To compute k_{max} leading eigenvectors of a matrix **M** the so called fixed-point algorithm is employed, as presented in Algorithm 1. It allows computation of k_{max} leading eigenvectors which are orthogonal. The main drawback is that the algorithm is iterative, although in practice in converges very fast. Apart from the input matrix **M** it requires setting of the orthogonality threshold ε as well as a maximal number i_{max} of allowable iterations to stop in case of lack of convergence.

Algorithm 1. The fixed-point algorithm for computation of the leading eigenvectors.

input A real *symmetric* matrix **M**
A number of maximum eigenvectors $1 \leq k_{\max} \leq rows(\mathbf{M})$
A maximal number of iterations i_{max}
An orthogonality threshold ε

output k_{max} leading eigenvectors of **M**

1: Random initialize vector $\mathbf{e}_0^{(0)}$
2: Set
3: $k \leftarrow 0$
4: **for** $k < k_{max}$
5: $i \leftarrow 1$
6: **do**
7: $\mathbf{e}_k^{(i)} \leftarrow \mathbf{M}\mathbf{e}_k^{(i-1)}$
8: Gram-Schmidt orthogonalization of $\mathbf{e}_k^{(i)}$ to eigenvectors $\{\mathbf{e}_{0\leq j<k}\}$:
$\mathbf{e}_k^{(i)} \leftarrow \mathbf{e}_k^{(i)} - \sum_{j=0}^{k-1}\left(\mathbf{e}_k^{T(i)}\mathbf{e}_j\right)\mathbf{e}_j$
9: Normalize $\mathbf{e}_k^{(i)}$:
$\mathbf{e}_k^{(i)} \leftarrow \mathbf{e}_k^{(i)} / \left\|\mathbf{e}_k^{(i)}\right\|_2$
10: $err = \left|\mathbf{e}_k^{T(i-1)}\mathbf{e}_k^{(i)} - 1\right|$
11: $i \leftarrow i+1$
12: **while** $err > \varepsilon$ **and** $i < i_{max}$
13: **end for**

In Eq. (4) we need a transposed matrix $\mathbf{S}_3$, so the eigenvectors computed with the outlined method become rows of this matrix. However, to compute (4) we need full size matrix $\mathbf{S}_3^T$. Hence, our idea is to choose a significant number of important components, i.e. eigenvalues, and the rest fill with 0. Under these conditions, $\mathbf{S}_3^T$ becomes

$$\mathbf{S}_3^T = \begin{bmatrix} \mathbf{e}_1^T \\ \mathbf{e}_2^T \\ \cdots \\ \mathbf{e}_{k_{\max}}^T \\ \mathbf{0}^T \end{bmatrix}, \tag{8}$$

where $\mathbf{e}_i^T$ denotes an ith eigenvalue of $\mathbf{T}_{(3)}\mathbf{T}_3^T$, and k_{max} stands for the chosen number of important eigenvectors.

The method outlined in Algorithm 1 draws from a number of known methods of eigenvalue computation. Step <7:> in Algorithm 1 is associated with the power method [7]. However, it allows computation of only eigenvector corresponding to the largest eigenvalue. To compute other eigenvalues, a deflation scheme based on the Gram-Schmidt orthogonalization (decorrelation) is applied in step <8:> of the algorithm. Having estimated $k - 1$ orthonormal eigenvectors, the kth is obtained by subtracting all projections of the previous $k - 1$ eigenvectors onto the kth one. As will be shown in the nexte section, the fixed-point algorithm allows much faster computation of the dominating tensor subspaces which greatly shortens training time.

4 Experimental Results

The presented method was tested with help of the handwritten digits database USPS [9, 18], which is a common test-set for comparison of different versions of tensor based classifiers [2, 3, 14, 16]. The dataset contains chosen and preprocessed handwritten digits from the envelopes of the U.S. Postal Service. These were converted to the common size of 16×16 pixels with gray level values. The numbers of training and testing patterns—7291 training and 2007 testing patterns, respectively—are presented in Table 1. The patterns are relatively difficult for machine classification and the reported human error is 2.5 %.

Table 1 Structure of the USPS database used in experiments

Training pattern	0	1	2	3	4	5	6	7	8	9
No of training patterns	1194	1005	731	658	652	556	664	645	542	644
No. of test patterns	359	264	198	166	200	160	170	147	166	177

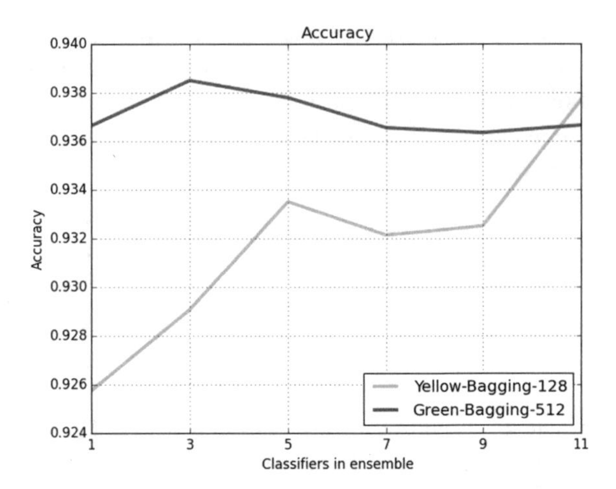

Fig. 1 Accuracy for different numbers of the HOSVD classifiers in the ensemble and two different bagging options —128 data and 512 data

The presented system was implemented in C++ with the Microsoft Visual 2013. Presented experiments were run on the laptop computer with the Intel® Core™ i7-4800MQ CPU @2.7 GHz, 32 GB RAM, and OS 64-bit Windows 7.

The first experiment was conducted to assess an overall accuracy of different configurations of classifiers and different amounts of data used in the bagging process [8]. Results are presented in Fig. 1. The output of the ensemble was estimated based on the majority voting scheme [3, 17]. Also, in all experiments the number of important factors was set to 16, which resulted in the best accuracy.

We can observe that bagging with larger number of elements per bag provides higher accuracy. Also interesting is to observe an influence of a number of base classifiers in ensembles. Accuracy always is higher for ensembles, that is if there is more than 1 classifiers. However, soon after adding new classifiers to the pool, accuracy does not grow as rapidly as expected. The reason is lack of diversity, since there are the same types of classifiers in the group. Also interesting to notice is that lower numbers of training data but with higher number of base classifiers compensate lower number of classifiers and high amount of bagging data [2, 3].

However, the main aspect we wish to test is a speed-up factor in the training process due to the new tensor decomposition method, as proposed in this paper, in respect to the accuracy of the final classification system. For this purpose we chose larger ensembles of classifiers, starting from 7 up to 11 members, as well as four different bagging assignments, 128, 256, 384, and 512 data objects for classifiers, respectively. The reason is to measure speed-up factor on ensemble setups which require significant amount of computations for training. Hence, in the presented experiments 12 different ensembles were used. Their structure is presented in Table 2. More details on ensemble construction are described in [2].

Table 2 Configurations of the ensembles of classifiers used in experiments

Ensemble No.	1	2	3	**4**	5	6	7	**8**	9	10	11	**12**
Number of classifiers	7	7	7	7	9	9	9	9	11	11	11	11
Size of bagging set	128	256	384	512	128	256	384	512	128	256	384	512

In bold font configurations which result in the highest speed-up factor due to large amount of processed data

Figure 2 presents timings obtained for different groups of ensemble classifiers and with different bagging sets. Using the fast eigenvalue computation, proposed in this paper, a significant gain in training speed is obtained. For some configurations, which in Table 2 are printed with bold font, this is larger than 5 times.

Figure 3 depicts bar graphs of showing accuracy for the same tensor ensemble configurations as used in Fig. 2. It can be observed that differences in accuracy between standard SVD and the proposed fast eigenvalue computation are negligible.

Thus, we conclude that the proposed method of fast training of the HOSVD classifiers allows even 5 times speed-up retaining the same accuracy as compared to the complete SVD decomposition.

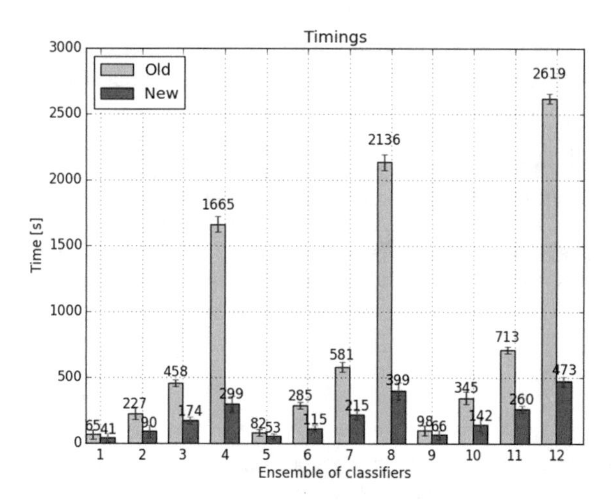

Fig. 2 Timings for different ensemble members and different bagging sets. Using the fast eigenvalue computation a significant gain in training speed is obtained. For some configurations this is larger than 5 times

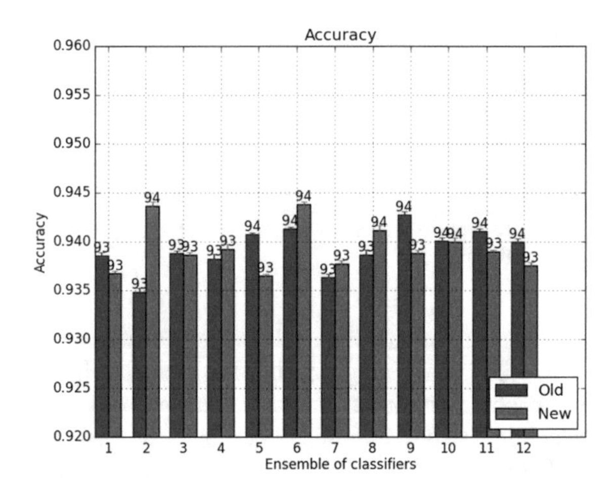

Fig. 3 Accuracy plots for different ensemble structure and different bagging sets. Differences in accuracy between standard SVD and the proposed fast eigenvalue computation are negligible

5 Conclusions

In this paper an improvement on fast computation of the HOSVD decompositions for the HOSVD based classifiers is proposed and analyzed. For this purpose the fast fixed-point method of eigenvector computation is employed. This iterative procedure allows for up to 5 times speed-up factor, as shown in experiments with handwritten data. It was also shown that the speed-up in the training stage does not incur any significant accuracy drop in the recognition stage. Further research will be focused on data preprocessing for further increase of the accuracy of the proposed method.

Acknowledgement This work was supported by the Polish National Science Centre under the grant no. DEC-2014/15/B/ST6/00609. This work was supported by EC under FP7, Coordination and Support Action, Grant Agreement Number 316097, ENGINE—European Research Centre of Network Intelligence for Innovation Enhancement (http://engine.pwr.wroc.pl/). All computer experiments were carried out using computer equipment sponsored by ENGINE project.

References

1. Bingham E., Hyvärinen A.: A fast fixed-point algorithm for independent component analysis of complex valued signals. Int. J. Neural Syst. **10**(1) (2000). World Scientific Publishing Company
2. Cyganek, B.: Ensemble of Tensor Classifiers Based on the Higher-Order Singular Value Decomposition. HAIS 2012, Part II, LNCS, vol. 7209, pp. 578–589. Springer (2012)
3. Cyganek B.: Embedding of the Extended Euclidean Distance into Pattern Recognition with Higher-Order Singular Value Decomposition of Prototype Tensors. In: Cortesi, A., et al. (eds.) IFIP International Federation for Information Processing, Venice, Italy CISIM 2012, Lecture Notes in Computer Science LNCS, vol. 7564, pp. 180–190. Springer (2012)

4. Cyganek, B.: Object Detection and Recognition in Digital Images: Theory and Practice. Wiley (2013)
5. Cyganek B., Krawczyk B., Woźniak, M.: Multidimensional data classification with chordal distance based kernel and support vector machines. Engineering Applications of Artificial Intelligence, Part A, vol. 46, pp. 10–22. Elsevier (2015)
6. Cyganek, B., Woźniak, M.: An improved vehicle logo recognition using a classifier ensemble based on pattern tensor representation and decomposition. New Gener. Comput. Springer **33** (4), 389–408 (2015)
7. Demmel J.W.: Applied Numerical Linear Algebra. Siam (1997)
8. Grandvalet, Y.: Bagging equalizes influence. Mach. Learn. **55**, 251–270 (2004)
9. Hull, J.: A database for handwritten text recognition research. IEEE Trans. Pattern Anal. Mach. Intell. **16**(5), 550–554 (1994)
10. Kolda, T.G., Bader, B.W.: Tensor decompositions and applications. SIAM Rev. 455–500 (2008)
11. Krawczyk, B.: One-class classifier ensemble pruning and weighting with firefly algorithm. Neurocomputing **150**, 490–500 (2015)
12. de Lathauwer, L.: Signal Processing Based on Multilinear Algebra. Ph.D. dissertation, Katholieke Universiteit Leuven (1997)
13. de Lathauwer, L., de Moor, B., Vandewalle, J.: A multilinear singular value decomposition. SIAM J. Matrix Anal. Appl. **21**(4), 1253–1278 (2000)
14. LeCun, Y., Bottou, L., Bengio, Y., Haffner, P.: Gradient-Based learning applied to document recognition. In: Proceedings of IEEE on Speech & Image Processing, vol. 86, No. 11, pp. 2278–2324 (1998)
15. Marot J., Fossati C., Bourennane S.: About advances in tensor data denoising methods. EURASIP J. Adv. Sig. Process. (2008)
16. Savas, B., Eldén, L.: Handwritten digit classification using higher order singular value decomposition. Pattern Recogn. **40**, 993–1003 (2007)
17. Woźniak, M., Grana, M., Corchado, E.: A survey of multiple classifier systems as hybrid systems. Inf. Fusion **16**(1), 3–17 (2014)
18. www-stat.stanford.edu/~tibs/ElemStatLearn/

Model Checking Mutual Exclusion Algorithms Using UPPAAL

Franco Cicirelli, Libero Nigro and Paolo F. Sciammarella

Abstract This paper proposes an approach to modelling and exhaustive verification of mutual exclusion algorithms which is based on Timed Automata in the context of the popular UPPAAL toolbox. The approach makes it possible to study the properties of a mutual exclusion algorithm also in the presence of the time dimension. For demonstration purposes some historical algorithms are modelled and thoroughly analyzed, going beyond some informal reasoning reported in the literature. The paper also proposes a mutual exclusion algorithm for $N \geq 2$ processes whose model checking confirms it satisfies all the required properties.

Keywords Mutual exclusion algorithms · Model checking · Timed automata · UPPAAL

1 Introduction

Mutual exclusion is the well-known problem faced by a collection of $N \geq 2$ asynchronous processes sharing some data variables. To avoid interference and ultimately unpredictable behavior of shared data, processes should access one at a time the common variables, i.e., they should execute one at a time their critical sections.

F. Cicirelli · L. Nigro (✉) · P.F. Sciammarella
Department of Informatics, Modelling, Electronics and Systems Science (DIMES), University of Calabria, Rende, CS, Italy
e-mail: l.nigro@unical.it

F. Cicirelli
e-mail: f.cicirelli@dimes.unical.it

P.F. Sciammarella
e-mail: p.sciammarella@dimes.unical.it

R. Silhavy et al. (eds.), *Software Engineering Perspectives and Application in Intelligent Systems*, Advances in Intelligent Systems and Computing 465,
DOI 10.1007/978-3-319-33622-0_19

Different solutions are described in the literature ranging from hardware to software based solutions. Software solutions can depend on blocked-queue semaphores or on a monitor structure. More challenging are solutions which make use of a few weak and unfair semaphores [1, 2] or they can be *pure*-software solutions, i.e., based on some communication variables and synchronization protocols regulating the *try* (or *enter*) and the *exit* sections which respectively precede and follow an execution of the critical section (CS). A process engaged into the try-protocol competes for entering its CS. A process not interested in entering its CS, e.g., because it has just finished executing its CS and the exit-protocol, runs its so called non critical section (NCS).

Methods for proving the correctness of mutual exclusion algorithms include proof-theoretic approaches [1] and model checking [3, 4]. Model checking can be preferable because it ensures formal modelling and automates the analysis activities. In particular, the use of Timed Automata [5] in the context of the popular and efficient UPPAAL toolbox [6] is advocated in this work, because it permits an in depth exploration of the behavior of an algorithm also along the time dimension, which can be difficult in a proof-theoretic approach.

This paper proposes a modelling approach based on UPPAAL and applies it to property checking of two known algorithms and to a new developed one. The afforded analysis enables to go beyond the informal reasoning often reported in the literature (see, e.g., the different indications about the worst case waiting time of competing processes in the $N > 2$ scenario of the Peterson algorithm [7, p. 101]). All of this testifies the known difficulties in mastering a concurrent solution due to complex action interleavings.

The paper argues that the proposed approach is of interest today in the software engineering domain because it helps the development of correct concurrent systems which can exploit the execution performance of multi-core machines.

The remainder of this paper is organized as follows. First the basic concepts, concurrency model and specification language of UPPAAL are summarized. Then the proposed modelling approach for mutual exclusion algorithms is presented. After that the Dekker's algorithm for 2 processes [8], the Peterson's algorithm for $N \geq 2$ processes [9] and a new algorithm for $N \geq 2$ processes are modelled and thoroughly analyzed. Finally, conclusions are presented with an indication of on-going and future work.

2 An Overview to UPPAAL

A timed automaton [5, 6] is a finite automaton augmented with a set C of real-valued variables named *clocks*. Clocks model the time elapsing and are assumed to grow synchronously at the same pace of the hidden system time. Constraints, of the form $x \sim k$ or $x - y \sim k$ where x and y are clocks, k is a non-negative integer and $\sim \in \{\leq, <, =, >, \geq\}$, are called *clock constraints* and can be introduced to restrict the behavior of the automaton. The set of all possible constraints over a set of clocks C is denoted by $B(C)$. A set of clock constraints used to label an edge it is called a *guard*. Clock constraints of the type $x \sim k$ can also be used to label locations and are called *invariants*. An automaton can stay in a location as long as the clocks satisfy the location invariant. Additionally, edges can be labeled by a set of clocks which are reset as the corresponding transition is taken, and by an *action* label ranging over a finite alphabet Σ.

Formally, a timed automaton A is as a tuple (L, l_0, E, I), where:

- L is a finite set of locations,
- $l_0 \in L$ is the initial location,
- $E \in L \times B(C) \times \Sigma \times 2^C \times L$ is a set of edges and
- $I: L \rightarrow B(C)$ is the invariant function.

The notation $l \rightarrow^{g,a,r} l'$ stands for $(l, g, a, r, l') \in E$, where g is a guard, a is an action, r is a set of clocks to be reset. The state of a timed automaton is a pair (l, u) where l is a location and u is a clock valuation, i.e., a function that associates a non-negative real value to each clock, and $(l_0, 0_{|C|})$ is the initial state. Let u be a clock valuation on a set of clocks C, and $d \in R_+$ a delay; $u + d$ denotes the clock assignment that maps all $x \in C$ to $u(x) + d$ and $r \subseteq C$, $[r \mapsto 0]u$ denotes the clock assignment that maps to 0 all clocks in r and agree with u for the other clocks. The semantics of a TA is defined by two state transition rules, namely *delay* and *action* transitions:

- $$(l, u) \rightarrow^d (l, u + d) \text{ if } (u + d) \in I(l) \text{ for any } d \in R_+$$

- $$(l, u) \rightarrow^a (l', u') \text{ if } l \rightarrow^{g,a,r} l', u \in g, u' = [r \mapsto 0]u \text{ and } u' \in I(l').$$

In the general case, for a given state (l, u) there are a continuous infinity values of d for which the first rule can be applied and hence, while the first state component l can assume only a finite set of values, the possible clock valuations are continuous infinity. Therefore, the state-space of a TA is infinite and uncountable. Despite this, reachability analysis of TA is decidable [5] because the infinite states of a TA can be partitioned into a finite set of equivalence classes called *zones*. A zone is a solution of a set of clock constraints.

TA can be composed to form a network of concurrent TA whose semantics depends on action interleavings and hand-shake synchronizations. UPPAAL adopts the notion of a *channel* for input and output action synchronization and uses a CSP-like notation. The edge of automaton labeled with `ch!` (output action), where `ch` is a channel, matches with an edge of another automaton labeled with `ch?` (input action). At a given time it may exist more than one pair of enabled and matched edges in which case a choice is made non-deterministically. Taking a transition (edge) in an automaton denotes an *atomic action* in the TA concurrent model. Moreover, the update of a sender is executed *before* that of a receiver.

The UPPAAL model-checker generates *on-the-fly* the zone graph of a network of TA for checking a subset of TCTL (Timed Computation Tree Logic) formulas [10] as in the following:

- $E <> \varnothing$ (Possibly $\varnothing$, i.e., a state exists where $\varnothing$ holds)
- $A[]\,\varnothing$ (Invariantly $\varnothing$, equivalent to: $not\, E <> not\, \varnothing$)
- $E[]\,\varnothing$ (Potentially Always $\varnothing$, i.e., a state path exists over which $\varnothing$ always holds)
- $A <> \varnothing$ (Always eventually $\varnothing$, equivalent to: $not\, E[]\, not\, \varnothing$)
- $\varnothing - - > \psi$ ($\varnothing$ always *leads-to* ψ, equivalent to: $A[]\, (\varnothing\, imply\, A <> \psi)$)

where $\varnothing$ and ψ are state properties (formulas), e.g., clock constraints or boolean expressions over predicates on locations. Verification of properties expressed by the above formulas reduces to reachability analysis, which is accomplished by traversing the zone graph associated with a TA network, and intersecting the zone associated with the formula with the zone of visited state nodes.

To facilitate the modelling task, integer variables with a bounded set of values and array of integers, clocks or channels can be introduced. A notion of automata *templates* which can be instantiated with different values for their parameters is supported. Integer variables, clocks and channels can be declared globally into a TA network, locally to a template, or used as template parameters. Locations can also be labeled as being *committed* (C) or *urgent* (U) both of which must be abandoned with no time passing. The exits from simultaneous committed locations are interleaved to one another, as well as the exits from urgent locations are interleaved each other, but committed locations have precedence over urgent locations. Channels can be declared to be urgent. An enabled synchronization on a urgent channel is required to occur without time passage.

UPPAAL consists of an editor, a simulator and the model checker. It is worth mentioning the possibility of building a counterexample (or diagnostic trace) of a not satisfied property, which can be analyzed in the simulator. The diagnostic trace furnishes evidence of a sequence of transitions bringing the model in a state not fulfilling the property.

3 Modelling Approach Based on UPPAAL

The common structure of processes regulated by a given mutual exclusion algorithm is assumed to be the following:

`process(i) = loop NCS; try-protocol; CS; exit-protocol; endloop.`

The parameter i identifies the generic instance of the process abstracted by an UPPAAL template. The critical section is modelled by a normal location (*CS*) with an associated clock invariant to enforce a maximum duration (C constant). The used clock is a local variable of the process. The non critical section is mapped to a normal location (*NCS*) which acts as the initial one for the process automaton. The absence of a clock invariant for *NCS* ensures an arbitrary time can elapse before the next arrival of the process competing for entering its *CS*.

A critical point of process modelling is the representation of the actions of thc try and exit protocols. To reproduce action interleaving and non determinism among processes, each elementary action (a variable assignment or a single condition test) is mapped onto an atomic action of the UPPAAL model, i.e., an update or a guard of a command of an edge exiting from a location. In order to ensure the *finite-delay* or *weak-fairness* property of the concurrent model [1], that is an action which is continuously enabled eventually fires, one could attach to the source location of an action a clock invariant mirroring its maximal duration. For simplicity, though, without any generality loss, the source location of an action is modelled as an urgent one. Therefore, each basic action is supposed to consume a negligible time with respect to the *CS*. However, to enable time advancement of a process, e.g., in the try-protocol, a critical point is the modelling of busy-waiting loop structures of the type: `while (cond) body`, where `cond` can be a complex condition and `body` can be void. Such loops are achieved by using a normal location (where time is allowed to pass) whose exiting is forced by an edge with guard `!cond` (`cond` is thus evaluated atomically and can exploit a user-defined function or the basic operators `exists`, `forall`, and `sum` which UPPAAL offers for checking arrays) and an urgent channel synchronization (see the unicast channel `synch` in Fig. 4). To preserve action interleaving and non determinism of processes during a complex condition evaluation, though, the exiting from a busy waiting location represents a *tentative* exit, thus it is followed by a detailed evaluation of `!cond` split into its component parts. During the detailed evaluation, the control can be transferred again to the busy-waiting location would `cond` be found still satisfied.

The worst case waiting time for a competing process can then be evaluated by observing the (hopefully bounded) number of *CS* s which are executed on behalf of competing processes, before the current one is enabled to enter its *CS*. Such a number (overtaking factor) can be determined either by counting the number of by-passes experimented by a competing process or by bounding the process clock during the try protocol.

The following properties will be verified on a modelled mutual exclusion algorithm.

Mutual exclusion (safety)–One and only one process can enter its *CS* at one time.

Deadlock free (safety)–The execution of the try/exit protocols in no case would induce a deadlock among processes.

Progress (liveness)–A process executing into its *NCS* would not forbid other processes to enter their *CS*.

Starvation free (bounded liveness)–A process competing for entering its *CS* eventually succeeds; that is, the number of by-passes of other competing processes is bounded.

4 Dekker's Algorithm

It has been the first algorithm proposed in 1962 for solving the mutual exclusion problem [8] for $N = 2$ processes. Three globals are used: boolean $b[0], b[1]$ and the integer k. Figure 1 shows the code of process i ($j = 1 - i$ denotes the partner process). $b[i] = true$ expresses willingness of process i to entering its *CS*. Process i can actually enter its *CS* when the turn variable k evaluates to i. The initial value of k can be either 0 or 1.

Figures 2, 3 and 4 show a corresponding UPPAAL model achieved according to the approach described in Sect. 3. The template process in Fig. 3 is named *Process* and admits one single *const* parameter i of type *pid*, the integer subrange type of process identifiers. Clock x serves to measure the elapsed time during the *CS*, or the waiting time during the try-protocol. Global declarations of the model and the local declarations of the $Process(i)$ template are collected in Fig. 2. Also the system configuration with implicit instantiation controlled by the *pid* parameter of *Process* (the two instances have names $Process(0)$ and $Process(1)$) is portrayed in Fig. 2. The *Synch* automaton is depicted in Fig. 4. It is always ready to send a signal on the urgent *synch* channel.

Absence of deadlocks was checked by the query $A[]!deadlock$ (*deadlock* is an UPPAAL keyword) which is satisfied. Mutual exclusion was checked by the query: $A[]forall(i\!: pid)forall(j\!: pid)Process(i).CS \,\&\&\, Process(j).CS\ imply\ i == j$ which is satisfied (would two processes be simultaneously in their *CS*, then the two processes are necessarily the same process). The progress property was checked by a query like $E <> Process(0).CS \,\&\&\, Process(1).NCS$ which is satisfied (it should be noted that being identical the two processes, one query suffices).

General liveness of the model was checked by query: $Process(0).while --> Process(0).CS$ which is *not* satisfied. This in turn mirrors the model (and the algorithm) has a zeno-cycle, which means any process can

Fig. 1 Dekker's algorithm

```
while( true ){
    NCS
    b[i] = true;
    while( b[j] ){
      if( k == j ){
        b[i] = false;
        while( k == j );
        b[i] = true;
      }
    }
    CS
    k = j;
    b[i] = false;
}
```

```
//Global declarations
const int C=4;
const int N=2;
typedef int[0,N-1] pid;
bool b[pid]={false,false};
pid k=0; //or 1
urgent chan synch;

//Process local declarations
clock x;
const pid j=1-i;

// List one or more processes to
// be composed into a system.
system Process,Synch;
```

Fig. 2 UPPAAL declarations

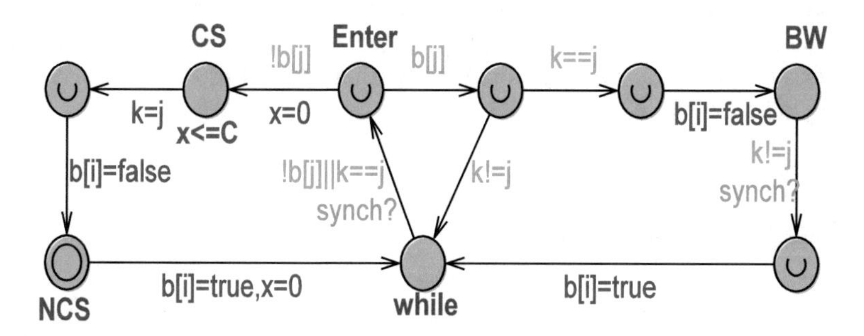

Fig. 3 Dekker's *Process*(i) automaton

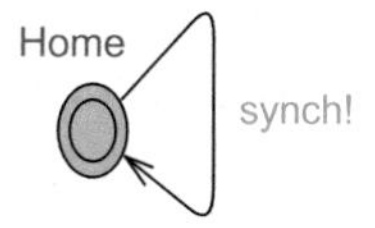

Fig. 4 The Synch automaton

enter/exit its critical section consecutively an infinite number of times and in 0 time (both *CS* and *NCS* are exited in 0 time). Therefore, the overtaking factor is theorically infinite. The zeno-cycle disappears by assuming the critical section has a duration strictly greater than zero (time-dependent behavior). As a consequence, the guard $x > 0$ was added to the model in Fig. 3 on the edge exiting the *CS* location and the remaining properties were verified on this modified model.

The starvation-free property was assessed by finding the supreme value of the clock x using the query: $\sup\{Process(0).while\}: Process(0).x$. Clock x is reset when the process starts competing, i.e., it reaches the *while* location. Similarly, it is reset when entering the *CS* location, i.e., abandoning the *Enter* location with $b[j] = = false$. The maximum value of the waiting time was found to be $2{*}C$, i.e., there are, in the worst case, two by-passes of $Process(1)$ whereas $Process(0)$ is competing and vice versa.

The worst case can happen when both processes are in the *while* location and $b[0] = true \wedge b[1] = true \wedge k = 1$. As a consequence $Process(0)$ reaches the BW location waiting for $k! = 1$. One time $Process(1)$ enters its *CS* and consumes *C* time units. On exiting the *CS* location, it executes $k = j$ (where *j* denotes $Process(0)$) and resets its $b[i]$ then enters its *NCS* which is immediately abandoned, i.e., in 0 time, then starts again competing by raising $b[i] = true$. Always in 0 time $Process(1)$ reaches again *Enter* and being $b[j] = = false$ it re-enters *CS*, thus consuming another critical section. The problem is that, for non determinism, $Process(0)$ in BW cannot immediately benefit of the update $k = j$ executed by $Process(1)$. But before exiting the second *CS* certainly $Process(0)$ reaches its *while* location. Now being $k = 0$ the next access to critical section will be granted only to $Process(0)$. The above behavior was confirmed by UPPAAL by answering to the query $E < > Process(0).while \,\&\&\, Process(0).x = = 2{*}C$ and by generating a corresponding diagnostic trace. It is worth noting that the worst case waiting time of processes also holds on the original model in Fig. 3.

5 Peterson's Algorithm

The algorithm was proposed in [9] to handle both the case $N = 2$ and the more general case of $N > 2$ processes. Figure 5 portrays the general version of the algorithm and the declarations of a corresponding UPPAAL model.

Processes are numbered $1, 2, \ldots, N$. They have to climb a ladder in order to enter the *CS*. The ladder has *N*-1 levels numbered from 1 to *N*-1. The process who reaches the *N*-1 highest level enters its *CS*. In the case more processes try to step a same level, at least one of them stops moving. Two arrays are used: $q[pid]$ and $turn[level]$. $q[i]$ contains the level occupied by $Process(i)$. $turn[j]$, where *j* is a level, stores the identifier of a process at level *j*.

The UPPAAL model of Peterson's algorithm was studied for $N \in [2..5]$. First of all the query $Process(1).for - - > Process(1).CS$ was found not satisfied, testifying, as in the Dekker's algorithm, the existence of a zeno-cycle which disappears by adding the guard $x > 0$ to the edge exiting the *CS* location (see Fig. 6). Using queries similar to the Dekker's model, it was found the Peterson's model satisfies the mutual exclusion and the progress properties. In addition the model is both deadlock and starvation free. The overtaking factor was determined using the query:

Fig. 5 Peterson's algorithm and UPPAAL declarations

```
while( true ){
  NCS
  for( j=1; j<=L; j++ ){
    q[i]=j;
    turn[j]=i;
    wait until(
    (for all k≠i,(q[k]<j))or
    (turn[j]≠i) );
  }
  CS
}

//Uppaal global declarations
const int C=4;
const int N=…;
const int L=N-1; //levels
typedef int[1,N] pid;
typedef int[1,L] level;
int[0,N] q[pid];
pid turn[level];
urgent chan synch;

//Process local declarations
int[1,L+1] j;
int[1,N+1] k;
clock x;
```

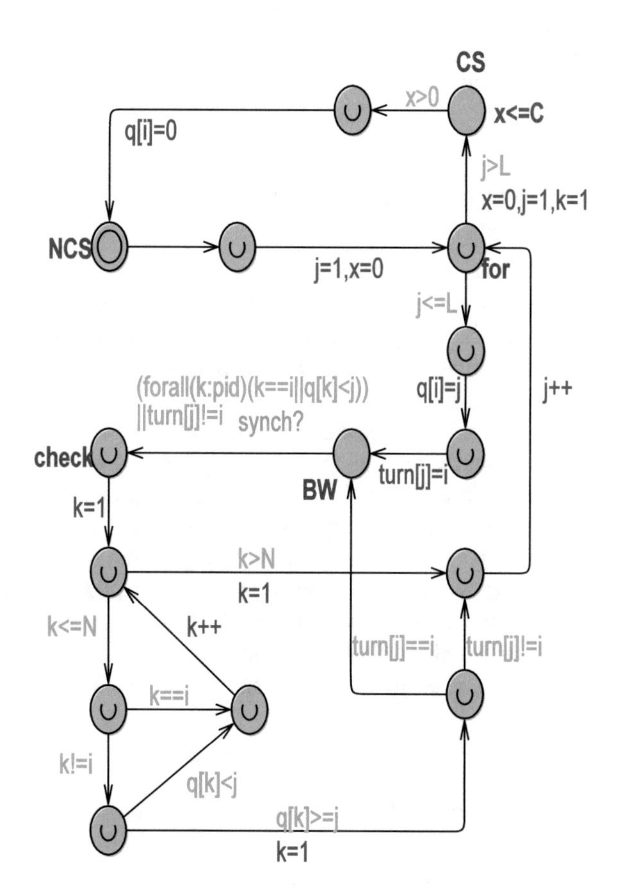

Fig. 6 Peterson's *Process*(i) automaton

Table 1 Overtaking factor versus *N*

N	*WCWT*	*OF*
2	4	1
3	12	3
4	24	6
5	40	10

$\sup\{Process(1).for\}: Process(1).x$. Table 1 shows the collected results for $N \in [2..5]$. *WCWT* denotes the emerged worst case waiting time for a trying process in the *for* location, $OF = WCWT/C$ is the corresponding overtaking factor, where $C = 4$ time units is the maximum duration of the critical section.

It emerged that in the Peterson's algorithm a trying process has a worst case overtaking factor of $((N-1)*N)/2$. This result agrees with the calculation reported in [11] but contrasts with the indications contained in [12, 13] which predicted a worst case number of by-passes of *N*−1.

The size and possible values of the arrays $q[]$and$turn[]$ critically affect the dimension of the state graph and the model checking activities. To favor the model checker, provisions were taken in Fig. 6 to assign a default value to a variable as soon as it gets unused (see the integer variables *j* and *k*). Checking the overtaking factor in the case $N = 5$ required about 31 min of wall clock time with a peak of used memory of about 19 GB. Experiments were carried out using UPPAAL version 4.1.19 64bit on a Linux machine, Intel Xeon CPU E5-1603@2.80 GHz, 32 GB.

6 Proposed Algorithm

As a further demonstration of the application of the proposed modelling and verification approach, the following describes an example of a new algorithm[1] for $N \geq 2$ processes (see Figs. 7, 8, 9) which uses fewer variables and it is more efficient than, e.g., the Peterson's algorithm. The processes are numbered $0, 1, \ldots, N-1$.

The algorithm uses an array $b[pid]$ of *N* booleans, two boolean variables *lock* and *cs_busy*, and a *turn* integer variable. Initially all the booleans are set to *false* and $turn = 0$. To access *CS*, $Process(i)$ must "climb a ladder" of three steps.

$Process(i)$ starts competing by putting *true* in $b[i]$. Climbing can fail at each of the first two steps (denoted respectively by the actions $turn = i$ and $lock = true$) thus forcing $Process(i)$ to restart trying (from the label *L*). The last step is signaled by the action $cs\_busy = true$. After *CS*, $Process(i)$ puts *false* in $b[i]$, then it makes a modular search in the *b* array looking for the first process *j*, if there are any, which is trying. The critical section grant is then transferred to it by assigning *false* to $b[j]$,

[1]The contribution of Domenico Spezzano to the design of this algorithm is acknowledged.

Fig. 7 Proposed algorithm

```
while( true ){
    NCS
    b[i]=true;
L:  while((turn!=i||lock)&&b[i]);
    if(!cs_busy) turn=i;
    if((turn!=i||lock)&&b[i]) goto L;
    lock=true;
    if((turn!=i||cs_busy)&&b[i]){
      if(!cs_busy) lock=false;
      goto L;
    }
    cs_busy=true;
    CS
    b[i]=false; j=(i+1)%N;
    while(j!=i&&!b[j])j=(j+1)%N;
    if(j==i){
      cs_busy=false;
      lock=false; turn=0;
    }
    else b[j]=false;
}//while
```

Fig. 8 UPPAAL declarations

```
//Global declarations.
const int C=4;
const int N=4;
typedef int[0,N-1] pid;
bool b[pid]=
{false,false,false,false};
bool lock=false;
bool cs_busy=false;
pid turn=0;
urgent chan synch;

//Process local declarat.
pid j;
clock x;
```

otherwise the trying protocol is re-initialized. Figure 9 portrays the *Process*(i) template automaton.

The model was verified for a number of processes N ranging in the interval [2..5]. It satisfies *all* the required properties without assumptions about the *CS* duration. The algorithm has no zeno-cycle and the overtaking factor is linear (although waiting processes are not *FIFO* managed) in the number of processes,

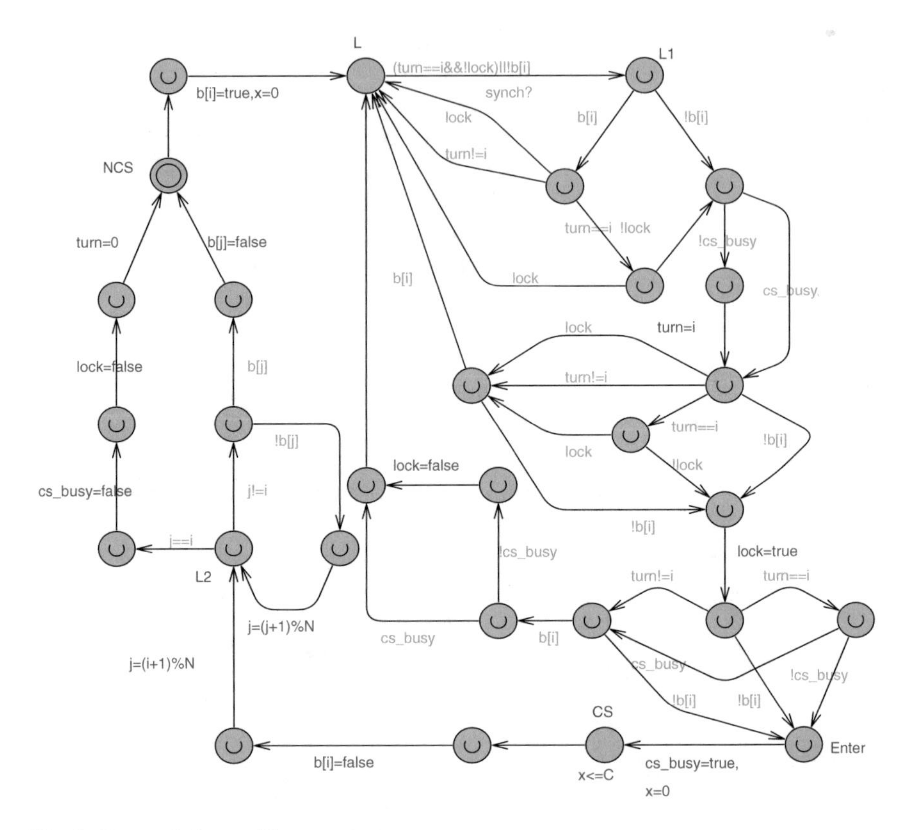

Fig. 9 Proposed $Process(i)$ automaton

i.e., in the worst case, a competing process has to wait for one by-pass of each of the remaining N–1 processes. For $N = 5$, a verification query is responded in about $1min$ of wall clock time with a memory peak of about $950MB$. Starvation-free behavior was checked by the satisfied queries: $A[]Process(0).Enter$ $implyProcess(0).x < = (N-1)*C$ and $Process(0).L - - > Process(0).CS$.The first query fails when the clock $Process(0).x$ is compared with $(N-1)*C-1$ or a lower quantity. The second query ensures bounded waiting for a trying process, and absence of zeno-cycles.

7 Conclusions

The analysis of concurrent systems is notoriously hard for the complex action interleavings occurring among the involved processes. All of this can introduce subtle errors in a concurrent program which cannot be dominated by informal reasoning or by program testing.

Mutual exclusion algorithms represent an active area of concurrency since the sixties, and some errors are famous about the characterization of their properties [7]. Nowadays, the availability of mature and powerful model checkers like UPPAAL [6] has the potential to improve the situation by enabling formal modeling and exhaustive exploration of a concurrent program.

This paper proposes a model checking approach based on Timed Automata and UPPAAL for the systematic analysis of mutual exclusion algorithms. With respect to proof-theoretic approaches (e.g., [1]) the proposal facilitates an in-depth verification of an algorithm also in the presence of the time dimension.

The approach was successfully applied to many existing algorithms of which known properties were confirmed and new properties disclosed in doubt situations. The article contributes to a better understanding of two historical algorithms like Dekker [8] and Peterson [9]. The approach was also exploited in the design and analysis of a new algorithm for $N \geq 2$ processes, which fulfils all the required properties. Prosecution of the work is geared at improving the approach and to experiment with its application to other algorithms.

References

1. Hesselink, W.H., IJbema, M.: Starvation-Free Mutual Exclusion With Semaphores. Formal Aspects of Computing (2011). DOI 10.1007/s00165-011-0219-y
2. Cicirelli, F., Nigro, L.: Modelling and verification of starvation-free mutual exclusion algorithms based on weak semaphores. Proc. FedCSIS **2015**, 785–791 (2015)
3. Clarke, E.M., Grumberg, O., Peled, D.A.: Model Checking. MIT Press (2000)
4. Cicirelli, F., Furfaro, A., Nigro, L.: Model checking time-dependent system specifications using time stream petri nets and UPPAAL. Appl. Math. Comput. **218**, 8160–8186 (2012)
5. Alur, R., Dill, D.L.: A theory of timed automata. Theoret. Comput. Sci. **126**, 183–235 (1994)
6. Behrmann, G., David, A., Larsen, K.G.: A tutorial on UPPAAL. In: Bernardo, M., Corradini, F. (eds.) Formal Methods for the Design of Real-Time Systems, Lecture Notes in Computer Science, vol. 3185, pp. 200–236. Springer (2004)
7. Alagarsamy, K.: Some myths about famous mutual exclusion algorithms. ACM SIGACT News **34**(3), 94–103 (2003)
8. Dijkstra, E.W.: Cooperating sequential processes. In: Genuys, F. (ed.) Programming Languages, pp. 43–112. Academic Press, New York (1968)
9. Peterson, G.L.: Myths about the mutual exclusion problem. Inf. Process. Lett. **12**(3), 115–116 (1981)
10. Alur, R., Courcoubetis, C., Dill, D.L.: Model-checking for Real-time systems. In: Proceedings of Seventh Annual IEEE Symposium on Logic in Computer Science, pp. 414–425. IEEE Computer Society Press (1990)
11. Raynal, M.: Algorithms For Mutual Exclusion. MIT Press (1986)
12. Kowalttowski, T., Palma, A.: Another solution of the mutual exclusion problem. Inf. Proce. Lett. **19**(3), 145–146 (1984)
13. Hofri, M.: Proof of a mutual exclusion algorithm—a 'class'ic example. ACM SIGOPS OSR **24**(1), 18–22 (1990)

Software Usability Evaluation Based on the User Pinpoint Activity Heat Map

Nikita Danilov, Tatiana Shulga, Natalya Frolova, Nina Melnikova, Nataliia Vagarina and Elena Pchelintseva

Abstract The article focuses on the software product's usability evaluation based on the user activity data. A heat map model is to be used as the user activity mathematical model. The authors propose a formal method for creating a heat map based on the point and click data of the applications user activity. The method considers both the data density and expert-defined parameters.

Keywords User activity · Visualization method · Heat map · Usability · User interface · Software

1 Introduction

The growing complexity of computer systems raises the standards for their usability i.e. the degree of efficiency, ease and user satisfaction with which a product can be used in particular context and for particular purposes [1]. The usability enhancement lies in maximizing the specified qualities, since each one of them plays an important

N. Danilov (✉) · T. Shulga · N. Frolova · N. Melnikova · N. Vagarina · E. Pchelintseva
Yuri Gagarin Saratov State Technical University,
77 Politechnicheskaya street, Saratov, Russia 410054
e-mail: Nikita_Danilov@outlook.com
URL: http://en.sstu.ru/

T. Shulga
e-mail: shulga@sstu.ru

N. Frolova
e-mail: natalya-fr@yandex.ru

N. Melnikova
e-mail: melnikovani@gmail.com

N. Vagarina
e-mail: v-n-s@yandex.ru

E. Pchelintseva
e-mail: Helen_Pchelintseva@mail.ru

R. Silhavy et al. (eds.), *Software Engineering Perspectives and Application in Intelligent Systems*, Advances in Intelligent Systems and Computing 465,
DOI 10.1007/978-3-319-33622-0_20

role both in the commercial sector and non-profit one [2]. For example, the less time a cashier spends on processing the purchases, and the fewer mistakes he makes, the more customers he will be able to serve throughout the day, and thus bring greater profit to the store.

Usability evaluation methods are based on understanding what kind of mistakes users typically make, and how they generally behave when interacting with the interface [3]. In order to gain such understanding, it is necessary to analyze the actual user activity data when working with the software product through the user interface.

There are various types of user activity, but the most curious are those which can be visualized in most graphic and easy-to-analyze way. These include mouse input data, such as mouse clicks and mouse movements, that is normally presented in the form of so-called heat maps. However, no methods for making such maps have been found in scientific sources. There are several software products implementing such methods, but they are non-free [3]. In this article, the authors look into ways of visualizing the user activity and offer a formal method for making a user activity heat map.

2 User Activity Visualization

When analyzing the user-application interaction, any information on user activity may be utile [3]. The more information an expert has, the more factors he will be able to take into account. However, this information must be suitable for further processing and use. For example, sometimes when testing the usability, the users behavior and things happening on the screen are recorded. This allows to keep track of virtually all types of activity, but the subsequent processing of such videos is extremely time-consuming, not to mention the difficulties with preparing the testees. Video recording ensures the completeness of the data but it provides no possibility of keeping it as statistical data, which can be used in further operations.

The possibility of collecting user activity data outside the laboratory is also important. In 2004, Katherine Thompson, together with her colleagues, published a research called “Here, there, anywhere: Remote usability testing that works” [4], which highlights the difference between laboratory research and collecting statistical data under real-life conditions. The research showed that, even with the most thorough preparation, the testees might act differently in lab conditions other than they would in real life, which makes some errors hard to detect.

There are various methods of user activity visualization. The choice depends on their structure and characteristics, as well as the clarity and detalization requirements necessary for the analysis. Tables, for example, are one of the simplest and most commonly used ways of presenting structured data, but they do not provide sufficient clarity.

Most popular as means of user activity visualizing are the so-called heat maps, which are graphical representations of the data, where matrix-like values are displayed in color. Similar systems of hierarchical color-coding are used in fractal images and other data representation systems. The term "heat map" was trademarked by a software developer and entrepreneur, Cormac Kinney, in 1991. He used the term to describe 2D-images depicting real time financial market information. Interestingly, in 1998 the trademark was acquired by SS&C TECHNOLOGIES, INC., but the company did not extend the licence, so it was annulled in 2006 [5].

Heat maps are often associated with cartograms i.e. a method of cartographic imaging, visually showing the intensity of a variable within the area on the map. The data can be displayed on the map using shading of different density, coloring of certain brightness (background cartogram) or points (point cartogram).

Biological heat maps are commonly used in molecular biology and medicine to provide data on gene expression in different samples, for instance, derived from different patients, or under different conditions from one patient. They are put in a table in which the color of the square indicates the level of expression, whereas the columns and rows show various genes or patterns, hierarchical organization of which can be displayed as a tree in the margins.

The basic principle, incorporated in all areas of application and the methods of creating heat maps, is to present different values using color, which provides a high level of visibility and speeds up the analysis process.

Traditionally, heat maps were used in the areas of science where the source data allowed one to easily determine the color of a specific cell (gene expression levels, stock indexes), area (cartograms) or point/pixel (brain tomography in medical research, temperature map in meteorology). However, with respect to usability, the standard methods of creating heat maps do not provide the proper level of clarity for all data types. Some types of user activity, such as mouse clicks and cursor movements, can be called "pinpoint", because they are actually associated with a specific point (pixel) on the screen which appears to be too small an area for detailed analysis compared to the entire interface. Unfortunately, searching scientific literature for the accurate methods of building heat maps, based on point and click data, brought no results.

In the early 21st century, the web industry began to actively use heat maps to display data on clicks, mouse movements and user clicks on the links. Currently, it is difficult to determine who was the first to propose collecting and using this data for further analysis. There is a patent CN 1949259 "Method for point contacting information of collecting web page by embedding code in web page", registered in 2006 under China Merchants Technology Holdings Co., Ltd. [6]. With no reference to any specific technology, it describes the principle of adding user clicks tracking logic to the web page code and sending the data to a special server. However, in the same year, an Israeli Internet company ClickTale appeared, beginning to offer similar web service of collecting user click data. Later on, various other similar patents and web services emerged due to the growth of interest of large companies towards such user activity. In Russian World Wide Web segment, Yandex.Metria [7] is the most popular one. The service offers a wide range of opportunities for the analysis. This tool allows to build different kinds of heat maps. Figure 1 shows an example of a user click heat map.

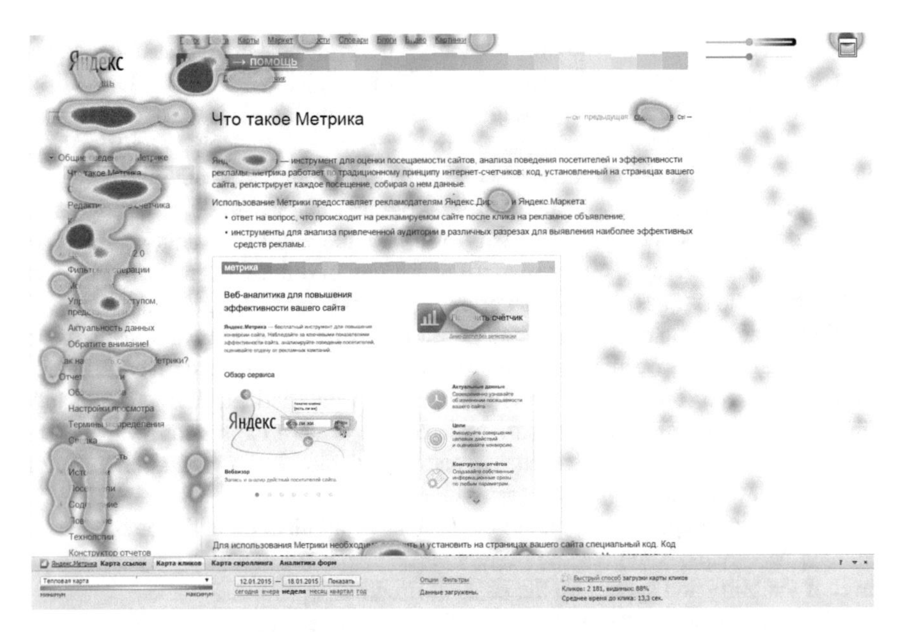

Fig. 1 Click heat map created on Yandex.Metrica web service

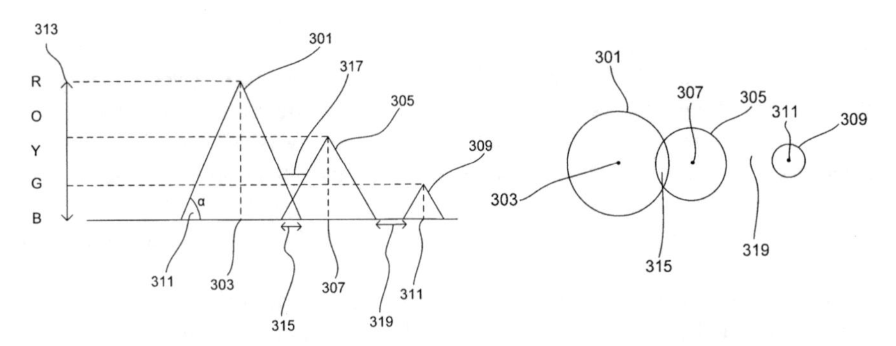

Fig. 2 Example of a heat map created with a uniform color gradient and partial intersection of the circles

Patent "Data visualization methods" [8] relates to a method of creating a graphical representation of data in the form of a heat map. It describes the process of creating heat maps with no reference to usability and user activity analysis, its main area of application is numeric data representation. It details the principle of presenting separate data elements in the form of circles, with a gradient color intensity from the center outward, and the principle of summing up the intensity values when the circles intersect. Figure 2 illustrates how the method works using a uniform color gradient and partial intersection of the circles.

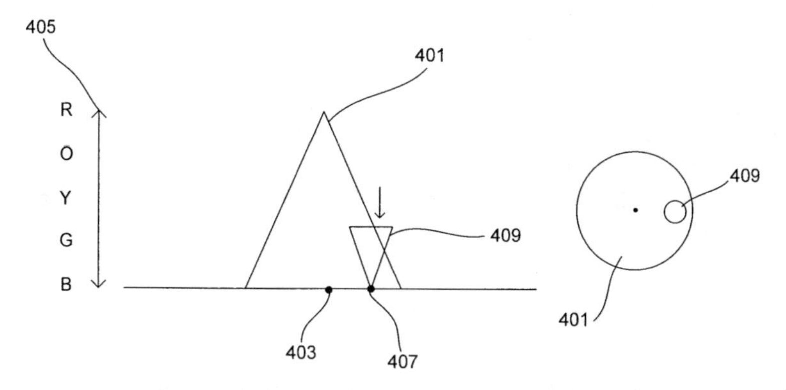

Fig. 3 The patent for point and click data visualization as a heat map with overlapped points

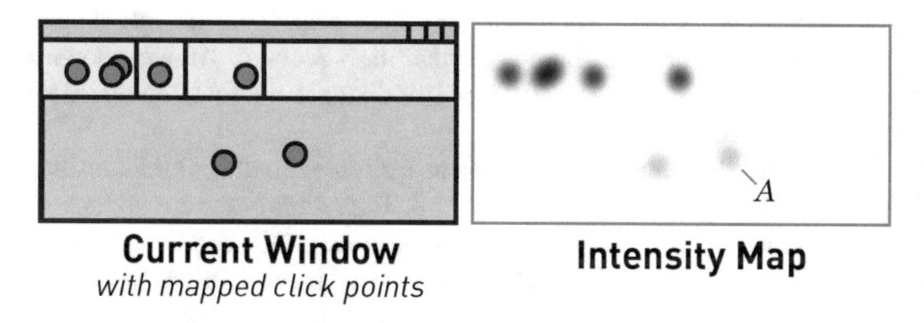

Fig. 4 Patina: intensity map creation

The patented method is not completely suitable for use in the analysis of the user activity point and click data, because it does not consider some crucial aspects. Firstly, it doesn't indicate how the initial intensity is chosen, as it is supposed to be used for specific numeric data. Secondly, in case if circles fully overlap each other, the cumulative intensity rather than its reverse distribution for the overlapped circle is displayed, so that it shows through the overlapping circle as shown on Fig. 3. This reduces the visibility when analyzing point user activity, since we aim to focus on the total summary intensity in particular.

Research results "Patina: Dynamic Heatmaps for Visualizing Application Usage" [9], presented at CHI 2013, contain description of the *Patina*—application independent system for collecting and visualizing software application usage data. Among other things in this article authors describe intensity map creating and heatmap coloring methods. Presented methods especially adapted for the user activity visualization and consider some aspects of this area. As for example, if there are many overlapping click points, the opacity of each point can be reduced to prevent the heatmap from becoming oversaturated. Another idea invented by authors is that the opacity of individual click points depends on the region size, reduced for large regions, principle is shown on Fig. 4.

Unfortunately in the article does not described in details possibilities for methods configuration, but it's important aspect. Different kinds of usability research require different visulization settings, such as click point size, point intensity opacity and heatmap color palette (which also may depends from the individual color vision deficiency of the researcher and target audience). As follows from the article, in the Patina just size of the intensity circle are configurable for intensity map creating.

3 Heat Map Creation Method

Authors propose a generalized method for making the user activity point and click data heat maps, that complements approaches presented above. The basic idea of the proposed method is that, when making a heat map, both the data density and the expert-defined parameters should be taken into account. An expert can determine such parameters as the intensity gradient distance and overlap distance. Lets elaborate on these concepts.

Each point activity element is presented as a circle with the color intensity gradient linearly decreasing from the center outward. The radius of the circle (the intensity area) is determined by the Intensity Gradient Distance (IGD) and is set by an expert. The specific color of each point on the heat map is determined by the value of its cumulative intensity, the sum of all intensity values of the covering areas, and the chosen color scheme (graphic palette). If the cumulative intensity value is more than 1 (one), it must be set to 1 (one).

The initial value of intensity gradient I_i is calculated as follows: $I_i = \frac{1}{MON}$, where MON is the Maximum Number of Overlapping circles on the entire data area. Two or more circles are considered overlapping when the distance between the pairs is less than the Overlap Distance (OD), set by an expert. The end value of the intensity gradient Ie always equals to 0 (zero).

Calculating I_i based on MON ensures the preservation and reliability of visual data. With the high density of point and click data, certain areas, surrounded by these points, will have the maximum intensity and not those areas where the maximum cumulative intensity would have formed. For instance, Fig. 5 shows two fields of intensity. For the sake of convenience, the circles are shown as triangles on Cartesian plane where the horizontal axis features the coordinate (position) of the point, and the vertical axis shows the intensity.

While the distance between the two points is more than half of the IGD, the cumulative intensity is less than 1 (one), since the intensity is calculated using the linear gradient. When the distance becomes less than or equal to half of the IGD, the cumulative intensity becomes either greater than or equal to 1 (one), respectively.

Depending on OD value set by an expert, visual data loss is considered acceptable to a certain extent, since we want to focus on the cumulative intensity of some interface area. For example, clicks on a button or the area around it, rather than a specific spot on it. If the points are at a distance less than or equal to the value of OD, as shown in Fig. 6, the initial intensity for all points is recalculated.

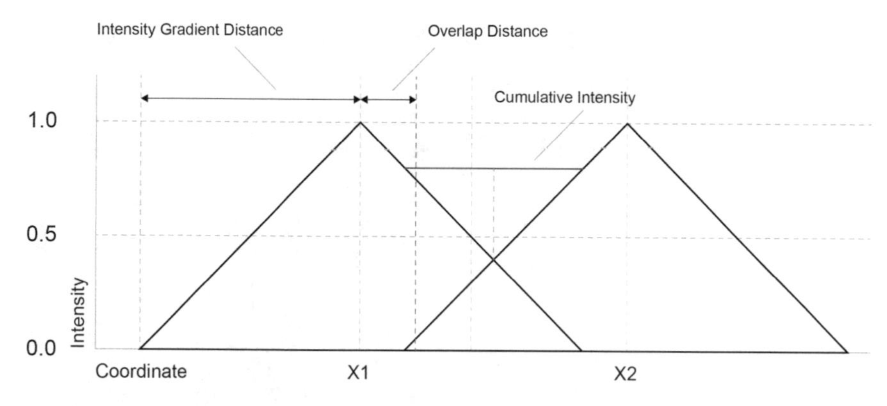

Fig. 5 Intersection of the points intensity areas

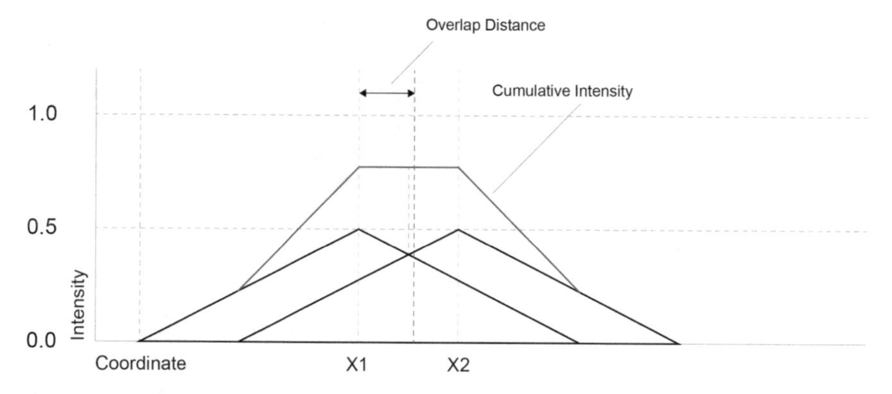

Fig. 6 Overlapping of the points intensity areas

Based on the intensity values and the chosen color scheme (graphic palette), various heat, transparency and other kinds of maps can be made. The choice depends on the specifics of the analysis. Transparency map, for instance, refers to imaging, in which the areas with lowest activity are rendered in black opaque color, with an increase in transparency in areas with greater intensity. Possible color schemes and their variations are not considered in the article.

Thus, by the user activity point and click data heat map we understand the imaging of users pinpoint activity, based on the calculated values of the cumulative intensity for each point using a certain color scheme.

Lets describe the method of making the heat maps of the user's point and click activity using the above-mentioned concepts of IGD, OD, MON, the initial and end value of the intensity gradient.

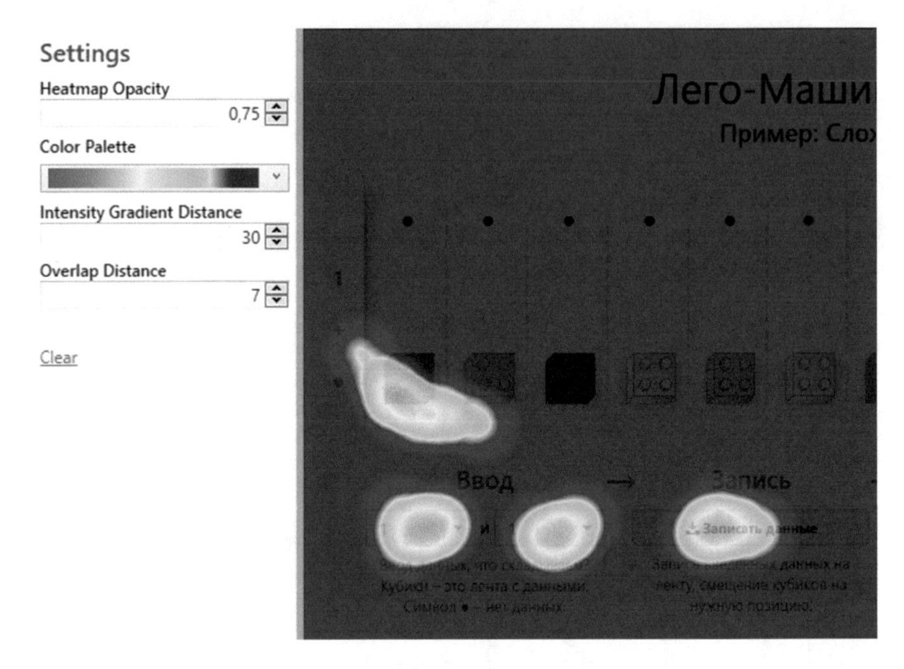

Fig. 7 Heat map created in accordance with suggested method

The method of creating a heat map based on the user activity point and click data.

Input: user activity point and click data (a set of points with their coordinates on the plane), IGD value, OD value, graphic palette.

Output: user activity point and click data heat map.

Step 1. Get the input data. User activity point and click data can be generated or collected in any manner, for example by embedding a special logic that monitors the mouse clicks and stores this information in a file. IGD value, OD value, graphic palette are defined by the expert and depend on the specifics of the analysis.

Step 2: Calculate MON.

Step 3: Calculate I_i by the formula: $I_i = \frac{1}{MON}$.

Step 4. For each user activity data item build the intensity circle centered at the specified point. Determine the radius equal to IGD and make a linear gradient intensity from the center outwards. The initial value of the gradient should be equal to I_i, the end value should be set to 0 (zero).

Step 4.1. If the intensity areas of two or more circles intersect, calculate the cumulative intensity for each point in the intersection, summing up the intensities of all the areas covering the point.

Step 4.2. If the cumulative intensity value is more than 1, set it to 1 (one).

Step 5. Visualize the heat map based on the intensity values for each point and the specified palette.

Step 6. Produce a heat map.

An example of the heat map built in accordance with this method is shown in Fig. 7.

Expert can adjust such parameters as IGD and OD. This makes it possible to achieve the desired level of visual detalization in accordance with the analysis, being conducted. For example, sometimes its a better idea to render each point activity using circles of small radius if the work with particular buttons is being examined, whereas sometimes knowing what areas of the interface are more commonly used is more important, so the IGD value may be set high enough.

4 Conclusion

The developed method of creating a heat map, based on the user activity point and click data, allows to take into account both the data location density and the expert-defined parameters, namely the intensity gradient distance and overlap distance. The proposed method can be used for the analysis and evaluation of the user interface usability, as well as for conducting research in this area.

References

1. ISO 9241-11:1998, ISO 9241-11:1998: Ergonomic requirements for office work with visual display terminals (VDTs)—Part 11: Guidance on usability. https://www.iso.org/obp/ui/#iso:std:iso:9241:-11:ed-1:v1:en
2. Bias, R.G., Mayhew, D.J.: Cost-Justifying Usability, 2nd edn. Morgan Kaufmann, San Francisco (2005)
3. Danilov, N., Shulga, T.: Method for constructing a heat map based on the point data of the application users activity. J. Appl. Inf. (Prikladnaya informatika) **10**(2) (56), 4958 (2015) (in Russian)
4. Thompson, K.E., Rozanski, E.P., Haake, A.R.: Here, there, anywhere: remote usability testing that works. In: 5th Conference on Information Technology Education, SIGITE 2004, pp. 132–137. ACM, Salt Lake City (2004)
5. SS&C TECHNOLOGIES, INC.: HEATMAPS. Trademark, US Serial Number: 75263259, US Registration Number: 2140964. Accessed 3 Mar 1998
6. Tianyun, M., Yan, L., China Merchants Technology Holdings Co., Ltd.: Method for point contacting information of collecting web page by embedding code in web page. Application, Publication Number CN1949259A, Application Number CN 200610002956. https://patents.google.com/patent/CN1949259A. Accessed 18 Apr 2007
7. Yandex.Metrica: https://metrica.yandex.com/
8. Cardno, A.J., Ingham, P.S., Singh, A.K.: Data visualization methods. Application, Publication Number US20110141118 A1, Application Number US 12/866, 842 https://patents.google.com/patent/US20110141118A1. Accessed 16 Jun 2011
9. Matejka, J., Grossman T., Fitzmaurice G., Fitzmaurice, G.: Patina: dynamic heatmaps for visualizing application usage. In: CHI 2013, SIGCHI Conference on Human Factors in Computing Systems, pp. 3227–3236. ACM, New York (2013)

Inflated Power Iteration Clustering Algorithm to Optimize Convergence Using Lagrangian Constraint

Jayalatchumy Dhanapal and Thambidurai Perumal

Abstract Spectral clustering is one of the machine learning techniques based on graph theory. It requires finding eigen values and eigen vector of large matrices consuming much time and space. Often, finding an optimal eigen value is sufficient to cluster data. Power Iteration Clustering algorithm (PIC) replaces the eigen values with pseudo eigen vector. Though PIC is fast and scalable it causes inter collision problem when dealing with larger datasets. Accuracy is also a concern. To solve the optimal Eigen value problem, in this paper we proposes an Inflated Power Iteration Clustering algorithm. It uses the modified lagrangian constraint which induces an exponential inflationary growth by increasing the rate of convergence of eigen vector to get an ideal optimal solution. This algorithm is validated by experimenting on various real and synthetic dataset of varying size. On validation it has been found that the performance of the algorithm has been improved in terms of speed up and efficiency. The speed up has been improved by 31 % when compared to PIC and almost 40–42 % with K-means algorithm. The accuracy has been improved by 7 % than k means and 4 % than PIC. The proposed algorithm is highly scalable and can be used for clustering Big Data.

Keywords Power iteration clustering · Inflation · Eigen value · Eigen vector · Lagrangian constraint

J. Dhanapal (✉)
Department of CSE, PKIET, Karaikal, India
e-mail: djlatchumy@gmail.com

T. Perumal
PKIET, Karaikal, India
e-mail: pdurai158@gmail.com

R. Silhavy et al. (eds.), *Software Engineering Perspectives and Application in Intelligent Systems*, Advances in Intelligent Systems and Computing 465,
DOI 10.1007/978-3-319-33622-0_21

1 Introduction

Clustering is an important technique used in data mining algorithm. It is a method of grouping objects which are similar within the same cluster and are different for varying clusters. It is most widely used in various fields of social network analysis and telecommunication. Clustering algorithms are mainly classified into Hierarchical based, Partitional based and Density based algorithm [1, 2]. Partitional based clustering outputs the result directly to a single level solution. There are many partitioning algorithm among which K-means seems to be the best. But K-means is not efficient as it converges slowly at local minimum and also, the number of cluster should be specified.

Spectral Clustering (SC) overcomes these limitations. K-means work directly on data points but in Spectral clustering the data points are converted into affinity matrix. The eigen vector are obtained from laplacian matrix using matrix decomposition. Hence, the effort for computing eigen vectors is $O(n^3)$. To overcome this problems Frank Lin Proposed Power Iteration Clustering (PIC) [3] that performs matrix vector multiplication. Instead of finding all the eigen vectors PIC computes an pseudo eigen vector which is a linear combination of vectors. Though, there is no data loss using PIC, it has limitations when dealing with multi class datasets.

In this paper, Inflation based PIC algorithm is proposed that has better execution time and accuracy than PIC. In this algorithm, the lowest eigen vector is found and inflated using lagrangian multiplier [4]. Nearby eigen pairs are found and the rate of convergence is calculated. Hence, it grows fast relative to its neighbor. Moreover, the computational time for finding the eigen vector is small when compared to PIC.

The rest of the paper is organized as follows. Section 2 presents the background of Spectral Clustering and Power Iterative Clustering; Sect. 3 explains the proposed algorithm. Section 4 examines the performance. Section 5 concludes the paper highlighting its future direction.

2 Background Work

- ***Spectral clustering (SC)***: It is a clustering technique based on graph theory. It constructs the unweighted graph with each point as vertex and the similarity between two points being the weights connecting the two vertices [5]. First, a similarity matrix is constructed using the given datasets. It is then converted into a laplacian matrix. The eigen values and eigen vectors are obtained using eigen decomposition of the laplacian [1]. Finally, the clusters are found using the K means algorithm. The drawback of SC is that the computational cost of finding the eigen vectors of a large matrix is very high. It takes $O(n^3)$. It consumes much time and space [3, 6, 7]. PIC is the adherent of spectral clustering which tries to solve this problem by finding only pseudo eigen vector.

- ***Power Method***: It is a technique to compute matrix decomposition. It finds only one eigen value and converges slowly [8]. This method starts with a vector v_0 and iterates further. The vector is multiplied with the matrix and the length of resultant vector is calculated and normalized for next iterations. The power method is used when A^{n*n} has n linearly independent eigen vectors. The eigen values can be ordered in magnitude where the first vector λ_1 is the leading eigen value of the matrix.
- ***Power Iterative Clustering (PIC)***: PIC is an algorithm that clusters data using power method which is used to find the largest of eigen vector a combination of the eigen vectors in a linear manner. The algorithm uses the matrix vector multiplication where the matrix W is combined with the vector v_0 to obtain Wv_0. PIC is an iterative process in which the vector gets updated and normalized to avoid becoming too large using $\frac{v_t}{||v_t||}$. The largest eigen vector using power method is

$$v^{t+1} = cWv^{t-1} \text{ where } c = \frac{1}{||Wv^t||}$$

For a set of data points $\{x_1, x_2, \ldots, x_n\}$, where x_i is a d-dimensional vector the similarity between the matrices is found using the Euclidian distance given as [6]

$$s(x_i, x_j) = \exp(-\frac{||x_i - x_j||_2^2}{2\sigma^2})$$

where σ is the scaling parameter that controls the kernel width [3, 6]. The algorithm for PIC is shown in Algorithm 1.

Begin

Read the dataset

Construct the similarity matrix $A \in R^{n*n}$

Obtain a row normalized affinity matrix W

Pick initial vector v_0

Repeat

$v^t = \gamma W v^{t-1}$ & $\delta^{t+1} = |v^{t+1} - v^t|$

where $\gamma = \frac{1}{||Wv^{t+1}||}$

Increment t

Until $|\delta^\wedge t - \delta^\wedge(t-1)| = 0$

Cluster point on v^t using K-means and

Output the clusters.

End

Algorithm 1. Power Iteration Clustering

Power Iterative Clustering works well for larger datasets. The computational complexity is given as $O(n^2)$ for Spectral Clustering and O(n) for PIC [3, 6, 7]. Vector <v^t> converges to local centre that forms the cluster of the datasets [3]. It converges more quickly than power iteration because, it stops when <v^t> stop accelerating.

3 The Proposed Inflated Power Iterative Clustering (IPIC)

Spectral Clustering consumes more time as it computes many eigen values and vectors. Though PIC finds only one pseudo eigen vector it causes inter collision problem for larger datasets. In certain conditions, using only one pseudo eigen vector does not yield best results. Sometimes finding the lowest or largest eigen values and eigen vectors yields a better solution. There are many methods to find these values which include lanczos algorithm, Jacob/Davidson technique and so on. Helle has presented in the paper [4] a way to solve it using modified lagrangian constraint. This constraint induces an inflation to obtain an optimal solution. Lagrangian multiplier is used to regulate the inflation where the lowest eigen vector grows exponentially fast relative to their neighbors. The lagrangian approach is given by

$$L = \sum_i x_i^2 - \sum_{ij} x_i A_{i,j} x_j + \lambda\left(\sum_i x_i^2 - 1\right)$$

where, λ is the Lagrangian multiplier that enforces normalization. The corresponding Euler equation is

$$\ddot{x}_i = -\sum_j A_{i,j} x_j + \lambda(t) x_i, \quad \text{i} = 1 \ldots \text{N}.$$

On integrating with time δt gives

$$x_i(t+\delta t) = x_i(t) + p_i(t+\delta t)\delta t$$

where,

$$p_i(t+\delta t) = p_i(t) - \sum_j \left[A_{i,j} - \lambda(t)\delta_{ij}\right] x_j(t)\delta t$$

Let λ(t) be a constant for small time step we have,

$$\xi_i(t) = a_i \cos(w_i t + \delta_i),$$

If $e_i > \lambda$, where δ_i is a real phase shift & $w_i = \sqrt{\lambda - e_i}$ for low eigen value $e_i < \lambda$, they coordinate as

$$\xi_i(t) = (a_i' e w_i'^t + b_i' e w_i'^t)$$

The eigen modes with eigen values $e_i < \lambda(t)$ gets inflated exponentially at a time 't'. $\lambda(t)$ regulates inflation by managing the inflated and the non-inflated states. The need for eigen pair and accuracy can be adjusted based on the convergence factor [4]. Hence convergence of matrices is the key term behind this algorithm. In this paper the computational time and accuracy is improved based on this convergence factor. The Algorithm 2 gives the steps for Inflated PIC.

Input: Given a set of data points $\{(x_1, x_2, x_3 \ldots x_n) \mid x_i \Sigma R, i = 1 \ldots n\}$ and the number of k clusters.

Step (i) Construct a similarity function using fully connected graph based method like Gaussian similarity function on a given set $s(x_i, x_j) = \exp(-\frac{\|x_i - x_j\|_2^2}{2\sigma^2})$ where σ is the scaling parameter that controls the kernel width

Step (ii) Build the affinity matrix A with $a_{ij} = s(x_i, x_j)$ if $i \neq j$ and $a_{ij} = 0$ if $i = j$

Step (iii) Obtain the diagonal matrix $D_{ii} = \Sigma_j A_{ij}$

Step (iv) Normalize the obtained matrix by dividing by its row sum. Find the eigen vector with an arbitrary vector v^0, $v^0 = \frac{R}{\|R\|}$ where R is the row sum of W

Step (v) Calculate new vector and new velocity $v^t = \gamma W v^{t-1}$ & $\delta^{t+1} = |v^{t+1} - v^t|$

Step (vi) Inflate the vector and perform update through iteration and normalize it from getting large for the time δ^t

Step (vii) Calculate the rate of convergence of inflation method of the ground state ξ_0 relative to the first excited state ξ_1 denoted as $\frac{\xi_0(t)}{\xi_n(t)} \sim -e^{\sqrt{e_1 - e_0 t}}$ where $\xi_0 = e_1 - e_0$ obtained by $\lambda(t) = \lambda(t) + \xi_{01}$ where $\lambda(t)$ is the lagrangian multiplier at the time t is given as $\delta t \sim \sqrt{\frac{e_1 - e_0}{e_{max} - e_0}}$

Step (viii) Repeat step 6 on incrementing the value of t until the algorithm converges using the constraint $(|\delta^t - \delta^{t-1}| \simeq 0 \;\&\&\; \frac{v_{max} - v_0}{2} \; != 0)$

Step (ix). Cluster the points v_t on a k dimensional subspace using K-Means algorithm.

Output: Clusters$\{C_1, C_2, C_3 \ldots C_n\}$ with n_j points in C_j

---

Algorithm 2. Inflated Power Iteration Clustering

3.1 Analysis of IPIC's Convergence

The convergence of eigen values plays an active role. It is optimized by inflating the needed eigen mode quickly. To find this rate of convergence, the initial state ξ_0 and final state ξ_1 is considered. The inflation border h is equal to e_1 [4]. Now ξ_0 grows at a faster rate

$$\xi_0(t)/\xi_n(t) \sim e^{\sqrt{e_1 - e_{0t}}}.$$

If the convergence of the algorithm at time δt is larger, then it will decline leading to wrong mode. If δt obeys the condition $\delta t < 2/w_{max}$ then it is stable,

$$w_{\max}^2 = e_{\max} - e_o \text{ is an upper bound of } e_1 - \alpha.$$

The optimal rate of convergence in time δt is given by

$$\delta t^2 \sim \frac{e_1 - e_0}{e_{max} - e_0}$$

$$\text{(i.e.) } \delta t \sim \sqrt{\frac{e_1 - e_0}{e_{max} - e_0}}$$

The square root is a result of second order differential equation against first order. Using this condition of inflation, the proposed algorithm is made to converge on the conditions,
where

$$\delta_t != 0 \;\&\&\; \frac{v_{max} - v_0}{2} != 0$$

which is a difference between the current vector to the previous vector [4]. The choice of inflation can be in any mode with $e_1 > e_o + w$, which $\sim \sqrt{}\,(e_{\max} - e_o)/e_o$ steps for convergence.

4 Performance Analysis

The PIC algorithm has been validated by the work done by Frank Lin [3]. In this paper the quality of the clusters of IPIC is considered. In this section various experiments conducted on real and synthetic dataset are explained. The datasets [9] used for this experiment includes.

IRIS** consists of 50 samples from 3 species, the Iris Setosa, Iris Virgincia and Iris Versicolor and four features from each sample. The length and width of sepals and petal in centimeters are stored in a 150 × 4 ndarray. The rows are the samples* *and columns are the sepal length, sepal width, petal length and petal width.* ***MNIST *dataset of handwritten digits. It includes 10 handwritten digits that contain 60,000 training pattern and 10,000 test patterns of 784 dimensions. It has 10,000 instances of 748 attributes.* ***BREAST*** *dataset that contains the prediction of the patients. The training dataset has 78 patients, 34 who developed metastates and 44 healthy patients. It has 699 instance and 9 attributes namely the Id, Diagnosis and 30 real value input features.* ***BIRCH*** *dataset which is a synthetic 2d data with 10,000 vectors and 100 clusters.* ***HOUSE*** *dataset which is an image data with RGB values quantized to 5 bits per color. It has 34112 instances and 3 attributes. The RGB value also contains 8 bits per color.*

4.1 Parameters Considered

To evaluate the quality of clustering, various parameters have been used in this paper includes Entropy, F-measure, Accuracy, Normalized Mutual Information (NMI), and cluster purity [10, 11]. This parameter calculates the goodness or the quality of the clusters formed and the similarity between the algorithms including its performance. The external measures do not inherit the dataset whereas the internal measures are derived from the data directly.

- ***Cluster purity***: It is defined as the weighted sum of purity of the clusters (i.e.) how far the clusters are pure. It states the extent to which a cluster contains variables from one partition and is given as

$$\text{purity} = \frac{1}{\text{n}} \sum_{i=1}^{r} \max_{\text{j}=1}^{\text{k}} \{n_{ij}\}.$$

  If the cluster contains entities from only one partition then the value of purity is maximum. The maximum value of purity is 1 and minimum is 0. If the value of purity is towards 1 then it can be said that the performance is high.
- ***F-measure***: It is the harmonic mean of the precision and recall values of each cluster. The precision is the proportion of the predicted results that is correct and recall is the ratio of the results that is correctly identified. Thus the precision and Recall are defined as [1, 11]

$$\text{prec} = \frac{1}{\text{n}_\text{j}} \max_{\text{j}=1}^{\text{k}} \{n_{ij}\}$$

$$Recall = \frac{n_{ij_i}}{m_{j_i}}$$

The F-measure of the cluster is given as [1, 11]

$$F_i = \frac{2}{\frac{1}{prec_i} + \frac{1}{recall_i}} = \frac{2.n_{ij_i}}{n_i + m_{j_i}}$$

The maximum value of F measure is 1 and minimum is 0. Higher the value maximum is the quality of the clusters. It tries to balance precision and recall values.

- ***Entropy***: It measures the amount of ambiguity present in the clusters with respect to ground truth. The entropy of the clustering algorithm $H(C)$ is given as [10, 11]

$$H(C) = \sum_{i=1}^{r} pc_i \log_2 pc_i \text{ where } pc_i = \frac{n_i}{n} \text{ the probability of cluster } c_i.$$

The entropy of a partition $H(T)$ is given as

$$(\mathrm{T}) = -\sum_{j=1}^{k} \mathrm{pT_j} \log_2 \mathrm{pT_j} \text{ where } pT_j = \frac{m_i}{n} \text{ is the probability of partition.} T_j.$$

Hence the entropy of T over c is given as

$$H(T|C) = -\sum_{i=1}^{r} \sum_{j=1}^{k} p_{ij} \log_2 \frac{p_{ij}}{pc_i} \text{ where } p_{ij} = \frac{n_{ij}}{n}$$

- ***NMI***: The Normalized Mutual Information between cluster C and partition T is used to quantify the shared information. The NMI is given as [10–12]

$$\mathrm{NMI(C,T)} = \sqrt{\frac{\mathrm{I(C,T)}}{\mathrm{H(C)}} \cdot \frac{\mathrm{I(C,T)}}{\mathrm{H(T)}}} = \frac{I(C,T)}{\sqrt{H(C).H(T)}}$$

$$I(C,T) = \sum_{i=1}^{r} \sum_{j=1}^{k} p_{ij} \log_2 \frac{p_{ij}}{pc_i pT_j}$$

The NMI value is 1 if the clustering results perfectly match and close to 0 if the partition is randomized. The maximum the value the NMI better is the quality of the cluster.

- ***Speed up***: The computational time for k-means, PIC and IPIC. IPIC is found to be fastest of all the algorithms. The average computational time of IPIC is lower than that PIC, especially when its data size increases. K-means is slower than even PIC. PIC is found to be better than its version of SC like Ncut, NJW [3, 7]. PIC finds only one pseudo eigen vector hence it seems to be better than its version of SC [3, 6, 7]. It is proven that PIC is 50 times faster than SCIRAM in most datasets [7]. The time to compute pseudo eigen vector is same as that of time to compute PIC pseudo eigen vector. But the rate of convergence IPIC is quick (i.e.) IPIC converges faster than PIC using

$$|\delta^t - \delta^{t-1}| \underset{\approx}{=} 0 \;\&\&\; \frac{v_{max} - v_0}{2} \;! = 0$$

Hence we justify that IPIC has better computational time than PIC, K-means and all versions of SC for various datasets. The graph for the computational time is given in Fig. 1. From the graph it is inferred that the proposed algorithm has higher speed up. IPIC is 42 % faster than K-means, and 31 % times faster than PIC. It is noted that as the data size increases the convergence rate is faster. Hence, it is efficient for big data.

- ***Accuracy***: The quality of the cluster is measured using how accurate the clustered output is. The accuracy is calculated using the formula

$$Accuracy = \frac{\sum_{i=1}^{n} \delta(y_i.map(c_i))}{n}$$

The accuracy of IPIC is better than K-means and PIC. The accuracy of PIC is less comparable because its finds pseudo eigen vector which is a linear combination of k largest eigen vector. IPIC tries to achieve better accuracy because it finds the smallest eigen vector and inflates it using the Lagrangian constraint [4]. Hence the accuracy is found to be better than PIC. From Fig. 2 it is inferred that the accuracy though increases than K-means but it is only 50 % of the predicted results. IPIC algorithm performs well in terms of time for computation the accuracy is comparable but do not have *IPIC* much variation. The accuracy of IPIC is better than K-Means but almost equal to that of PIC. IPIC is 4 % better than PIC and 7 to 8 % better than k means. The graph for accuracy, entropy and NMI is given in Fig. 2.

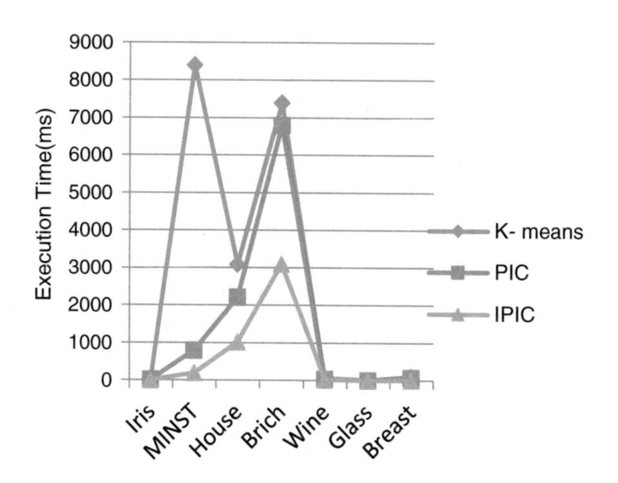

Fig. 1 Computational time of K-Means, PIC, IPIC

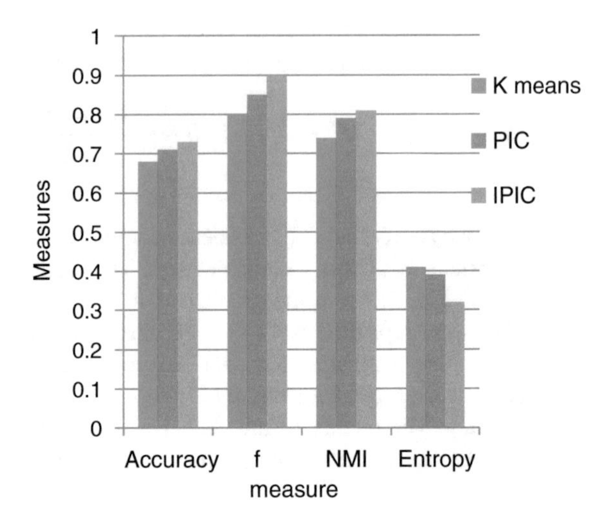

Fig. 2 Comparison chart accuracy, f-measure, NMI and entropy

The parameters to prove the effectiveness of the algorithm includes the Speedup, NMI, Accuracy and F-measure. The proposed algorithm converges using the constraint. Since the rate of convergence is fast the vectors formed meet the meeting criteria early and hence the clusters are formed fast. The accuracy is calculated based on true positive, rue negative, false positive and false negative. They are found for iris dataset and the accuracy is obtained. The proposed algorithm is better than K-means and PIC in terms of speedup and almost accuracy is the same to an extent. The parameters have been tested for various datasets like the IRIS, MNIST, BRICH, HOUSE C5 and BREAST.

5 Conclusion

In this paper an Inflated Power Iteration Clustering is proposed and implemented. It is concluded that PIC is not much accurate and causes intercollision problem for larger datasets. Also the execution time is to be improved. This drawback has been addressed in IPIC by inflating the eigen vectors. Experiments conducted using datasets of varying size justify that the proposed algorithm has a lesser execution time. Though, accuracy has not increased to a greater extent it is found to be better than PIC measured in terms of its quality of the clusters. In future, work has to be done to automate the value of k to find the clusters. Accuracy of the clustering algorithms also has to be concentrated since most of the clustering techniques obtain only 50–70 % of the predicted results.

References

1. Shah, G.H., Bhensdadia, C.K., Ganatra, A.P.: An empirical evaluation of density based clustering techniques. Int. J. Soft Comput. Eng. (IJSCE) **2**(1) (2012). ISSN: 2231-2307
2. Rokach, L., Maimon, O.: Clustering methods. In: Data Mining and Knowledge Discovery Handbook. Springer (2005). ISBN: 978-0-387-09822-7. e-ISBN: 978-0-387-09823-4. doi:10.1007/978-0-387-09823-4
3. Lin, F., William, W.: Power iteration clustering. In: International Conference on Machine Learning, Haifa, Israel (2010)
4. Heller, E.J., Kaplan, L., Polimann, F.: Inflationary dynamics for matrix eigen value problems. In: PNAS, vol. 105, no. 22, 7631–7635, June 2008. doi:10.1073/pnas.0801047105
5. Jia, H., Ding, H., Xu, X., Nie, R.: The latest research progress on spectral clustering. Neural Comput. Appl. **24**, 1477–1486 (2014). ISSN: 0941-0643. doi:10.1007/s0052-013-1439-2
6. Yana, W., et al.: p-PIC: parallel power iteration clustering for big data. Models and algorithms for high performance distributed data mining. J. Parallel Distrib. Comput. **73**(3) (2013)
7. The, A.P., Thang, N.D., Vinh, L.T., Lee, Y.-K., Lee, S.: Deflation based power iteration clustering. Appl. Intell. **39**, 367–385 (2013). doi:10.1007/s10486-012-0418-0
8. http://www.cs.cornell.edu/~bindel/class/cs6210-f09/lec26.pdf
9. http://cs.joensuu.fi/sipu/datasets/
10. http://www.cs.rpi.edu/~zaki/www-new/uploads/Dmcourse/Main/Lecture18.pdf
11. http://www.seas.gwu.edu/~bell/csci243/lectures/performance.pdf
12. Chen, W.Y., Song, Y., Bai, H., Lin. C.: Parallel spectral clustering in distributed systems. IEEE Trans. Patt. Anal. Mach. Intell. **33**(3), 568–586 (2011)

Virtualization of Operating System Using Type-2 Hypervisor

Jiri Vojtesek and Martin Pipis

Abstract A virtualization is one of the very progressive fields in information technologies. It is widely used especially in the server virtualization because it offers better usage of hardware resources by using one physical hardware for various virtual host servers. The history of the virtualization starts nearly before fifty years and the development runs from the virtualization of the mainframe to the virtualization on the hardware level. The second IT field which uses a virtualization is a testing and an education which uses a specialized software for virtualization of the operating system. The task of this contribution is to compare a desktop virtualization software from the performance and the hard-disk drive usage point of view. The practical part is focused mainly on the comparison of the VMware Player and the Oracle VM VirtualBox as one of the best-known free software for the desktop virtualization.

Keywords Virtualization · Hypervisor · VMware player · Oracle VM VirtualBox · Performance test · HDD test

1 Introduction

A computer virtualization of operating systems belongs to the one of the most progressive spheres in the information technology. The basic idea is based on the virtual sharing of physical sources of a hardware into more virtual hosts [1].

One of the main field for a virtualization is a server hosting [2]. It is known, that major group of servers has mean usage in the range 5–15 % depending for example on the attractiveness of the server, day time etc. Everybody can guess that it means that the server is inefficiently used. As a result, a virtualization in this case provides possibility to share the rest of the capacity to the other virtual servers. As a results,

J. Vojtesek (✉) · M. Pipis
Faculty of Applied Informatics, Tomas Bata University in Zlin,
Nam. T.G.Masaryka 5555, 760 01 Zlín, Czech Republic
e-mail: vojtesek@fai.utb.cz
URL: http://www.utb.cz/fai

R. Silhavy et al. (eds.), *Software Engineering Perspectives and Application in Intelligent Systems*, Advances in Intelligent Systems and Computing 465,
DOI 10.1007/978-3-319-33622-0_22

we can have one instead of three or four servers in the company which means big cost savings to the hardware and other operating costs like electric energy costs, server cooling costs etc. [3]

Other benefits can be found in the better administration of the server. The installation of the new software is also much more efficient. Finally, the restoration after the crash, a backup or a security generally is better for the virtualized servers [4].

The start of the virtualization can be found in 1960's when IBM tries to invent the concept which uses full computation power of computers [5]. The example could be found in the experimental system IBM M44/44X which runs virtualization of multiple virtual computers IBM 7044 in one computer. The virtualization was these years the domain of mainframe computers till late 1990's [6]). The role of x86 servers increases after 1990 with the client-server configuration which reduces costs and makes x86 architecture as an industrial standard.

American company VMware, Inc. [7], one of the main leaders in the virtualization field, was invented in 1998 and it offers first virtualization software VMware Workstation ported on Microsoft Windows in 1999. Microsoft also adopts virtualization technology when they buy company Connectix with their product named Virtual PC.

The hardware virtualization [8] grows after the year 2005 because both main producers of computer processors, AMD and Intel, focus and develop their technologies AMD-V and Intel VT that supports virtualization on the hardware level. This shifts a virtualization again in the next level.

There are numerous fields of usage of the virtualization. It was already mentioned that the server virtualization is widely used in the server or the hosting rental companies [9]. The virtualization is also very nice tool if you want to run a program which is not ported in the actual operating system or it is not supported for the newest versions of operating systems. It can be also used in the situation if we want to test software or operating system. In this case,the restoration after the crash from the backup is much simple then installing the operating system with all programs and settings from the zero again [10]. That is why we also use a virtualization software as a learning tool for example in the subject Operating systems and other subjects at Faculty of Applied Informatics, Tomas Bata University in Zlin [11, 12].

The paper is focused mainly on the simplest version of the virtualization a local virtualization of the operating system in the desktop programs *VMware Player* and *Oracle VM VirtualBox*. Both programs uses Type-2 Hypervisor which runs as a special virtualization software according to the Type-1 Hypervisor which is focused on the hardware virtualization.

2 Virtualization

As it was already mentioned, the virtualization means the use of the one physical hardware (Central Processing Unit—CPU, Random-access Memory—RAM etc.) for various smaller independent virtual hosts (i.e. virtual hosts or guests). Each guest could have its own operating system with the use of the shared CPU and RAM of the host machine.

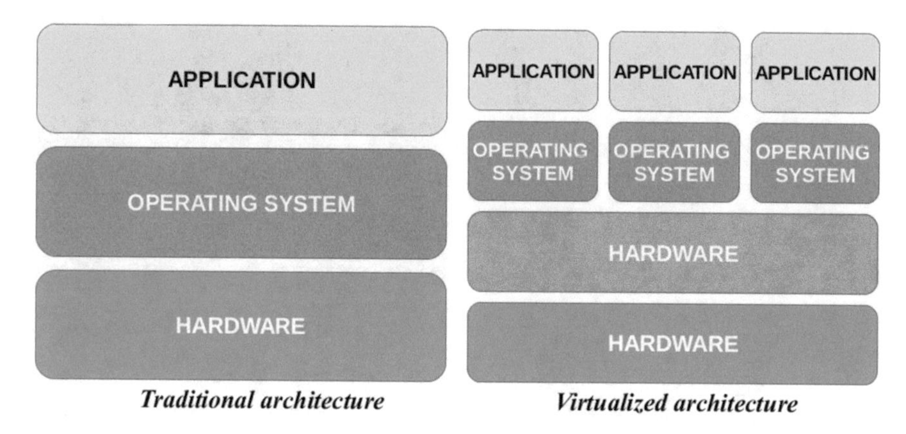

Fig. 1 Comparison of the traditional and virtualization computer architecture

The comparison of the traditional computer architecture and the architecture used for the virtualization is shown in Fig. 1.

A *Hardware virtualization* [13] runs on the real computer and guest machines uses hardware resources separately. It is common, that virtual hosts are separated from hardware resources. Advantage of this type of virtualization is that the operating system which runs in the virtual machine does not need any modifications as the virtual machine acts as a real computer. Special type of hardware virtualization is *paravirtualization* where not all virtual components are emulated. There is a program interface which uses a real hardware and it is common, that some parts of the virtual machine are equal to the real hardware. As a result, the paravirtualization could be more efficient because there is no need for emulating of all hardware components. On the other hand, operating system could detect that it is used in virtual machine and that is why it needs to be modified.

An *Application virtualization* [13] on the contrary means that desktop or server applications use local resources but runs in a special virtual machine which means that they do not need to be installed in the local operating system. The application then runs in a thin virtual environment that has only components needed for starting and acts as a layer between an application and an operating system.

One of the main components of the virtualization is hypervisor or Virtual Machine Monitor (VMM). It could be a software, a firmware or a hardware depending on the type of hypervisor and its main task is to create virtual machine (VM). The operating system which carries on a hypervisor is denoted as a host machine and the virtual system that runs in the virtualization software is called a guest machine.

The task of the hypervisor is also to separate individual VM with a different access to hardware sources for each VM.

There are two main types of hypervisors based on the underlying in the system—Type-1 and Type-2 hypervisor [14].

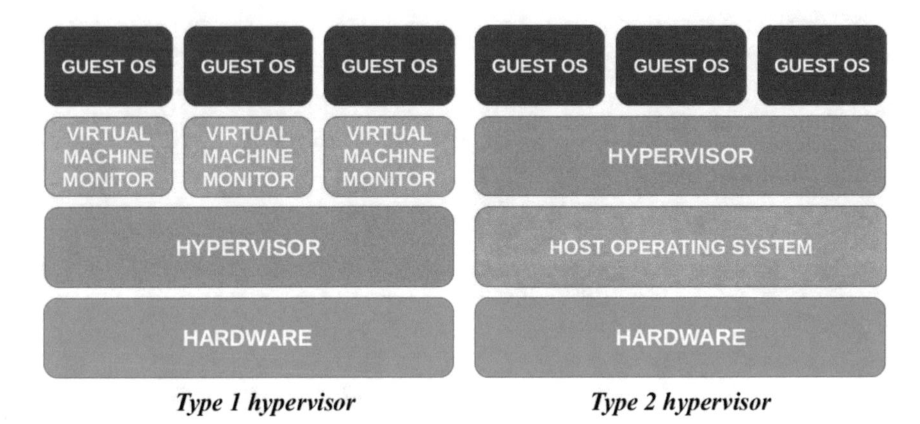

Fig. 2 Comparison of Type-1 and Type-2 hypervisor

2.1 *Type-1 Hypervisor*

This type of a hypervision is also described as *a native* or *bare-metal hypervision* because it runs directly as a code on the host's hardware. A native hypervisor does not need a host machine and we can imagine it as a thick software layer over the hardware which manage VM—see Fig. 2. As a result, this type of virtualization has much better performance than Type-2 hypervisor. The big advantage of this type of hypervisor is that individual VMs are independent and crash of the one of the VM does not affect others. It is also more secure because of this. Disadvantage could be found in the worse hardware support and more complicated installation.

Typical software which uses native hypervision are VMServer ESX/ESXi, Citrix XenServer, Microsoft Hyper-V, Oracle VM Server x86 etc.

2.2 *Type-2 Hypervisor*

The second type of the hypervisor is called *hosted hypervisor* and it runs as a program in the guest operation system—see Fig. 2. Guest virtual machines are then critically dependent on the hosted OS and each problem of this OS affects all guest machines. On the other hand, hosted hypervisor software is much accessible for ordinary, not very experienced, computer users and it has a big hardware support.

The virtualization software is very popular with typical members VMware workstation, VMware player [15], Oracle VM VirtualBox [16], Microsoft Virtual PC [17]. The most of these software has also free version for the non-commercial use.

3 Software for Virtualization

This article deals with local virtualization of the operating system using Type-2 hypervisor because it is simpler and in some cases, such as teaching purposes, with sufficient results.

There exists various software which can be used. We will focus on software which will be under the freedom license or free for the non-commercial use.

If we talk about the virtualization, we must mention an American company VMware which is specialized on a virtualization and a cloud computation software since 1998 [7]. VMware provides multiplatform virtualization software for desktops or servers which can run as computer programs or bare-metal hypervisors. The main and a well-known product is a VMware workstation which can be used for running multiple instances with virtual hosts on single host machine. The special version of the VMware Workstation used for the presentation and demonstration purposes is a VMware Player [15]. This software is used in this work, especially VMware Player 6 32-bit. The latest version in January 2016 is named VMware Workstation Player 12 but it runs only in the 64-bit operating systems.

The second virtualization software tested here is also a multiplatform program Oracle VM VirtualBox [16]. VirtualBox was developed in a German company Innotek GmbH which was bought in 2008 by Sun Microsystems and then by Oracle in 2009. The virtualization software used in this work was an Oracle VM VirtualBox, version 4.3. The newest version from December 2015 is an Oracle VM VirtualBox 5.0. It provides similar functionality as a VMware Workstation but it is in GNU v2 licence.

We can find also other free software for virtualization like XEN, KVM, Microst Virtual PC [17], Qemu etc. but these are not tested here.

4 Tests for Local Virtualization

The practical part of this contribution test performance of the hosted virtualization with the Oracle VM VirtualBox and VMware Player. Both programs are free for non-commercial use.

Parameters of virtual machines for both cases are the same for better comparability of the results.

The VM has following configuration:

- Single-core processor,
- 1 GB of RAM,
- 10 GB of virtual hard disk drive,
- The operating system is Windows XP SP3 32-bit.

The operating system Windows XP is chosen because of quick installation and less disk space and RAM requirements. It has still 10 % of market share in the desktop OS . Both VM uses clean Windows XP OS, without any additional software and improvements.

The host machine was notebook MSI with Intel Core i3 330M 2.13 GHz processor (CPU), 4 GB of RAM, ATi Mobility Radeon HD 5470 1 GB DDR3 graphical card, standard 2.5 500 GB SATA 5400 rpm hard-disk drive (HDD) and operating system Windows 7 Professional.

There was done performance and HDD tests on both VM and results are shown in following subchapters.

4.1 Performance Test

The goal of this test is to test the performance of the virtualized OS and how the virtualized system uses the hardware resources from the host OS. Three benchmark test were used in this part.

The first benchmark software is a PiTest which evolutes π number and computes first 10 000 numbers after the decimal separation of the π number. The performance of the CPU is then illustrated by the time when CPU reaches this computation in ms. The lowest value is the best.

The similar performance test is a Prime Benchmark which computes primes for one minute and the result is number n of computed primes. The bigger value is better in this case.

The last program, a Geekbench 2, combines a CPU performance and a RAM memory benchmark test. The result of this test is a score which reflects the performance and throughput of the memory. The bigger value is again better.

Results of tests are shown in Table 1 and Fig. 3.

Presented results are mean values after four measurements for each test. They have shown, that both virtualization software have similar results, but VMware player got a little bit better results in all three tests. It computes 10 000 evolutions of the π number in PiTest in approximately 44 s and VirtualBox in 48.6 s. VMware player computes more primes in one minute and it has also a better score in Geekbench2 test.

Table 1 Results of performance tests

Virtualization software	PiTest [ms]	Prime Benchmark [n]	Geekbench2 [-]
Oracle VM VirtualBox	48 599	7 393	2 336
VMware player	43 924	7 880	2 399

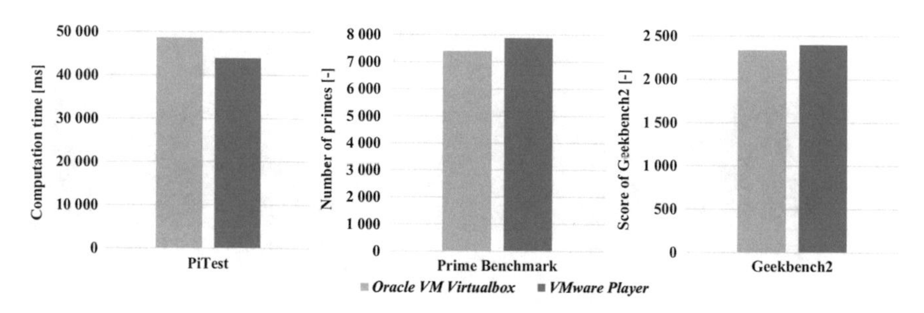

Fig. 3 Results of performance tests—PiTest, Prime Benchmark, Geekbench2

4.2 HDD Test

The performance of the VM also depends on the read and write speed of the HDD. The speed of reading and writing data from or to the HDD could be also one of parameters which affect the speed of the virtual machine. That is why is the second test is focused on this feature again for both virtualization software VMware Player and Oracle VirtualBox.

There were tested two read and write techniques—a sequential and a random data access. The sequential read and write speed test was evaluated on big files and the random access read and write speed was tested on small files.

Used program Parkdale measures read and write speeds for both, the sequential and the random access and results are shown in Tables 2, 3 and Fig. 4.

Presented results are again mean values from four measurements. These values are very tentative, because results varies very rapidly in this case. We can generally say, that VMware Player has again better results because we obtained bigger read and write speeds in three cases from four tests. The VirtualBox is better only in the sequential read. The sequential reading speed of big files is around 50–60 MB/s and write speed is in the range 35–45 MB/s. On the other hand, the speed of small files read and written randomly is predictably smaller, only 300–400 kB/s for reading and 600–1 000 kB/s for writing.

Table 2 Results of sequential read and write HDD test

Virtualization software	Sequential read [MB/s]	Sequential write [MB/s]
Oracle VM VirtualBox	61.6	36.6
VMware player	54.1	43.4

Table 3 Results of random read and write HDD test

Virtualization software	Random read [kB/s]	Random write [kB/s]
Oracle VM VirtualBox	291.8	623.8
VMware player	381.3	955.3

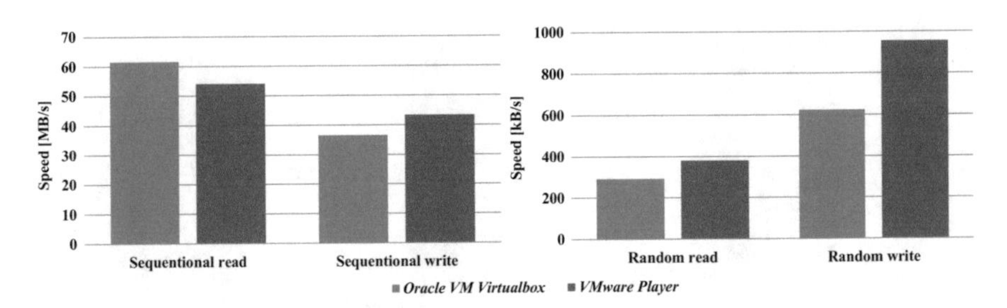

Fig. 4 Results of HDD tests—Sequential and random read/write

5 Conclusion

The paper deals with the local virtualization of the operating system. There was used Type-2 hypervision where the virtualization runs as a special program in the host operating system. The VMware Player and Oracle VM Virtualbox were chosen as a sample representative software because both are multiplatform programs ported to the main operating systems Microsoft Windows, Linux or iOS and both can be downloaded as a free software. The main task was to compare performance and HDD operations in Windows XP as a virtualized operating system in both chosen virtual machines. Results are comparable but VMware Player is a bit better in both of the tested criteria. We can generally say, that both software—VMware Player and Oracle VM VirtualBox have very good results and it is up to the user which of them he or she will use. The advantage of the VirtualBox can be found in more options accessible for free because VMware Player is limited version of the commercial product VMware Workstation.

References

1. Virtualization on Wikipedia. https://en.wikipedia.org/wiki/Virtualization
2. Barrett, D.M., Kipper, G., Liles, S.: Virtualization and forensics: a digital forensic investigator's guide to virtual environments. xvii, 254 s. Amsterdam, Elsevier (2010). ISBN: 978-1-59749-557-8
3. Berl, A.: Energy-efficient cloud computing. Comput. J. **53**(7) (2010). ISSN: 0010-4620
4. Virtualization Basics: How Virtualization Works. VMware. https://www.vmware.com/virtualization/how-it-works
5. History of Virtualization: Everything VM. http://www.everythingvm.com/content/history-virtualization
6. History of Virtualization: InfoBarrel. http://www.infobarrel.com/History_of_Virtualization
7. History of VMware: Welcome to vSphere-Land! http://vsphere-land.com/vinfo/history-of-vmware
8. x68 Virtualization: Wikipedia. https://en.wikipedia.org/wiki/X86_virtualization
9. Introduction to the Benefits of Local Desktop Virtualization: Enterprise Management Associates. http://download.microsoft.com/download/5/7/0/570BECAE-0F3E-4E4C-8DC1-8DD1711933ED/EMA_Microsoft-MED-V_B%20-%2020090331a.pdf

10. Hess, K., Newman, A.: Practical virtualization solutions: virtualization from the trenches. Prentice Hall/Pearson Education, xxiii, p. 304. Upper Saddle River, NJ (2010). ISBN: 0137142978
11. Blik, M.: Virtualization technologies as an e-learning support in the academic environment. In: The 18th International DAAAM Symposium Zadar Croatia (2007)
12. Vojtesek, J., Matusu, R., Bliznk, M.: The use of virtualization in the IT courses. In: Proceedings of the 17th International Conference on Process Control 09, pp. 498–505. trbsk Pleso, Slovakia (2009)
13. Understanding Hardware-Assisted Virtualization: Admin magazine. http://www.admin-magazine.com/Articles/Hardware-assisted-Virtualization
14. Hypervior: Wikipedia. https://en.wikipedia.org/wiki/Hypervisor
15. VMware Player: VMware. http://www.vmware.com/products/player
16. ORACLE: VirtualBox. https://www.virtualbox.org/
17. Widows Virtual PC: Wikipedia. https://en.wikipedia.org/wiki/Windows_Virtual_PC

A New Game-Theoretical Approach in Network Routing: Algorithms and Their Performance Analysis in OMNeT++

Serap Ergün and Tuncay Aydoğan

Abstract In this paper, we generally study performance of some network routing algorithms. These are Kruskal's, Prim's and Sollin's algorithms as tree algorithms that are used in minimum spanning trees problems. Further, we propose new algorithms that are modeled by game-theoretical approach. Mathematical models have been used to solve complex problems such as those in social sciences, economics, psychology, politics and telecommunication. In this context, game theory can be defined as a mathematical framework consisting of models and techniques analyzing the behavior of individuals concerned about their own benefits. Game theory deals with multi-person decision making, in which each decision maker tries to maximize own utility or minimize own cost and is applied to networking, in most cases to solve routing and resource allocation problems in a competitive environment. Modeling the network scenarios with the game-theoretical approach is one of the pioneering aims of the study. The algorithms for performance analysis are carried out OMNeT++, which is a network simulation program. Finally, the results are compared with each other and the literature.

Keywords Network routing · Algorithms · Game theory · OMNeT++ · Algorithm · Performance

1 Introduction

Networking is an exciting field that brings together what many students, practitioners, and researchers like best about the mathematical and computational sciences. It couples deep intellectual content with a remarkable range of applicability,

S. Ergün (✉)
Department of Computer Engineering, Süleyman Demirel University, Isparta, Turkey
e-mail: serapbakioglu@sdu.edu.tr

T. Aydoğan
Department of Software Engineering, Süleyman Demirel University, Isparta, Turkey

R. Silhavy et al. (eds.), *Software Engineering Perspectives and Application in Intelligent Systems*, Advances in Intelligent Systems and Computing 465,
DOI 10.1007/978-3-319-33622-0_23

covering literally thousands of applications in such wide-ranging fields as computer networking, scheduling and routing, telecommunications, and transportations.

Game theory is a field of applied mathematics for analyzing complex interactions among entities. It is basically a collection of analytic tools that enables distributed decision process. In the early to mid-1990, game theory was applied to networking problems including flow control, congestion control, routing and pricing of Internet services. More recently, there has been growing interest in adopting game-theoretic methods to model today's leading communications and networking issues, including power control and resource sharing in wireless and peer-to-peer networks [1].

One of the classical problems in Operational Research is the problem of finding a minimum cost spanning tree in a connected network. An important part of the correlation between cooperative games and Operational Research systems, from the basic structure of a graph, network or system that underlies various types of combinatorial optimization problems. If one assumes that at least two players are located at the system, then a cooperative game can be associated with this type of optimization problem [2].

In this study, we provide an integrative view of cooperative game theory, minimum cost spanning tree algorithms, and performance applications in OMNeT++. For more information about how using OMNeT++, the studies [3–10] can be examined.

The rest of paper is organized as follows: Game theory approach and usage in network areas are discussed in Sect. 2. In Sect. 3, the algorithms that are studied in minimum cost spanning trees are given and additionally, the performance analysis between new algorithms modeled by cooperative game theory and minimum cost spanning tree algorithms is introduced in Sect. 4. Therefore, this method is presented in more detail. Finally, Sect. 5 gives a short summary and an outlook on future work.

2 Basics of Game Theory

Game theory is a field of applied mathematics that describes and analyzes interactive decision situations. It provides analytical tools to predict the outcome of complex interactions among rational entities, where rationality demands strict adherence to a strategy based on perceived or measured results [11, 12]. The main areas of application of game theory are economics, political science, biology and sociology. From the early 1990s, engineering and computer science have been added to this list [11, 13, 14].

Game theory is concerned with decision-making in strategic settings, where you must factor the preferences and rational choices of other players into your decision to make the best choice for yourself. In many such settings, you're on your own: the choice you must make is yours and yours alone, because cooperation with other players is either impossible to implement or without any possible benefits.

However, in some situations it is both possible and fruitful to cooperate with other players. Where players can make binding agreements with each other, and there is some added value available by cooperating with others, then it can make sense for players to form coalitions that will work together to mutual advantage. Formal legal contracts are the most obvious mechanism available in the real world for implementing binding agreements. The field of cooperative game theory studies strategic decision-making in settings where binding agreements are possible and where agents can therefore act collectively [12, 15].

Game theory has two major subdivisions as non-cooperative and cooperative game theory. In non-cooperative games individual players make their own decisions about the strategy and the target of each player is to maximize his or her payoff regarding one's own preferences. This is not the case in cooperative games. In cooperative games the players can form coalitions which can make binding agreements about the distribution of payoffs and also about the strategies to be chosen [16]. We limit our discussion to cooperative models.

Cooperative game theory has been enriched in the last recent years with several models which provide decision-making support in collaborative situations. Additionally, it provides a wealth of tools that can be applied to the design and operation of communications system.

For over a decade, game theory has been used as a tool to study different aspects of computer and telecommunication networks, primarily as applied to problems in traditional wired networks. In the past three to four years there has been renewed interest in developing networking games, this time to analyze the performance networks. Since the game theoretic models developed for networks focus on distributed systems, results and conclusions generalize well as the number of players (nodes) is increased [11].

2.1 Preliminaries

An (undirected) graph is a pair $\langle V, E\rangle$, where V is a set of vertices or nodes and E is a set of edges of the form $\{i,j\}$ with $i,j \in V, i \neq j$. The complete graph on a set V of vertices in the graph $\langle E, E_V\rangle$, where $E_V = \{\{i,j\} \mid i,j \in V \text{ and } i \neq j\}$. A path between i and j in a graph in a graph $\langle V, E\rangle$ is a sequence of nodes $i = i_0, i_1, \ldots, i_k = j, k \geq 1$ such that all the edges $\{i_s, i_{s+1}\} \in E$, for $s \in \{0, \ldots, k-1\}$, are distinct. A cycle in $\langle V, E\rangle$ is a path from i to i for some $i \in V$. Two nodes $i, j \in V$ are connected in $\langle V, E\rangle$. A connected component of V in a graph $\langle V, E\rangle$ is a maximal subset of V with the property that any two nodes in this subset are connected in $\langle V, E\rangle$.

A minimum cost spanning tree (mcst) situation is a situation, where $N = \{1, \ldots, n\}$ is a set of agents who are willing to be connected as cheaply as possible to a source (i.e., a supplier of a service) denoted by 0 or * [17].

Formally, an mcst problem is triple $\Gamma = (N, *, t)$, where $N = \{1, \ldots, n\}$ the player set is, * is the source and $t\colon E_{N^*} \to \mathbb{R}_+$ is the nonnegative cost function. E_S is defined as the set of all edges between pairs of elements of $S \subset N^*$ so that (S, E_S) is the complete graph on S

$$E_S = \{\{i,j\} \mid i,j \in S, i \neq j\}.$$

Because of connection costs are non-negative; it is obvious that a minimal cost graph that connects all players to the source is indeed a tree, which explains the name of the problem.

A cooperative game in characteristic function form is an ordered pair $\langle N, c\rangle$ consisting of the player set N and the characteristic function $c\colon 2^N \to \mathbb{R}$ with $c(\varnothing) = 0$. The real number $c(S)$ can be interpreted as the maximal worth or cost savings that the numbers of S can obtain when they cooperate. Often we identify the game $\langle N, c\rangle$ with its characteristic function v [18].

3 Minimum Spanning Tree

In this section we motivate several algorithms for solving minimum spanning tree problem. The minimum spanning tree problem arises in a number of applications, such as designing physical systems, optimal message passing, all-pairs mini-max path problem, reducing data storage, cluster analysis, routing scenarios, etc. The minimum spanning tree problem is important not only because it serves as a valuable prototype model in combinatorial optimization that has stimulated many lines of inquiry [19].

A spanning tree T of G is a connected acyclic sub-graph that spans all the nodes and every spanning tree of G has $n-1$ arcs. Given an undirected graph $G = (N, A)$ with $n = |N|$ nodes and $m = |A|$ arcs and with a *length* or *cost* c_{ij} with each arc $(i,j) \in A$, we wish to find a spanning tree, called a minimum spanning tree, that has the smallest total cost (or length) of its constituent arcs, measured as the sum of costs of arcs in the spanning tree [19, 20].

The spanning tree algorithm examines the edges in any arbitrary sequence and decides whether each edge will be included in the spanning tree. When an edge is examined, the algorithm simply checks if the edge under consideration forms a cycle with the other edges already assigned to the tree. If so, then the edge under consideration is excluded from the tree; otherwise, it is assigned to the tree [20].

Let us clarify the situation with an example where 3 players 1, 2, and 3 want to be connected directly or indirectly with a source 0 and where the cost situation is represented in Fig. 1.

Take the ordering $\sigma = \{1, 2, 3\}$. In step 1, player 1 constructs and pays the cheapest of the edges (1, 0), (1, 2), (1, 3) which is edge (1, 3) with cost 3. Then player 2 constructs and pays edge (1, 0) with cost 5 and this is the cheapest of his

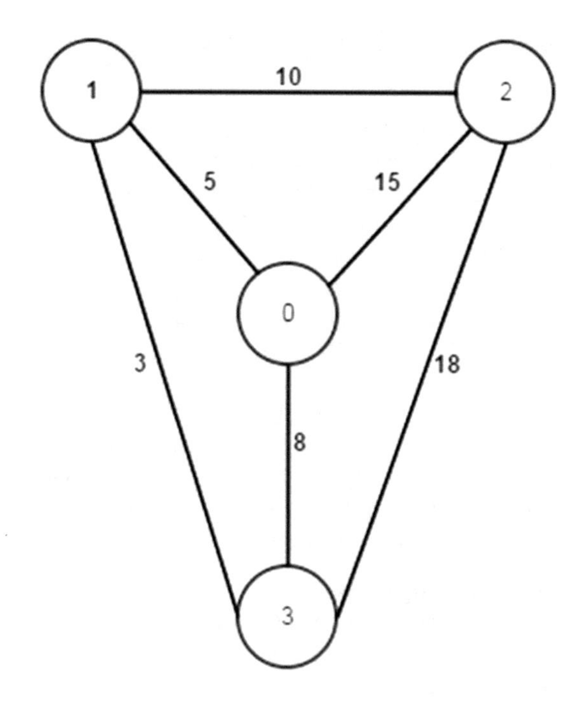

Fig. 1 An example of minimum cost spanning tree

Table 1 Construct and charge results for the mcst situation in Fig. 1

	Constructed edges by			Costs for		
σ	1	2	3	1	2	3
(1, 2, 3)	(1, 3)	(1, 0)	(1, 2)	3	5	10
(1, 3, 2)	(1, 3)	(1, 2)	(1, 0)	3	10	5
(2, 1, 3)	(1, 0)	(1, 3)	(1, 2)	5	3	10
(2, 3, 1)	(1, 2)	(1, 3)	(1, 0)	10	3	5
(3, 1, 2)	(1, 3)	(1, 2)	(1, 0)	3	10	5
(3, 2, 1)	(1, 2)	(1, 3)	(1, 0)	5	3	5

allowed edges (2, 1), (1, 3), (3, 0). Finally player 3 chooses edge (1, 2) which is the cheapest of his allowed edges (3, 1), (1, 0), (0, 2). Note that (2, 0) is not allowed for player 3 because it generates a cycle. The result is the mcst with edges (1, 3), (1, 0), (1, 2) and the cost share vector (3, 5, 10). In Table 1, the construct and charge results for all orderings of the players can be seen.

Now, Let us construct the game where 3 players 1, 2, and 3, which is represented in Fig. 1. Let $\langle N, c\rangle$ be a cooperative mcst game with $N = \{1, 2, 3\}$ and $c(\{1\}) = 5, c(\{2\}) = c(\{1, 2\}) = 15$, $c(\{3\}) = c(\{1, 3\}) = 8$, $c(\{2, 3\}) = 23$, $C(\{1, 2, 3\}) = 28$ and $c(\varnothing) = 0$.

3.1 Minimum Spanning Tree Algorithms

The resulting algorithms seem simple, although implementing them efficiently requires considerable care and ingenuity. The tree algorithms we consider in this study, Kruskal's algorithm [21], Prim's algorithm [22], and Sollin's algorithm [23], all share one characteristic: They are greedy algorithms in the sense that each step they add an arc of minimum cost from a candidate list, as long as the added arc does not form a cycle with the arcs already chosen. All three algorithms maintain a forest containing arc already chosen and then they add one or more arcs to enlarge the size of forest. For Kruskal's algorithm, the candidate list is the entire network; for Prim's algorithm, the forest is a single tree plus a set of isolated nodes and the candidate list contains all the arcs between the single tree and the nodes nor in the tree; Sollin's algorithm is a hybrid approach that maintains several components in the forest, as in Kruskal's algorithm, but then adds several arcs at each iteration, choosing (like Prim's algorithm) the minimum cost arc connecting each component of the forest to the nodes not in that component [19, 20].

Kruskal's Algorithm

The path optimality conditions immediately suggest the following straightforward algorithm for solving the minimum spanning tree problem. Let us start with any arbitrary spanning tree T satisfies this condition, it is an optimal tree; otherwise $c_{ij} > c_{kl}$ for some non-tree arc (k, l) and some tree arc (i, j) contained in the unique path in T connecting nodes k and l. Thus situated, adding arc (k, l) to T in place of arc (i, j) gives us a spanning tree with a lower cost. Repeating this step will give us

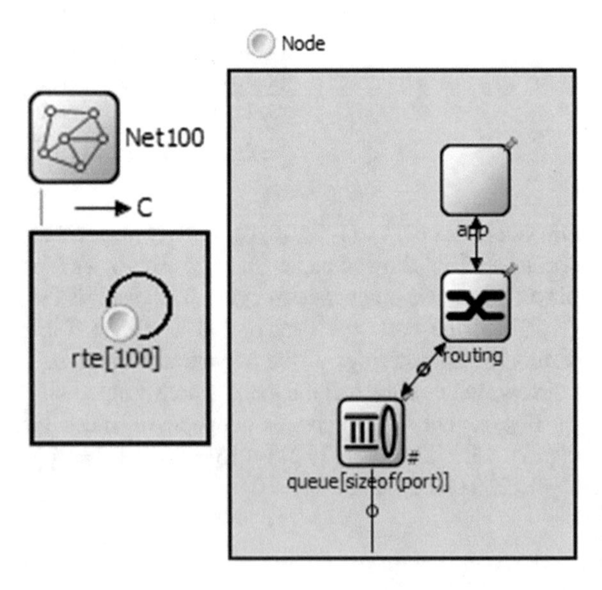

Fig. 2 The construction of nodes

a minimum spanning tree within a finite number of iterations. For more detail of algorithm, we refer to [19–21, 24] and we leave the rest to the reader.

Prim's Algorithm
Another simple algorithm for the minimum spanning tree problem, knows as Prim's algorithm. This algorithm builds a spanning tree from scratch by fanning out from a single node and adding arcs one at a time. It maintains a tree spanning on a subset S of nodes and adds a nearest neighbor to S. The algorithm does so by

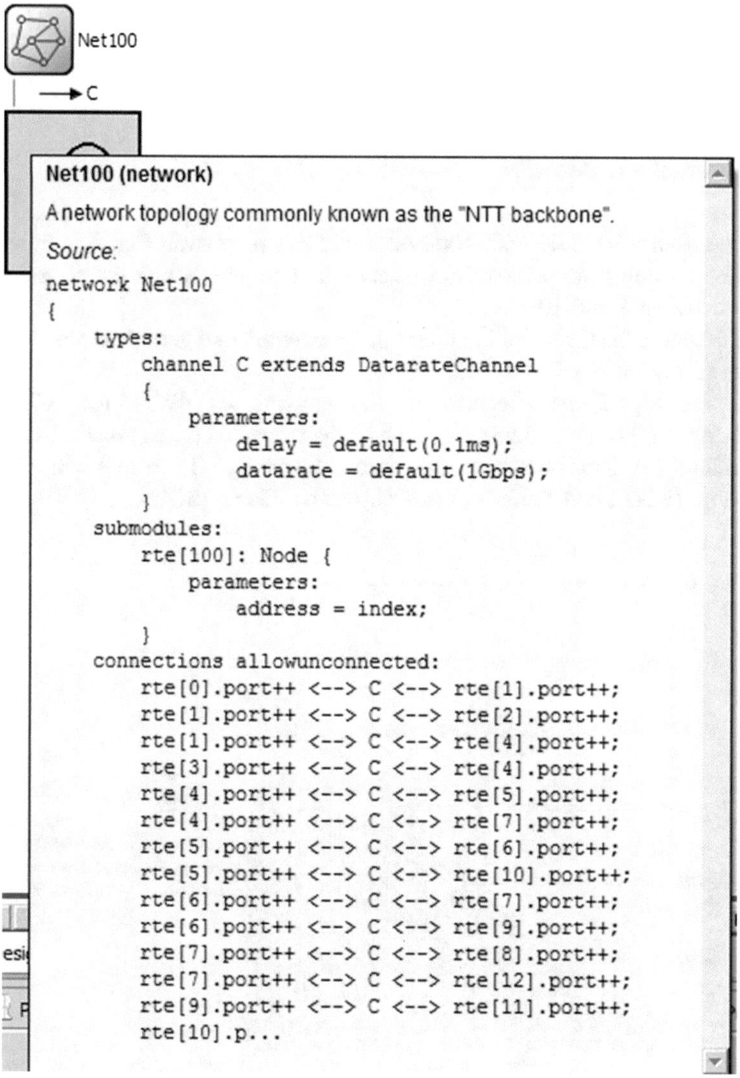

Net100

→C

Net100 (network)

A network topology commonly known as the "NTT backbone".

Source:

```
network Net100
{
    types:
        channel C extends DatarateChannel
        {
            parameters:
                delay = default(0.1ms);
                datarate = default(1Gbps);
        }
    submodules:
        rte[100]: Node {
            parameters:
                address = index;
        }
    connections allowunconnected:
        rte[0].port++ <--> C <--> rte[1].port++;
        rte[1].port++ <--> C <--> rte[2].port++;
        rte[1].port++ <--> C <--> rte[4].port++;
        rte[3].port++ <--> C <--> rte[4].port++;
        rte[4].port++ <--> C <--> rte[5].port++;
        rte[4].port++ <--> C <--> rte[7].port++;
        rte[5].port++ <--> C <--> rte[6].port++;
        rte[5].port++ <--> C <--> rte[10].port++;
        rte[6].port++ <--> C <--> rte[7].port++;
        rte[6].port++ <--> C <--> rte[9].port++;
        rte[7].port++ <--> C <--> rte[8].port++;
        rte[7].port++ <--> C <--> rte[12].port++;
        rte[9].port++ <--> C <--> rte[11].port++;
        rte[10].p...
```

Fig. 3 The simple codes of construction of network

identifying an arc (i,j) of minimum cost in the cut $[S,\overline{S}]$. It adds arc (i,j) to the tree, node j to S, and repeats this basic step until $S = N$. For more detail of algorithm, we refer to [19–21] and we leave the rest to the reader.

Sollin's Algorithm

Sollin's algorithm is an algorithm for the minimum spanning tree problem. This algorithm can be viewed as a hybrid version of Kruskal's and Prim's algorithm. As in Kruskal's algorithm, Sollin's algorithm maintains a collection of trees spanning the nodes $N_1, N_2, N_3, \ldots,$ and adds arcs to this collection. However, at every iteration, it adds minimum cost arcs emanating from these trees, an idea borrowed from Prim's algorithm. For more detail of algorithm, we refer to [19–21] and we leave the rest to the reader.

4 Simulation Results

We have used a scenario with 100 nodes that can be seen in Fig. 2.

Figure 3 shows the parameters of network. The node's link costs created randomly between 1 and 10.

We compare the timing of the nodes in the network and get the results which can be seen in Fig. 4.

The six algorithms executed in the network as 49.348 ms, 45.032 ms, 37.665 ms, 47.348 ms, 35.658 ms, 33.670 ms respectively, Kruskal's (K), Prim's (P), Sollin's (S), Kruskal's with Cooperative Game (KCG), Prim's with Cooperative Game (PCG), and Sollin's with Cooperative Game (SCG).

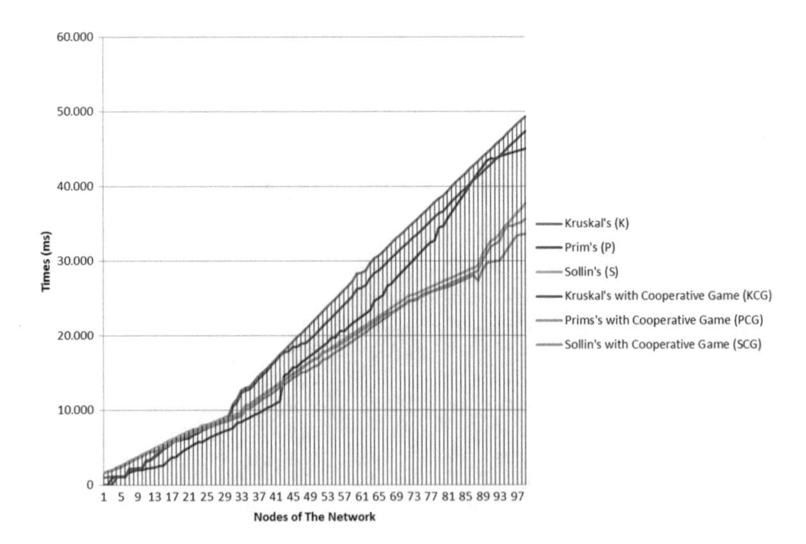

Fig. 4 The time performances of the algorithms' runtime

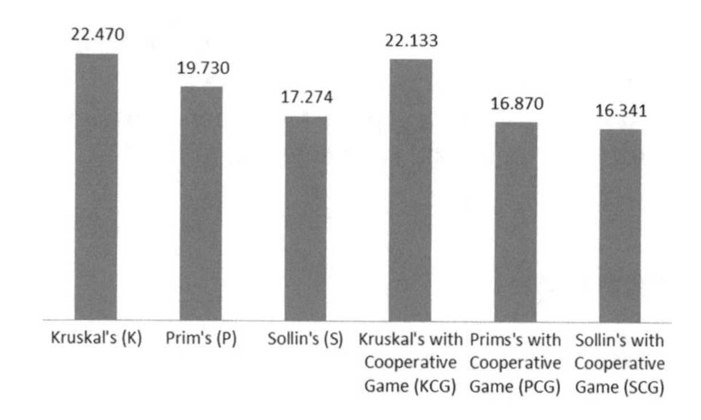

Fig. 5 Time performances based on construction of minimum spanning trees

Table 2 The comparisons of six algorithms

Time performances of algorithms' runtime*	$SG < PG < S < P < KG < K$
Time performances based on construction of minimum spanning trees	$SG < PG < S < P < KG < K$
* : $A < B$ means: A is faster than B	

Figure 5 gives the performances of construction of minimum spanning trees. In Table 2, a brief comparison of six algorithms in network is provided.

5 Conclusion and Outlook

In this paper, we bring together main topics of cooperative game theory, minimum cost spanning tree algorithms and OMNeT++ network simulator. Game theory proves to be a powerful tool for modeling various aspects of networking, such as routing algorithms, optimization problems or energy saving. We have showed how our algorithms can be used to simulate the network. Using the approach of this study, optimization algorithms for networks and graphs can be modeled by cooperative game theory. We believe the efficiency of cooperative game theory will create a new breath to the network researchers.

The selection procedure is based on cooperative game theory. We proposed new algorithms based on cooperative game approach in order to solve the minimum spanning tree problems. We compared the performance of algorithms in OMNeT++; it provides us a good workspace and result information. OMNeT++ also supports parallel distribution simulation. Maybe, the next study may have in this regard.

Future work involves extending our algorithms to not only route network, find optimal path, but also using in the other network types as optimization algorithms such as scheduling.

Before closing, we note that the research areas of networking and game theory are both extensively studied fields with many problems and solutions. Yet, the cross-over between cooperative game theory, algorithms and performance analysis was surprisingly very small. The importance of this study is given by that is a pioneering and ongoing project and offer to the Operational Research and Network community in Europe and the world.

References

1. Leino, J.: Applications of game theory in ad hoc networks. UNe **100**, 2–5 (2003)
2. Borm, P., Hamers, H., Hendrickx, R.: Operations research games: a survey. Top **9**(2), 139–199 (2001)
3. Diaz-Estebaranz, J., Portilla-Figueras, J. A., Salcedosanz, S., Faro-Rivas, M., Esteve-Asensio, G.: Using an OMNET++ network based simulator as test-bed for network design algorithms. In: SMO'08: Proceedings of the 8th Conference on Simulation, Modeling and Optimization, Sept 2008
4. INET Framework Documentation. http://www.omnetpp.org/doc/INET
5. Mayer, C.P., Gamer, T.: Integrating real world applications into OMNeT++. Institute of Telematics, University of Karlsruhe, Karlsruhe, Germany, Technical report TM-2008-2 (2008)
6. OMNeT++ Discrete Event Simulator. https://omnetpp.org/ (2015)
7. Varga, A.: The OMNeT++ discrete event simulation system. In: Proceedings of the European Simulation Multiconference (ESM'2001), vol. 9, no. S 185, p. 65, June 2001
8. Xian, X., Shi, W., Huang, H.: Comparison of OMNET++ and other simulator for WSN simulation. In: 3rd IEEE Conference on Industrial Electronics and Applications, ICIEA 2008, pp. 1439–1443. IEEE, June 2008
9. Mallanda, C., Suri, A., Kunchakarra, V., Iyengar, S.S., Kannan, R., Durresi, A., Sastry, S.: Simulating wireless sensor networks with OMNeT++. Submitted to IEEE Computer (2005)
10. Varga, A., Hornig, R.: An overview of the OMNeT++ simulation environment. In: Proceedings of the 1st International Conference on Simulation Tools and Techniques for Communications, Networks And Systems and Workshops, p. 60. ICST (Institute for Computer Sciences, Social-Informatics and Telecommunications Engineering) Mar 2008
11. Srivastava, V., Neel, J.O., MacKenzie, A.B., Menon, R., DaSilva, L.A., Hicks, J.E., Gilles, R. P.: Using game theory to analyze wireless ad hoc networks. IEEE Commun. Surv. Tut. **7**(1–4), 46–56 (2005)
12. Peleg, B., Sudhölter, P.: Introduction to the Theory of Cooperative Games, vol. 34. Springer Science & Business Media (2007)
13. Osborne, M.J., Rubinstein, A.: A Course in Game Theory. MIT press (1994)
14. Vazirani, V.V.: In: Roughgarden, T., Tardos, E. (eds.) Algorithmic Game Theory, vol. 1. Cambridge University Press, Campbridge (2007)
15. Chalkiadakis, G., Elkind, E., Wooldridge, M.: Cooperative game theory: basic concepts and computational challenges. IEEE Intell. Syst. **3**, 86–90 (2012)
16. Savunen, T.: Application of the cooperative game theory to global strategic alliances. Doctoral dissertation, Helsinki University of Technology (2009)
17. Moretti, S., Gök, S.Z.A., Branzei, R., Tijs, S.: Connection situations under uncertainty and cost monotonic solutions. Comput. Oper. Res. **38**(11), 1638–1645 (2011)
18. Branzei, R., Dimitrov, D., Tijs, S.: Models in Cooperative Game Theory, vol. 556. Springer Science & Business Media (2008)
19. Ahuja, R. K., Magnanti, T. L., Orlin, J.B.: Network flows (No. MIT-WP-2059-88). Alfred P Sloan School of Management, Cambridge, MA (1988)
20. Evans, J.: Optimization Algorithms for Networks and Graphs. CRC Press (1992)

21. Kruskal, J.B.: On the shortest spanning subtree of a graph and the traveling salesman problem. Proc. Am. Math. Soc. **7**(1), 48–50 (1956)
22. Prim, R.C.: Shortest connection networks and some generalizations. Bell Syst. Tech. J. **36**(6), 1389–1401 (1957)
23. Borůvka, O.: O jistém problému minimálním (1926)
24. Chartrand, G., Oellermann, O. R.: Applied and Algorithmic Graph Theory (1993)

Using Analytical Programming for Software Effort Estimation

Tomas Urbanek, Zdenka Prokopova, Radek Silhavy and Ales Kuncar

Abstract This paper evaluates the usage of analytical programming for software effort estimation. Analytical programming and differential evolution generate regression models. The new model was generated by analytical programming and it was tested and compared with Karner's model to assess insight to its properties. Mean Magnitude of Relative Error and k-fold cross validation were used to assess the reliability to this experiment. The experimental results shows that the new model generated by analytical programming outperforms the Karner's equation about 12 % MMRE. Moreover, this work shows that analytical programming method is viable method for calibrating Use Case Points method. All results were evaluated by standard approach: visual inspection and statistical significance testing.

Keywords Analytical programming · Differential evolution · Effort estimation · Use case points

1 Introduction

Effort estimation is defined as the activity of predicting the amount of effort required to complete a development of software project [1].

There are two major groups of methods which could be used for effort estimation. The first group is algorithmic methods and the second group is non-algorithmic methods. The algorithmic methods use mathematical formula for prediction. It is very common that this group is also depending on the historical data. The most famous example of algorithmic methods are COCOMO [2], FP [3] and last but not least UCP [4]. However, there are a many others algorithmic methods. The main feature of the non-algorithmic methods is expert judgment or using analogy for effort estimation. The most common example can be method called Delphi [5]. It

T. Urbanek (✉) · Z. Prokopova · R. Silhavy · A. Kuncar
Faculty of Applied Informatics, Tomas Bata University in Zlin, Nad Stranemi, 4511 Zlín, Czech Republic
e-mail: turbanek@fai.utb.cz

R. Silhavy et al. (eds.), *Software Engineering Perspectives and Application in Intelligent Systems*, Advances in Intelligent Systems and Computing 465,
DOI 10.1007/978-3-319-33622-0_24

is essential that the calculation of effort estimation should be completed in early stage of software development cycle. The best case is if these calculations are known during the requirement analysis [4]. The Use Case Points method used in this article could predict the effort required to complete the development in early stage. The accurate and reliable effort estimates are the crucial factor for the proper development cycle. These estimates are used for effective planning, monitoring and controlling the process of the software development. Due to the more accurate effort estimates the software engineers could be more accurate in their decisions. The prediction of effort estimations in software engineering is complex and complicated process. The main reason is that there are a lot of factors which influencing the final prediction. For instance, these factors are the size of development team, programming language use, the complexity of requirements and other factors. One of the most substantial factor is human factor. For this reason, the artificial intelligence could compensate the prediction error calculated by software engineer. In this point of view, the usage of artificial intelligence can be a promising way for effort estimation in software engineering. Nowadays, using of artificial intelligence is very common in this research field. In this article is investigated the method of analytical programming for effort estimation in software engineering. The analytical programming can be seen as a regression function generator.

In recent time we published study that examined a selection of fitness function for analytical programming and it was found that the best fitness function is MSE and very common MMRE [6]. In published study we compared a lot of models statistically; however there was no presented a specific model for examination. This article is evolution of the published study and in this article we investigated the one specific model from the previously generated population. In our best knowledge no previous study has investigated the model (equation) produced by analytical programming when Use Case Point method and k-fold cross validation was used. Therefore, this study makes a major contribution to research of Use Case Points method, while analytical programming method is used.

1.1 The Use Case Points Method

This effort estimation method was presented in 1993 by Gustav Karner [4]. It is based on a similar principle to the function point method. Project managers have to estimate the project parameters to four tables. These tables are as follows:

- Unadjusted Use Case Weight (UUCW)
- Unadjusted Actor Weight (UAW)
- Technical Complexity Factor (TCF)
- Environmental Complexity Factor (ECF)

Due to the aims of this paper, the detailed description of well known Use Case Point method basic principles is insignificant and hence omitted. Please refer to [4, 6] for more detailed description of the Use Case Point method.

1.2 Differential Evolution

Differential evolution is an optimization algorithm introduced by Storn and Price in 1995, [7]. This optimization method is an evolutionary algorithm based on population, mutation and recombination. Differential evolution is easy to implement and has only four parameters which need to be set. The parameters are: Generations, NP, F and Cr. The Generations Parameter determines the number of generations; the NP Parameter is the population size; the F Parameter is the weighting factor; and the Cr Parameter is the crossover probability [8]. In this research, the differential evolution is used as an analytical programming engine.

1.3 Analytical Programming

Analytical programming (AP), is a symbolic regression method. The core of analytical programming is a set of functions and operands. These mathematical objects are used for the synthesis of a new function. Every function in the analytical programming set core has its own varying number of parameters. The functions are sorted according to these parameters into General Function Sets (GFS). For example, GFS_{1par} contains functions that have only 1 parameter e.g. *sin*(), *cos*(), or other functions. AP must be used with any evolutionary algorithm that consists of a population of individuals for its run [9, 10]. In this paper, Differential evolution (DE) is used as an analytical programming evolutionary algorithm.

2 Related Work

Despite of a lot of effort of scientists and software engineers, there is still no optimal and effective method for every software project. Some work has been done to enhance the effort estimation based on the Use Case Points method. These enhancements cover the review and calibrating the productivity factor such as the work of Subriadi et al. [11]. Another enhancement could be the construction investigation and simplification of the Use Case Points method presented by Ochodek et al. [12]. The recent work of Silhavy et al. [13] suggest a new approach "automatic complexity estimation based on requirements", which is partly based on Use Case Points method. Or using fuzzy inference system approach to improve accuracy of the Use Case Points method [14]. Very promising way is a research of Kocuganeli et al. [15], this paper shows, that ensemble of effort estimation methods could provide better results then a single estimator. The work of Kaushik et al. [16] and Attarzadeh et al. [17] uses neural networks and COCOMO [2] method for prediction. COCOMO method is widely used for testing and calibrating in cooperation with artificial intelligence or with fuzzy logic [18]. Because it is very difficulty to obtain a reliable dataset in case of Use Case Points method, this paper shows results on a dataset from Poznan University of Technology [12] and from this paper [11].

3 Problem Statement

This section presents the design of the research questions we carried out to get an insight in the use of analytical programming for effort estimation. The research questions of our study can be outlined as follows:

- RQ-1: Analyzing the differences between Karner's model and new model.
- RQ-2: There is an evidence that new model is more accurate than Karner's model

The first research question (RQ-1) aims to get an insight on the estimation accuracy of analytical programming and understand the actual effectiveness of this technique with respect to the estimates by standard Use Case Points method. For this reason, we first set the productivity factor to the standard value of 20 and then, we try to outperformed this estimates by the method of analytical programming. To address research question (RQ-2) we experimented with built model as reported and discussed in experiment planning section. To asses the performance of fitness function we used paired t-test and exploratory statistical analysis.

4 Experiment Planning

The proposed experiment can be seen in the Fig. 1. In this experiment we used 10-folds cross-validation. In one loop was generated one equation which was then verified on the rest of the dataset. The process begins with a cycle that loops through the number of folds. In the data preparation loop, the 10-fold cross-validation was used to split the dataset into two distinct sets. The dataset is depicted in Table 1. Then, there is a second loop. In this loop, the differential evolution process starts to generate an initial population. Analytical programming then uses this initial population to synthesize a new function. After that, the new function is evaluated by the mean magnitude of relative error. If the termination condition is met, one can assume that one has an optimal predictive model, and this model is then evaluated by the calculation of the mean magnitude of relative error (MMRE) on the testing set. Then, the results are saved to file for further analysis.

4.1 Dataset

The data for this study was collected using document reviews. The Use Case Points method dataset was obtained from Poznan University of Technology [12] and from Subriadi's paper [11].

Table 1 displays the Use Case Points method data from 24 projects. Only the Use Case Points method data with transitions were utilized in this paper in the case of the Poznan University of Technology dataset. There are 5 values for each software

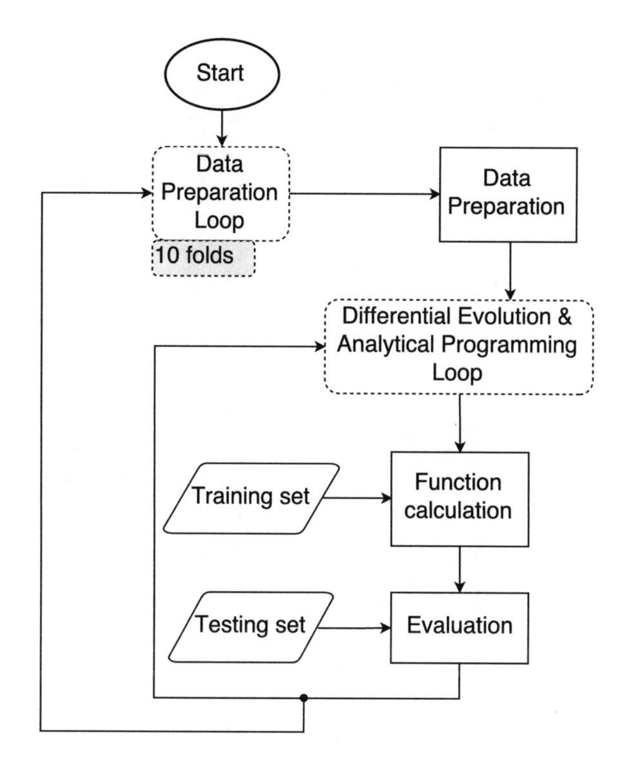

Fig. 1 Diagram of proposed experiment

project: UUCW, UAW, TCF, ECF and actual effort. Software projects 1–14 are from Poznan University of Technology. The rest are from Subriadi's paper. As can be seen Subriadi's data are quite consistent in actual effort. The possible reason is that these projects are related to one context, respectively linked to the web development software projects. The distribution of actual effort of this dataset can be seen on Fig. 2. From the figure can be seen, that the most of the software projects was completed from 1250 to 3250 man/hour.

Table 2 shows the analytical programming set-up. The number of leafs (functions built by analytical programming can be seen as trees) was set to 30, which can be recognized as a relatively high value. However, one needs to find the model that will be more accurate than the Karner's model. There is no need to generate short and easily memorable model, but rather, model that will be more accurate.

Table 3 shows the set-up of differential evolution. The best set-up of differential evolution is the subject of further research.

Table 1 Data used for effort estimation

ID	UUCW	UAW	TCF	ECF	Actual effort [man/hours]
1	195	12	0.780	0.780	3037
2	80	10	0.750	0.810	1917
3	75	6	0.900	1.050	1173
4	130	9	0.850	0.890	742
5	85	12	0.820	0.790	614
6	50	9	0.850	0.880	492
7	50	6	0.780	0.510	277
8	305	14	0.940	1.020	3593
9	85	12	1.030	0.800	1681
10	130	12	0.710	0.730	1344
11	80	9	1.050	0.950	1220
12	70	12	0.780	0.790	720
13	30	4	0.960	0.960	514
14	100	15	0.900	0.910	397
15	355	15	1.125	0.770	3684
16	145	18	1.080	0.770	1980
17	325	12	1.095	0.935	3950
18	90	6	1.085	1.085	1925
19	125	9	1.025	0.980	2175
20	120	9	1.115	0.995	2226
21	200	12	1.000	0.920	2640
22	175	9	0.950	0.920	2568
23	245	12	0.890	1.190	3042
24	140	6	0.965	0.755	1696

4.2 Fitness Function

The new model built by the analytical programming method contains the following parameters: UUCW, UAW, TCF and ECF. There is no force applied to the analytical programming that the models built by the analytical programming method have to contain all of these parameters.

$$MMRE = \frac{1}{n}\sum_{i=1}^{n}\frac{|y_i - \hat{y}_i|}{y_i}, \tag{1}$$

where n is equal to the number of projects in training set, $\hat{y}_i$ is prediction, y_i is actual effort

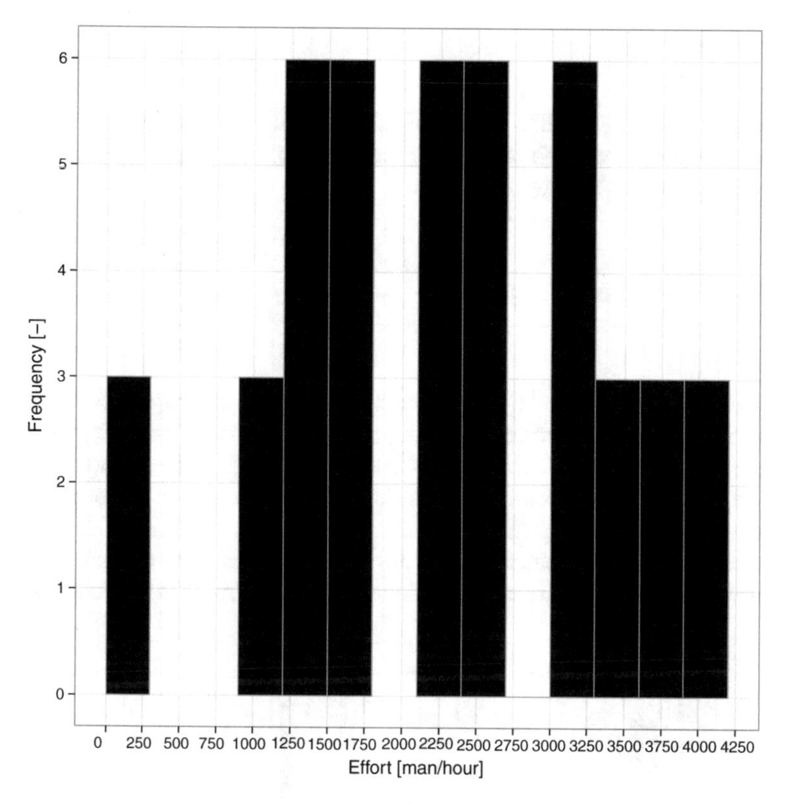

Fig. 2 The distribution of actual efforts in dataset

Table 2 Set-up of analytical programming

Parameter	Value
Number of leafs	30
GFS-functions	Plus, subtract, divide, multiply, tan, sin, cos
GFS-constants	UUCW, UAW, TCF, ECF, K

Table 3 Set-up of differential evolution

Parameter	Value
NP	40
Generations	60
F	0.7
Cr	0.4

The Eq. 1 is used for optimization task. When the Mean Magnitude of Relative Error result is closer to zero, then the accuracy of the proposed model is higher.

5 Results

In this section, we present the result of our study. Exploratory statistical analysis was utilized to describe research results. All the calculations was performed by 10-fold cross validation on 24 software projects.

$$UCP = UAW * csc\left(\frac{sin\left(e^{-sin\left(cos\left(e^{TCF}\right)\right)}\right)}{UUCW}\right) \tag{2}$$

The Eq. 2 was calculated by the process described in this paper. As can be seen the complexity of calculated equation is higher then the Karner's equation. Moreover trigonometric functions was preferred. On the other hand, Karner's model assumes linear dependencies. We could also see that the ECF parameter is not used in the new equation.

Figure 3 shows the calculation least absolute deviation error of each project for both models. In general Karners's equation had greater error than new equation (this

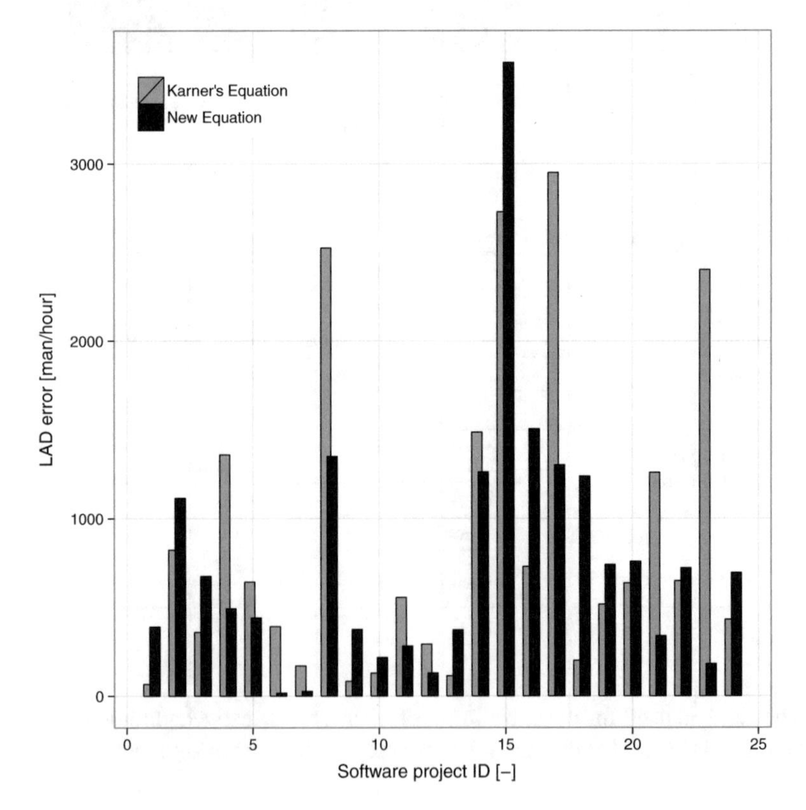

Fig. 3 Least absolute deviation error of each project for both models

Table 4 Accuracy measure

	MMRE [%]	LAD [man/hour]
Karner's Equation	62.34	21494.0
New equation	50.02	18199.4

can be seen in Table 4). As can be seen the project 15 was highly overestimated by both models. This project have a very high value of UUCW.

Table 4 provides the calculation of mean magnitude of relative error and least absolute deviation. As can be seen the new model had lower relative error rate by approximately 12 % and the mean error on one software project is 758 man/hour. Karner's equation had mean error on one project 895 man/hour.

Table 4 provides the graphical comparison of new equation with Karner's equation and actual effort. As can be seen the median of new equation had lower value than actual effort. On the other hand, the Karner's equation had median value higher than actual effort. The shape of the box plot of new equation is more similar to actual effort. Both new equation and Karner's equation had greater maximum value than actual effort.

5.1 *Test*

Paired t-test was conducted to provide the evidence that the new equation had statistically lower error than Karner's equation.

Paired t-test
data: karLADvalues newLADvalues
$t = -0.8776$, $df = 23$, p-value $= 0.3893$
alternative hypothesis: true difference in means is not equal to 0
95 % confidence interval:
−460.866 186.318
sample estimates:
mean of the differences −137.274

As can be seen the p-value of the t-test has been 0.3893. The null hypothesis that the Karner's equation and new equation had the same true difference in mean is accepted. Hence, there is no statistical evidence that the new equation had statistically lower error than Karner's equation.

6 Discussion

The study started out with a goals of answering the question of whether analytical programming technique outperformed the standard UCP equation. This question is

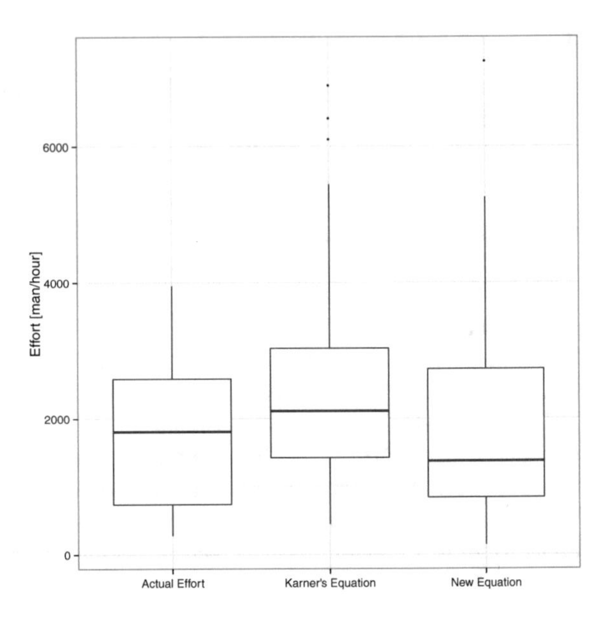

Fig. 4 Graphical comparison of new equation with Karner's equation and actual effort

answered in the result section. If the productivity factor and the whole UCP method is set to default values, there is a possibility, that model built by analytical programming outperform the standard UCP equation.

There is also a another question (RQ-2), which must be answered. For answering this question we need to study Table 4 and Fig. 4 and statistical t-test from result section. From Table 4 could be seen that, the new equation generated by analytical programming have had slightly smaller LAD than Karner's equation. Also we could seen on the Fig. 4 that the overall shape of the boxplot of new equation is more similar to actual effort than the Karner's equation. The statistical test proves that the true mean of the new equation is the same as Karner's equation on the level of significance 5 %. Nevertheless the MMRE of the new equation was about 12 % better and on the long run the new equation provides lower error than Karner's equation.

Evidences provided by these statements could be probably false, when the productivity factor will be set to the optimal value. However the optimal value for productivity factor is not known and therefor in this paper we used standard productivity factor. When the productivity factor is set to standard value the new model outperform the Karner's equation. Both Karner's equation and new equation generate some outliers as can be seen on Fig. 4. This could be caused by inaccurate estimation of UCP parameters at the estimation process. We could also see that the ECF parameter is not used in the new Eq. 2. The reason for this is probably that the ECF parameter provides no additional information for more accurate model.

7 Conclusion

The current study found that the prediction accuracy measurement by analytical programming method can be seen as a viable method for effort estimation. However, this is true if and only if the UCP method is not optimized. The presented model outperform the Karners's equation about 12 % MMRE. The findings of this study have a number of important implications for future research of the using of analytical programming as an effort estimation technique. More research is required to determine the efficiency of analytical programming for this task. It would be interesting to find the best set-up for analytical programming.

Acknowledgments This study was supported by the internal grant of TBU in Zlin No. IGA/FAI/2016/035 funded from the resources of specific university research.

References

1. Keung, J.W.: Theoretical maximum prediction accuracy for analogy-based software cost estimation. In: Software Engineering Conference, 2008. APSEC '08. 15th Asia-Pacific, pp. 495–502 (2008)
2. Boehm, W.: Software engineering economics. IEEE Trans. Softw. Eng. **SE-10**, pp. 4–21 (1984)
3. Atkinson, K., Shepperd, M.: Using function points to find cost analogies. In: 5th European Software Cost Modelling Meeting, pp. 1–5. Ivrea, Italy (1994),
4. Karner, G.: Resource estimation for objectory projects. Objective Systems SF AB, pp. 1–9 (1993)
5. Rowe, G., Wright, G.: The Delphi technique as a forecasting tool: issues and analysis. Int. J. Forecast. **15**, 353–375 (1999)
6. Urbanek, T., Prokopova, Z., Silhavy, R., Vesela, V.: Prediction accuracy measurements as a fitness function for software effort estimation. SpringerPlus (2015)
7. Storn, R., Price, K.: Differential evolution—a simple and efficient adaptive scheme for global optimization over continuous spaces, vol. 11. Technical Report TR-95-012 (1995)
8. Storn, R.: On the usage of differential evolution for function optimization. In: Proceedings of North American Fuzzy Information Processing, pp. 519–523 (1996)
9. Zelinka, I., Davendra, D., Senkerik, R., Jasek, R., Oplatkova, Z.: Analytical programming-a novel approach for evolutionary synthesis of symbolic structures. InTech, Rijeka (2011)
10. Oplatkova, Z.K., Senkerik, R., Zelinka, I., Pluhacek, M.: Analytic programming in the task of evolutionary synthesis of a controller for high order oscillations stabilization of discrete chaotic systems. Comput. Math. Appl. **66**, 177–189 (2013)
11. Subriadi, A.P., Ningrum, P.A.: Critical review of the effort rate value in use case point method for estimating software development effort. J. Theor. Appl. Inf. Technol. **59**(3), 735–744 (2014)
12. Ochodek, M., Nawrocki, J., Kwarciak, K.: Simplifying effort estimation based on Use Case Points. Inf. Softw. Technol. **53**, 200–213 (2011)
13. Silhavy, R., Silhavy, P., Prokopova, Z.: Algorithmic optimisation method for improving Use Case Points estimation. PLOS ONE **10** (2015)
14. Nassif, A.B., Capretz, L.F., Ho, D.: Estimating software effort based on use case point model using sugeno fuzzy inference system. 2011 23rd IEEE International Conference on Tools with Artificial Intelligence (ICTAI), pp. 393–398 (2011)
15. Kocaguneli, E., Menzies, T., Keung, J.W.: On the value of ensemble effort estimation. IEEE Trans. Softw. Eng. **38**(6), 1403–1416 (2012)

16. Kaushik, A., Soni, A.K., Soni, R.: An adaptive learning approach to software cost estimation. 2012 National Conference on Computing and Communication Systems (NCCCS), pp. 1–6 (2012)
17. Attarzadeh, I., Ow, S.: Software development cost and time forecasting using a high performance artificial neural network model. Intell. Comput. Inf. Sci. pp. 18–26 (2011)
18. Reddy, C., Raju, K.: Improving the accuracy of effort estimation through fuzzy set combination of size and cost drivers. WSEAS Trans. Comput. **8**(6), 926–936 (2009)

Intra-frame Prediction Mode Decision for Efficient Compression and Signal Quality

N.S. Pradeep Kumar and H.N. Suresh

Abstract Accomplishing superior version of compression with proper data integrity is a challenging task when it comes to fast intra-prediction for multimedia. We have reviewed various existing techniques of fast intra-prediction mode decision techniques and found a significant trade-off between inequilibrium between compression and signal quality. Hence, we present a novel and simple algorithm that can perform fast intra-prediction mode decision using H.264 standard. The presented paper also uses multiple iterations of enhancement to each prediction model during encoding process considering frame and block sizes. The study also maintains the highest integrity of the meta-data information during frame transition. The outcome of the study is found to outperform the existing similar technique on compression and signal quality.

Keywords Power transmission system · Stochastic modelling · Fault tolerance · Outage · Prediction

1 Introduction

The development of fast and efficient internet and of growing applications of media has given many applications in the image compression by which we can create the original images and also the web slides, posters and slides. H.264/AVC is the good

N.S. Pradeep Kumar (✉)
Department of Electronics & Communication Engineering,
South East Asian College of Engineering & Technology, Bangalore, India
e-mail: pradeepkumarnsvtu@gmail.com

H.N. Suresh
Department of Electronics & Instrumentation Engineering,
Bangalore Institute of Technology, Bangalore, India

R. Silhavy et al. (eds.), *Software Engineering Perspectives and Application in Intelligent Systems*, Advances in Intelligent Systems and Computing 465,
DOI 10.1007/978-3-319-33622-0_25

image coding standard in the image compression, but they cannot compress the web images, slides etc. [1, 2]. These images can be compressed by using H.264/AVC intra coding. The H.264 is useful in the image compression of level 4–10X times of MPEG format. The H.264 compresses the image and reports only the differences with pre-compressed image. The H.264 divides the image by 4 parts to store the moving image volume. Also compression takes place within the camera by which bandwidth on Image Acquiring Network (IAN). In case of still images the storage volume will be more. The H.264 has the 4×4 and also 16×16 intra prediction block sizes. The block size 4×4 has nine different intra prediction modes and also exhibits 16 pixels of arrangement a, b, c, …., p. Commonly, the H.264 intra prediction is carried out spatial domain. The intra prediction is performed by considering the previously coded adjacent blocks (Left/above the block). In case of the intra prediction of luma samples, blocks are formed for 8×8 or 4×4 or 16×16. The 8×8 and 4×4 exhibit nine more extra prediction modes, while for 16×16 exhibit only 4 extra prediction modes. The optimization technique of rate distortion (r-d) can be used to achieve better coding mode in nine prediction modes to increase the quality of coding and reduce the bit-rate. i.e. the encoder has to be performed for the code in all combinations of modes [3, 4]. The mode with minimum r-d cost is considered as the better mode. The r-d cost can be executed by, (i) performing same forward and backward transform or quantization operation, (ii) Flowingly the repeated entropy coding has to performed. The entire process will derive the complexity in the r-d cost computation; hence the execution complexity for the encoder will increase randomly. Many prediction modes are explained and implemented in this paper. Section 3 describes problems explored after reviewing the literatures. Discussion of the proposed system using H.264/AVC approach is presented in Sect. 4 followed by research methodology to accomplish the stated goal of study in Sect. 5. Algorithm description is discussed in Sect. 6 followed by Sect. 7 that demonstrates the outcomes of the proposed study followed by summary of paper in Sect. 8.

2 Related Work

This section discusses about the studies being carried out by the significant authors pertaining to Intra-Frame Prediction Mode Decision using H.264/AVC encoding standards over multimedia. We have already reviewed some of the existing technique [5]; however, some more techniques are discussed here. Most recently, Balaji and Thyagharajan [6] have presented a fast mode decision technique using H.264/AVC using scalable video coding for minimizing the computational complexity as well as better controlling of the encoding time. Elyousfi [7] have presented a subsampling techniques as well as rate distortion technique using

H.264/AVC. Park et al. [8] have extended the same technique using the latest version of encoding i.e. high efficiency video coding. Similar line of research has also been carried out by Zhang et al. [9] who have developed a three dimensional encoder for estimating the disparity. Zhao et al. [10] have also used the similar encoding technique for developing an efficient coding tree unit. A unique direction of the study was carried out by Zhu et al. [11] focusing on a mode decision technique using correlation-based approach. The study also presents a unique block mode decision over compressive sensing frames using cross diamond. Schier et al. [12] have developed a prioritization technique based on H.264/AVC for enhancing the quality of video. The outcome of the study was evaluated with respect to PSNR and loss rate. Palomino et al. [13] have presented a technique for fast intra mode decision module developed over FPGA using H.264 encoders. Bharanitharan et al. [14] have investigated about the temporal and spatial homogeneity of the signal blocks for minimizing the quantity of the modes with an aid of Rate-Distortion theory. Xin et al. [15] have presented a study focusing on a unique transcoding scheme of intra-coded Multimedia focusing on the minimizing of the complexity and faster response time.

Hence, it can be seen that there are various research work being carried out using H.264 for intra-prediction mode decision. The next section discusses about the limitations associated with the existing system.

3 Problem Description

As majority of the existing studies were more focused on performing compression using H.264, the retention of Intra-Frame Prediction Mode Decision was less emphasized on the existing system. Although usage of H.264/AVC ensures potential performance of rate-distortion compared to its legacy version, but it doesn't guarantee an efficient Intra-Frame Prediction Mode Decision. We also find that there are various forms of the mode decision techniques and algorithms presented by various authors. Unfortunately, the entire mode decision techniques designed till date seriously lacks multi-level optimization, which will mean that the existing algorithm can offer one round of better solution and cannot go beyond that. Usage of conventional optimization techniques are also highly resource bound and is expensive in nature. Hence, the next section discusses about proposed system to overcome this issues.

4 Proposed System

This work is a continuation of our prior study [16]. The prime aim of the proposed study is to design and develop an algorithm for Intra-Frame Prediction Mode Decision based on multi-mode scheme for H.264. The study also aims for evolving up with a technique to balance between mode decision technique and data quality using the recent H.264 standard. The prime contributions of the presented study are:

- To develop a novel encoding algorithm that can perform extensive intra-frame prediction mode decision of video file using H.264/AVC standard.
- To develop a condition that can perform computational enhancement of video multimedia encoding cost effectively.
- To present a technique that can perform an effective management of metadata of the unit frame ensuring no loss of data during video streaming.
- To assess the performance with respect to size of compressed bits and data quality.

5 Research Methodology

The proposed study has considered explanatory research strategy and is a continuation of our earlier studies [5, 16]. The study includes performing quick intra prediction mode choice of video record utilizing H.264. This segment talks about the methodology received for building up the proposed compression procedure. Not quite the same as traditional video processing methods, the proposed framework utilizes preparing the video data in form of NTSC standard. The following phase of creating novel encoder fundamentally performs improving the resolution by 16 times of the current pixels for quick intra forecast mode choice. Numerous conditions were additionally formulated for guaranteeing if the framework truly requires an optimizing (if there should arise an occurrence of any variance in stature and width from standard configuration). The third phase of the study is centered around outlining of auxiliary encoding framework. This configuration organize basically performs a second layer of encoding to perform compression. The framework at first registers Lagrangian multiplier taking into account quality parameter. So as to keep up meta-information of the pixel enhancement during intra-fast prediction mode decision, it performs obstructing of different sizes taken after by different macroblocks of the image. These strategies definitely minimizes the size furthermore keep up a decent measure of image meta-information as header, which is iteratively kept up at each level of compression in a different lattice. At last, we ascertain least cost of Lagrangian multiplier to show signs of improvement upgraded estimations of compacted bits.

The last phase of the examination approach manages planning final encoding System. The whole idea of this encoding plan starts with the novel concept that arrangements with both present and neighboring pixels. The framework keeps up projection of the pixels in light of frame size so as to guarantee that streaming of video turns out to be to a great degree smooth during transition. This is one of the novel alternative techniques that we have used in replacement of temporary memory for maintaining bitstreams. Another special part of this study is it can proficiently keep up two forms of memory (one for picture quality and other for exceedingly diminished size of packed bits). At each phase of H.264-based encoding, we include header that guarantees no pixels related data is lost amid the compression. At long last, the framework figures the computational multifaceted nature by extricating the base Lagrangian cost. Another essential fuse is reliance of essential and auxiliary encoding plan on performing tertiary encoding plan. At long last, the result is subjected to whole number based changes took after by quantization to achieve the packed bits. The result of the study was at long last surveyed utilizing size of packed bits and Peak-Signal-to-Noise-Ratio.

6 Algorithm Implementation

The data to the calculation can be any types of video document v of as often as frequently utilized formats. We utilize the Matlab inbuilt method mmreader to peruse the video and performs extraction of begin and end frames of the video. In spite of the fact that, we do the preparatory analyses on a scope of casings, however later we performed the examination for the complete video outlines. The center calculation is in charge of performing essential encoding plan for compressed video utilizing H.264 standard. The calculation separates the height h and width w data from the edge f. Attributable to upgrade the intra-mode prediction, the proposed framework will perform encoding to expand 16 times in each passes of h and w. We select another parameter i and j which represents further 4 bits of augmentation in pixel determination for both h and w separately. We additionally apply three distinct conditions for further improving the results considering both the estimations of i and j. If both are found to be similar to unity than it doesn't represents a case of even 1080p, henceforth, we reject such cases to be subjected for streamlining. On the off chance that the estimation of i is observed to be unity, it speaks to a situation where level directionalities of the pixels required to be amplified. The vertical expansion of the pixels is finished by checking the estimation of j to be unity. Finally, the system performs integer-based transform and some conventional steps of quantization and entropy encoding mechanism to further leverage primary encoding in video compression using H.264 standard.

Algorithm for Encoding with H.264 standard

Input: *h* (height), *w* (Width), *f* (frame sequence), ρ (quantization).

Output: Encoding of frame

Start:

1. define (h, w)→size(f)
2. for m=1:16:h
3. for n=1:16:w
4. bits_frame→encode(binary)
5. If (i==1 && j==1)
6. No Optimization
7. elseif (i==1)
8. Horizontal optimization
9. elseif(j==1)
10. Vertical optimization.
11. Apply Lagrangian multiplier (λ_{me})

$$\lambda_{me} = ((0.85) * 2^{\frac{Q_P - 12}{13}})^5$$

12. mat→prior motion vectors
13. for i=1:b_i:h
14. for j=1:b_i:w
15. Write the reconstructed 16x16 macroblock to the output
16. Add header for a quadruplet split
17. Extract minimum Lagrangian cost
18. Get value of S_r, mse, R w.r.t b_i.
19. End
10. End
20. Perform integer-based transform, quantization, and entropy
21. Perform encoding.

End

Algorithm for secondary encoding is carried out for further generating balance between H.264 based video compression and retention of high resolution. The algorithm takes the input of frame f and quantized parameter ρ and applies the principle of Lagrangian multiplier considering the quality parameter Q_P. The prior motion vectors are reposited on the matrix *mat*. The algorithm further generates reconstructed macro-blocks of the frames of size 16 × 16 for any sizes of blocks (b_i, where suffix i was taken as 4, 8, and 16). The algorithm further add header for retaining more pixel information making the size lowered down to a extremely large extent. The compression technique was further more optimized by evaluating the minimum Lagrangian cost that is calculated by multiplying absolute difference between the old and new block position, Lagrangian multiplier, and encoded bits using JPEG 2000 standards. Finally, the algorithm extracts the value of mean squared error (*mse*), compressed present frame (S_r), and bitrate (R). The algorithm for tertiary encoding mechanism highly depends on the primary and secondary encoding mechanism applicable on different sizes of block. This algorithm also considers performing orthographic projections considering multiple frame size of the projected pixels. The prime purpose is to maintain high end compression using H.264 with maximum remnant pixel information on defined image plane. The algorithm computes the row and column of the block. It than extracts the header information and forward it to the bitstreams thereby maintains a superior quality of frame transition without much loss of motion vectors. The algorithm than performs encoding of the I-Frame and append it to the frame header followed by extraction of the I frame. The size is further reduced to unsigned integer followed by applying primary encoding algorithm. The received frame is stored in a temporary matrix for the purpose of appending I-frame along with the bit stream. The process is repeated for all the frames in a given video clip which is followed by appending of P-frame in the header. Finally, the secondary encoding technique is applied and entire bitstream is encoded. This technique ensures maintaining all the metadata information of the pixels along with superior reduction of sizes of video packets while streaming.

7 Result Discussion

This section discusses about the outcomes being accomplished from the presented study. We present the accomplished result in two ways i.e. we discuss the individual outcomes as well as comparative analysis outcomes. For discussion of individual outcomes, we use the performance parameter e.g. compressed bits, PSNR, and algorithm processing time to assess the outcomes. The evaluation was carried out over 10 samples of video files of few seconds in multiple video formats in order to assess the quality of the outcomes. For the purpose of the comparative analysis, we compare the recent work being done by Kim et al. [17].

Figure 1 shows that proposed system highlights the superior data quality in comparison to existing approach. The technique implemented by Kim et al. [17] uses complex mechanism of encoding process that results in more storage of buffer in frame transition that finally results in not more than 44 dB of PSNR. Moreover, the study didn't attempt to enhance the resolution of the signal. The study on the other hand has focused on data quality using basic test cell. However, due to lack of optimization, the trend of the curve is found to be maximum 46 dB. The prime reason for superior outcomes of proposed system was due to multi-level optimization technique and efficient management of meta-data.

Figure 2 shows the outcome of compression in terms of percentage, which shows that proposed system is capable of performing compression of the data to a larger extent as compared to the existing approaches. The approach of Kim et al. [17] maintains almost a stable and quite predictable performance of compression even with increasing number of processing frames. The trend of the curve is more or less same even if the frames are increased. Although, this compression

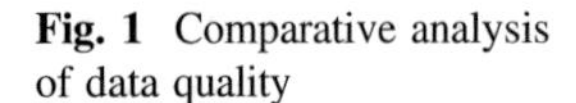

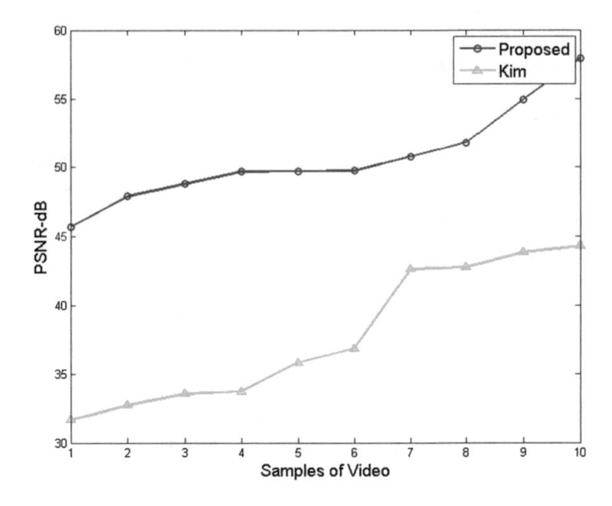

Fig. 1 Comparative analysis of data quality

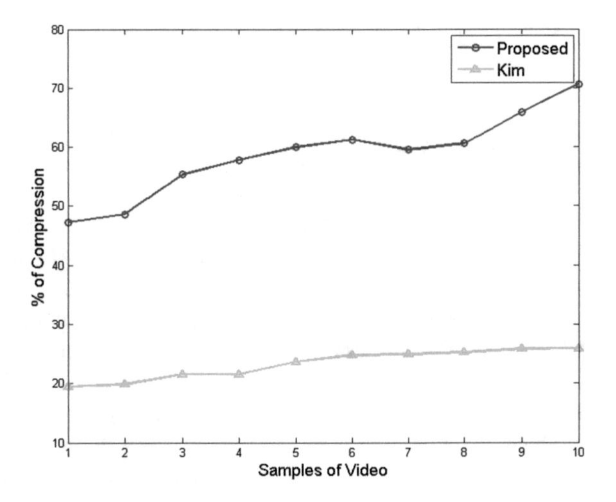

Fig. 2 Comparative analysis of compression (%)

performance is good for static video and dataset, it cannot be used for streaming massive bitstreams. The scheme introduced by proposed system actually performs multiple level of optimization of the video frames to ensure efficient Intra-Frame Prediction Mode Decision.

8 Conclusion

The paper has emphasized on usage of the recent video compression standard called as H.264/AVC. Our theoretical study findings on this protocol shows that it has superior compression performance and can offer better resolution standards too.

However, we have found that there is a trade-off in hardware supportability with compression cum data quality. Hence, we reviewed some of the existing techniques on usage of H.264 and found that majority of the research implements it for compression with less focus on visual quality. Although, there are many research work found to discuss about connection of H.264 with high resolution, but till date there is no technical implementation to prove it. Hence, we propose a novel approach where we use multiple video samples to be compressed using H.264. However, we have not only perform H.264/AVC encoding, but we have redefined the encoding pattern by incorporating primary, secondary, and tertiary encoding scheme with multi-level optimization. The outcome of the study was found to posses a well balance between the video compression and retention of superior data quality.

References

1. Tian, X., Le, T.M., Lian, Y.: Entropy Coders of the H.264/AVC Standard: Algorithms and VLSI Architectures. Springer, Berlin (2010)
2. Juurlink, B., Mesa, M.A., Chi, C.C.: Scalable Parallel Programming Applied to H.264/AVC Decoding. Springer, New York (2012)
3. Angelides, M.C., Agius, H.: The Handbook of MPEG Applications: Standards in Practice. Wiley, Chichester (2010)
4. Kondoz, A.: Visual Media Coding and Transmission. Wiley (2009)
5. Pradeep Kumar, N.S., Suresh, H.N.: Studying an effective contribution of techniques of video compression. Int. J. Electron. Commun. Comput. Eng. **6**(1) (2015)
6. Balaji, L., Thyagharajan, K.K.: H.264 SVC complexity reduction based on likelihood mode decision. Sci. World J. (2015). Hindawi Publishing Corporation
7. Elyousfi, A.: An improved fast mode decision method for H.264/AVC intracoding. Adv. Multimedia (2014). Hindawi Publishing Corporation
8. Park, C.-S., Hong, G.-S., Kim, B.-G.: Novel intermode prediction algorithm for high efficiency video coding encoder. Adv. Multimedia (2014). Hindawi Publishing Corporation
9. Zhang, Q., Li, N., Gan, Y.: Low complexity mode decision for 3D-HEVC. Sci. World J. (2014). Hindawi Publishing Corporation
10. Zhao, T.: Flexible mode selection and complexity allocation in high efficiency video coding. IEEE J. Sel. Top. Sig. Process. **7**(6) (2013)
11. Zhu, J., Cao, N., Meng, Y.: Adaptive multi-hypothesis prediction algorithm for distributed compressive video sensing. Int. J. Distrib. Sens. Netw. (2013). Hindawi Publishing Corporation
12. Schier, M., Welzl, M.: Optimizing selective ARQ for H.264 live streaming: a novel method for predicting loss-impact in real time. IEEE Trans. Multimedia **14**(2) (2012)
13. Palomino, D., Correa, G., Diniz, C.: Algorithm and hardware design of a fast intra frame mode decision module for H.264/AVC encoders. Int. J. Reconfig. Comput. (2012). Hindawi Publishing Corporation
14. Bharanitharan, K., Liu, B.-D., Yang, J.-F.: Classified region algorithm for fast intermode decision in H.264/AVC encoder. EURASIP J. Adv. Sig. Process. (2010). Hindawi Publishing Corporation
15. Xin, J., Vetro, A., Sun, H., Su, Y.: Efficient MPEG-2 to H.264/AVC transcoding of intra-coded video. EURASIP J. Appl. J. Appl. Sig. Proc. (2007). Mitsubishi Electric Research Laboratories

16. Praveen Kumar, N.S., Suresh, H.N.: Multi-mode schema for an efficient intra-frame prediction technique for H.264. Am. J. Comput. Sci. Inf. Technol. (2015). ISSN: 2349-3917
17. Kim, T., Hwang, U., Jeong, J.: Efficient block mode decision and prediction mode selection for intra prediction in H.264/AVC high profile. In: IEEE-International Conference on Digital Image Computing: Techniques and Applications (2011)

Multidimensional Design of OLAP System for Context-Aware Analysis in the Ambient Intelligence Environment

Jan Tyrychtr, Martin Pelikán, Hana Štiková and Ivan Vrana

Abstract Ambient Intelligence (AmI) is currently a perspective area of development intelligent systems that react on human presence, their behavior and AmI adapts to requirements based on contextual knowledge. The important issue in the study of AmI is thinking about context-aware preference. In the context of ubiquitous computing technologies there is not any access for users to the system at one point, but in different contexts. This creates need for context-sensitive preferences. The aim of context reasoning is getting new knowledge, so that systems or services were more intelligent. This process is not a trivial problem, so that we propose multidimensional view on context-aware knowledge for support of contextual reasoning. In this paper we introduce a new TCAP procedure for transformation context-aware preference through OLAP in the AmI. The OLAP technology enables us better analyzing contextual dependence on preferences and choosing relevant content for users in the AmI environment.

Keywords OLAP · Ambient intelligence · Multidimensional data model · Context-aware · Knowledge · Decision support

1 Introduction

Currently the fastest growing area of intelligent systems is AmI [1, 11, 12, 24, 31] as multidisciplinary technological paradigm. AmI is set of processes, applications and technologies designed to efficiently and effectively support human needs in many areas (in household, at business, in hospitals, in automotive industry etc.).

J. Tyrychtr (✉)
Faculty of Economics and Management, Department of Information Technologies, Czech University of Life Sciences in Prague, Prague, Czech Republic
e-mail: tyrychtr@pef.czu.cz

M. Pelikán · H. Štiková · I. Vrana
Faculty of Economics and Management, Department of Information Engineering, Czech University of Life Sciences in Prague, Prague, Czech Republic

R. Silhavy et al. (eds.), *Software Engineering Perspectives and Application in Intelligent Systems*, Advances in Intelligent Systems and Computing 465,
DOI 10.1007/978-3-319-33622-0_26

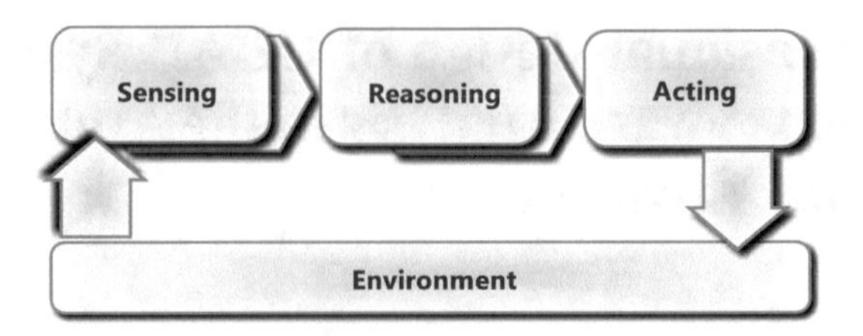

Fig. 1 Process of AmI systems, according to [6]

The purpose of AmI is to create adaptive, friendly, individualized digital environment which understands, learns, reasons and interacts with user needs [4].

AmI involves a large number of technologies for sensing, reasoning and acting systems. *Sensing systems* enable perceived user status and environments thanks to sensors, *reasoning systems* use data for support making decisions how to act upon the environment and finally, *acting systems* perform these decisions [6]. Figure 1 illustrates this process of AmI.

AmI is sometimes confused with similar concepts: *Smart Environments* [9, 10, 14], *Ubiquitous or Pervasive Computing* [18, 25, 27], but they aren't equivalent. AmI gives bigger importance for users so that AmI system could foresee and adapt to users behavior. This human aspect is essential for AmI. It is very important to pay attention to learning and understanding human needs and their environment during design of these systems. Learning is that the environment has to gain knowledge about preferences, needs and habits of users to be able help them with their daily human life in AmI environment [17, 21]. It is very important to consider context-awareness as one of function AmI for these purposes. *Context* is typically the location, identity, and state of people, groups, and computational and physical objects [15].

1.1 The OLAP

OLAP (On-Line Analytical Processing) describes an approach for the decision support which aims to gain knowledge from the data warehouse or data marts [2]. OLAP permits to aggregate data (due to SQL aggregate operator, such as SUM, COUNT, AVG etc.) and look at the indicators from different points of view. OLAP gained aggregated indicators by grouping various relational data from multidimensional database. *The multidimensional databases* are suitable for storing large amounts of analytical data over which analyses and surveys are performed used to support decision making. The actual method of organizing data in a multidimensional database is enabled through data cubes.

Data cube is a data structure for storing and analyzing large amounts of multidimensional data [22]. Data cube allows to utilize benefits of multidimensional view on data and through OLAP operators, such as roll-up, drill-down, slice-dice, pivoting etc., perform OLAP questions.

There are many approaches to the formal definition of operators' data cubes (a comprehensive overview is available in [30]). Generally, the data cube consists of dimensions and measures. *Dimension* is a hierarchical set of dimensional values that provide categorical information characterizing a particular aspect of the data [23]. Dimension hierarchy specifies aggregation levels and granularity. For example: *day* → *month* → *quarter* → *year* is a hierarchy on dimension Time [26]. *Measures* (monitored indicators) of the cubes are mainly quantitative data that can be analyzed.

1.2 The OLAP in the AmI Environment

Current approaches [5, 16, 29, 32] primarily deal with ways of gaining, saving and specification of context-aware information. OLAP as decision-making processes support tool gives AmI applications possibility to dynamically analyze large data volume which originated from contextual data. We can consider using OLAP technology in following areas:

1. OLAP as a target AmI tool—analyses and OLAP outputs will be adapted to the specific needs of users.
2. OLAP as a component of AmI systems:
   - OLAP for source data analysis (from sensors [33], wireless network, smart systems [7], etc.)
   - OLAP for knowledge analysis (context-aware [3, 34], user preferences etc.)

There are not OLAP approaches used for analysis user preferences in the AmI. The aim of this paper is to fill in this gap and suggest OLAP for analysis of knowledge context-aware preferences.

2 Methods

In this paper, we adopt the formal apparatus of the data cube according to [13]. Let us have 6-tuple $<D, M, A, f, V, g>$, where four component indicate properties of data cube. Those properties are:

1. The set of n dimensions $D=\{d_1, d_2, \ldots, d_n\}$, where each d_i is the name of dimension from particular domain $dom_{\text{dim}(i)}$.
2. The set of k measures $M=\{m_1, m_2, \ldots, m_k\}$, where each m_i is the name of measure from particular domain $dom_{\text{measure}(l)}$.
3. The set of dimension names and measures is disjoint; i.e. $D \cap M = 0$.

4. The set of t attributes $A = \{a_1, a_2, \ldots, a_t\}$, where each a_i is the name of attribute from particular domain $dom_{\mathrm{attr}(r)}$.
5. The one-to-many mapping $f : D \to A$ exists for every dimension and set of attributes. The mapping is such that attribute sets corresponding are pair wise disjoint, i.e. $\forall i, j, i \neq j, f(d_i) f(d_j) = 0$.
6. The set V represents a set of values that were used to materialize data cube. Therefore every element $v_i \in V$ is k-tuple $< \mu_1, \mu_2, \ldots, \mu_k >$, where μ_i is instance of i-th measure m_i.
7. The g represents a mapping g: $dom_{\mathrm{dim}(1)} \times dom_{\mathrm{dim}(2)} \times \cdots \times dom_{\mathrm{dim}(n)} \to V$. Thus, intuitively g mapping indicates which values are associated with a particular 'cell'. *Cells* are measures or values based on a set of dimensions.

The definition presented above is the abstract structure for multidimensional paradigm. We represent knowledge elements of context-aware preference as matrix values (similarly to [19]).

Let K be the set of all *knowledge elements* in a particular domain, i.e.

$$K = \{k_1, k_2, \ldots k_n\}, \quad (1)$$

where $k \in C(Context) \vee k \in P(Preference)$. The set of Context-Aware Preference View $(CAPV)$ is defined as the set of matrices, in which each matrix represents context-aware preference values of a particular users in the AmI environment. A *CAPV* is defined as:

$$CAPV = \{[cv_1], [cv_2] \ldots [cv_m]\}, \quad (2)$$

where each $[cv]$ is a 2-dimensional matrix, as follow:

$$[cv] = \begin{array}{c} \\ p_1 \\ p_2 \\ \vdots \\ p_n \end{array} \begin{array}{c} \begin{array}{cccc} c_1 & c_2 & \ldots & c_m \end{array} \\ \begin{bmatrix} cv_{11} & cv_{12} & \ldots & cv_{1m} \\ cv_{21} & cv_{22} & \ldots & cv_{2m} \\ \vdots & \vdots & \ddots & \vdots \\ cv_{n1} & cv_{n2} & \ldots & cv_{nm} \end{bmatrix} \end{array} \quad (3)$$

where each p represents *preference element* from P, each c represents *context element* from C and each cv is a particular measurable *context value*.

We consider knowledge rules of context-aware preferences in the form IF-THEN as *IF context THEN preference*. Each knowledge element k is represented as a 3-tuple in the form $<object>(<attribute>.<value>)$. Design of our new Transformation Context-Aware Preferences (TCAP) procedure to capture the context-aware knowledge of agricultural activities will serve as an illustration. We use the databases of the Farmer's Portal run by the Czech Ministry of Agriculture. These databases are basically implemented using standard relational tables containing set of registered animals, medicaments, etc. (see Table 1).

Table 1 Example of relational table containing a set of registered cows

Ear code	Birth date	Sex	Breed	Category
CZ000562606042	17.04.2014	Male animal	Limousin	B7-24M
CZ000562607042	19.04.2014	Male animal	Charolais	B7-24M
CZ000562613042	13.05.2014	Male animal	Other beef breeds	B7-24M
CZ000562608042	21.04.2014	Male animal	Blonde d'aquitaine	B7-24M

This paper is based on author's previous research and is a follow-up to their previously published work about transformation of knowledge rules [28]. The conceptual schema is designed using the star schema [20]. The prototype of the OLAP for context-aware preference analysis is created using the ROLAP (Relational OLAP) technology [8] and Microsoft Power Pivot software.

3 Results

In this section we propose new TCAP procedure for transformation context-aware preferences to multidimensional database. First, we propose technological aspects of OLAP focused on context-aware analyses in the AmI. Subsequently we present TCAP procedure.

3.1 The OLAP for Analysis Context-Aware Preferences

There is a need to pay attention to technological aspects of such a solution for designing OLAP for analysis of context-aware preferences. We include OLAP technologies into the process of AmI systems to avoid loss of knowledge in the process knowledge-to-data conversion and conversely, Fig. 2 illustrates the relationships among these technologies.

Contextual knowledge is stored in knowledge base from which it is loaded into a multidimensional (MD) database. The multidimensional database (based on relational technology) serves as a basic construct for creating data cube and context-aware OLAP. Analytical outputs provided to the OLAP technology further enables to consider context and create a new knowledge that is stored in the knowledge base or used for a decision-making processes. We propose transforming knowledge base of contexts-aware preferences into the multidimensional database via the TCAP procedures in the next section.

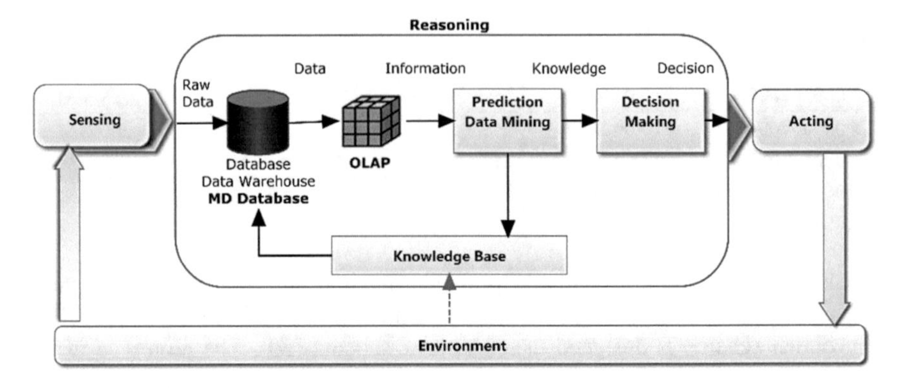

Fig. 2 OLAP context-aware technologies

3.2 The TCAP Procedure

Further we describe our approach of modeling OLAP system for analysis of context-aware preferences in Agribusiness context, then we set rules of transformation, we suggest multidimensional schema and we create OLAP prototype of context-aware preferences. For instance, suppose a user has one preference.

Example 1 The executive farmer prefers to sell cattle. This is a young bull with ear code CZ000562606042, who was born April 17, 2014, it is Limousin breed with the market price of meat 88 CZK/kg. The bull is in good health and weighs 250 kg. Thus, he expresses the following preferences:

Context: *AnimalSpecies* (*Name.Cattle*) ^ *Animal*(*Breed.Limousin* ^ *Category.B7-24M* ^ *Sex.MaleAnimal* ^ *BirthDate*.4-17-2014 ^ *EarCode.CZ*00562606042) ^ *Market*(*Price*.88.CZK/kg) ^ *Condition*(*Treatment.None* ^ *weight*.250 *kg*)
Preference: *Preference*(*InterestOf.*Sell)

Above we use only one context and preference knowledge. There can be a large number of these rules in the real environment of agribusiness. In order that AmI system could best adapt to users preferences we transform these knowledge elements into multidimensional paradigm in following steps:

Step 1: Transformation Context-Aware Preferences
First we set following rules of transformation of knowledge elements to create OLAP system for CAPV analysis in the AmI environment:

1. Each <object> from k_n is transformed to dimension d_n.
2. Each <attribute> from k_n is transformed to attribute a_t.
3. Each <value> from k_n is transformed to values v_i.

Table 2 Description of the results of the transformation of the context-aware preferences

Knowledge elements	Dimension	Attribute
Context	AnimalSpecies	Name
	Animal	Breed
		Catcgory
		Sex
		BirthDate
		EarCode
	Market	Price
	Condition	Treatment
		Weight
Preference	Preference	InterestOf

We apply the rules above and transform knowledge elements from Example 1. Transformation of the context-aware preferences leads to identification of dimensions and attributes for designing a star conceptual schema (the whole diagram contains only one fact table).

The result (Table 2) of the transformation of knowledge elements from context-aware preferences allows creation of conceptual schema for designing OLAP system in the AmI environment.

Step 2: **Multidimensional Design**

We suggest multidimensional data model through conceptual and logical design in this step. In the conceptual design we associate the identified dimensions with fact table from results in Table 2. All attributes are also added to dimension tables. It is important to implement analytic processing in the short and long term for context-aware analysis. That's the reason why we add time dimension to the whole conceptual model. The Time dimension contains attributes that will allow analysis both in the short term (Month, Week, Day and Hour) and in the long term (Period, Year). There is used a basic aggregation operator Count for analyzing and preferences measurement. If the context-aware preferences were based on quantitative evaluation of preferences they could be used for next operators as average, standard deviation, etc. Conceptual model is expandeartlogd to include the necessary logic design. The primary keys are added into all dimensions and a fact table. The foreign keys are added into the fact table to create 1:N relation. The basic logical model is extended by the additional dimension attributes. The diagram of the resulting multidimensional design is shown in Fig. 2.

Step 3: Creation of the Prototype

Creating prototype OLAP application is verified that whole multidimensional model for analysis context-aware preferences is applicable for analytical processing in AmI environments. We apply the whole solution through relational databases. First we create table with columns according to multidimensional model in the Fig. 3. There are added records with values of knowledge elements into rows. Then this table elaboration is imported into software MS Power Pivot where we associate relationship between fact tables and dimensions (according to Fig. 3). There we

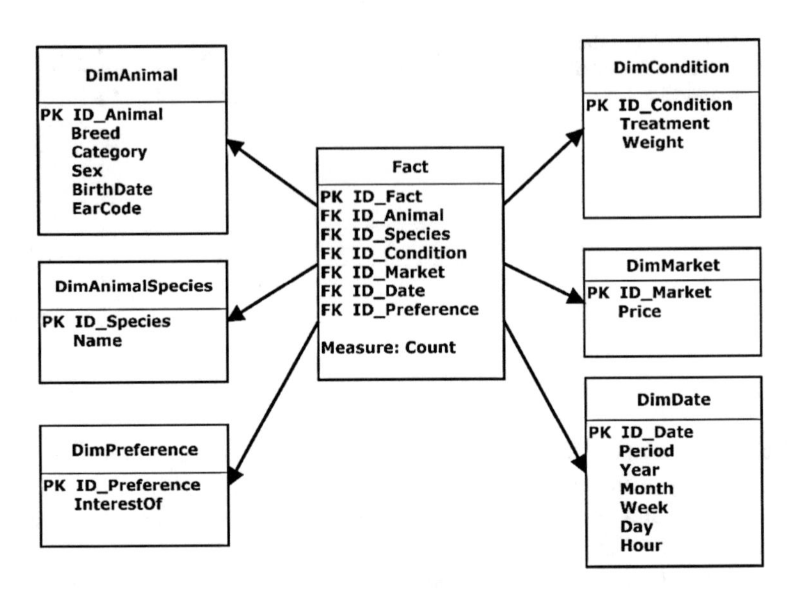

Fig. 3 Multidimensional data model context-aware preferences

create analytic output in the form of contingent table for different analytical context view in the Power Pivot Dashboard. For example, to answer the following analytical questions:

- How many times did the user prefer a sale during the autumn in years 2012–2015?
- How did change preferences of sales breed Charolaise in comparison with other breeds?
- How much has changed preferences of sales with young bulls in last year?
- Etc.

In the physical design, the measure is selected so that it will be displayed inside the pivot table and dimensions are placed in rows and columns where their values are calculated within. For example: there will be calculated a measure for Count, dimensions will be Animal and Time at the last question.

4 Conclusion

The paper presented approach to support for analytical processing of context-aware preferences in AmI environments. We have proposed a TCAP procedure for transforming knowledge base of contexts-aware preferences into the multidimensional database and we have provided a mechanism for CAPV in the multidimensional paradigm. Preferences are store in data cube for using OLAP techniques. Our approach brings advantages for next design AmI systems, such as:

- Analytical processing of context-aware preferences in the AmI environments.
- Design of a more adaptive system that will automatically select the most relevant content according to preferences and present it to the user.

Presented results are only the first step of a more complex research. One of the planned research phases also concerns automated transformation context-aware preferences into multidimensional data model.

Acknowledgements This work was conducted within the project DEPIES—Decision Processes in Intelligent Environments funded through the Czech Science Foundation, Czech Republic, grant no. 15-11724S.

References

1. Aarts, E., Wichert, R.: Ambient Intelligence, pp. 244–249. Springer, Berlin (2009)
2. Abelló, A., Romero, O.: On-line analytical processing. In: Encyclopedia of Database Systems, pp. 1949–1954. Springer, US (2009)
3. Adomavicius, G., Tuzhilin, A.: Context-aware recommender systems. In: Recommender systems handbook, pp. 217–253. Springer, US (2011)
4. Aly, S., Pelikán, M., Vrana, I.: A generalized model for quantifying the impact of Ambient Intelligence on smart workplaces: Applications in manufacturing. J. Ambient Intell. Smart Environ. **6**(6), 651–673 (2014)
5. Athanasopoulos, D., Zarras, A.V., Issarny, V., Pitoura, E., Vassiliadis, P.: CoWSAMI: interface-aware context gathering in ambient intelligence environments. Pervasive Mob. Comput. **4**(3), 360–389 (2008)
6. Aztiria, A., Izaguirre, A., Augusto, J.C.: Learning patterns in ambient intelligence environments: a survey. Artif. Intell. Rev. **34**(1), 35–51 (2010)
7. Cho, D.Y., Lee, S.P.: Ubiquitous data warehouse: integrating RFID with mutidimensional online analysis. J. Korea Soc. IT Serv. **4**(2), 61–69 (2005)
8. Colliat, G.: OLAP, relational, and multidimensional database systems. ACM Sigmod Record **25**(3), 64–69 (1996)
9. Cook, D., Das, S.: Smart Environments: Technology, Protocols and Applications, vol. 43. Wiley (2004)
10. Cook, D.J., Das, S.K.: How smart are our environments? An updated look at the state of the art. Perv. Mob. Comput. **3**(2), 53–73 (2007)
11. Cook, D.J.: Multi-agent smart environments. J. Ambient Intell. Smart Environ. **1**(1), 51–55 (2009)
12. Cook, D.J., Augusto, J.C., Jakkula, V.R.: Ambient intelligence: Technologies, applications, and opportunities. Perv. Mob. Comput. **5**(4), 277–298 (2009)
13. Datta, A., Thomas, H.: The cube data model: a conceptual model and algebra for on-line analytical processing in data warehouses. Decis. Support Syst. **27**(3), 289–301 (1999)
14. Das, S.K., Cook, D.J.: Designing smart environments: a paradigm based on learning and prediction. In: Pattern Recognition and Machine Intelligence, pp. 80–90. Springer, Berlin (2005)
15. Dey, A.K., Abowd, G.D., Salber, D.: A conceptual framework and a toolkit for supporting the rapid prototyping of context-aware applications. Hum. Comput. Interact. **16**(2), 97–166 (2001)
16. Feng, L., Apers, P.M., Jonker, W.: Towards context-aware data management for ambient intelligence. In: Database and Expert Systems Applications, pp. 422–431. Springer, Berlin (2004)

17. Galushka, M., Patterson, D., Rooney, N.: Temporal data mining for smart homes. In: Designing Smart Homes, pp. 85–108. Springer, Berlin (2006)
18. Lyytinen, K., Yoo, Y.: Ubiquitous computing. Commun. ACM **45**(12), 63–96 (2002)
19. Memon, T., Lu, J., Hussain, F.K., Rauniyar, R.: Subject-oriented semantic knowledge warehouse (SSKW) to support cognitive DSS. In: On the Move to Meaningful Internet Systems: OTM 2013 Conferences, pp. 291–299. Springer, Berlin (2013)
20. Mendoza, M., Alegría, E., Maca, M., Cobos, C., León, E.: Multidimensional analysis model for a document warehouse that includes textual measures. Decis. Support Syst. **72**, 44–59 (2015)
21. Nugent, J.C.A.C.D.: Designing smart homes. LNAI **4008**, 109–131 (2006)
22. Pedersen, T.B.: Cube. On-line analytical processing. In: Encyclopedia of Database Systems, pp. 538–539. Springer, US (2009a)
23. Pedersen, T.B.: Dimension. On-line analytical processing. In: Encyclopedia of Database Systems, pp. 836. Springer, US (2009b)
24. Remagnino, P., Foresti, G.L.: Ambient intelligence: a new multidisciplinary paradigm. IEEE Tran. Syst. Man Cybern. Part A: Syst. Hum. **35**(1), 1–6 (2005)
25. Saha, D., Mukherjee, A.: Pervasive computing: a paradigm for the 21st century. Computer **36** (3), 25–31 (2003)
26. Sarawagi, S.: Indexing OLAP data. IEEE Data Eng. Bull. **20**(1), 36–43 (1997)
27. Satyanarayanan, M.: Pervasive computing: vision and challenges. IEEE Pers. Commun. **8**(4), 10–17 (2001)
28. Tyrychtr, J., Brožek, J., Vostrovský, V.: Multidimensional modelling from open data for precision agriculture. In: Enterprise and Organizational Modeling and Simulation, pp. 141–152. Springer (2015)
29. van Bunningen, A.H., Feng, L., Apers, P.M.: A context-aware preference model for database querying in an ambient intelligent environment. In: Database and Expert Systems Applications, pp. 33–43. Springer, Berlin (2006)
30. Vassiliadis, P., Sellis, T.: A survey of logical models for OLAP databases. ACM Sigmod Record **28**(4), 64–69 (1999)
31. Weber, W., Rabaey, J., Aarts, E.H. (eds.): Ambient Intelligence. Springer (2005)
32. Wu, S., Chang, A., Chang, M., Liu, T.C., Heh, J.S.: Identifying personalized context-aware knowledge structure for individual user in ubiquitous learning environment. In: Fifth IEEE International Conference on Wireless, Mobile, and Ubiquitous Technology in Education, 2008. WMUTE 2008, pp. 95–99. IEEE (2008)
33. Xu, M.: Experiments on remote sensing image cube and its OLAP. In: 2004 IEEE International Geoscience and Remote Sensing Symposium Proceedings 2004. IGARSS'04, vol. 7, pp. 4398–4401. IEEE (2004)
34. Xu, K., Zhu, M., Zhang, D., Gu, T.: Context-aware content filtering & presentation for pervasive & mobile information systems. In: Proceedings of the 1st International Conference on Ambient Media and Systems, p. 20. ICST (Institute for Computer Sciences, Social-Informatics and Telecommunications Engineering) (2008)

A Process for Creating the Elicitation Guide of Non-functional Requirements

Andreia Silva, Plácido Pinheiro, Adriano Albuquerque and Jônatas Barroso

Abstract Non-functional requirements play a crucial role in the process of software development because they correspond to the characteristics and restrictions on which the software must running. The earlier this information is available, the faster the decision process of the solution that will best meet the customer need. However, the elicitation of these requirements is not an easy task. This paper presents a process that aims at the construction of a guide that assists analysts in identifying the requirements for the customer software context. The first run of the process showed that the results are motivators.

Keywords Software process · Elicitation · Non-functional requirements

1 Introduction

The non-functional requirements (NFR) must be known and properly treated since the beginning of the development life cycle because they can influence the selection of the technologies used, the standards to be adopted for the development of software, defining allocation strategies hardware resources, the security mechanisms and how to license and distribute the software.

A. Silva (✉) · P. Pinheiro · A. Albuquerque
University of Fortaleza, Av. Washinton Soares, 1321, BL J, SL 30,
Fortaleza, Ceara 60833-155, Brazil
e-mail: andrearsp@gmail.com

P. Pinheiro
e-mail: placido@unifor.br

A. Albuquerque
e-mail: adrianoba@unifor.br

J. Barroso
Dataprev, Technical Architecture, Fortaleza, Brazil
e-mail: jonatas.barroso@dataprev.gov.br

R. Silhavy et al. (eds.), *Software Engineering Perspectives and Application in Intelligent Systems*, Advances in Intelligent Systems and Computing 465,
DOI 10.1007/978-3-319-33622-0_27

Also, knowing the NFR in the early stages of development can help improve estimates of size, effort and cost of the project, and can avoid rework. A NFR discovered late can significantly increase the effort and cost of the project. It will probably have to be adapted to meet the new feature. According to Boehm [1], without a well-defined set of NFR and its proper fulfillment, software projects are vulnerable to failure. In addition, errors caused by inadequate treatment or lack of treatment of these requirements are appointed among the most expensive and difficult to correct [2].

However, although the NFR are of great importance for the development of the software, there has been an unbalance between the importance dedicated to these requirements and functional requirements. Software development methods, despite incorporating the NFR in the requirements phase, do not present details of how to treat them [3]. This is a point that deserves attention because, as [4], specify and formally characterize NFR is a complex task.

Not least is the fact that the NFR be characteristics related to several specialist areas and hardly a systems or business analyst know all the requirements for any context. Also, a common problem according to [5, 6] is that, in many cases, the customer is not aware of their needs related to the NFR and considered to be difficult understand details about them.

An important source of knowledge and reference for the NFR elicitation can be found in the international standard ISO 25010 [7]. The standard describes eight product characteristics that define the quality of software. Importantly, the standard establishes the quality requirements, but does not determine how the requirements should be elicited nor establishes a written standard. In addition, some other studies have been developed to support the process of NFR elicitation [1–6, 8–10], but the industry seems still find barriers for the treatment of these requirements.

According to [11], it is likely that the reasons for the inadequate treatment of NFR are related to the high level of abstraction and lack of understanding of their applications, operating scopes and features. Moreover, as [6], there is a lack of well-formatted definitions for NFR.

In order to facilitate and improve the quality of non-functional specifications of software, this paper proposes a process for creating a elicitation guide NFR applicable to software development and maintenance. The use of this process will support companies in defining the fundamental requirements for each type of software developed in addition to providing a mechanism that facilitates the elicitation of these requirements with customers.

This article consists of five sections: the following are the main concepts of NFR. Section 3 contains the importance of process in the guide definition. Then, in Sect. 4 describes the proposed process. Finally, Sect. 5 presents the conclusions.

2 Non-functional Requirements

The requirements of a system correspond to descriptions of the services that the software must provide and the restrictions on its operation. They are divided into functional and non-functional requirements.

Functional requirements describe the functions that a system or software component must provide [12]. The NFR, on the other hand, represent the system constraints. According to [13], the NFR may be associated with emerging properties of the system such as: reliability, response time, protection or availability. In general, specify or restrict the characteristics of the system as a whole.

The NFR are also known for quality requirements [14] and have significant impact on the final product [15]. The term software quality, before assigned to those systems that met all their functional requirements, is changing as the NFR are increasingly being demanded by users [3].

Likewise, the NFR strongly impact the total cost and time required to develop a software [16]. Not least, the errors originated by the inadequate treatment of NFR or their failure to observe are identified as the most cost and difficulty of correction [2]. This is a point that deserves attention because, as Rahman et al. [6], the failure rate of projects is increasing due to the inadequate and insufficient treatment of the NFR at the proper time of the project life cycle. This has considerably increased customer dissatisfaction.

In addition, NFR are often more critical than the individual functional requirements because, in general, system users can find a way around a system function that does not meet their needs, as opposed to a NFR which, if not met, can come to disable the system as a whole [13].

Similarly, Ebert [17] says that overlook the NFR can bring more serious consequences than omit functional project requirements. Some of the problems that can be generated for not observing these requirements are well known software development, such as cost and schedule deviations, system interruptions for the inability to use, in addition to customers and users dissatisfaction [18].

However, the identification of NFR is not an easy task. Unlike functional requirements which already have well developed elicitation techniques, there is still no consensus on appropriate mechanisms and techniques to these requirements elicitation. This is one of the challenges of requirements analysis [6].

According Slankas and Williams [19], all systems have NFR but, not always, are explicitly defined in a formal specification. To Cysneiros and Leite [3], several causes may contribute to the neglect of these requirements, among which may be emphasized that despite being this developing methods, these requirements are only mentioned and not effectively treated.

Also, often functional and non-functional requirements are mixed and the process to differentiate them raises doubts and ambiguities [6]. Another point to consider is the fact that many analysts are not convinced of the importance of these requirements or do not have sufficient skills to elicits them.

Some factors contribute directly to this scenario. Firstly, the NFR are very diverse. This makes it difficult an analyst master all kinds applicable to a given context of software. Customers, in turn, do not always know the non-functional needs of their software or do not know how to explicit them. In addition, the various sources of knowledge, such as standards and norms or relevant bibliographic references related to NFR, do not specify in what situations a requirement should be elicited, or define a standard format for its definition.

While there have been proposed work dealing NFR elicitation, yet there is no evidence that indicate the most appropriate method for eliciting these requirements. None of these methods has been adopted as standard by the engineering community requirements [6].

Thus, knowing the requirements elicitation is one of the most critical activities and requires greater degree of knowledge in the software development process [20] and, in addition to the difficulty in identifying and quickly categorize NFR available from several sources, this work proposes a process to support organizations in the definition of a group of NFR that represent the context of their applications.

The motivation for creating a process that supports the guide definition to elicit NFR is described below.

3 Process Importance in Guide Definition

A process consists of a systematic approach to creating a product or performing a particular task [21]. According to ISO/IEC 15504 standard [22], process is a set of interrelated activities that aims to transform inputs into outputs.

The main motivation for the definition of processes is the quest for improved the embodiment of a work [23]. According to Humphrey [24], with the thought in an organized way process, it acquired the ability to anticipate problems and take actions that allow preventing them or solve them. Moreover, without the existence of a minimally defined process it is impossible to carry out assessments aimed at promoting improvements in these processes [25].

In the context of this work, the motivation for defining a process to support the creation of the NFR elicitation guide was given the following factors:

- Each company can work with a specific type of software: web, mobile, data warehouse applications, among others. The creation of a single guide would not meet all contexts.
- Select the requirements for software context will be developed from the several available sources, as well as time consuming, it can be considerably complex task. It is believed that the use of the process remove the complexity of the moment of the elicitation of requirements.
- Using the process will be possible to spread the knowledge of the different specialists departments of the company with software development teams.

- The process can contribute to maintaining the continuous improvement of elicitation guide in future releases.

The description of the process is presented in detail below.

4 Process for Creating the Elicitation Guide

The description of the proposed process establishes the consumed and generated artifacts, input and output criteria for each activity and those responsible for performing these activities. The graphical process description was made using the standard BPMN for modeling business processes [26].

The defined process involves executing nine activities, as illustrated in Fig. 1 and detailed below.

1. **Define application context**: the purpose of this activity is to establish the type of software application to be analyzed to define the applicable non-functional requirements. For example: software application for web, mobile and data warehouse.
    a. *Required Artifacts*: List of products developed by the company.
    b. *Produced Artifacts*: Definition of the type of application being treated.

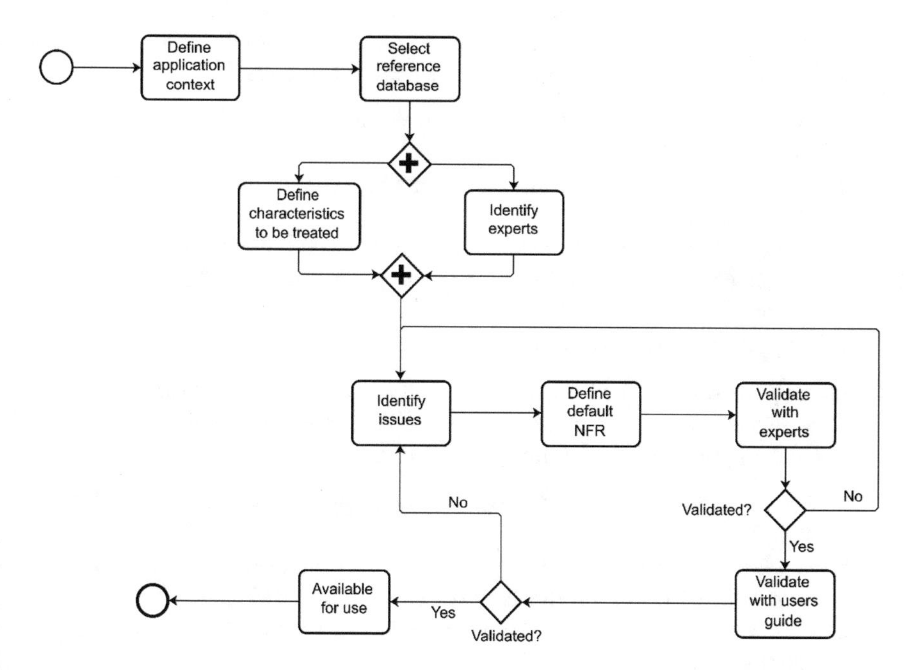

Fig. 1 Process for creating the elicitation guide of NFR

   c. *Input Criteria*: Not applicable.
   d. *Output Criteria*: Application types defined.
   e. *Responsible*: Processes team.

2. **Select reference database**: the purpose of this activity is to identify, among the reference sources on NFR, the one that most adequate for the company context and the type of project in question. Examples of reference sources for NFR can be found in [7, 8, 13].

   a. *Required Artifacts*: Standards and bibliographic references related to NFR.
   b. *Produced Artifacts*: Definition of the reference source selected.
   c. *Input Criteria*: List of standards and bibliographic references related to NFR finalized.
   d. *Output Criteria*: Reference source defined.
   e. *Responsible*: Processes team.

3. **Identify experts**: the purpose of this activity is to identify the organization's employees with in-depth knowledge of the subject being treated.

   a. *Required Artifacts*: List of employees and their skills.
   b. *Produced Artifacts*: Definition of experts for each NFR category.
   c. *Input Criteria*: List of employees and their competencies finalized.
   d. *Output Criteria*: Identification of NFR experts completed.
   e. *Responsible*: Processes team.

4. **Define characteristics to be treated**: the purpose of this activity is to define, among the different categories, types or characteristics of NFR, those that must be treated in the context of the type of software being studied.

   a. *Required Artifacts*: Reference source selected in the activity "Select reference database".
   b. *Produced Artifacts*: Definition of categories, types or characteristics of NFR to be treated.
   c. *Input Criteria*: Activity "Select reference database" complete.
   d. *Output Criteria*: Categories, types or characteristics of NFR to be treated defined.
   e. *Responsible*: Processes team and experts.

5. **Identify issues**: the purpose of this activity is to define, for each category of NFR, questions that can be asked for the customer to identify the requirements that will apply to the application context in development.

   a. *Required Artifacts*: List of categories, types or characteristics of NFR to be treated.
   b. *Produced Artifacts*: Checklist with questions directed to the customer for each selected feature in the activity "Define characteristics to be treated".
   c. *Input Criteria*: Activities "Define characteristics to be treated" and "Identify experts" concluded.

d. *Output Criteria*: Checklist with questions directed to the customer finalized.
e. *Responsible*: Processes team.

6. **Define default NFR**: the purpose of this activity is to establish, for each question, a corresponding NFR for cases in which the questioning of the customer response is positive.

   a. *Required Artifacts*: Checklist with questions directed to the customer obtained through the activity "Identify issues".
   b. *Produced Artifacts*: Checklist update to include NFR corresponding to each question defined in the activity "Identify issues".
   c. *Input Criteria*: Activity "Identify issues" concluded.
   d. *Output Criteria*: Checklist updated with the corresponding NFR to each question.
   e. *Responsible*: Processes team.

7. **Validate with experts**: the purpose of this activity is to validate with the experts if the list of questions and defined requirements are within the required quality standards.

   a. *Required Artifacts*: Checklist with questions directed to the customer and the requirements corresponding to each questioning.
   b. *Produced Artifacts*: Checklist quality evaluation report.
   c. *Input Criteria*: Activity "Define default NFR" concluded.
   d. *Output Criteria*: Checklist quality evaluation report filled.
   e. *Responsible*: Experts.

8. **Validate with users guide**: the purpose of this activity is to validate with elicitation guide users if the list of questions and requirements defined are within the required quality standards.

   a. *Required Artifacts*: Checklist with questions directed to the customer and the requirements corresponding to each questioning.
   b. *Produced Artifacts*: Checklist quality evaluation report.
   c. *Input Criteria*: Activity "Validate with experts" concluded.
   d. *Output Criteria*: Checklist quality evaluation report filled.
   e. *Responsible*: NFR elicitation guide users.

9. **Available for use**: the purpose of this activity is to make available for the entire organization so that the guide can be applied in their new projects.

   a. *Required Artifacts*: Checklist quality evaluation report obtained through the realization the activity "Validate with users guide".
   b. *Produced Artifacts*: Guide (checklist) for NFR elicitation.
   c. *Input Criteria*: Activity "Validate with users guide" concluded.
   d. *Output Criteria*: Guide (checklist) for NFR elicitation available for the entire organization.
   e. *Responsible*: Processes team.

5 Final Considerations

Although the NFR is of great importance for the development of the software, there has been an unbalance between the importance dedicated to these requirements and functional requirements. For many years, much of the attention in software engineering was focused on notations and techniques to define functional requirements and the many software development methods, despite incorporating the NFR in the requirements phase, do not present details of how to handle them.

This neglect of the NFR can generate significant losses for software projects. In addition to influencing the cost and time to develop the software, the NFR are essential for the process of deciding which solution will best meet customer needs. For this reason, it is essential that the NFR be known as soon as possible.

Thus, considering the wide range of existing NFR, the difficulty inherent in the elicitation activity of these requirements, together with the required level of knowledge and the absence of a method or standard technique, has motivated this work. The objective of the proposed process is, above all, to build a guide to assist business and system analysts to identify the requirements for the customer software. The process execution results in a set of questions that are directed to customers, along with a set of NFR already defined and associated with these issues.

Furthermore, the process presented here also contributes to reduce complexity and optimize the time of elicitation activity of the NFR, disseminate knowledge of experts to the several members of the organization, optimize the analysis time of the solution that best fits the needs of customer and improve estimates of time and cost of the project.

Currently, the process has been applied in a large governmental software development company and the results obtained so far are very positive, particularly as regards the involvement of the different departments of the company in the creation of the guide and the opinion of the analysts who participated the first evaluation of the final product.

The next steps of this research aim at applying the guide in different software development projects and evolve the process with the user experience. Furthermore, to improve the utilization of the guide is being proposed to develop a tool to support the NFR elicitation process.

Furthermore, applying a multicriteria model for selection of most important NFR the customer's point of view is being studied. The application of this methodology in the context of software process demonstrated good results, as shown in [27–29].

Acknowledgments The second author is thankful to National Counsel of Technological and Scientific Development (CNPq) via Grants #475239/2012-1.

References

1. Boehm, B., In, H.: Identifying quality-requirement conflicts. IEEE Softw. **13**, 25–35 (1996)
2. Cysneiros, L.M.: Requisitos Não Funcionais: Da elicitação ao Modelo Conceitual. Ph.D. Thesis, Dept. de Informática PUC-Rio, Feb 2001
3. Cysneiros, L.M., Leite, J.C.S.P.: Definindo Requisitos Não Funcionais. In: XI Simpósio Brasileiro de Engenharia de Software, 49–54. Anais, Outubro 1997
4. Cysneiros, L.M., Yu, E.: Non-functional requirements elicitation. In: Perspectives on Software Requirements, vol. 753, pp. 115–138. Springer, US (2004)
5. Galster, M., Bucherer, E.A.: Taxonomy for identifying and specifying non-functional requirements in service-oriented development. In: IEEE Congress on Services-Part I, pp. 345–352 (2008)
6. Rahman, M.M., Ripon, S.: Elicitation and modeling non-functional requirements-a POS case study. In: International Journal of Future Computer and Communication, vol. 2, pp. 485–489 (2013)
7. ISO 25001. http://www.abntcatalogo.com.br/norma.aspx?ID=38835
8. Chung, L., Nixon, B., Yu, E., Mylopoulos, J.: Non-functional Requirements in Software Engineering. Kluwer Academic Publishers (1999)
9. Al Balushi, T.H., Sampaio, P.R.F., Dabhi, D., Loucopoulos, P.: ElicitO: a quality ontology-guided NFR elicitation. In: 13th International Conference on Requirements Engineering Foundations for Software, Trndheim, Norway (2007)
10. Cleland-Huang, J., Settimi, R., Zou, X., Solc, P.: Automated classification functional requirements. Requirements Eng. **12**(2), 103–120 (2007)
11. Hasan, M.M., Loucopoulos, P., Nikolaidou, M.: Classification and qualitative analysis of non-functional requirements approaches. In: Enterprise, Business-Process and Information Systems Modeling, pp. 348–362. Springer, Berlin, Heidelberg (2014)
12. Thayer, R., Dorfman, M.: System and Software Requirements Engineering. IEEE Computer Society Press (1990)
13. Sommerville, I.: Software Engineering, 8ª edition. Addison Wesley (2007)
14. Chung, L.: Representing and using non-functional requirements: a process oriented approach, Ph.D. Dissertation, Department of Computer Science. University of Toronto, June 1993
15. Boehm, B.W., McClean, R.K., Urfrig, D.B.: Some experience with automated aids to the design of large-scale reliable software. IEEE Trans. Softw. Eng. **1**, 125–133 (1975)
16. Abdulmonem, R., Olga, O., Russell, W.: Ontology-based classification of non-functional requirements in software specifications: a new corpus and svm-based classifier. In: IEEE 37th Annual Computer Software and Applications Conference (COMPSAC), pp. 381–386 (2013)
17. Ebert, C.: Dealing with nonfunctional in large software systems. In: Annals of Software Engineering, pp. 367–395 (1997)
18. Neto, J.M.S., Leite, C.S.P., Cysneiros, L.M.: Non-functional Requirements for Object-Oriented Modeling (2000)
19. Slankas, J., Williams, L.: Automated extraction of non-functional requirements in available documentation. In: 1st International Workshop on Natural Language Analysis in Software Engineering (NaturaLiSE), pp. 9–16. IEEE (2013)
20. Gottesdeiner, E.: Requirements by Collaboration. Addison-Wesley (2002)
21. Osterweil, L. Software processes are software too. In: International Conference on Software Engineering, pp. 2–13 (1987)
22. ISO/IEC-15504. Information Technology—Software Process Assessment. Parts 1–9. The International Organization for Standardization and the International Electrotechnical Commission (2004)
23. Barreto, A.S.: Uma Abordagem para Definição de Processos Baseada em Reutilização Visando à Alta Maturidade em Processos. Diss. Universidade Federal do Rio de Janeiro (2011)
24. Humphrey, W.S.: Managing the Software Process. Addison-Wesley (1989)

25. Wang, Y., King, G.: Software Engineering Processes: Principles and Applications. ICRC Press (2000)
26. Object Management Group. Business Process Modeling Notation (BPMN), Version 2.0. OMG Specification. OMG. http://www.omg.org/spec/BPMN/2.0 (2011)
27. Rodrigues, A., Pinheiro, P.R.; Rodrigues, M.M., Carvalho, A., Gonçalves, F.M.: Applying a multicriteria model for selection of test use cases: a use of experience. In: International Journal of Social and Humanistic Computing (Print). vol. 1, pp. 246–260 (2010)
28. Rodrigues, A., Pinheiro, P.R.; Rodrigues, M.M., Albuquerque, A.B., Gonçalves, F.M.: Towards the selection of testable use cases and a real experience. In: Communications in Computer and Information Science (Print). vol. 49, pp. 513–521 (2009)
29. Sampaio, M., Donegan, P., Castro, A.K., Pinheiro, P.R., Carvalho, A., Belchior, A.D.: Multicriteria model for selection of automated system tests. In: Tjoa, A.M, Xu, L., Sobail, C. (eds.) (Org.). IFIP International Federation for Information Processing, 1st edn, vol. 205, pp. 777–782. Springer, New York, US (2006)

A Data-Centric Algorithm for Identifying Use Cases

Erki Eessaar

Abstract Creating use case models is a popular method of requirements analysis. Unfortunately, guidelines for the identification of use cases are quite general. Thus, inconsistencies with state machines that model lifecycles of entity types can easily appear. Traditionally, one writes use cases as requirements and uses these as an input for creating analysis models. We propose a data-centric algorithm that partially turns this process upside down. It takes a set of system actors, a set of their informational requirements, an initial conceptual data model, and state machines of the main entity types as an input. It produces a set of use case descriptions, each of which contains the name of the use case, its primary actors, and general understanding of its functionality. One would use this model for the elaboration of use cases and writing their textual specifications.

Keywords Use case · Requirements · State machine · UML · System analysis

1 Introduction

Use cases allow us to describe system (or business) functionality by concentrating to its responsibilities from the viewpoint of its actors instead of its internal algorithms. Use cases are popular for capturing requirements to systems and developers can later use these for planning iterations, driving the design, testing the system, and writing user manuals. Jacobson [1] comments that use case model is like the hub of a wheel that binds together all the activities within a software project. Each use case model can contain a visual (one or more diagrams) and a textual part. UML specification [2] describes abstract and concrete syntax of use case diagrams. It

E. Eessaar (✉)
Department of Informatics, Tallinn University of Technology,
Akadeemia tee 15A, 12618 Tallinn, Estonia
e-mail: Erki.Eessaar@ttu.ee

R. Silhavy et al. (eds.), *Software Engineering Perspectives and Application in Intelligent Systems*, Advances in Intelligent Systems and Computing 465,
DOI 10.1007/978-3-319-33622-0_28

acknowledges that one may specify use cases in various formats such as natural language, tables, trees, etc. However, [2] does not specify the structure of textual descriptions of use cases as well as ways how to identify use cases. One can specify use cases with different levels of granularity. Cockburn [3] suggests three levels of use cases—summary-level, user goal-level, and subfunction-level. Larman [4] thinks that for the requirements analysis of a computer application, one should focus to use cases at the level of elementary business processes (EBPs). These corresponds to the user goal-level. Larman [4] defines EBP as "A task performed by one person in one place at one time, in response to a business event, which adds measurable business value and leaves the data in a consistent state."

We claim (see Sect. 2) that literature offers quite vague guidelines of how to identify use cases. Thus, there are no clear criteria to decide and explain as to whether the identified use cases are necessary and sufficient. Similarly, it is hard to decide and explain as to whether the decomposition of the responsibilities in terms of use cases has a good quality. We have taught use case modeling to undergraduate students in the context of database design courses. In parallel with collecting requirements to the database, one must collect requirements to the functionality of the system and synchronize the requirements. In student projects, we have encountered models that map each entity type or only each main entity type to either a *Manage <X>* use case or to four (create, read, update, delete—CRUD) use cases. A variation of the latter is a use case model that for each main entity type has one additional use case for conducting the state changes according to the lifecycle of the corresponding main entity type (for instance, *Change the state of an order*). A uniform structure tempts its users to copying and pasting of textual descriptions that could again lead to errors. Use cases that correspond to the read operation might say that the system has to present one or more reports without sufficiently explaining their content. Sometimes modelers have forgotten altogether use cases that describe responsibility of the system to present predefined reports. In the student projects and elsewhere, we have encountered consistency problems of use case models and state machines. We have seen state transitions in the state machines that do not have corresponding use cases to express the need that the system has to initiate or react itself to such transitions or make possible initiation of or reacting to such transitions. For instance, according to the lifecycle of orders, which one presents in terms of the state machine of orders, it is possible to reject an order but no use case describes the functionality. We have seen use case models that do not pay attention to the interactions of lifecycles of different entity types. For instance, cancelling of a show means that its organizers must refund tickets or replace these with tickets to another show and the system must provide functionality for that. Yet another problem is that some use cases remain at the summary-level even after the detailed analysis of requirements.

Fixing these problems always takes many resources. One could accept this kind of models by arguing that there are no right or wrong models but more or less useful models and that writing textual descriptions has at least lead writers to think about the system. However, still the models are not as good as they might be and considerable effort of making them is bigger than the advantage that they offer to their readers.

Our proposed algorithm of finding use cases appeared because of the practical need to solve the previously mentioned problems. It must employ quite easily explainable rules, must ensure consistency with the models that depict lifecycles of entity types, and must facilitate clear expression of requirements to the reporting functionality. It must lead to the creation of use cases that are at the user goal-level. It must be possible to automate it with the help of an extension of a UML CASE tool. We think that such algorithm is not only useful for novices or students but to system analysts in general.

The method of the research and the basis of structuring the paper is *action design research* (ADR) [5]. The *goal* of the work is to both address the problems that we encountered in teaching in university as well as to produce an IT artifact (in this case an algorithm). Building of the algorithm involved continuous presentation of new versions to students who at the same time made their projects and making improvements in the algorithm based on the feedback of students as well as problems in their models. This paper is an attempt to formalize the learning outcomes of this cyclic process.

We organize the paper as follows. Sections 1 and 2 correspond to the ADR problem formulation phase. Description of related works in Sect. 2 describes existing theories in order to explain the research opportunity and address our claim about identifying use cases. Section 3 presents the algorithm itself. Section 4 discusses application of the algorithm. Finally, we conclude and point to the future work with the current topic.

2 Related Works

Shu [6] comments that Object-Oriented Software Engineering "does not provide clear guidelines about how to identify use cases." Among others, Cockburn [3] and Larman [4] propose us to firstly identify primary actors, next identify their goals, and then map each goal to a certain use case that helps the actor to achieve the goal. Larman [4] offers a possibility to find actors, goals, and use cases by firstly finding external events. Ferg [7] suggests us to use domain model as the infrastructure for the systematic definition of use cases. For the entity types that are described in this model, one has to find events in the lives of their entities (essentially describe lifecycles) and map these to use cases. The current paper actually operationalizes this approach. The Rational Unified Process [8], which is a variant of the Unified Process, provides a set of questions that suggest us to look the system from the perspective of actors and their tasks as well as informational needs. The document also points to four types of events that analysts often overlook and thus do not have triggered use cases in place. Although both Larman and Cockburn want us to create user-goal level use cases, the former suggests us to try to create them directly but the latter by reconsidering and refining initial summary-level use cases. Jacobson [1] offers a numeric measure by suggesting that a large system that supports one business process should have at most twenty concrete (real) use cases.

Columns of the Zachman Framework ™ [9] define different aspects (phenomena) that one must use to describe and understand systems. Use case models combine all these aspects. The courses of action in the text of a use case describe a process that the system and actors conduct together. It correspond to the "How" column. Other columns of the framework correspond to the primary actors ("Who" column), goals that the use case helps its primary actors to achieve ("Why" column), events that trigger the use-case ("When" column), locations where the use case actions take place ("Where" column), and the entity types that corresponding data the use case process creates, updates, deletes, or reads ("What" column). These other columns describe aspects that determine the context for the process. Thus, in general, we can say that different authors suggest us to use different context-setting aspects as aids for discovering use cases.

There is quite a lot of research about how to derive use cases, which describe the system, from the models that describe goals and processes of the business that needs the system. For instance, one could present the transformed business process descriptions with the help of UML activity diagrams [10], Business Process Model and Notation, or Petri nets. If one has described the functions and processes of the system by using business use cases, then it is possible to map these to system use cases [8]. Deriving use cases from the source code to facilitate maintenance and evolution of the system is yet another example of transforming other artifacts to use cases [11].

Many author acknowledge that use cases offer a good and flexible way to capture requirements of stakeholders and integrate these to the requirements gathering process. However, there are also references to problems of use case modeling that include:

- inconsistencies and discrepancies between use cases that are caused by their definition in isolation [12, 13] or redundant descriptions of the same functionality;
- lack of precision due to little-restricted semi-structured textual nature of use cases [12, 14–16], which complicates interpretation and finding anomalies [12];
- multitude of guidelines by different authors (see, for instance, [3, 4]) what and how to represent in use case descriptions together with the lack of a standard [14];
- coupling of functional and non-functional concerns in use case descriptions [16];
- lacking explicit information about what is the order of executing use cases (and thus the lack of information about the global behavior) or how actors interact outside use cases [14]. Limited support to showing intra-use case dependencies in UML [14].

All this makes it difficult to detect inconsistencies, incompleteness, redundancy, or ambiguity in use case descriptions. Thus, there is a lot of research about structuring use cases as well turning their textual specifications into representations that one could use for analysis, validation, simulation, and generation of further artifacts. Larman [4] comments that although the Unified Process does not specify "state model" it is possible to investigate the dynamic behavior of any element in any model (including use cases) by using state machines. Yue et al. [17] provides an extensive and systematic literature review of approaches for generating analysis models from use case models. The former present requirements to the system in more structured and less ambiguous manner compared to use case model. For instance, one could represent the analysis models in terms of different types of visual models. Quite often, these approaches attempt to produce state machines based on use cases (see [12–15, 18]). The listed approaches propose the creation of state machines to validate written requirements [12–14], be a part of conceptual design schemas [18], or generate test cases [15]. In case of validating requirements, some of the approaches [12] only produce state machines that one can use for manual analysis. On the other hand, some [13, 14] produce executable simulations that frees analysts from the need to build prototypes. A commonality of the approaches [12–15] is to restrict the processed use case text according to some rules to increase the precision of written specifications. Some of the approaches produce intermediate results, like an instance of an intermediate metamodel [15] or conceptual predesign schema [18], during the transformations.

To summarize, none of the approaches uses state machines as the basis for identifying use cases, whereas we propose the use of state machines as a basis to identify use cases and their scope. If [12] applies use cases to detect possible anomalies in requirements, then our algorithm tries to reduce these in the first place. A difference of approaches is that [12, 13] create a state machine for every use case whereas our algorithm leads to the creation of multiple use cases based on one state machine. On the other hand, [18] proposes the creation of state machines to thing-types, which represent important domain concepts, and correspond thus to the main entity types in our algorithm. The approach in [14] makes simulations of use cases and their dependencies based on the states of actors, whereas we use states (in terms of lifecycle) of the main entity types to determine use cases. The transformation rules in [15] take into account pre- and post-conditions of use cases. Because these could refer to the steps in the lifecycles of the main entity types, one can say that the use of the conditions is a reversal of an idea of our proposed algorithm. The algorithm offered in our paper does not restrict structure and language of use cases that one can use to write them down after identification. The algorithm also does not produce an intermediate artifact.

The idea of using lifecycles of the main entity types to describe processes is not new. Artifact-centric business process modeling that has gained traction during the last decade specifies systems and their processes in terms of business artifacts as well as their lifecycles, services, and associations. Estañol et al. [19, 20] propose how to use UML for artifact-centric process modeling. They model lifecycles of artifacts, which one can identify based on the main entity types, by using state machines. For modeling the associations between services, they suggest us to use activity diagrams, not use case models. Each goal-level use case specifies a localized process conducted by one actor in one place at one time to achieve a goal whereas activity diagrams allow us to model processes that span multiple goal-level use cases. The latter gives bigger flexibility in modeling the associations between services. In [20] they propose modelers to create a separate activity diagram based on each state transition. Our algorithm proposes modelers to present in general one use case for each state transition. Thus, the approaches lead to the same granularity of processes that we represent as a logical whole.

We have proposed a set of practices for the development of data-centric information systems [21]. The set of system actors gives us decomposition of the system in terms of areas of competence. The practices describe how to use the set of the main entity types for the functional and data-centric decomposition of systems into functional subsystems and registers, respectively. There are techniques for the *quick* discovery of the main entity types. Specification of each register and functional subsystem must contain the state machine of the corresponding main entity type and use case model, respectively. One needs actors and state machines as an input of the algorithm that we present here. Because the required inputs are not the exclusive results of the data-centric development practices and because the practices themselves are usable in the context of different methodologies and processes, we think that the use of the algorithm is yet another data-centric practice, which is not restricted with specific methodology or process.

3 The Proposed Algorithm

The algorithm needs the following information as the input.

- A set of actors (roles) that representatives will use the system.
- For each actor, a set of predefined informational requirements (reports) that its representatives want to get from the system.
- State machines depicting lifecycles of the main entity types that corresponding data the system has to manage. In case of using UML, the state machines would be protocol state machines that we use to specify lifecycles of classifiers from the external perspective [2]. The algorithm considers only elements that are possible in protocol state machines—initial pseudostate, (protocol) state, (protocol) transition, and final state. Each state machine must only consider the periods in the life of entities that are interesting and relevant to the organizations that need the information system. Therefore, for instance, a transport company owns trucks. The life of a truck entity might end in a car-scrapping yard. However, because the transport company does not own the scrapping yard, then from the perspective of the company the life of the truck ends with selling it to some other party.
- For each state transition, a set of actors who trigger it in the sense that they are the ones who should register information about the new state in the system.

- An initial version of the conceptual (aka logical [22]) data model of the system.
- For each state transition, at least initial understanding of its pre- and post-conditions in terms of the elements of the conceptual data model.

The algorithm allows us to identify use cases with the precision of name, associated primary actors, and general understanding of functionality that the system offers to the primary actors. It also allows us to establish links between use cases and model elements that are the basis of their creation. In case of each main entity type, the algorithm takes the following consecutive steps.

Step 1: In general, create a use case for each state transition that changes the state of the entities of the main entity type. There are the following exceptions.

- If there are multiple state transitions from different states to the same state and the transitions have the same post-conditions and are triggered by the same set of actors *A*, then create one use case based on all these transitions and connect it with *A*. If, in addition, there is some other set of actors *B*, who could trigger only a proper subset S' of these transitions, then create for *B* a separate use case based on S'.
  - For instance, if managers can discard active and inactive products, then create the use case "Discard a product" and connect it with the actor "Manager". If product administrators can discard only inactive products, then add the use case "Discard an inactive product" and connect it with the actor "Product administrator".
  - However, if in the lifecycle of products there is the state "Active" that has two incoming transitions—one from the initial pseudostate and another from the state "Inactive", then this will not lead to the creation of one use case. Even if the same actor triggers both transitions, the transitions have different post-conditions. The first have post-conditions that describe initial registration of the product whereas the second only requires changing the state of an existing product instance.
- There could be dependencies between the lifecycles of different main entity types in the sense that a transition in the lifecycle of an entity from one main entity type triggers an *automatic* transition in the lifecycles of entities that belong to other entity types. We should do it in the context of one use case.
  - For instance, a business process permits reservation of a resource before committing to its actual usage. *Reservation* and *Usage* are two different transactional main entity types. If one creates a usage based on a reservation, then the use case "Register a usage" must also change the state of the reservation based on that one created it. It is a matter of choice as to whether to create only one use case "Register a usage" or to add the use case "Register a usage based on a reservation".

- It is possible that one does not use in database state classifiers to make possible recording of the current states of the main entities. Instead, one may use values of attributes as well as the existence or lack of relationships between entities to calculate the current state. If a state transition of a main entity always happens because time passes a recoded timestamp and there is no need to trigger some action because of that, then we do not define a use case for this transition.
  - For instance, a session goes to the state "passed" after the end of its period.
- If there are two or more outgoing transitions from the same state that are triggered by the same set of actors and the actors have to decide what path to take, then there is a possibility to present it in one use case. In this use case, the most probable path corresponds to the main scenario and less probable paths correspond to alternate scenarios. However, we advise against creating the single use case due to resulting large textual description and coupling of concerns.
  - For instance, if a manager can accept an order (more probable) or reject an order, then there is a possibility to create the use case "Accept an order" that is connected with the actor "Manager" and that has an alternate scenario for rejecting orders.

Step 2: Create zero or more use cases for changing the data corresponding to the main entity type in some other aspect than registering the transition of a main entity from one lifecycle step to another. This may include updating facts (true propositions) about the main entities as well as inserting, updating, or deleting facts about their related non-main entities. For each actor, one has to determine the subset of facts that its representatives can change. For identifying the subset, one has to look vertically across different attributes and relationships as well as horizontally by specifying Boolean expressions that determine facts that the actor can insert, update, or delete. The expressions could take into account the current state of main entities. Create a use case for each unique subset and connect with it all the actors that can change exactly this subset. One can indicate the need of such use cases with the help of self-transitions in the state machine.

- For instance, if a client can update his/her own submitted active orders but a manager can update all the active orders, it will require two different use cases. Similarly, if a manager can update the delivery addresses and change the set of ordered products of active orders but a secretary can only update the delivery addresses of active orders, it will lead to two different use cases.

Step 3: For each set of actors, who are interested about exactly the same view to the data corresponding to the main entity type and its related non-main entity types, create a use case describing the view as well as the process of getting the view.

– For instance, in an e-shop a manager of orders could look the detailed information of all the orders, a client could look the detailed information of his/her own orders, and a general manager could look the total number of orders as well as the number of orders by each lifecycle step. It means that one has to define in the use case model three different use cases for that purpose.

Step 4: If there are informational requirements (reports) that the previously described use cases do not satisfy, then for each such requirement, create a separate use case. There is a possibility to group the reports that each actor needs together to one use case. However, we advise against it because it violates the separation of concerns principle meaning that the use case has multiple change drivers [16]. Moreover, if multiple actors need the same report, then it could lead to the duplication of the report description.

Step 5: Although some authors like [3] consider the use cases "Log in" and "Log out" as subfunction-level use cases that the relevant goal-level use cases should

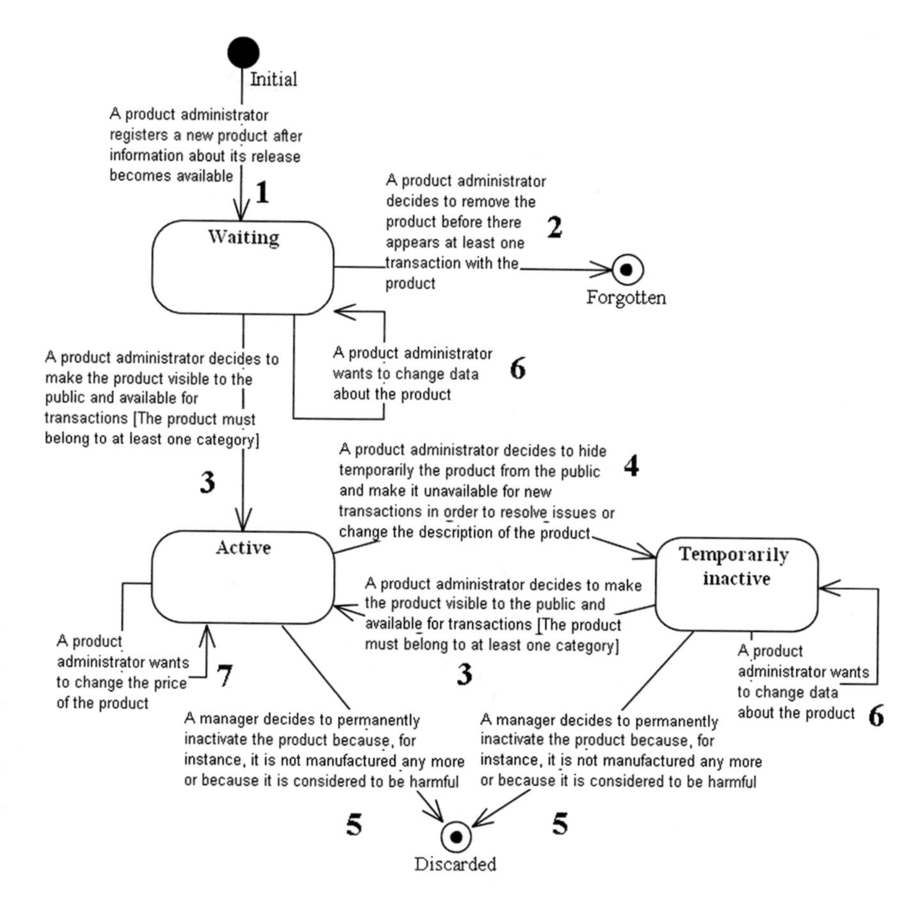

Fig. 1 A possible state machine of products in an e-shop system

invoke, we propose to define "Identify a user" as a goal-level use case. We do it because ensuring confidentiality of data and avoiding unauthorized modification is a goal of its own, which violation leads to the loss of business value.

Figure 1 presents a possible state machine of products in an e-shop system. Figure 2 presents the use case model that we have partly created based on the state machine. We have annotated state transitions in the state machine with numbers. Numbers in the use case model indicate, which use cases have been found based on which state transitions.

The state machines that we use as a basis of finding the use cases contain additional information that we can reflect in the textual specifications of use cases. Triggers (events) that initiate state transitions initiate also the corresponding use cases. The preconditions of these use cases must contain the logical conjunction of information about the origin states and pre-conditions (guards) that are associated with the transition.

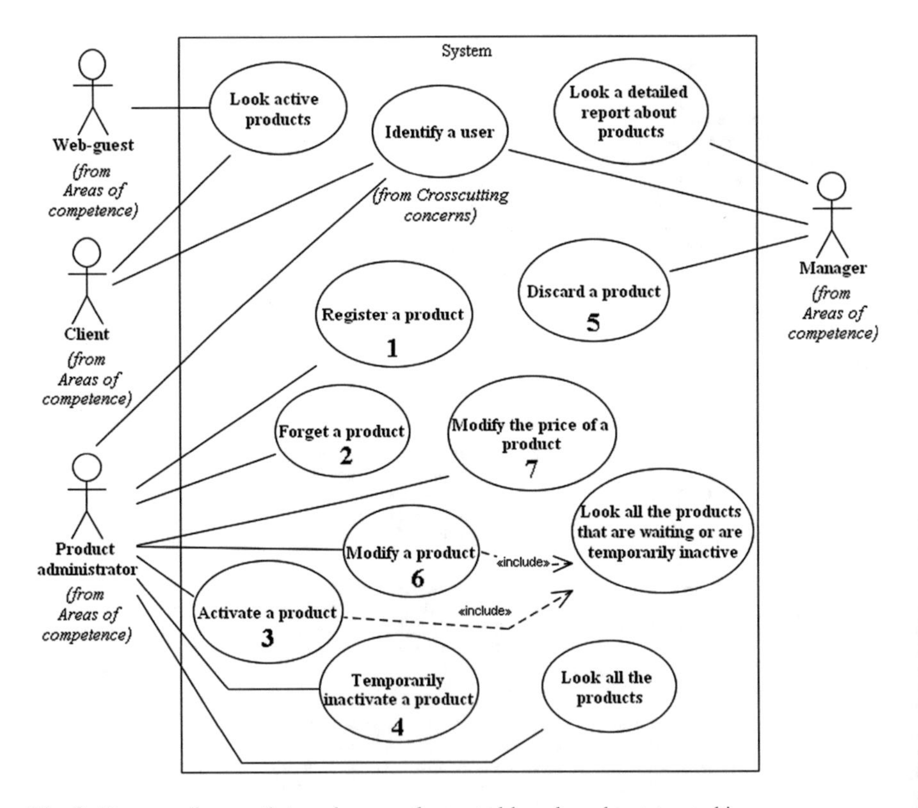

Fig. 2 Use case diagram that we have partly created based on the state machine

4 Discussion

The algorithm produces use cases that represent CRUD operations on the main entity types. They are not overly generic meaning that they conform better (than examples in the Sect. 1) to the *separation of concerns* principle. This property also means that it is easier for us to achieve *separation of duties*, meaning that the management of data of an entity through its entire lifecycle requires more than one person in order to reduce the possibility of fraud or error [23].

Next, we outline a possible process of collecting requirements and analyzing the system by using our algorithm. It should be iterative, meaning that thinking about the system and modeling it leads continuously to the need to update the artifacts that we next mention. During all the steps, collaboration between the stakeholders of the system and system analysts is possible and recommended. In our experience, interacting with stakeholders during a workshop and at the same time creating models so that all the participants of the session can see it right away is an effective way of working.

The goal of the first steps of the process is quickly and with acceptable quality reach to the initial business architecture of the system. It will support and guide elaboration of requirements. One has to find a set of actors who are going to use the system and for each such actor a set of goals that the system must help its representatives to achieve. One could present the goals as a textual list or with the help of specialized goal models. Moreover, one has to identify a set of the main entity types that corresponding data the system has to manage. This set is the input to identifying initial set of predefined reports that the system must provide. Most of the reports need data that corresponds to multiple main entity types. One has to produce for each identified main entity type the state machine depicting all the possible lifecycles of its entities (instances) in the context of the organization. Moreover, one needs the first version of conceptual data model.

The produced artifacts are inputs to the process that identifies initial set of use cases, which one could depict on use case diagrams. Thus, we produce one or more use case diagrams for each main entity type. One can characterize the approach as "faking" of the "ideal" development process that starts with use cases and proceeds with the creation of analysis models [24]. What matters is the result and less when and how we achieved it. Because each main entity type has a corresponding functional subsystem and register in the business architecture of an information system [21], each such set of use case diagrams declares the responsibilities of a particular functional subsystem. The union of all the use cases that we connect with a particular actor declare the responsibilities of the actor in the context of the entire information system (its area of competence). The use case diagrams will be an input for the writers of textual descriptions of use cases. State machines exhibit the sequence of states. Thus, in case of using our algorithm, these give information about the order of executing some of the use cases. In case of each state in the state machine, one has to ask, who and in what extent can modify data corresponding to the main entity type and its related entity types.

Each functional goal should have a corresponding use case depicting the responsibilities of the system and its users in terms of achieving the goal. One should evaluate the produced set of use cases against the goals of actors. It could lead to both modification of the set of goals as well as the set of use cases.

The guidelines from [21] should help us to increase the completeness of the set of the main entity types. The higher it is, the higher will be the completeness of process descriptions derived from the specifications of the lifecycles of the main entity types.

A weakness of the proposed approach is that clients and inexperienced analysts might be unable to create meaningful state machines (for instance, most of their models could consist of states "active" and "inactive"), thus leading to unsatisfactory use case models. Similarly, there could be mistakes in state machines that lead to the set of incorrect use cases. For instance, if one represents a final state twice, once as a (protocol) state and then as a final state, then the depicted transition between these will lead to the specification of an unnecessary use case. It could also be that modelers forget to depict the possibility of forgetting a main entity (meaning deleting all its data from the system) with the help of a final state (for instance, "Deleted").

With the help of more experienced modelers and perhaps also with the help of patterns of state machines, it could be possible to quickly and with a good quality reach to a set of use cases that could be used as a basis to elaborate requirements. In principle, Burns [22] offers a similar approach. He suggests quick creation of the first version of logical data model. One can do it based on universal models and patterns. Requirements analysts could use the model as a basis for collecting requirements in general and requirements about data that the system needs in particular. The first quick version of logical data model should be refined and further artifacts generated from it in an incremental manner. To support a similar approach in case of use case modeling, it would be useful to have universal state machines for entity types that could play a role in many different organizations and their information systems (agreement, document, party, etc.). In principle, it means universal models and patterns about business processes.

5 Conclusions and Future Work

Writing use cases is a popular requirements analysis method. Unfortunately, the guidelines for finding use cases are quite vague, leading to inconsistencies with analysis artifacts. We started the work from a practical need to explain to students how to find use cases in case of data-centric systems and how not to forget some use cases. We proposed an algorithm that uses the set of actors, their informational requirements, the state machines of the main entity types, and conceptual data model as the input. It produces a set of use cases. For each use case, it helps us to identify name, primary actors, and general functionality. Thus, it produces an initial framework based on that to describe more precisely the desired system behavior. It turns partly upside down the traditional approach that firstly produces use case models and then starts to create analysis artifacts based on these. Use case models created in our way are analysis artifacts that specify system functionality. With the emphasis to the lifecycles of the main entity types, the algorithm has a common ground with the artifact-centric business process modeling.

Future work must include evaluation of the algorithm in real world projects and comparing the results with other methods of identifying use cases. One should implement the algorithm in a CASE tool. In case of UML, we need a modeling profile to show interactions of lifecycles, relationships between actors and state transitions, as well as data needs and responsibilities of actors. To facilitate the use of the algorithm, we need universal business process models represented as interacting state machines.

References

1. Jacobson, I.: Use cases—Yesterday, today, and tomorrow. Softw. Syst. Model. **3**, 210–220 (2004)
2. OMG Unified Modeling Language ™ (OMG UML) Version 2.5
3. Cockburn, A.: Writing Effective Use Cases. Addison Wesley (2001)
4. Larman, C.: Applying UML and Patterns. An Introduction to Object-Oriented Analysis and Design and the Unified Process, 2nd edn. Prentice Hall, Upper Saddle River (2002)
5. Sein, M.K., Henfridsson, O., Purao, S., Rossi, M., Lindgren, R.: Action design research. MIS Q. **35**, 37–56 (2011)
6. Shu, X.: Fitting design patterns into object-oriented methods. PhD thesis, Graduate College of the University of Illinois at Chicago (1996)
7. Ferg, S.: What's Wrong with Use Cases? (2003)
8. Rational Unified Process. http://sce.uhcl.edu/helm/rationalunifiedprocess/
9. Zachman, J.A.: The Zachman Framework for Enterprise Architecture ™. The Enterprise Ontology, Version 3.0. http://www.zachman.com/about-the-zachman-framework
10. Dijkman, R.M., Joosten, S.M.M.: An algorithm to derive use case diagrams from business process models. In: 6th International Conference on Software Engineering and Applications, pp. 679–684 (2002)
11. Zhang, L., Qin, T., Zhou, Z., Hao, D., Sun, J.: Identifying use cases in source code. J. Syst. Softw. **79**, 1588–1598 (2006)
12. Tiwari, S., Gupta, A.: Statechart-based use case requirement validation of event-driven systems. In: 27th ACM Symposium on Applied Computing, pp. 1091–1093. ACM, New York (2012)
13. Somé, S.S.: Supporting use case based requirements engineering. Inf. Softw. Technol. **48**, 43–58 (2006)
14. Kanyaru, J.M., Phalp, K.T.: A lightweight state machine for validating use case descriptions. Technical report, ESERG, Bournemouth (2004)
15. Yue, T., Ali, S., Briand, L.: Automated transition from use cases to UML state machines to support state-based testing. In: France, R.B., Kuester, J.M., Bordbar, B., Paige, R.F. (eds.) ECMFA 2011. LNCS, vol. 6698, pp. 115–131. Springer, Heidelberg (2011)
16. Verelst, J., Silva, A.R., Mannaert, H., Ferreira, D.A., Huysmans, P.: Identifying combinatorial effects in requirements engineering. In: Proper, H., Aveiro, D., Gaaloul, K. (eds.) EEWC 2013. LNBIP, vol. 146, pp. 88–102. Springer, Heidelberg (2013)

17. Yue, T., Briand, L.C., Labiche, Y.: A systematic review of transformation approaches between user requirements and analysis models. Requirements Eng. **16**, 75–99 (2011)
18. Fliedl, G., Kop, C., Mayr, H.C., Salbrechter, A., Vöhringer, J., Weber, G., Winkler, C.: Deriving static and dynamic concepts from software requirements using sophisticated tagging. Data Knowl. Eng. **61**, 433–448 (2007)
19. Estañol, M., Queralt, A., Sancho, M.R., Teniente, E.: Artifact-centric business process models in UML. In: La Rosa, M., Soffer, P. (eds.) BPM 2012 International Workshops. LNBIP, vol. 132, pp. 292–303. Springer, Heidelberg (2013)
20. Estañol, M., Sancho, M.R., Teniente, E.: Reasoning on UML data-centric business process models. In: Basu, S., Pautasso, C., Zhang, L., Fu, X. (eds.) ICSOC 2013. LNCS, vol. 8274, pp. 437–445. Springer, Heidelberg (2013)
21. Eessaar, E.: A set of practices for the development of data-centric information systems. In: 22nd International Conference on Information Systems Development, pp. 73–84. Springer (2014)
22. Burns, L.: Building the Agile Database. How to Build a Successful Application Using Agile Without Sacrificing Data Management. Technics Publications, LLC, New Jersey (2011)
23. Sandhu, R.S.: Separation of duties in computerized information systems. In: IFIP WG11.3 Workshop on Database Security, pp. 179–190 (1990)
24. Parnas, D.L., Clements, P.C.: A rational design process: how and why to fake it. IEEE Trans. Softw. Eng. **SE-12**, 251–257 (1986)

Algorithm to Balance Compression and Signal Quality Using Novel Compressive Sensing in Medical Images

M. Lakshminarayana and Mrinal Sarvagya

Abstract Usage of compressive sensing plays a highly contributory role in compression, storage, and transmission in medical images even in presence of inherent complexities associated with radiological images. After reviewing the existing system, we found that existing techniques are less focused on medical images ignoring the complexities associated with it. Hence, this paper presents a very simple and novel transform-based technique where the performance of compressive sensing is enhanced using novel parameters of linear approximation, index ordering, along with number of low pass coefficient, and auxiliary measurement. The algorithm formulated by the proposed system is purely capable of minimizing L1-minimization. The outcome of the proposed system shows well balance between the compression ratio and signal quality in contrast to the existing technique of compressive sensing in medical images.

Keywords Compressive sensing · Compression ratio · Discrete wavelet transform · Medical images · Peak signal-to-noise ratio

1 Introduction

The mechanism of compressive sensing allows the linear projection of the image or video or any form of signal whose size is much less than input source signal. It also ensures that the recovered signal also undergo extremely less extent of data loss.

M. Lakshminarayana (✉)
Department of Electronics & Communication Engineering,
VTU Research Center, Belgaum, India
e-mail: lakshminarayana.m.2015@ieee.org

M. Sarvagya
School of Electronics & Communication Engineering, REVA University,
Bengaluru, India
e-mail: mrinalsarvagya@gmail.com

R. Silhavy et al. (eds.), *Software Engineering Perspectives and Application in Intelligent Systems*, Advances in Intelligent Systems and Computing 465,
DOI 10.1007/978-3-319-33622-0_29

The compressive sensing technique also allows integrating the acquisition of the signal and then minimizes the dimension of the matrix in one step. Therefore, it is said that compressive sensing is highly computationally cost effective process with highly reduced storage requirements and faster processing time. Moreover the compressed sensed signal finds suitable to be transmitted to any low-bandwidth signal. Therefore, compressive sensing may explore its valuable contribution in medical image compression [1, 2]. Usually, the radiological images are captured from multiple devices which differentiate from its data capturing mechanism. Owing to presence of external artifacts, sometimes there are huge amount of noisy or spurious data which is really not required to be acknowledged during diagnostic process. Hence, if a medical image is simply subjected to compression that an image as a wholesome will be compressed along with unwanted data. Therefore, usage of compressive sensing will allow only the information of clinical importance to be decomposed first using discrete wavelet transform [3] and is then subjected to L1-minimization, total variation, and computational cost. Hence, we present a novel technique that can perform compressive sensing considering the complexities of the medical images. The paper is organized as follows Sect. 2 highlights prior research work pertaining to compressive sensing. Section 3 describes problems explored after reviewing the literatures. Discussion of the proposed system using novel compressive sensing approach is presented in Sect. 4 followed by research methodology to accomplish the stated goal of study in Sect. 5. Algorithm description is discussed in Sect. 6 followed by Sect. 7 that demonstrates the outcomes of the proposed study followed by summary of paper in Sect. 8.

2 Related Work

Our prior review has already discussed various existing system to carry out compressive sensing over multimedia [4] with a clear description of the research gap. This section, we review some more relevant research work towards compressive sensing considering more complexities.

Zhou et al. [5] have carried out a study of compressive sensing using image fusion and formulation of sparse matrix. Trocan et al. [6] have presented a predictive technique for enhancing the quality for reconstructed data taking as case study of multi-view multimedia files. Zhu et al. [7] have presented a compressive sensing technique to mainly focus on total variational problems by using a dual mechanism of sparsity functions. Usage of matrix decomposition of order to low rank is carried out by Ren et al. [8] for accomplishing compressive sensing. The authors have used optimization mechanism for nonlinear types for obtaining superior quality of image.

Focus on performing compression of medical image was seen in the study of Smith et al. [9], where the authors have used compressive sensing using graphical processing unit and conventional split Bregman technique. Similar direction of work is also carried out by Kim et al. [10], where the authors have used the technique of compressive sensing in three dimensional orders over multi-cores. Puy et al. [11] have used spread spectrum and recommended the consideration of modulation of wide bandwidth of the signal before performing projection in compressive sensing. Tzagkarakis et al. [12] have developed a Bayesian technique for computing the amount of noise to be removed from the signal. The authors have used Gaussian Scale Mixture approach. Gurbuz [13] have developed a compressive sensing technique considering the predefined knowledge of speed of the moving target using spatial-based approach.

Hence, it can be seen that there are various techniques being introduced in recent time for evolving up with new strategies of compressive sensing. All the techniques discussed have significant beneficial point of learn and adopt while associated with limitations and constraints too. The problems pertaining to existing studies are discussed briefly in next section.

3 Problem Description

This section presents the problems that have been identified after reviewing the existing techniques on compressive sensing. It has been found that focuses of existing techniques on compressive sensing in medical images are quite less. Although, there are some significant studies being carried out considering medical images, but the complexity of the medical images (e.g. need of normalizing and processing the medical images to suit the compression) is less focused on. The next bigger problem is less efficient balance between compression ratio and PSNR to be seen in the existing studies. Therefore, there is a requirement of a cost efficient compressive sensing technique that is suitable to address the complexity of medical images and can maintain a well balance between size and signal quality. Hence, the next section discusses about proposed system to overcome this issues.

4 Proposed System

The proposed work is a continuation of our recent studies [14, 15]. The idea of the proposed system is to develop a mechanism of compressive sensing motivated by the work carried out by Sevak et al. [16], who have used compressive sensing using wavelet transform. The major goal of the proposed system is to enhance it to accomplish better compression performance and signal quality with cost effective computational performance too. Hence, our contribution/novelty laid in this research manuscript are as follows:

- **Minimized Wavelets and maximized sampling**: Usage of discrete wavelet transform is restricted to only 2 level and focus was more on maximizing the sampling performance with an aid of its generated coefficients in proposed system.

- **L1-minimization**: Different from approach of [16], the proposed system will perform linear approximation mechanism in order to generate index ordering. The algorithm also performs zigzag ordering in order to accomplishing faster response time during decoding process.
- **Inclusion of new parameters**: Different from approach of [16], the study uses number of low pass coefficient and number of auxiliary measurements to perform compressive sensing.
- **Balance between compression ratio and signal quality**: The proposed system is not only focus on accomplishing higher compression ratio but also it is capable of maintaining a superior signal quality measured using peak signal-to-noise ratio.

The next section discusses about the research methodology adopted for designing the proposed system.

5 Research Methodology

The development of the proposed study is carried out using analytical research methodology. The adopted schema used in the methodology is highlighted in Fig. 1. The normalized medical image is subjected to 2D transformation followed by compressive sensing operation with an aid of number of low pass coefficient and number of auxiliary measurements. The proposed study carry out compressing sensing using linear approximation in order to generate index ordering as well as JPEG zigzag order. Finally, the low pass discrete cosine transformation is combined with noise lets to accomplish compressed bits which are further subjected to quantization and entropy encoding. The decompression stage is reverse of it. In the entire process, design emphasis was laid upon solving L1-minimization problem in compressive sensing.

6 Algorithm Implementation

The proposed system is implemented in Matlab considering medical image dataset. Although, the prime purpose of the algorithm design is to perform compressive sensing but emphasize was majorly laid upon two aspect i.e. (i) reduction of computational complexity and (ii) accomplishing superior reconstructed image. The algorithms description is as follows:

Algorithm for performing compressive sensing
Input: I_{orig} (original image), k_1 (number of low pass coefficient), k_2 (number of auxiliary measurements), n (side length), b (block-size)

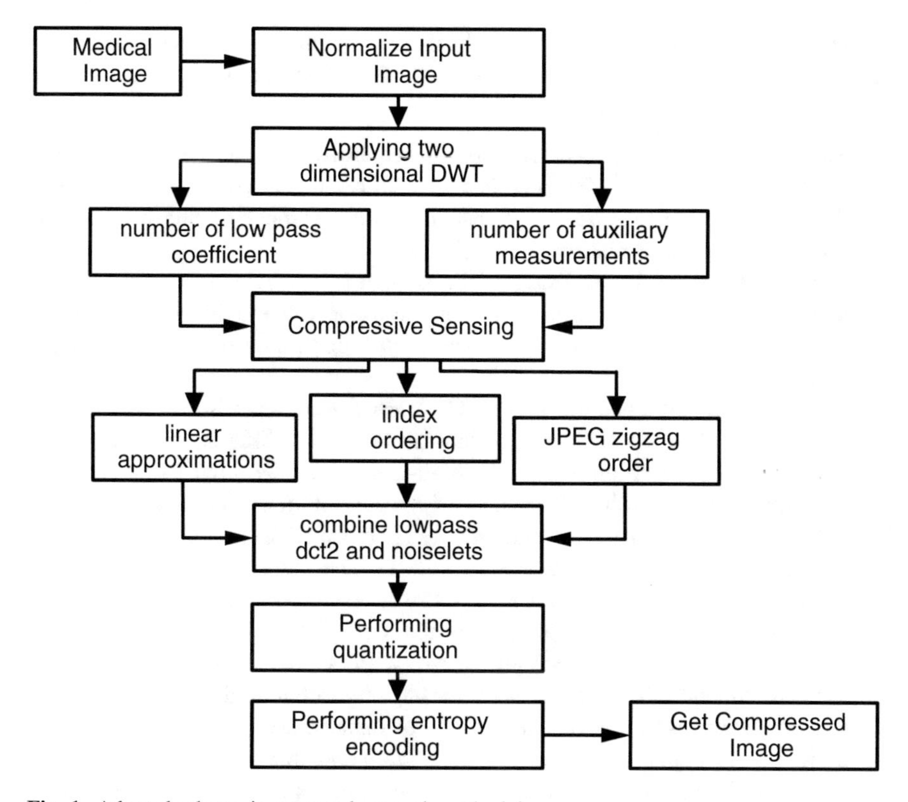

Fig. 1 Adopted schema in proposed research methodology

Output: Compressed bits
Start

1. init k_1, k_2
2. eval_bits(I_{orig})*8
3. $I_{orig} = [I_{orig}/arg_{max}(I_{orig})*255]$
4. Apply 2D-DWT
5. Struct → {lpf, hpf}
6. lo = perform linear approximation
7. OM1 = lo(1:k1)//Apply ordering
8. Apply Mersenne Twister
9. OM2 = rand_perm(1:k2)
10. Combine lowpass dct2 and noiselets
11. Perform quantization of all coordinates
12. Perform Huffman entropy
13. Save compressed data

End

The algorithm takes the input of medical image, performs compressive sensing, followed by quantization and entropy encoding. However, the process can be descriptively discussed as follows:

- **Taking input of image**: The algorithm takes medical image as an input, which is further checked for its RGB contents which is later converted to greyscale. In this process itself, the algorithm calculates total number of bits of the input image considering 8 bits per pixels (line-2).
- **Normalization of input Image**: The digitized image is now enhanced for increasing its precision level to double. We apply line-3 to normalizing the image. It will further assist in compressing the appropriate contents of the input image.
- **Applying two dimensional DWT**: The normalized image is subjected to two dimensional Haar discrete wavelet transforms to get 4 different coefficients viz. LL (low-low), HH (high-high), HL (high-low), HH (high-high) (Line-3).
- **Measurement Sampling**: This part of the algorithm considers two different types of compression parameters viz. k_1 (number of low pass coefficient), k_2 (number of auxiliary measurements) (line-1). The algorithm considers storing the outcome of DWT in a specific structure *struct*. Finally, the algorithm implements compressive sensing algorithm considering the 3 input arguments i.e. (i) respective coordinate from the *struct*, (ii) k_1, and (iii) k_2. We develop a separate function for compressive sensing algorithm with an aid of linear approximation using the blocked matrix of discrete cosine transform. Considering the new input arguments of side length n and size of block, the algorithm checks of the modulus of this new input argument is non-zero number. A new block b is generated by dividing side length n with size of block. A zigzag mode of accessing the block is done considering DCT coefficient. Finally implementation of linear approximation leads to generation of order of the compressive sensing. We develop two different orders in line-7–9. A random permutation for a squared matrix is created. Finally, the low pass dct as well as noiselets are combined to get the compressed bits.
- **Performing quantization & entropy encoding**: The compressed bits is than subjected to quantization considering the user-defined values of the quantization. The quantized coordinates are collected and rounded after quantization that is finally stored back to structure *struct*. The outcome of the quantized image is than subjected to entropy encoding using Huffman dictionary and Huffman encoding. The total compressed bits are then computed by adding the prior compressed bits with length of Huffman code. The compression ratio is than calculated using total number of the bits of original image divided by compressed bits received in the end stage of compressive sensing.

The above mentioned steps are the stage for performing compressive sensing that finally results in generation of compressed bits. In order to perform decompression, the compressed bits as well as similar values of k1 (number of low pass coefficient), k2 (number of auxiliary measurements) will be used. It will be followed by performing dequantization with an aid of similar user-defined value that has been used in compression stage. This stage leads to generation of bits and dictionaries from the structure. Which is subjected to Huffman decoding. Finally, the decoded signals are stored back in the structure.

This stage is also followed by L1-minimization, which is very much dependent on k1 (number of low pass coefficient), k2 (number of auxiliary measurements) as well as the structure. The dequantized vectors are accomplished and sizes of the coefficients are extracted. The reconstruction process is soon followed in this process that takes the input arguments of (i) recently accomplished dequantized vector, (ii) size of coefficient, (iii) k1 (number of low pass coefficient), and (iv) k2 (number of auxiliary measurements). The outcome is reshaped in forms of squared dimension. The outcome of the L1-minimization is subjected to inverse transformation using same two dimensional Haar approach. The final outcome of the two dimensional Haar based discrete wavelet transformation that results in reconstructed image. Finally, the proposed system is checked for its signal quality in the form of peak signal to noise ratio that considers the original image before compression and reconstructed image after compression. The complete process takes around 0.256 s to execute for a normal medical image on core i3 processor. An interesting aspect of the proposed algorithm is that it doesn't store any runtime information except the generated compressed bit resulting in considerable storage of memory.

7 Results and Discussion

This section discusses about the result analysis of the proposed as well as existing system. The outcome of the proposed system is compared with that of work carried out by Sevak et al. [16] considering the performance parameter of peak signal-to-noise ratio and compression ratio (Table 1).

The above Table 1 shows that peak signal to noise ratio of proposed system is approximately found to be 36.02 dB, whereas for the Sevak's approach [16], it is found to be 25.59 dB. The analysis was carried out for 20 sample medical images. Although the approach of Sevak [16] is similar to ours, but the approach doesn't uses the linear approximation method for performing compressive sensing. The proposed mechanism not only develops significant algorithm where the prime focus was laid over L1-minimization process. The usage of ordering and application of Mersenne twister as in the process of randomization allows the system to perform faster computation while consideration of k1 (number of low pass coefficient) and k2 (number of auxiliary measurements) in the compressive sensing retains better

Table 1 Numerical analysis of comparative study (PSNR)

Image sample	Sevak's approach	Proposed approach
Img_1	25.782	36.834
Img_2	20.766	38.652
Img_3	28.541	39.112
Img_4	23.241	32.677
Img_5	23.366	36.775
Img_6	20.182	31.061
Img_7	23.761	32.432
Img_8	27.021	38.954
Img_9	25.552	35.987
Img_10	26.528	37.411
Img_11	26.233	35.252
Img_12	28.767	32.176
Img_13	25.034	36.878
Img_14	30.981	35.622
Img_15	24.551	37.318
Img_16	23.781	34.865
Img_17	25.389	38.598
Img_18	26.768	35.422
Img_19	28.061	37.016
Img_20	27.676	37.385

signal quality. We also carry out analysis of the structural similarity for both Sevak et al. [16] and proposed method, which provides better assessments in contrast to peak signal-to-noise ratio. For better evaluation process, we estimate Structural Similarity Index (SSIM) as well as Feature-based Structural Similarity (FSIM) motivated from the work done by Zhang et al. [17]. The authors have mechanized a new technique that can assess similarity between two signals based on its features.

The closer look into the above numerical outcomes in Table 2 will show that proposed system has enhancement of 15.88 % with respect to SSIM value (SSIM for proposed = 68.51 % and Sevak et al. = 52.63 %). The numerical outcome of the proposed system is also found to have 19.49 % of improvement with respect to FSIM value (FSIM for proposed = 90.83 % and Sevak et al. = 71.35 %). Hence, higher values of FSIM of proposed system (i.e. 90.83 %) shows that final reconstructed image have better retention of data quality, which is one of the prime attribute for medical image compression when compression is carried out in bulk manner.

The above Table 3 shows the trend of the compression ratio of the Sevak's [16] approach as well as proposed technique. The compression ratio was evaluated by calculation the uncompressed bits by compressed bits. The compression ratio for

Table 2 Numerical analysis of structural similarity

Image sample	SSIM		FSIM	
	Sevak's approach	Proposed	Sevak's approach	Proposed
Img_1	0.510	0.613	0.847	0.987
Img_2	0.578	0.692	0.524	0.941
Img_3	0.502	0.522	0.846	0.925
Img_4	0.618	0.738	0.826	0.964
Img_5	0.712	0.854	0.891	0.607
Img_6	0.454	0.837	0.872	0.849
Img_7	0.588	0.819	0.867	0.944
Img_8	0.634	0.618	0.914	0.971
Img_9	0.417	0.698	0.703	0.921
Img_10	0.428	0.749	0.819	0.945
Img_11	0.534	0.701	0.918	0.973
Img_12	0.519	0.692	0.738	0.901
Img_13	0.460	0.872	0.687	0.879
Img_14	0.417	0.826	0.745	0.829
Img_15	0.479	0.720	0.671	0.933
Img_16	0.701	0.321	0.381	0.876
Img_17	0.591	0.671	0.437	0.899
Img_18	0.399	0.619	0.521	0.968
Img_19	0.476	0.297	0.745	0.947
Img_20	0.510	0.844	0.317	0.908

Sevak's approach [16] was found to be 3.4322 while that of proposed technique was found to be 7.3551. Hence, there is a superior compression ratio being established for the proposed system. The prime reason behind this is that we apply dual steps of ordering in linear approximation principle of proposed compressive sensing. The first ordering uses linear approximation of k1 (number of low pass coefficient) while the second ordering uses random permutation of k2 (number of auxiliary measurements). This results in significant compression performance that is further enhanced by combining low pass discrete cosine transformed value with noiselets. However, Sevak's approach doesn't perform this step that result in compression ratio of only 3.4322.

Reviewing the accomplished outcome, it can be stated that proposed system have maintained a well balance between the signal quality and compression performance. The compression performance is collectively improved to 40 % while signal quality is improved to approximately 10 % from the existing system. Therefore, a better performance of compressive sensing can be witnessed in this paper.

Table 3 Numerical analysis of compression ratio

Image sample	Sevak's approach	Proposed approach
Img_1	2.145	6.177
Img_2	2.765	5.726
Img_3	3.146	4.723
Img_4	3.157	5.172
Img_5	3.786	7.242
Img_6	4.121	6.781
Img_7	3.128	7.863
Img_8	3.785	7.562
Img_9	2.677	8.242
Img_10	4.321	9.162
Img_11	2.333	8.726
Img_12	3.918	7.246
Img_13	3.176	6.421
Img_14	4.186	7.862
Img_15	4.231	8.242
Img_16	3.541	5.202
Img_17	3.721	7.866
Img_18	4.788	8.966
Img_19	3.555	9.203
Img_20	2.165	8.718

8 Conclusion

Compressive sensing not only performs superior compression of signals but also saves the cost of using other network-related resources like bandwidth and buffer. However, study on compressive sensing towards medical image processing are less focused on computational excellence and was more inclined towards accomplishing higher compression. This paper presents a technique where the performance of the compressive sensing is more enhanced by adopting linear approximations and various new set of parameters. The outcome of the study was evaluated with respect to performance parameters e.g. compression ratio, SSIM, FSIM and peak signal-to-noise ratio to find that proposed system is found to have better computational performance in comparison to existing similar techniques.

References

1. Majumdar, A.: Compressed Sensing for Magnetic Resonance Image Reconstruction. Cambridge University Press, Computers (2015)
2. Carmi, A.Y., Mihaylova, L.S., Godsill, S.J.: Compressed Sensing and Sparse Filtering. Springer Science & Business Media, Technology & Engineering (2013)

3. Boche, H., Calderbank, R., Kutyniok, G., Vybíral, J.: Compressed Sensing and its Applications. Birkhäuser, Mathematics (2015)
4. Lakshminarayana, M., Sarvagya, M.: Scaling the effectiveness of existing compressive sensing in multimedia contents. Int. J. Comput. Appl. **115**(9) (2015)
5. Zhou, X., Wang, W., Liu, R.: Compressive sensing image fusion algorithm based on directionlets. Springer-EURASIP J. Wirel. Commun. Netw. **19** (2014)
6. Trocan, M., Tramel, E.W., Fowler, J.E., Pesquet, B.: Compressed-sensing recovery of multiview image and video sequences using signal prediction. Springer-Multimedia Tools Appl. **72**(1), 95–121 (2014)
7. Zhu, Z., Wahid, K., Babyn, P., Cooper, D.: Improved compressed sensing-based algorithm for sparse-view CT image reconstruction, Hindawi Publishing Corporation. Comput. Math. Methods Med. **2013** (2013)
8. Ren, K., Xu, F., Gu, G.: Compressed sensing and low-rank matrix decomposition in multisource images fusion, Hindawi Publishing Corporation. Math. Prob. Eng. **2014** (2014)
9. Smith, D.S., Gore, J.C., Yankeelov, T.E., Welch, E.B.: Real-time compressive sensing MRI reconstruction using GPU computing and split Bregman methods, Hindawi Publishing Corporation. Int. J. Biomed. Imaging **2012** (2012)
10. Kim, D., Trzasko, J., Smelyanskiy, M., Haider, C.: High-performance 3D compressive sensing MRI reconstruction using many-core architectures, Hindawi Publishing Corporation. Int. J. Biomed. Imaging **2011** (2011)
11. Puy, G., Vandergheynst, P., Gribonval, R., Wiaux, Y.: Universal and efficient compressed sensing by spread spectrum and application to realistic Fourier imaging techniques. EURASIP J. Adv. Signal Process. **6** (2012)
12. Tzagkarakis, G., Milioris, D., Tsakalides, P.: Multiple-measurement Bayesian compressed sensing using GSM priors For DOA estimation. In: IEEE International Conference on Acoustics Speech and Signal Processing, pp. 2610–2613 (2010)
13. Gurbuz, A.C.: Analysis of unknown velocity and target off the grid problems in compressive sensing based subsurface imaging. In: 18th European Signal Processing Conference, Denmark, pp. 1087–1091 (2010)
14. Lakshminarayana, M., Sarvagya, M.: Random sample measurement and reconstruction of medical image signal using compressive sensing. In: IEEE International Conference on Computing and Network Communications (CoCoNet'15), pp. 261–268 (2015)
15. Lakshminarayana, M., Sarvagya, M.: Lossless compression of medical image to overcome network congestion constraints. In: Springer-Proceedings of Third International Conference on Emerging Research in Computing, Information, Communication and Application (ERCICA-15), vol. 01, pp. 305–311 (2015)
16. Sevak, M.M., Thakkar, F.N., Kher, R.K., Modi, C.K.: CT image compression using compressive sensing and wavelet transform. In: IEEE International Conference on Communication Systems and Network Technologies, pp. 138–142 (2012)
17. Zhang, L., Zhang, L., Mou, X., Zhang, D.: FSIM: a feature similarity index for image quality assessment. IEEE Trans. Image Process. **20**(8) (2011)

Application of Virtualization Technology for Implementing Smart House Control Systems

Maxim Polenov, Andrey Kostyuk, Evgenia Muntyan, Vyacheslav Guzik and Vladislav Lukyanov

Abstract This paper describes the rationale for a choice, basic features, and test results of the Smart House control system model that has been developed for small and medium size households. To resolve common problems found in the existing Smart House systems, the authors proposed installing a virtual machine monitor, also known as a hypervisor, on to a personal desktop computer. This approach centralized the entire system, simplified its installation and configuration processes, and increased its cost effectiveness. The paper provides a series of experiments on installing hypervisors known manufacturers to ordinary personal computers. Shows the result of the work of hypervisors with deployed virtual machines alight on water graphs showing consumption of resources, the table with the characteristics of physical and virtual machines.

Keywords Virtualization · Smart House · Hypervisor · Control · Citrix · Hyper-V · vSphere · Server · Physical computer

---

M. Polenov (✉) · A. Kostyuk · E. Muntyan · V. Guzik · V. Lukyanov
Department of Computer Engineering, Southern Federal University, Taganrog, Russia
e-mail: mypolenov@sfedu.ru

A. Kostyuk
e-mail: aikostyk@sfedu.ru

E. Muntyan
e-mail: ermuntyan@sfedu.ru

V. Guzik
e-mail: vfguzik@sfedu.ru

V. Lukyanov
e-mail: sith@pochta.ru

R. Silhavy et al. (eds.), *Software Engineering Perspectives and Application in Intelligent Systems*, Advances in Intelligent Systems and Computing 465,
DOI 10.1007/978-3-319-33622-0_30

1 Introduction

The "Smart House" is a home automation technology and it is intended to help household residents in various aspects of their living. This may include electronically controlled features for elderly and disabled people, as well as security systems for tracking pets, children, and contractors working on the house.

There are many companies on the market that offer various Smart House services. We reviewed related information posted on their web-sites [1–7] and proposed our solution to centralized management of all Smart House systems utilized on the property. This solution was based on the second method of centralized system management [8]. Our research team performed a variety of experiments on utilizing common personal computers as Smart House management servers.

Various ways of resolving conflicts between software and hardware were considered. A virtualization technology was used to address resource related issues and certain software incompatibility problems of running multiple applications concurrently on a single machine. This virtualization approach also allowed resolving the issue of underutilizing the server resources [9]. Several cost effective hypervisors that well handled managing multiple Smart House applications were considered.

2 Hardware Layer

Hypervisors are typically installed on servers. However, for the purpose of managing a Smart House system within a single small or medium-size cottage, hypervisors can be put onto an ordinary desktop computer. This computer will be the main platform for the installation and configuration of all necessary systems of Smart House.

For example, on a physical server it could be installed a hypervisor. On hypervisor you can create a necessary amount of virtual machines (VM). The server has to be connected to the network switch. On the switch two VPN could be configured, one is intended for tenants of the house, another for sensors and systems of video surveillance. The necessary software for sensors and video cameras is installed on virtual machines for first VPN. Computers and mobile devices are connected to another VPN. In order to test how well a desktop machine would handle this task, we utilized two personal computers. Their characteristics are listed in Table 1.

Table 1 Basic characteristics of the test personal computers

Resource	The first test computer	The second test computer
CPU	AMD FX-8350	AMD FX-4300
Motherboard	MSI970A-G43	Gigabyte GA 78LMT-S2PV
RAM size (GB)	8	8
The integrated network card	Realtek RTL8111E	Realtek GbE

Several supervisors were installed on both machines at different times. Each physical computer (the host) was configured with the same set of four virtual machines. These virtual machines had different amounts of host's resources allocated to them. Different versions of Windows and Linux operating systems were installed on these virtual machines.

2.1 First Series of Experiments

The VMware vSphere 4.0 hypervisor is installed on both test computers with integrated and removable network cards.

Just as majority of modern machines do, both test computers had integrated network cards. After VMware vSphere 4.0 hypervisor was installed on both test computers, none of their integrated network adapters were recognized by the operating systems. Therefore, D-Link, Star Tech, and TP-Link PCI network cards (Table 1) were purchased and installed on the test computers. However, these cards were not recognized as well. Finally, we tested IBM network cards that are shown in Table 1.

These cards were recognized by operating systems (OS) on both test computers. Please refer to Table 2 for the results of network card recognition.

2.2 Second Series of Experiments

The VMware vSphere 5.1 hypervisor installed on both test computers with integrated and removable network cards.

Because there was a newer version of VMware vSphere software also available on the market, next we performed similar tests with VMware vSphere 5.1. It was installed on the test computers listed in Table 1, but these computers did not have removable network cards installed. As it was observed during the first series of experiments, none of the integrated network cards were recognized on both test computers with installed VMware vSphere 5.1 hypervisor. Next, the IBM removable network cards (#3 and #4, see Table 2) were installed one after another on the first test computer. The hypervisor was running fine in both these cases, and it also recognized all the hardware except the integrated network card.

Table 2 Removable network adapters and results of their recognition

#	The model of removable network adapter	Was it recognized by OS?
1	D-Link DFE-520TX 10/100 MBps PCI	No
2	TP-Link TF-3239DL/TF-3283 10/100 MBps PCI, Low	No
3	IBM Intel Ethernet Dual Port I340-T2	Yes
4	IBM Broadcom NetXtreme I Quad Port GbE Adapter	Yes

Both IBM removable network cards (#3 and #4, Table 2) were then installed one at a time on the second test computer. The same results, as seen on the first test computer, were observed on the second test computer. The removable network adapter #3 was then used for further experiments on both test computers.

2.3 Third Series of Experiments

Four virtual machines were created and configured on the VMware vSphere 5.1 hypervisor that was installed on the second test computer. Basic characteristics of these virtual machines and operating systems installed on them are given in Table 3.

The hypervisor was configured to automatically increase amount of resources being used by virtual machines when required. This allowed providing virtual machines with available resources on demand. All virtual machines had no application software installed at that time. All four virtual machines were started and utilization of CPU and RAM resources was observed in Fig. 1.

As it is shown in Fig. 1, the four concurrently running virtual machines with only operating systems installed on them, utilized about 25 % of CPU resources and used almost a half of available RAM. It suggested that there were plenty of hardware resources available for normal functioning of the hypervisor and also for running additional software on these virtual machines.

Table 3 Virtual machine configuration

Virtual machine	Operating System installed on VM	Amount of RAM dedicated to VM	No. of CPU cores dedicated to VM
First VM	Windows Server 2003 R2	1 GB	One core
Second VM	Windows Server 2008	2 GB	Two cores
Third VM	Debian 7.5.0	512 MB	One core
Fourth VM	Ubuntu Server 14.5.0	1 GB	One core

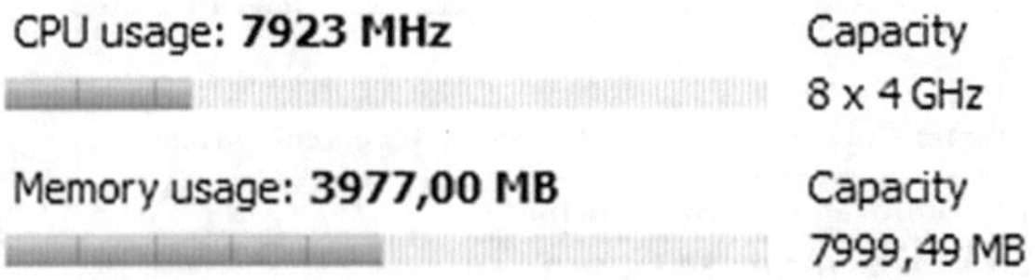

CPU usage: **7923 MHz** Capacity 8 x 4 GHz

Memory usage: **3977,00 MB** Capacity 7999,49 MB

Fig. 1 Utilization of CPU and RAM on the second test computer with installed VMware vSphere 5.1hypervisor when four virtual machines were running

At the next step each virtual machine was configured with a specific set of Smart House software applications as shown in Table 4.

After this software was installed on the four virtual machines, all these virtual machines were simultaneously started. The graphs of CPU and RAM utilization that were observed during these processes in the VMware vSphere 5.1 monitoring system are shown in Figs. 2 and 3.

The initial spikes of CPU and RAM utilization were related to concurrent installation of software applications on all four virtual machines. However, when this process was finished, CPU and RAM utilization stabilized. Insignificant

Table 4 Configuring virtual machines with Smart House applications

Virtual machine	Servers that were installed and configured on the virtual machine
First VM	A video surveillance server
Second VM	An authorization server with multiple emulated requests initiated by clients or sensors
Third VM	The Apache virtual web server that included PHP libraries, MySQL databases, phpMyAdmin application for administering MySQL databases, and also several virtual web-hosts
Fourth VM	An additional MySQL server

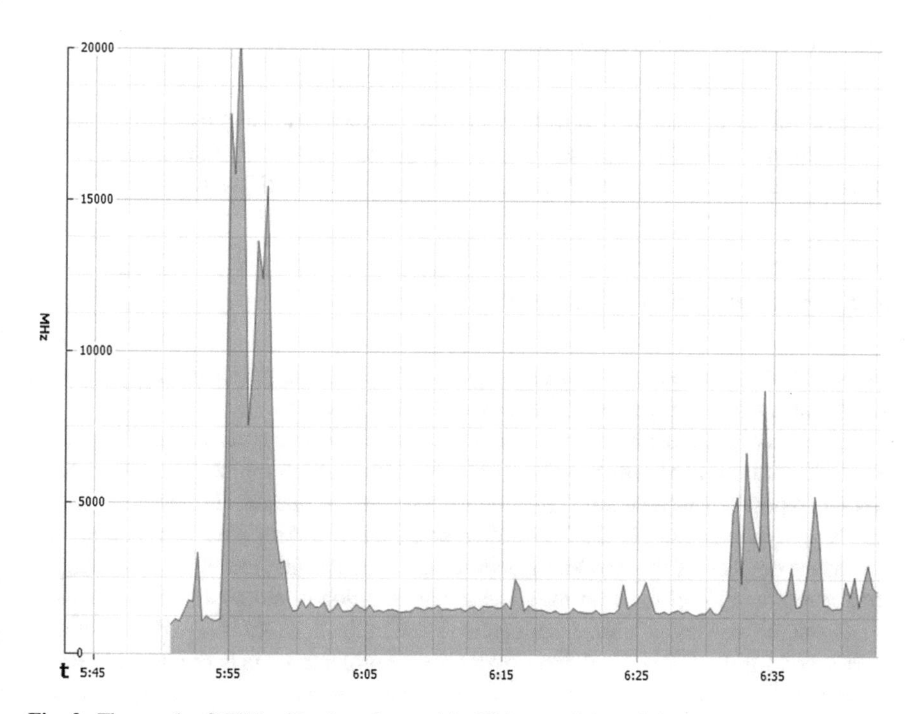

Fig. 2 The graph of CPU utilization observed in VMware vSphere 5.1

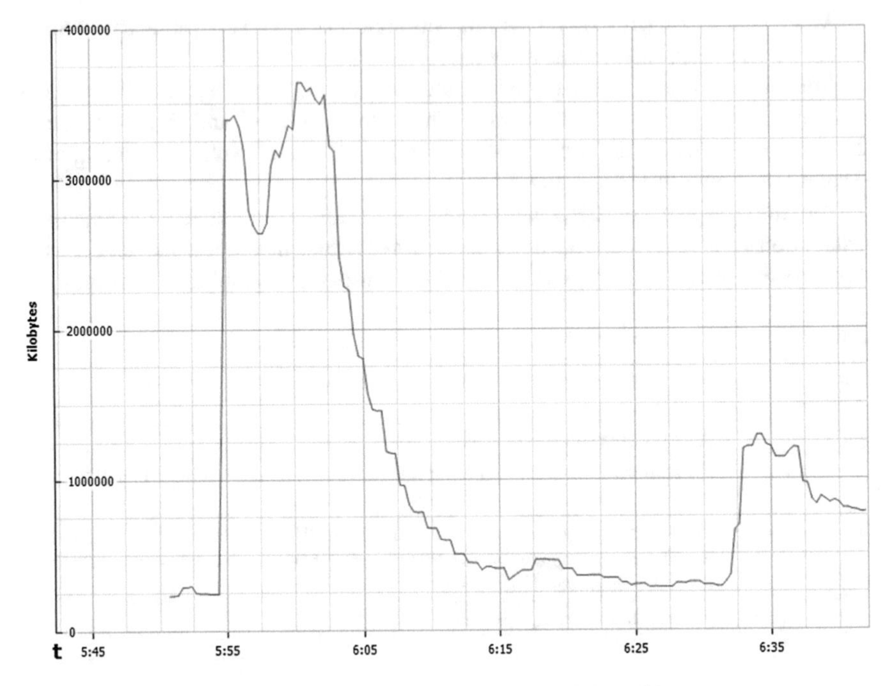

Fig. 3 The graph of RAM utilization observed in VMware vSphere 5.1

increase of CPU load and RAM usage was observed during five simultaneous active sessions on the second virtual machine that had an authorization server installed, and also during web browsing and using the MySQL database.

2.4 Fourth Series of Experiments

Four virtual machines were controlled by the XenServer hypervisor that is installed on the first test computer with the integrated network card.

After successful testing of the VMware vSphere 5.1 on the second test computer, the PCI network card was removed from the first test computer. The Citrix XenServer hypervisor was then installed on the first test computer. Unlike the VMware vSphere hypervisors, the Citrix XenServer hypervisor was able to recognize all hardware devices of the first test machine. Because the integrated network adapter was recognized, it eliminated the need for an additional, removable network card.

The same tests, which are described in detail in Sect. 2.3 of this paper, were also performed for the Citrix XenServer hypervisor. Utilization levels of CPU and RAM on the first test computer with four virtual machines that were running under the Citrix XenServer hypervisor were very similar to the VMware vSphere 5.1 resource utilization as shown on Fig. 3.

The graphs were observed for the Citrix XenServer hypervisor (shown in Figs. 4 and 5) during installing and running software applications that are shown in Table 4.

2.5 *Fifth Series of Experiments*

Four virtual machines running hypervisor Microsoft Hyper-V, set on the second test computer with built-in network card. In the third test machine was running Windows Server 2012 R2 with the established role of the hypervisor (Hyper-V). As is the case with the hypervisor company Citrix, Hyper-V was able to identify all the hardware devices in the test machine including a built-in network card. Schedule resources consumed Windows Server 2012 R2 with the established role of the hypervisor are shown in Figs. 6 and 7.

As follows from Figs. 6 and 7 the server without the deployed virtual machines uses 5 % of the CPU and 850 MB of RAM.

After you deploy and run four virtual machines (see Table 3), the amount of consumed resources has changed and results are shown in Figs. 8 and 9.

As can be seen in the work of all virtual machines with the installation of the necessary software used promptly volume rose to 5.4 GB. CPU utilization increased to 27 %, if necessary, the virtual machine can use more system resources than it was originally set.

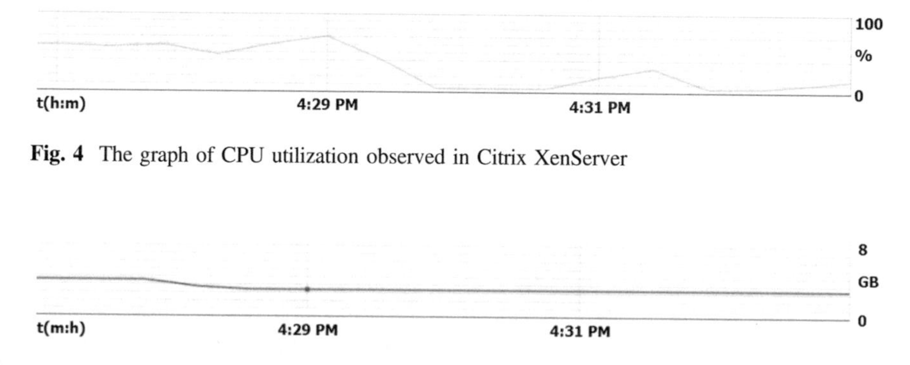

Fig. 4 The graph of CPU utilization observed in Citrix XenServer

Fig. 5 The graph of RAM utilization observed in Citrix XenServer

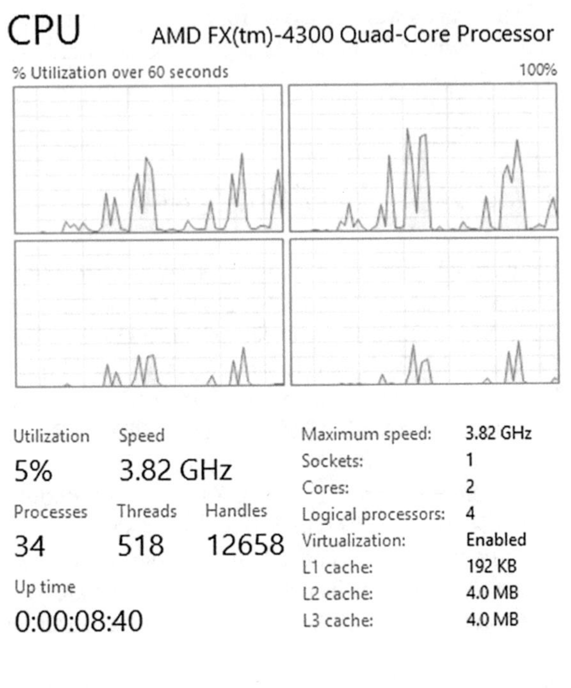

Fig. 6 The graph of CPU utilization observed in Hyper-V

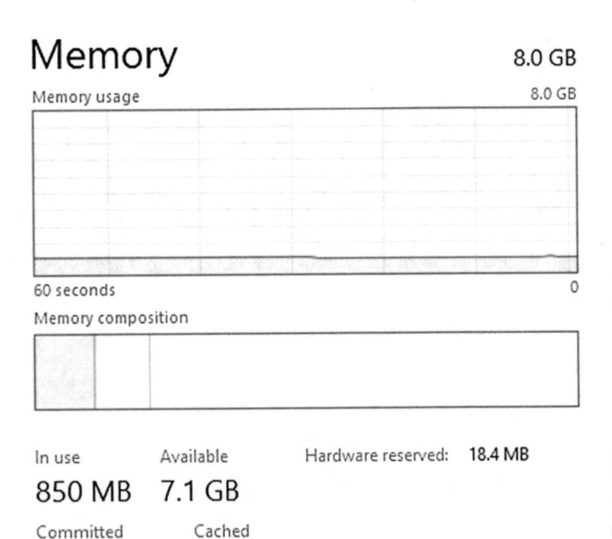

Fig. 7 The graph of RAM utilization observed in Hyper-V

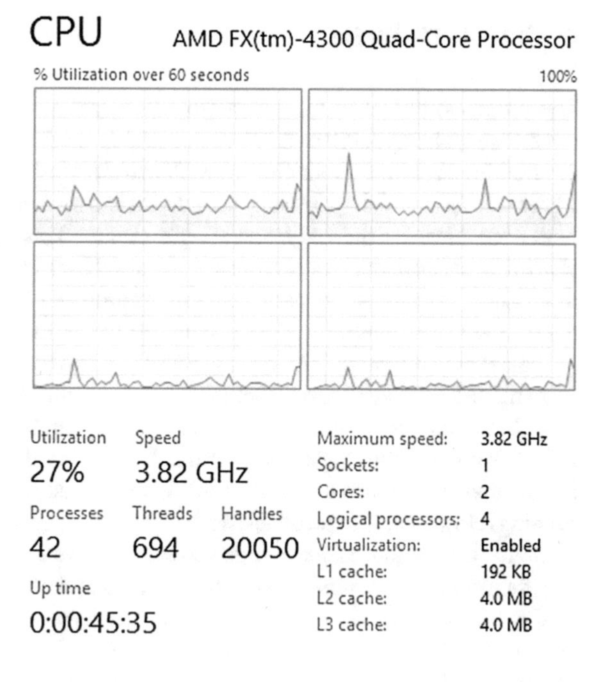

Fig. 8 Using the CPU Hyper-V virtual machines

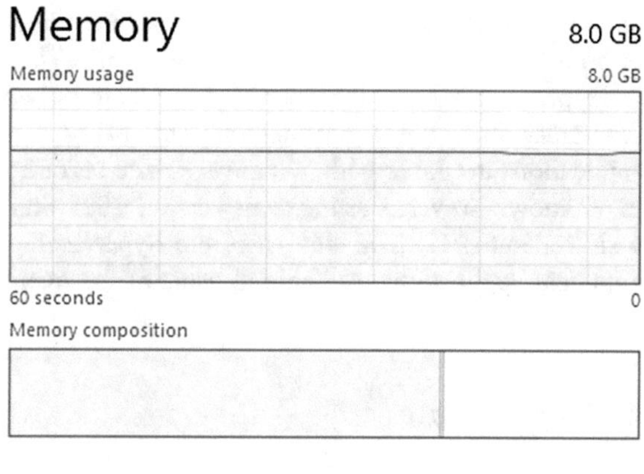

Fig. 9 Memory use Hyper-V virtual machines

3 Software Layer

The proposed model of the Smart House control system utilized a specific type of hypervisor. The hypervisor allows to centralize control systems of Smart House and to combine them into a single server. VMware vSphere 5.1, Citrix XenServer and Microsoft Hyper-V hypervisors include their own kernels and are installed on the host computers as the only operating systems. This eliminates expenses of purchasing a host operating system, as well as releases hardware resources for running the virtual machines. In addition, this approach provides effective and optimal distribution of available resources between the virtual machines.

To develop our model of a Smart House Control System, we used hypervisors that were made by the three most popular software manufacturers: VMware vSphere [10, 11], Citrix XenServer [12–15] and Microsoft Hyper-V [16]. The main advantages of the VMware vSphere hypervisor include flexible configuration of the backup system and increasing the size of virtual hard drives on the fly while the virtual machine is running. However, the use of VMware vSphere hypervisor is limited in the small and medium size residences due to high software license costs and additional expenses for user support and any separately purchased product or module.

On the other hand, the Citrix XenServer hypervisor has a free version. This version has limited capabilities, but the provided features are quite sufficient for private use. Other advantages of the Citrix XenServer hypervisor include memory optimization and configurable balancing of the virtual machine loads. The features of both hypervisors are described in detail in [17].

Hypervisor Microsoft Hyper-V is available in two versions. The first version is absolutely free, but controlled only by using the command line or graphical applications available with Windows 8. The second version is distributed as a role in Windows Server, starting with version 2008 and Microsoft Windows, starting with version 8 [18]. In this case, the hypervisor can be controlled by intuitive graphical applications available to users of the operating system.

4 Conclusion

Our test results supported the hypothesis of successful utilization of a personal desktop computer for configuring and running a Smart House control system. Using the VMware vSphere hypervisor called for some modification of hardware, such as installation of an additional network card, because the integrated network controller was not recognized. However, this was not necessary with a Citrix Xen Server hypervisor based system, because all hardware was recognized by XenServer. Both hypervisors and their virtual machines functioned well on personal desktop computers, although this use is limited to small and medium size households. Because of high resource demands, server platforms should be used for configuring Smart House control systems in large houses and apartment buildings. The hypervisor developed by Microsoft, does not require the installation of additional equipment and also showed good results during the experiment. However, using free version, the administrator may have difficulty with the command line hypervisor.

We used the Citrix XenServer hypervisor to develop a system that was based on the proposed conceptual model [19]. Convenience of using the recognized integrated network adapter and availability of a free version of the hypervisor motivated us to make this decision. The majority of features included in the paid Citrix XenServer software releases are intended for the use by companies and organizations.

In conclusion, it should be noted that the free edition of the Citrix XenServer hypervisor is a highly reliable and secure virtualization platform. It provides good application performance and stable operation of virtual machines.

Acknowledgments This work was financially supported by the Ministry of Education and Science of the Russian Federation under Project No. 2336 (Base part, State task 2014/174).

References

1. What is the "Smart House"? http://hmps.ru/smart_house/
2. The Smart Home: http://www.cleverism.com/smart-home-intelligent-home-automation/
3. What is Virtualization? http://www.ixbt.com/
4. The Purpose of Virtualization: http://www.itiks.ru
5. Virtualization Market: http://www.rg.ru
6. Market Assessment: http://www.forrester.com/home/
7. Top 10 Virtualization Technology Companies for 2016: http://www.serverwatch.com/server-trends/slideshows/top-10-virtualization-technology-companies-for-2016.html
8. Kostyuk, A., Lukyanov, V., Polenov, M., Muntyan, E.: A concept of utilizing a hypervisor in the Smart House control systems. In: Proceedings of the Congress on Intelligent Systems and Information Technology «IS&IT'14», vol. 2. pp. 103–107. Physmatlit, Moscow (2014)
9. Josyula, V., Orr, M., Page, G.: Cloud Computing. Automating the Virtualized Data Center, Cisco Press, Indianapolis (2012)
10. Mikheev, M.: Administration of VMware vSphere 5. DMK Press, Moscow (2012) (in Russian)
11. Langone, J., Leibovici, A.: VMware View 5 Desktop Virtualization Solutions. Packt Publishing (2012)
12. Chisnall, D.: Definitive Guide to XenHypervisor. Pearson Education, Prentice (2007)
13. Tosatto, D.: Citrix XenServer 6.0 Administration Essential Guide. Packt Publishing (2012)
14. Williams, D., Garcia, J.: Virtualization with Xen: Including Xenenterprise, Xenserver, and Xenexpress. Syngress (2007)
15. Semeonov, A.B.: Structured Cable Interactive Control Systems. IT Press, Moscow (2011) (in Russian)
16. Minasi, M., Greene, K., Booth, C., et al.: Mastering Windows Server 2012 R2. Wiley (2015)
17. Kostyuk, A., Polenov, M., Lukyanov, V., Muntyan, E.: Model of centralized control in smart house system with hypervisor usage. Izvestiya SFedU. Engineering Sciences, vol. 7, pp. 170–177. SFedU Publishing, Taganrog (2014) (in Russian)
18. Carvalho, L.: Windows Server 2012 Hyper-V Cookbook. Packt Publishing (2012)
19. Kostyuk, A., Polenov, M., Lukyanov, V., Muntyan, E., Nikolava, A.: Research of virtualization deployment possibility in Smart House control systems. Informatization and Communication, vol. 3, pp. 72–77. Moscow (2015) (in Russian)

Utilization of Motion Animation for Analysis of Basic Self-defense Techniques

Dora Lapkova, Lukas Kralik, Zuzana Oplatkova Kominkova and Milan Adamek

Abstract This paper is focused on possible utilization of an artificial neural network connected with biometric systems and motion animation for the purpose of training of self-defense techniques. The described experiment was performed in a specialized laboratory of university hospital in Brno with the help of VICON system. The aim was to obtain new inputs for artificial neural networks. Simultaneously, this research proceeds with previous project and research. This paper also contains the most interesting results from the experiment.

Keywords Self-defense · Animation · Biometric system · Striking techniques

1 Introduction

Biometric systems are used by human civilization in their most trivial form for at least 3000 years. The first preserved mention about biometrics can be found in an ancient Egypt. Egyptians used facial features for authorization of payments for services or for purchase of various products. It is the first type of biometric system

D. Lapkova (✉) · M. Adamek
Faculty of Applied Informatics, Department of Security Engineering,
Tomas Bata University in Zlin, Nam. T. G. Masaryka 5555, 760 01 Zlin,
Czech Republic
e-mail: dlapkova@fai.utb.cz

M. Adamek
e-mail: adamek@fai.utb.cz

L. Kralik · Z.O. Kominkova
Faculty of Applied Informatics, Department of Informatics
and Artificial Intelligence, Tomas Bata University in Zlin,
Nam. T.G. Masaryka 5555, 760 01 Zlin, Czech Republic
e-mail: kralik@fai.utb.cz

Z.O. Kominkova
e-mail: kominkovaoplatkova@fai.utb.cz

R. Silhavy et al. (eds.), *Software Engineering Perspectives and Application in Intelligent Systems*, Advances in Intelligent Systems and Computing 465,
DOI 10.1007/978-3-319-33622-0_31

based on recognition of face and its features, anthropometric characteristics (e.g. body height, weight) and fingerprints. [1] In 19th century Alfonse Bertillon laid the foundations of modern systems which are still used today. His methods were further developed by Russia and USA [2, 3].

The biometrics of move is based on walking. As the first scientifically based description of the walk is considered the research by Borelli from 1692 which describes the position of the center of gravity and defines walking as an alternating shift of support bases forward [4].

The greatest development of the recognition of people by their movement started in 1990s with expanding availability and power of computer systems [4].

Modern technologies allow detailed analysis of biomechanical movement and they can also analyze the related variables.

Self-defense is a field which is primarily focused on the legal protection of personal interests. It covers various areas—the theory and the practice of defence, attack and prevention, scientific disciplines such as tactics (e.g. skill in the counter attack), strategy (precautionary action) and operation (behaviour after a conflict situation). Moreover, it includes the knowledge of somatology and the chosen parts of crisis management, especially the phases of the conflict and solutions to conflict situations. [5, 6]. Striking techniques are very important part in majority of martial arts [6, 7] and this is the reason why our experiment was focused on them.

During our experiment, the system VICON was used for visualization of body motion in the moment of striking. This can be used in self-defense training, allowing trainees to see correct techniques and also their main mistakes. Also this system may help to recognize trained person on the first view.

In the past research [7, 8], it was observed that an artificial neural network is suitable for recognition of gender and training level of monitored subject in self-defense. The force profile and some statistical information such as mean and maximum of the force were used for input data during the training. The main motivation of this paper was to find other possible input data with the same or better results of classification. This experiment was mainly focused on identifying of significant points on human body during motion.

2 Motion Analysis

Biometric systems are experiencing a great boom in the last few years. One major area is the analysis of human motion [9]. Based on the method of collecting data from the sensors, these systems can be divided into three groups:

- Optical
- Electromagnetic
- Electromechanical

The most frequently used measuring devices are optical (more specifically optoelectronic) systems with utilization of video (stereometric methods). Among the devices operating on this principle belong VICON, ARIEL, OPTORAK, BTS SMART and QUALISYS. The motion scanning is performed in three planes:

- The sagittal plane
  - The median plane
- The frontal (coronal) plane
- The transverse plane (Figs. 1 and 2)

Fig. 1 Preview of BTS SMART [10]

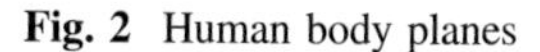

Fig. 2 Human body planes

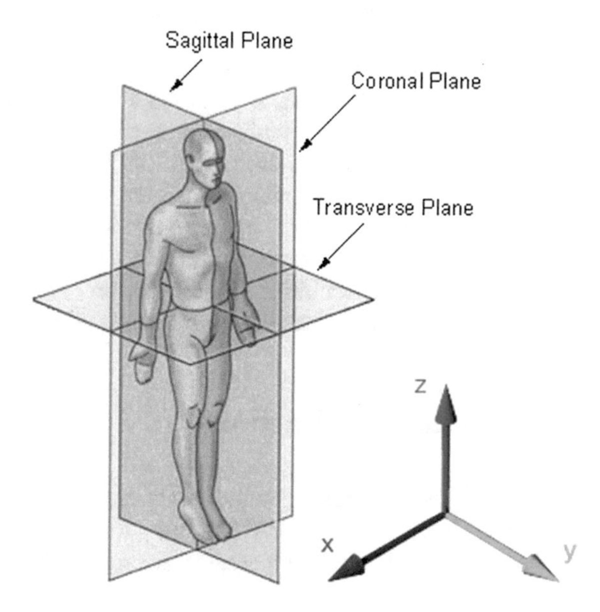

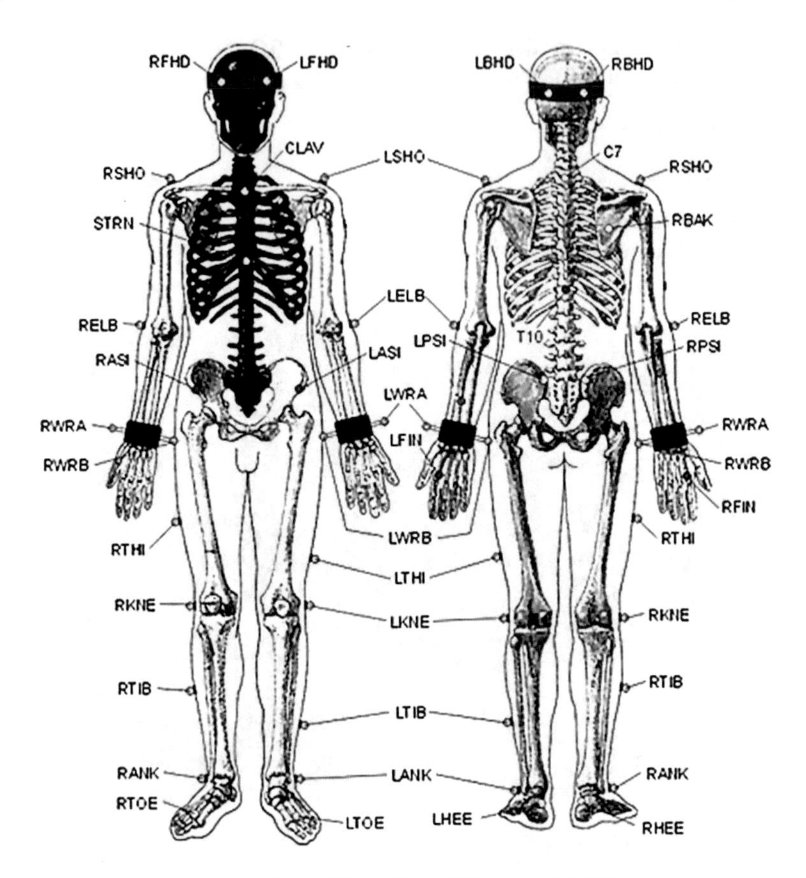

Fig. 3 Marker placement according to PlugInGaitFullBody model [10]

Special markers, which define selected body parts, are placed in these planes. All markers are placed according to validated available models (and then identified by specialized software) or customized by user's specific requirements (Figs. 3 and 4). At least two video cameras are needed for recording the motion and calculating coordinates in 3D. Aforementioned markers are used for the ease of measurement. The most often used markers are retro-reflexive markers (also known as passive markers) that excellently reflect infrared radiation from video cameras. This fact ensures fine level of visibility of these markers in the final digitized record. The newer systems use active markers which include infrared LEDs [9].

3 Artificial Neural Network

The aim of the paper is to analyze data and prepare them for input vector into Artificial Neural Networks (ANN) used for classification. The artificial neural networks are inspired in the biological neural nets and are often used for complex

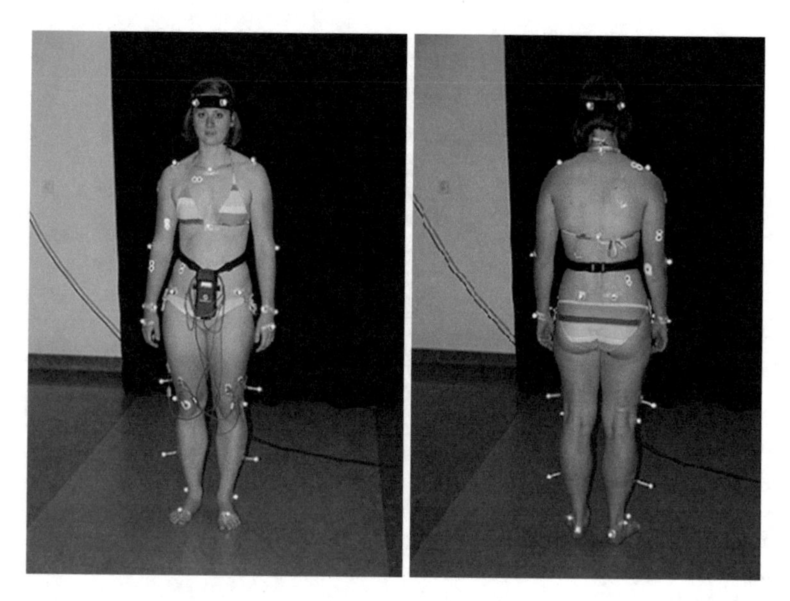

Fig. 4 Marker placement

and difficult tasks [11–14]. The most typical use is the classification of different objects. ANNs are capable of generalization and therefore the classification is natural for them. Some other possibilities are in pattern recognition, data approximation, control, filtering of signals and others.

There are several kinds of ANN. The most used ANN are with feedforward net structure with supervision. Often, Backpropagation and Levenberg-Marquardt algorithms are used for training [11]. ANN needs a training set of known solutions to be learned on them. Supervised ANN has to have input and also required output.

4 Experiment

The system VICON, that is located in university hospital in Brno, was used for this experiment. The laboratory is equipped with 8 video cameras (in height 1.4–2.5 m) and illuminated by rings radiating infrared light with wavelength of 780 nm. Retro-reflexive markers and infrared light allow scanning of the whole trajectory of motion with accuracy of few hundredths of a millimeter. The type of used cameras was MX20+ with the resolution of 1600 × 1280 pixels and frequency of 120 FPS (frames per second).

The experiment was performed with 21 participants: 15 men and 6 women in age of 19–30 years. Each basic self-defense technique was performed 10 times in a row. These basic techniques were selected:

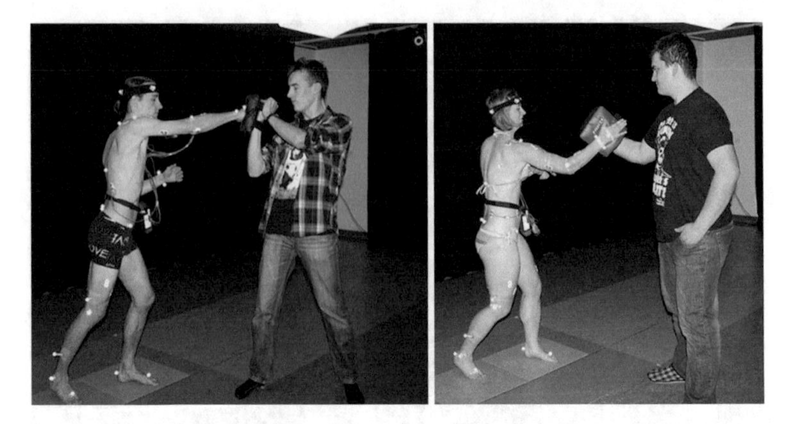

Fig. 5 Preview of direct punch (*left*) and slap (*right*)

- Direct punch (Fig. 5 left)
- Slap (Fig. 5 right)
- Direct kick
- Round kick

Data processing was divided into two main parts:

- Motion visualization
- EMG analysis of individual muscles.

Other areas (such as calculation of speed, visualization of trajectory of major joints, etc.) will be solved in future research. The main aim of the current research and this experiment was ascertaining if motion scanning systems are usable for analysis of striking techniques because primary focus of these systems is different. Also this causes data processing to be more complicated, because software for these systems is intended for analysis of human movement (walking).

8 groups were formed for the analysis. The division was based on skills (trained and untrained persons). Description of these groups is in Table 1. These pairs are described in more details in part 5 (Results).

Table 1 Compared groups

Trained woman—direct punch	Untrained woman—direct punch
Trained man—slap	Untrained woman—slap
Trained woman—direct kick	Untrained woman—direct kick
Trained man—round kick	Untrained woman—round kick

5 Results

This part describes only the most interesting results. The reason is simple. There is a great amount of collected data and a lot of possible comparisons.

The first compared group is composed of a trained and an untrained woman with the scanned marker placed on the back of the hand and in one case on the elbow. The process of direct punch was compared. As it is depicted in Fig. 6, there is a difference in arm trajectory. It is clear that trajectories for direct punch of the untrained woman are absolutely different. Also the marker on elbow shows visible round motion. The trained woman has both trajectories almost same and direct.

The next pair was a trained man and an untrained woman. The process of slap was compared. The main focus was placed on the marker on the back of a hand. The following figure (Fig. 7) shows that the curve of trajectory for the trained man is higher than for the untrained woman. The untrained woman has prepared her hand in a lower position. For an accurate hit, she had to move her hand higher. This caused the decomposition of force which affected the overall intensity of the slap. The trained man raised his hand at the beginning of the slap and the final move to the target is almost in one axis. The fall came almost at the end of trajectory; however, the weight of the hand was also used and then the final slap is more intense with a lower force reduction.

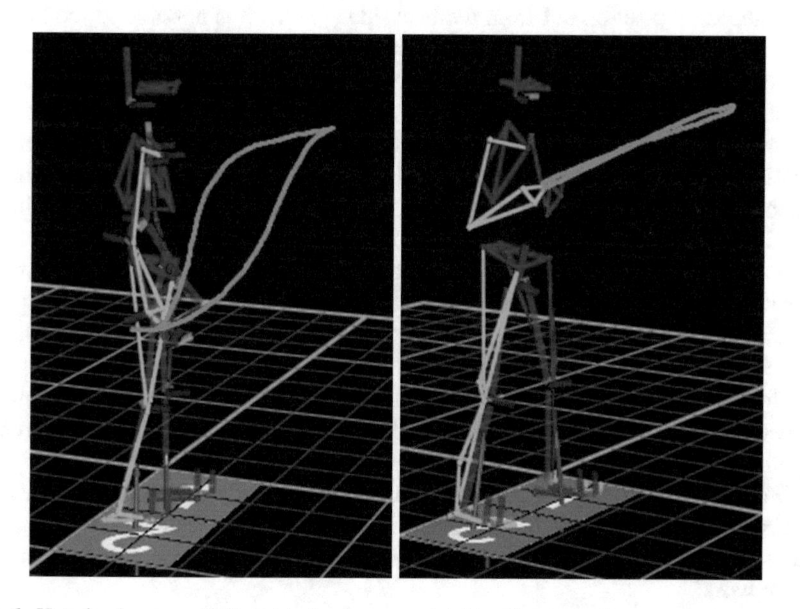

Fig. 6 Untrained woman (*left*) and trained woman (*right*)—direct punch

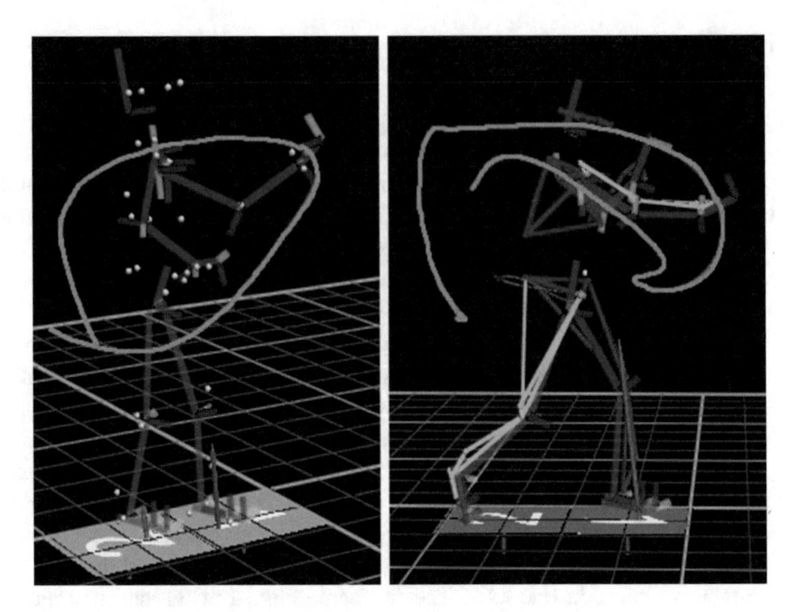

Fig. 7 Untrained woman (*left*) and trained man (*right*)—slap

The direct kick is very similar to the direct punch for untrained woman. She had leg in the right position but from the backside (Fig. 8) it is possible to see that she has round motion in her move while the trained woman lifted her knee at the beginning of the kick and then directed her kick in the straightest trajectory to the target.

From the side view, a performance of round kick is almost the same for an untrained woman and a trained man. The main difference is in positions of the support leg and arms. A better angle of view is from the backside. A significant round motion is visible for the trained man (Fig. 9). For this analysis, two markers were used. The first one was on a knee and the second was on an instep. This helped us with visualization of round kick motion for the untrained woman. Her instep was on a round trajectory; however, her knee was almost on a direct trajectory. The trained man rotated both these joints.

The differences between different untrained, trained, women and men are clearly visible. ANN would be suitable for the classification. The future step is to prepare input data which were measured by above described technique.

On the basis of human body motion, significant markers were analyzed as inputs for ANN. The results are listed in the following Table 2. The observed markers were depicted in Fig. 3. The identified suitable markers will be used in future work.

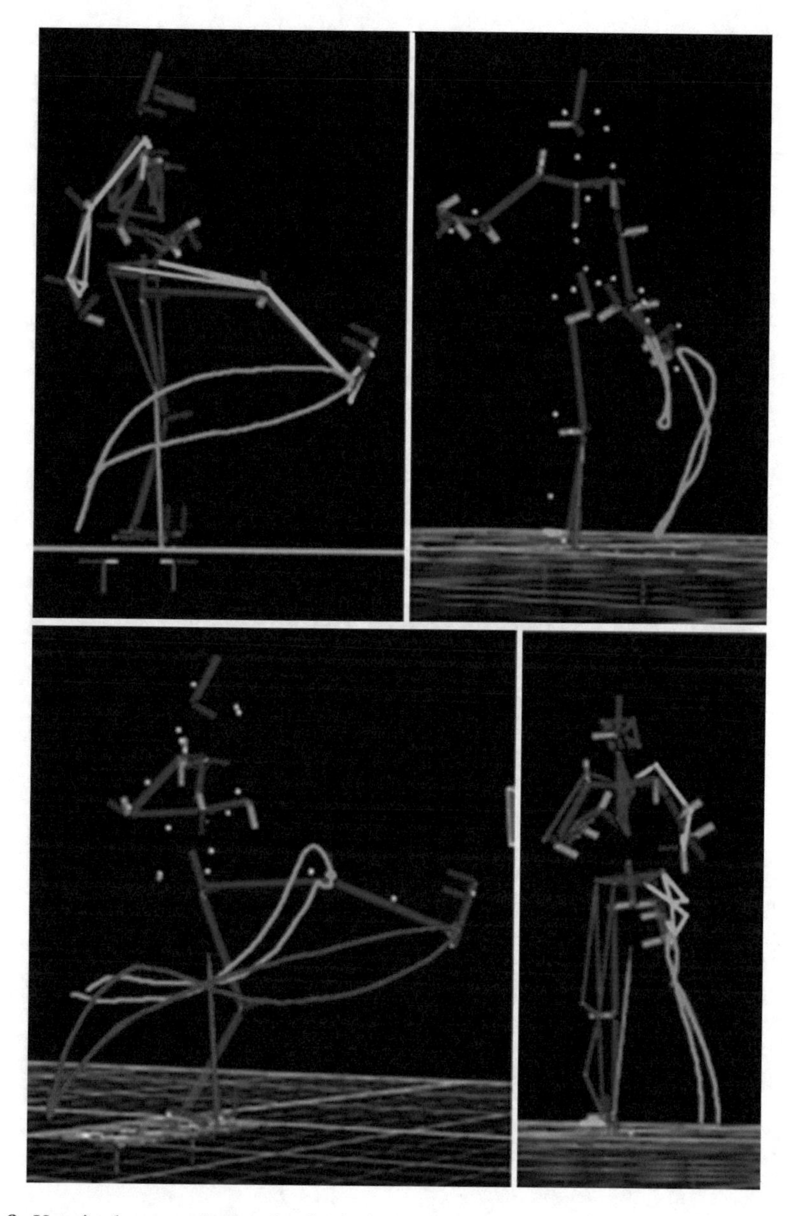

Fig. 8 Untrained woman (*top*) and trained woman (*bottom*)—direct kick

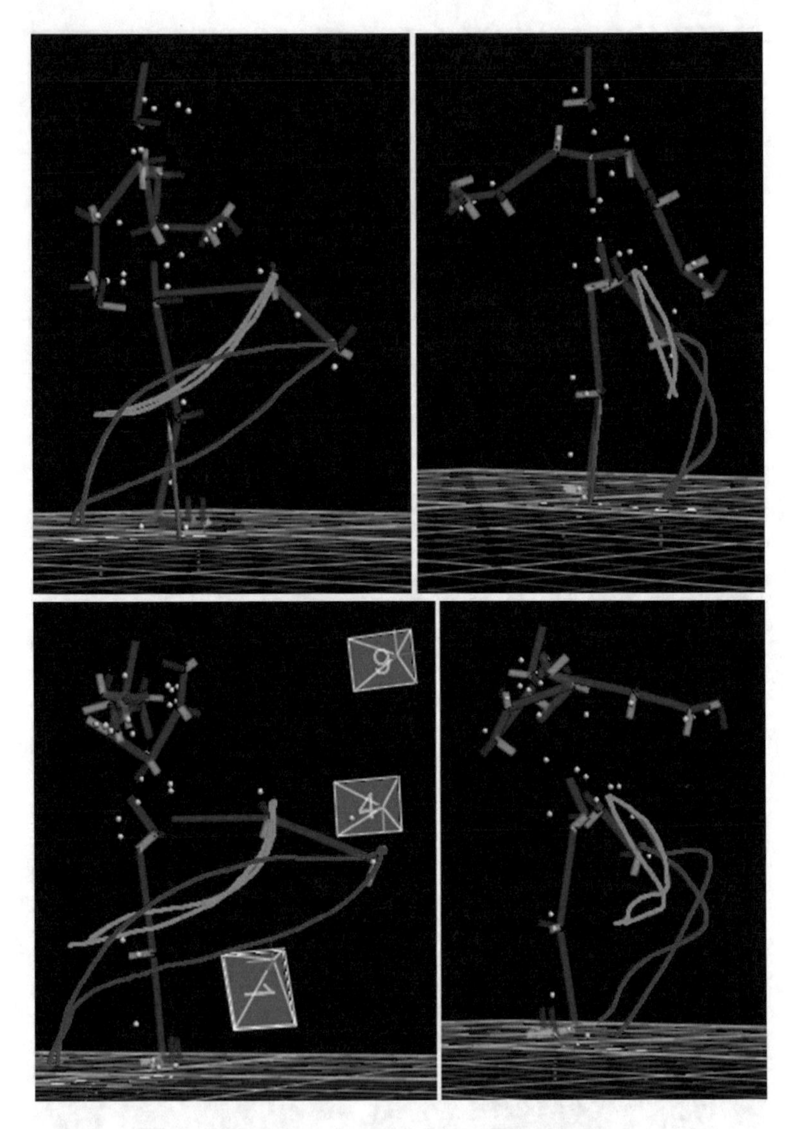

Fig. 9 Untrained woman (*top*) and trained man (*bottom*)—round kick

Table 2 Significant markers

Technique	Significant markers			
Punch	RFIN	RELB	RSHO	RPSI
Kick	RANK	RKNE	RPSI	LPSI

6 Conclusion

The paper deals with biometric motion analysis. Biometric motion system VICON has a great potential due to its clear visualization of motion. The analysis result could be used during the training of self-defense where trainees could see correct techniques and also their main mistakes. Also this system may help to recognize a trained person on the first view.

Differences in techniques according to the training level are very significant. Therefore the identified suitable markers will be used as an input feature vector into artificial neural networks in the automatic recognition system. In future research, other striking techniques will be compared and other pairs of persons will be determined too.

Acknowledgments This work was supported by by Internal Grant Agency of Tomas Bata University in Zlin under the project No. IGA/CebiaTech/2016/006.Further this work was supported by financial support of research project NPU I No. MSMT-7778/2014 by the Ministry of Education of the Czech Republic and also by the European Regional Development Fund under the Project CEBIA-Tech No. CZ.1.05/2.1.00/03.0089 and also supported by Grant Agency of the Czech Republic—GACR P103/15/06700S.

References

1. Talandova, H.: Study about application of biometric systems in the industry of commercial security. Zlin. Bachelor thesis. UTB in Zlin (2010)
2. Drahansky, M., Orsag, F.: Biometrics. 1. 294 s. (2011). [Brno: M. Drahanský]. ISBN: 978-80-254-8979-6
3. Rak, R., Matyas, V., Riha, Z.: Biometrics and identity of a person in forensic science and commercial applications. 1. Grada Publishing, Praha (2008). ISBN: 978-80-247-2365-5
4. Li, H., Li, L., Toh, K.-A.: Advanced Topics in Biometrics, xv, 500 s. World Scientific, New Jersey (2012). ISBN: 978-981-4287-84-5
5. Lapkova, D., Pospisilik, M., Adamek, M., Malanik, Z.: The utilisation of an impulse of force in self-defence. In: XX IMEKO World Congress: Metrology for Green Growth. Republic of Korea, Busan (2012). ISBN: 978-89-950000-5-2
6. Lapkova, D., Pluhacek, M., Adamek, M.: Computer aided analysis of direct punch force using the tensometric sensor. In: Modern Trends and Techniques in Computer Science: 3rd Computer Science On-line Conference 2014 (CSOC 2014), s. 507–514. Springer, Switzerland (2014). ISBN: 978-3-319-06739-1. ISSN: 2194-5357
7. Lapkova, D., Pluhacek, M., Oplatkova, Z.K., Adamek, M.: Using artificial neural network for the kick techniques classification—an inticial study. In: Proceedings 28th European Conference on Modelling and Simulation ECMS 2014, s. 382–387. Digitaldruck Pirrot GmbH, Germany (2014). ISBN: 978-0-9564944-8-1
8. Lapkova, D., Adamek, M., Oplatkova, Z.K.: Analysis of direct punch force in professional defence. In: Proceedings 29th European Conference on Modelling and Simulation ECMS 2015, s. 564–569. Digitaldruck Pirrot GmbH, Germany (2015). ISBN: 978-0-9932440-0-1
9. Sulovska, K., Adamek, M.: Research of biometric systems based on the recognition of human walking (Výzkum biometrických systémů založených na rozpoznávání lidské chůze). Jemná mechanika a optika, roč. 59, č. 10, s. 273–276 (2014). ISSN: 0447-6441

10. Sulovska, K.: On different approaches to human body movement analysis. In: Proceedings of the 2014 International Conference on Applied Mathematics, Computational Science and Engineering. Craiova: Europment, s. 264–274 (2014). ISSN: 2227-4588. ISBN: 978-1-61804-246-0
11. Fausett, L.V.: Fundamentals of Neural Networks: Architectures, Algorithms and Applications. Prentice Hall (1993). ISBN: 9780133341867
12. Gurney, K.: An Introduction to Neural Networks. CRC Press (1997). ISBN: 1857285034
13. Hertz, J., Kogh, A., Palmer, R.G.: Introduction to the Theory of Neural Computation, Addison-Wesley (1991)
14. Wasserman, P.D.: Neural Computing: Theory and Practice. Coriolis Group (1980). ISBN: 0442207433

EMG Analysis for Basic Self-defense Techniques

Dora Lapkova, Lukas Kralik and Milan Adamek

Abstract Electromyography (EMG) is an electro diagnostic medicine method used for measuring an electrical activity of skeletal muscles and nerves that control muscles. Beside the medicine, EMG is also used for measuring a local muscle load and for the purpose of this research EMG was used for determination of utilization of individual great muscles in self-defense techniques. The aim was obtaining new inputs for artificial neural networks. Simultaneously this research proceeds on previous project and researches.

Keywords Self-defense · EMG · Striking techniques

1 Introduction

This paper deals with the analysis of the utilization of the muscles during the self-defense techniques. The research uses electromyography for the description of the muscle motion. This paper introduces preliminary images of time series with patterns that will be able to use for automatic recognition system based on artificial neural networks in future.

D. Lapkova (✉) · M. Adamek
Faculty of Applied Informatics, Department of Security Engineering, Tomas Bata University in Zlin, Nam. T.G. Masaryka 5555, 760 01 Zlin, Czech Republic
e-mail: dlapkova@fai.utb.cz

M. Adamek
e-mail: adamek@fai.utb.cz

L. Kralik
Faculty of Applied Informatics, Department of Informatics and Artificial Intelligence, Tomas Bata University in Zlin, Nam. T.G. Masaryka 5555, 760 01 Zlin, Czech Republic
e-mail: kralik@fai.utb.cz

R. Silhavy et al. (eds.), *Software Engineering Perspectives and Application in Intelligent Systems*, Advances in Intelligent Systems and Computing 465,
DOI 10.1007/978-3-319-33622-0_32

Electromyography (EMG) is an electro diagnostic medicine method used for measuring an electrical activity of skeletal muscles and nerves that control muscles [1, 2]. Electrical bio potentials (electrical muscle activity) can be measured in two ways. The first way is using needle electrodes. These electrodes are placed directly into the muscle. Another option is the use of surface electrodes. This type of electrodes is stuck on the skin just above the muscle belly [3].

A result of EMG diagnostic is a graph. More specifically, this graph is called an electromyogram or also EMG diagram [2]. Diagnostic is carried out by a special device called electromyograph. This device registers action potential (an impulse) that is caused by intentional activation of skeletal muscle or peripheral nerve irritation (e.g. pain). Created impulse spreads over nerve fiber and activates muscle fibers. This causes muscle to twitch. These twitches are recorded by electrodes and then the signal is transmitted into device where signal is processed and displayed as EMG diagram. [1–4].

Beside the medicine, EMG is also used for measuring a local muscle load [4] and for the purpose of this research, EMG was used for determination of utilization of individual great muscles in self-defense techniques.

Self-defense is a field which is primarily focused on the legal protection of personal interests. It covers various areas- the theory and the practice of defense, attack and prevention, scientific disciplines such as tactics (e.g. skill in the counter attack), strategy (precautionary action) and operation (behavior after a conflict situation). Moreover, it includes the knowledge of somatology and the chosen parts of crisis management, especially the phases of the conflict and solutions to conflict situations. [5, 6] Striking techniques are very important part in majority of martial arts [6, 7] and this is the reason why our experiment was focused on them.

Our main hypothesis is that the level of training affects the utilization of individual muscles. The better trained person uses more muscles and makes the technique more effective.

In the past research [7, 8], it was observed that an artificial neural network is suitable for recognition of gender and training level of monitored subject in self-defense. The force profile and some statistical information such as mean and maximum of the force were used for input data during the training. The main motivation of this paper was to find other possible input data with the same or better results of classification. The experiment was mainly focused on identifying of significant EMG signal profile and patterns.

2 Artificial Neural Network

The paper prepares data for classification by means of artificial neural networks (ANN). Therefore it is useful to describe them shortly. ANN are inspired in the biological neural nets and are used for complex and difficult tasks [9–12]. The most often usage is classification of objects as also in this case. ANNs are capable of generalization and hence the classification is natural for them. Some other

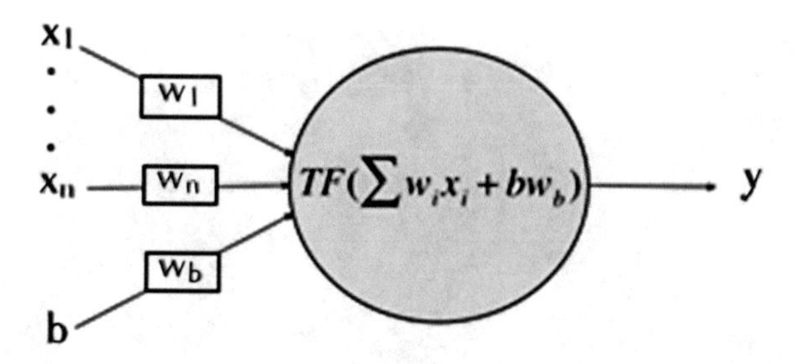

Fig. 1 A node model, where TF (transfer function like sigmoid), x1–xn (inputs to neural network), b—bias (usually equal to 1), w1–wn, wb—weights, y—output [7]

possibilities are in pattern recognition, control, filtering of signals and also data approximation and others.

There are several kinds of ANN. The most used ANN are with feedforward net structure with supervision. Often, Backpropagation and Levenberg-Marquardt algorithms are used for training [9]. ANN needs a training set of known solutions to be learned on them. Supervised ANN has to have input and also required output. ANN with unsupervised learning exists and there a capability of selforganization is applied.

The neural network works so that suitable inputs in numbers have to be given on the input vector. These inputs are multiplied by weights which are adjusted during the training. In the neuron (node) (Fig. 1), the sum of inputs multiplied by weights are transferred through mathematical function like sigmoid, linear, hyperbolic tangent etc. to the output from a neuron unit—node. The settings of the structure and transfer functions are tuned during the training on the specific task.

3 Experiment

The system VICON (Fig. 2), that is located in university hospital in Brno, was used for this experiment. The laboratory is equipped with 8 video cameras (in height 1.4–2.5 m) and illuminated by rings radiating infrared light with wavelength of 780 nm. Retro-reflexive markers and infrared light allow scanning of the whole trajectory of motion with accuracy of few hundredths of a millimeter. The type of used cameras was MX20 + with the resolution of 1600 × 1280 pixels and frequency of 120 FPS (frames per second).

The other used device was an eight-channel EMG (3000 Hz) which was connected to great muscles via electrodes. The monitored muscles were (Fig. 3):

- *M. palmaris brevis* (1)
- *M. biceps brachii* (2)
- *M. triceps brachii* (3)
- *M. deltoideus, spinal part* (4)

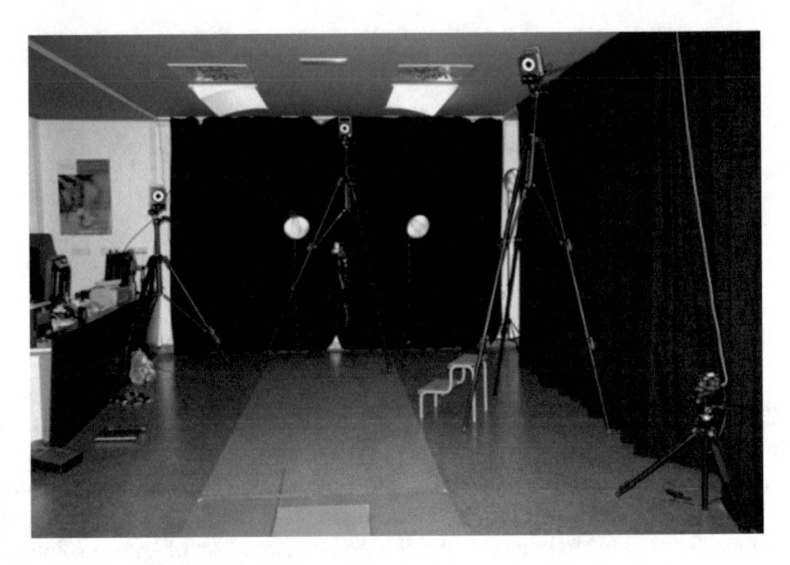

Fig. 2 Preview of VICON laboratory

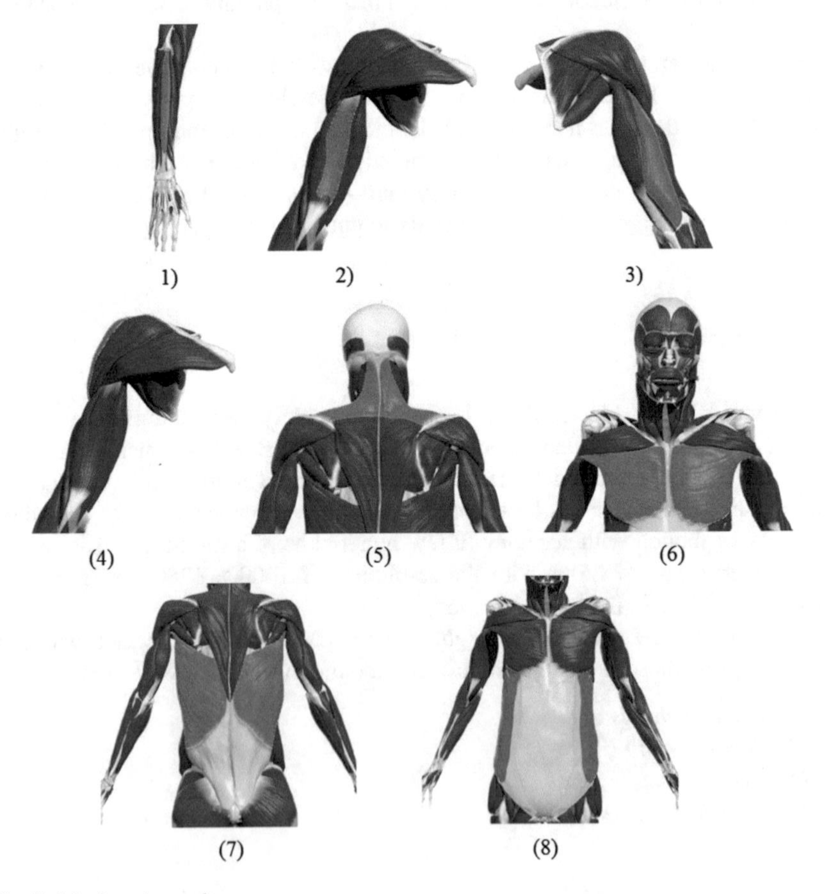

Fig. 3 Monitored muscles

- *M. trapezius* (5)
- *M. pectoralis major* (6)
- *M. latissimus dorsi* (7)
- *M. obliquus externus abdominis* (8)

This helped to discover utilization of muscles during individual self-defense techniques. Our main hypothesis was that the level of training affects the utilization of individual muscles. The better trained person uses more muscles and makes the technique more effective.

The experiment was performed with 21 participants: 15 men and 6 women in the age of 19–30 years. Each basic self-defense technique was performed 10 times in a row. These following basic techniques were selected for the experimentation: *direct punch and a slap.*

4 Results

This chapter presents results from the experiment in the image form mostly. This way is more clear and easy to understand. Pure text description would be a little bit confusing and thus difficult to understand. These images should also help to

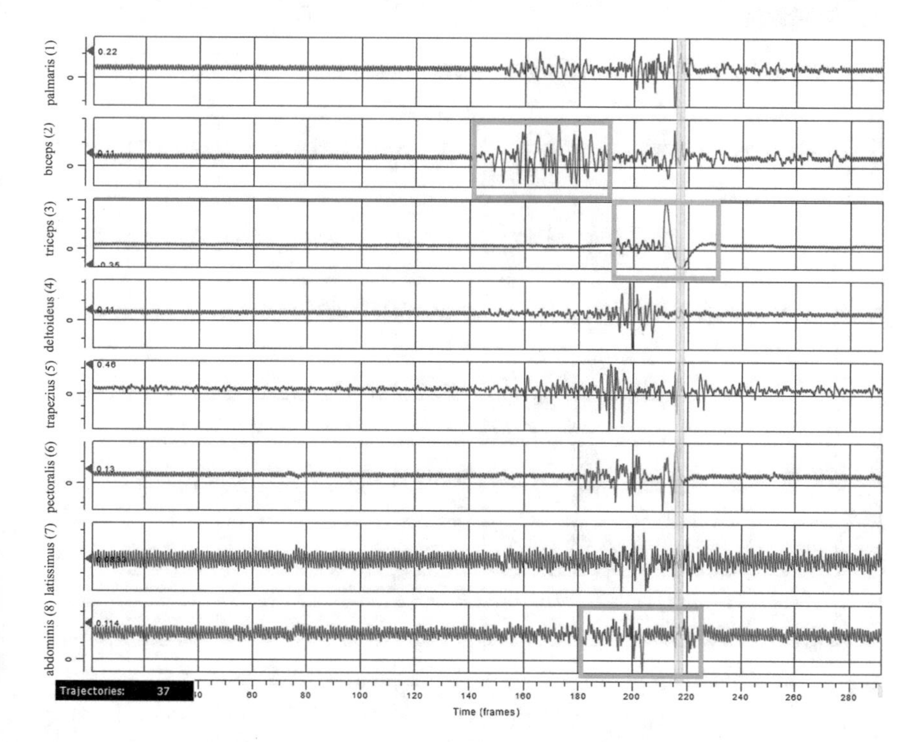

Fig. 4 EMG diagram for the untrained woman—direct punch

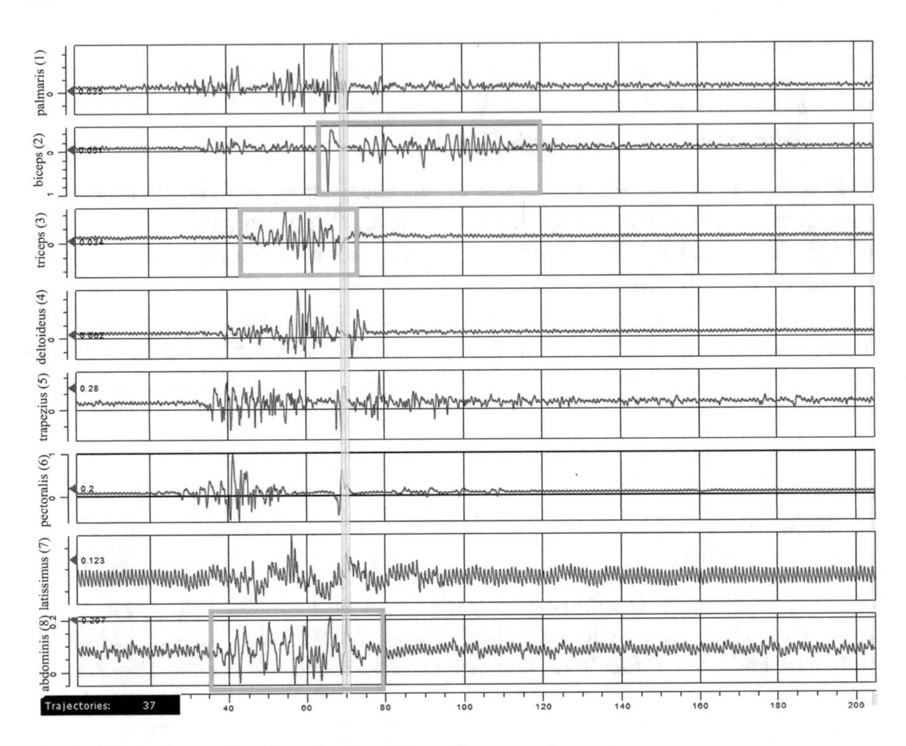

Fig. 5 EMG diagram for the trained woman—direct punch

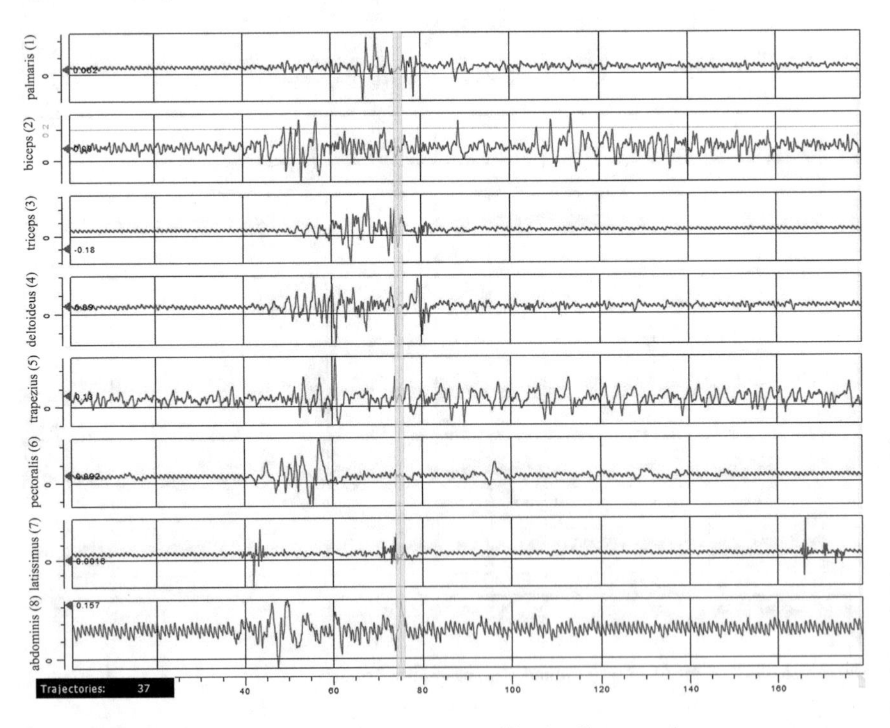

Fig. 6 EMG diagram for the untrained woman (the athlete)—direct punch

determine if ANN are able to use for classification later; the differences observed between categories have to be visible.

The first comparison is between a trained and an untrained woman. As following figures show, EMG diagram for both women is very similar (Figs. 4 and 5). The most visible difference is in utilization of abdominal muscles and triceps. The trained woman (Fig. 5) used these muscles more intensely than the untrained woman. The lesser difference is in the time of utilization of biceps that is shorter for the untrained woman (Fig. 4).

The order of muscles on all EMG diagrams is the same as in the list shown above and in Fig. 3.

The one untrained woman had very interesting values of utilization of monitored muscles (Fig. 6) because she is a professional athlete. It is clear that she is untrained in self-defense because her trajectory of direct punch is an arc. However, EMG diagram shows that she used all of her monitored muscles and in some cases, she used some muscles more effectively than trained women.

A correct slap is performed by slash with relaxed muscles and a firming comes at the end of the motion. This can be clearly visible on EMG diagrams. The untrained woman utilized all her muscles for almost all time of the motion (Fig. 7). On the other side, the trained man performed slap in an absolutely exemplary way (Fig. 8).

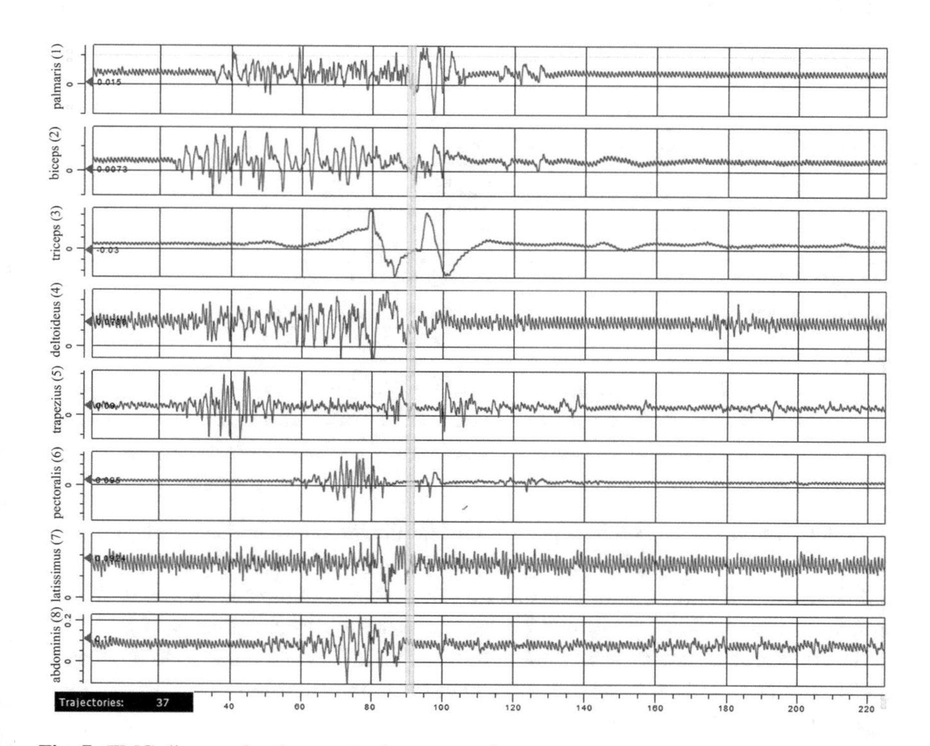

Fig. 7 EMG diagram for the untrained woman—slap

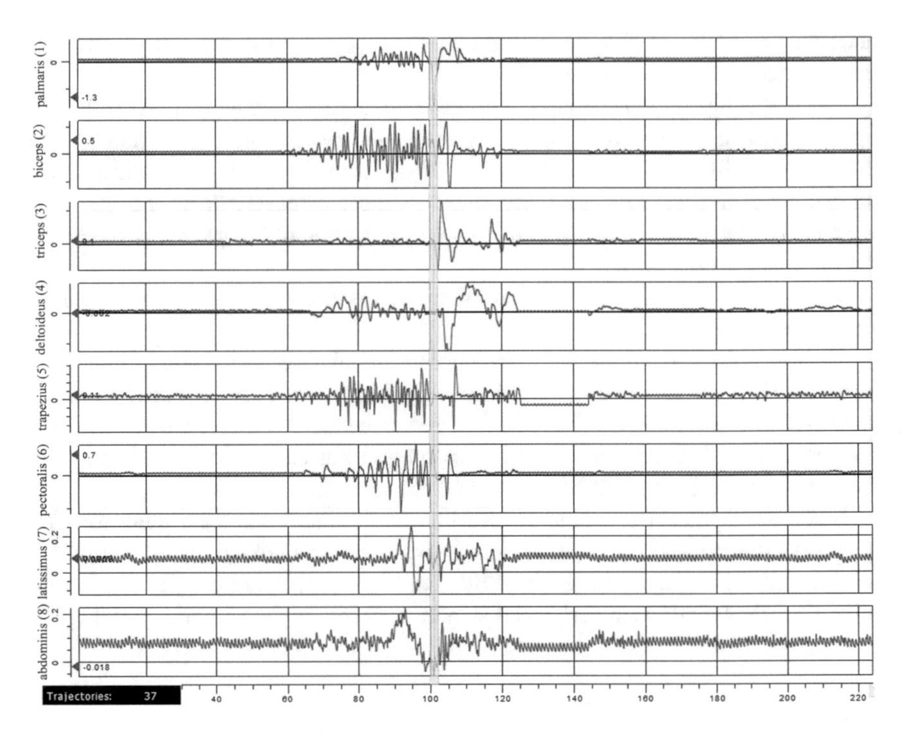

Fig. 8 EMG diagram for the trained man—slap

Table 1 Muscles with significant EMG profile

Direct punch	Slap
M. biceps brachii	M. palmaris brevis
M. triceps brachii	M. biceps brachii
M. latissimus dorsi	M. triceps brachii
M. obliquus externus abdominis	M. deltoideus, spinal part
	M. trapezius
	M. pectoralis major
	M. latissimus dorsi
	M. obliquus externus abdominis

Other compared pairs are very similar in the process of slap. All untrained people had all monitored muscles very firmed while trained people performed self-defense techniques with relaxed muscles. A firming came at the end of a technique which allows performing a technique with greater intensity.

The differences between different untrained, trained, women and men are clearly visible. ANN would be suitable for the classification. The future step is to prepare input data which were measured by above described technique.

On the basis of EMG signal, significant profile was analyzed as inputs for ANN. The results are listed in the following table (Table 1). Monitored muscles were depicted in Fig. 3.

For direct punch, only 4 muscles were selected to be ANN inputs because there were the greatest differences between untrained and trained subjects. When comparing a slap, all monitored muscles showed significant differences mainly in the time period.

5 Conclusion

Biometric motion system VICON has a great potential due to its clear visualization of motion and possibility of using another device such as EMG. This could be used for training of self-defense, where trainees could be shown correct techniques and also their main mistakes. Also this system may help to recognize a trained person on the first view.

Differences in techniques according to the training status are very significant. The interesting group is persons who do some kind of sport (football, gymnastic, etc.) on high level or even professionally. They know their bodies and can use mass, muscles and center of gravity for their profit. Then the main difference lies in quality of used technique, especially in trajectory of techniques.

On the basis of EMG signal, significant profile was analyzed as inputs for ANN. An analysis of results from experimentation shows that only 4 muscles are enough as an input vector for ANN in the case of direct punch. However, all muscles will be used with focusing on time period in the case of slap. The findings will be used in further work for assessment of gender and training level of a subject for the purpose of self-defense in a complex recognition system.

Acknowledgments This work was supported by by Internal Grant Agency of Tomas Bata University in Zlin under the project No. IGA/CebiaTech/2016/006.Further this work was supported by financial support of research project NPU I No. MSMT-7778/2014 by the Ministry of Education of the Czech Republic and also by the European Regional Development Fund under the Project CEBIA-Tech No. CZ.1.05/2.1.00/03.0089.

References

1. Michell A.: Understanding EMG, 288 s. OUP Oxford, Oxfrord, Great Britan (2013). ISBN: 9780191509407
2. Norris, F.H.: The EMG: a guide and atlas for practical electromyography. Grune & Stratton (2008)
3. Remijn, L., Groen, B.E., Speyer, R., van Limbeek, J., Nijhuis-van der Sanden, M.W.G.: Reproducibility of 3D kinematics and surface electromyography measurements of mastication. Physiol. Behav. **155**, 112–121 (2016)

4. Liu, J., Liu, Q.: Use of the integrated profile for voluntary muscle activity detection using EMG signals with spurious background spikes: a study with incomplete spinal cord injury. Biomed. Signal Process. Control **24**, 19–24 (2016)
5. Lapkova D., Pospisilik M., Adamek M., Malanik Z.: The utilisation of an impulse of force in self-defence. In: XXIMEKO World Congress: Metrology for Green Growth. Busan, Republic of Korea (2012). ISBN: 978-89-950000-5-2
6. Lapkova D., Pluhacek M., Adamek M.: Computer aided analysis of direct punch force using the tensometric sensor. In: Modern Trends and Techniques in Computer Science: 3rd Computer Science On-line Conference 2014 (CSOC 2014), s. 507–514. Springer (2014). ISBN: 978-3-319-06739-1. ISSN: 2194-5357
7. Lapkova D., Pluhacek M., Kominkova Oplatkova Z., Adamek M.: Using artificial neural network for the kick techniques classification– an inticial study. In: Proceedings 28th European Conference on Modelling and Simulation ECMS 2014, s. 382–387. Digitaldruck Pirrot GmbH, Germany (2014). ISBN: 978-0-9564944-8-1
8. Lapkova D., Adamek M., Kominkova Oplatkova Z.: Analysis of direct punch force in professional defence. In: Proceedings 29th European Conference on Modelling and Simulation ECMS 2015, s. 564–569. Digitaldruck Pirrot GmbH, Germany (2015). ISBN: 978-0-9932440-0-1
9. Fausett L.V.: Fundamentals of Neural Networks: Architectures, Algorithms and Applications. Prentice Hall (1993). ISBN: 9780133341867
10. Gurney K.: An Introduction to Neural Networks. CRC Press (1997). ISBN: 1857285034
11. Hertz, J., Kogh, A., Palmer, R.G.: Introduction to the Theory of Neural Computation. Addison–Wesley (1991)
12. Wasserman, P.D.: Neural Computing: Theory and Practice. Coriolis Group (1980). ISBN: 0442207433

Proposal of the Web Application for Selection of Suitable Job Applicants Using Expert System

Bogdan Walek, Ondrej Pektor and Radim Farana

Abstract Every company sometimes recruits new employees. Selecting suitable job applicants is often very complicated and there are many criteria which enter the hiring process. The paper proposes a web application for selection of suitable job applicants. An important part of the proposed web application is an expert system, which propose suitable job applicants based on their hard and soft skills and process of the job interview. We also suggest a structure of a database of job positions and a method for comparing job position requirements with job applicant's skills and evaluating suitability of job applicant due to job position requirements.

Keywords Web application · Job applicants · Job position · Job interview · Hiring process · Expert system · Fuzzy · Evaluation number

1 Introduction

Every company sometimes recruits new employees. Hiring of new staff and selecting suitable job applicants is often a very complicated process, because there are a lot of criteria which enter this process. There are also different hard and soft skills of each job applicant.

Currently, selecting suitable job applicants is usually solved by an HR department. If there is no HR department or HR manager in a company, this task is solved

B. Walek (✉) · O. Pektor · R. Farana
Department of Informatics and Computers, University of Ostrava,
30. dubna 22, 701 03 Ostrava, Czech Republic
e-mail: bogdan.walek@osu.cz

O. Pektor
e-mail: ondrej@pektor.cz

R. Farana
e-mail: radim.farana@osu.cz

R. Silhavy et al. (eds.), *Software Engineering Perspectives and Application in Intelligent Systems*, Advances in Intelligent Systems and Computing 465,
DOI 10.1007/978-3-319-33622-0_33

by the director or the owner of the company. But the selection process is the same or very similar. This paper focuses mainly on medium and big companies where the HR manager is responsible for the hiring process.

This paper will propose a web application for selection of suitable job applicants using an expert system. The paper is continuation of paper [1, 2].

2 Problem Formulation

As said above, hiring of new staff is a complicated process and an HR manager has a difficult task to select the most suitable job applicant. In medium and big companies, information about job applicants is stored in a database. Such a database typically contains CV of job applicants and their hard skills.

One of difficult tasks is how to store soft skills and subsequently process them appropriately. It is difficult to measure the level of each soft skill of a job applicant and compare it with soft skill levels of other job applicants. During the hiring process, the HR manager can recognize strong or weak soft skills but it might be difficult to describe and store all information about job applicant's soft skills in a database, including subjective impressions and evaluation of the HR manager.

Another problem is that there are many characteristics and properties of candidates for a specific job position. Moreover, each characteristic may affect the decision-making process during selecting the most suitable job applicant.

Another problem is that the HR manager's subjective evaluation of each job applicant, and a specific candidate can be prioritized because of his impression on the HR manager. Thus, it is not his hard and soft skills which are more important than job applicant's impression on the manager. The HR manager also can make mistakes, for instance: forgetting important characteristics during multi-round selection process, time pressure, misunderstanding of the importance of the requirements for the job position, etc.

Currently, there are open source systems for human resource management, such as Opencats or OrangeHRM. They are mainly focused on supporting the process of hiring new employees, but not on evaluating the most suitable employees or on decision-support of selecting the most suitable candidates for the job.

3 Problem Solution

Based on the above-mentioned reasons, we propose a web application with an expert system for the selection of the most suitable candidates for a given position.

The database of the proposed web application fulfils these roles:

- Storing information about job applicants and their hard and soft skills
- Storing information about job positions

- Storing information about specific selection processes (job interviews) and their evaluations

The proposed web application consists of two main evaluating parts. In the first part, based on characteristics (hard and soft skills) of job applicant and required properties for the specific job position, a special evaluation number is calculated. This special evaluation number is calculated for each job applicant and expresses the suitability of the candidate for the job position and how much the job applicant corresponds to the given job position. The highest evaluation number means that the job applicant is the most suitable for the job position, but only from the perspective of the conformity of his properties with the job requirements.

The output of the first evaluation part is the first input for the expert system. In the second part, during the selection process, the HR manager fills in a special questionnaire corresponding with the job interview. Based on the results of the questionnaire and the output of the first evaluation part, the expert system evaluates suitability of each job applicant. The knowledge base of the proposed expert system contains a set of IF-THEN rules, which is filled in by an expert on human resources.

The main parts of the proposed web application are shown in the following figure: (Fig. 1)

The following sub-chapters will define the main parts of the proposed web application:

3.1 *Definition and Storing Requirements for the Job Position*

Firstly, it is essential to define and store requirements for the job position. Based on information from typical job adverts, we propose categorization into three categories of requirements for the job position: mandatory; desired; is an advantage. Next categorization specifies the skill type of requirement: hard skills; soft skills.

For storing requirements into a database, we propose the following universal structure for all requirements:

- Category (possible values: mandatory-M, desired-D, is an advantage-A)
- Type (possible values: hard skill-H, soft skill-S)
- Quantification/Level
- Skill (expertise) description (name)
- Detailed description—optional
- Final description for requirement for job advert—consists of quantification, skill description and optionally detailed description, final description should be slightly modified for publishing in a job advert

An example of a few requirements for a specific job position called Mid-level Application Java Programmer are shown below:

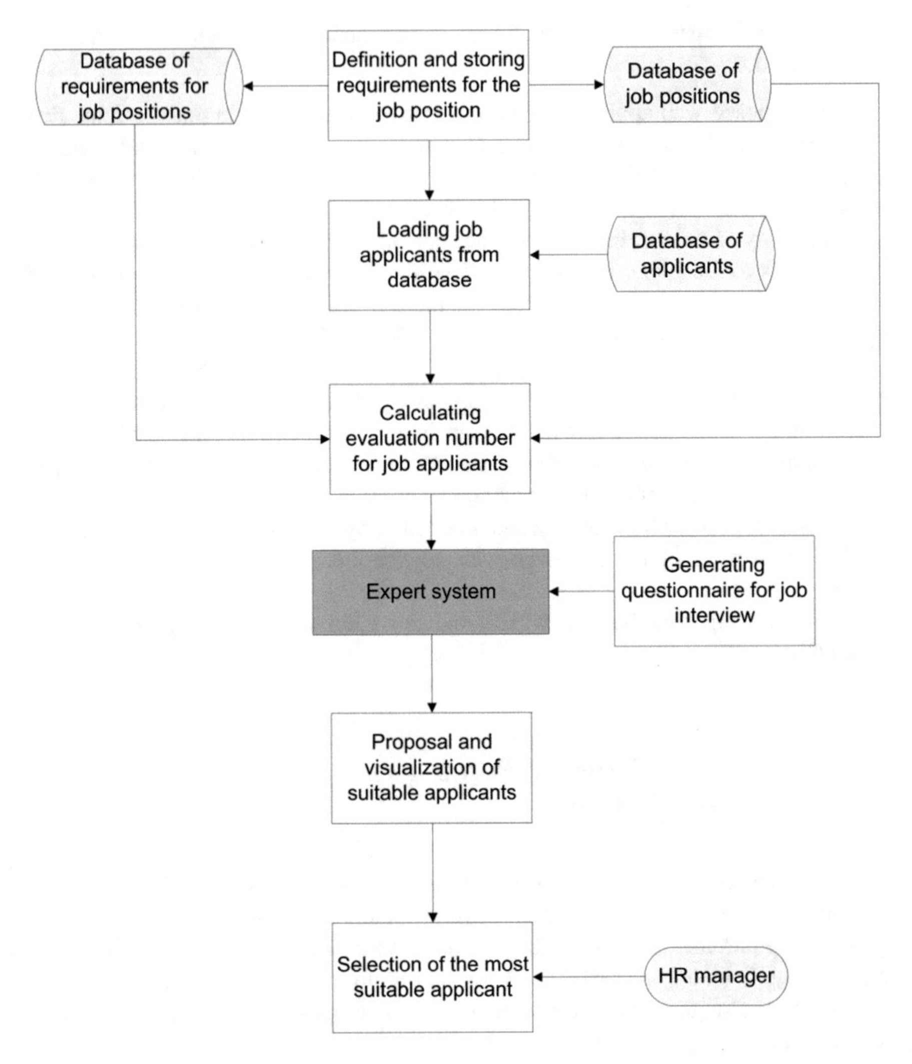

Fig. 1 Proposed web application with an expert system

For each requirement, the final description of a requirement (published in a job advert) consists of Quantification/Level, Skill description and optionally Detailed description. The HR manager may supplement this information for a job advert, but the main information is stored in three properties shown above. The final descriptions for the requirements in Table 1 are:

- 3+ years professional experience with Java EE
- 3+ years experience with multi-tier Enterprise Web Application development using Java Spring MVC framework
- 3+ years experience using UI Libraries such as AngularJS, hibernate

Table 1 Example of requirements for a job position

Cat.	Type	Quant./level	Skill desc.	Det. desc.
M	H	3 years	Java EE experience	–
M	H	3 years	Experience with multi-tier Enterprise Web Application development	Java Spring MVC framework
M	H	3 years	Experience using UI Libraries	AngularJS, Hibernate
M	H	Professional	Hibernate	–
M	H	Intermediate	Version control software	–
D	H	Professional	Oracle database design	–
D	H	Intermediate	Working on DoD systems	–
D	S	Excellent	Communication	–
D	S	Good	Teamwork	–
D	S	High	Responsibility	–
A	H	Intermediate	Experience with web development	–
A	H	Intermediate	Experience with jQuery	–

- Professional experience with Hibernate
- Intermediate experience with version control software
- Professional experience with Oracle database design
- Intermediate experience working on DoD systems
- Excellent communication skills
- Good ability to work in team
- High sense of responsibility
- Intermediate experience with web development
- Intermediate experience with jQuery

Each requirement requires its priority to be specified. It represents the importance of a specific requirement in relation to the job position. During the selection process (job interview), the HR manager knows which requirements from mandatory or desired requirement are more or less important. Possible priority values for requirements are: low; medium; high.

The process of defining requirements and their properties for a job position is a task for an expert in the company who knows everything necessary for the specific kind of the job position. Certain job requirements (especially soft skills) may be specified in cooperation with the HR manager.

3.2 *Loading Job Applicants from Database*

In the second step, job applicants are loaded from the database. The database of applicants consists of new candidates for the job position and current employees. Current employees are stored in the database of job applicants because we have information about the skills acquired from their CV before the job interview and current (continually updated) information about their skills. Another reason is that in the case of a new job position in the company (especially in medium and big company), we can include them into the selection process (they are suitable candidates, they want to change job position within the company).

In the case of new job candidates, the HR manager opens their CV and stores their hard and soft skills in the same structure as the job position requirements are stored. It is appropriate to store all important skills of new job candidates before the job interview. A profile of each entry in the job applicant database consists of:

- Personal information
- Relevant hard skills
- Relevant soft skills
- Other skills—for instance, a job applicant seeks for a job position Java programmer and he puts into his CV—skill "driver license for buses". This is an interesting skill but irrelevant to this job position. Though a hard skill, but the HR manager stores it into category Other skills

For the selection process, the HR manager has a few options how to load job applicants for the job interview (selection process):

- Load all stored job applicants and employees
- Load all job applicants registered for the current job interview
- Load all job applicants registered for the current job interview and selected employees

The output of these steps is a set of selected job applicants for the job interview.

3.3 *Calculating Evaluation Number for Job Applicants*

Based on a set of selected job applicants from the previous step and specific job position, we can calculate the "evaluation number" for each job applicant (from the selected job applicants for the job interview). Calculating of the evaluation number is a process of comparing requirements for the job position with skills of job applicants and determining which applicants match the specific job position better or worse. This part is a pre-selection of appropriate job applicants and we want to determine which job applicants match the requirements for the job position and which do not.

A formal definition of calculating the evaluation number EN for a specific job applicant J is:

$$EN[J] = (p_1 {*} P(r_1)) + (p_2 {*} P(r_2)) + \ldots + (p_n {*} P(r_n)), \quad (1)$$

where:

p – points acquired in process of compare founded skill of job applicant with requirement of job position r

r – requirement of job position
P(*r*) – priority of requirement of job position *r*

Calculating thc cvaluation number will be demonstrated on a simple example. Requirements for the job position are:

- r_1—3+ years professional experience with Java EE
- $P(r_1)$—high
- r_2—Professional experience with Oracle database design
- $P(r_2)$—medium
- r_3—Intermediate experience with jQuery
- $P(r_3)$—low

Quantification/level for r_1 has the the following values and points. In the case of required minimal length of practice, we can use an evaluation system with penalties, which allows to select a candidate who does not meet this requirement but has good evaluation in other requirements and therefore is a very promising candidate. An example of two types of evaluation systems is shown in the table below (Table 2).

Quantification/level for r_2 and r_3 has these values and points:

- Unknown/Not founded—0 points
- Beginner—2 point
- Intermediate—4 points
- Professional—6 points

Priorities of requirements have these coefficients:

- Low—0.2
- Medium—0.5
- High—0.8

Table 2 Example of quantification/level values

Scoring system without penalties	Scoring system with penalties
Unknown/Not founded—0 points	Unknown/Not founded—0 points
1 year—1 point	1 year—0.25 point
2 years—2 points	2 years—1 point
3 years—3 points	3 years—3 points
4 years—4 points	4 years—4 points
5 years—5 points	5 years—5 points
6 years—6 points	6 years— points
7 years—7 points	7 years—7 points
8 years—8 points	8 years—8 points
9 years—9 points	9 years—9 points
10 years—10 points	10 years—10 points

Priorities can be determined depending on the place occupied by expert group using the methods of multi-criteria analysis like Fuller's triangle or Saaty's method [3–5] or using fuzzy approach [6, 7].

Job applicants are:

- J1
  - r_1 = 3 years professional experience with Java EE—**3 points**
  - r_2 = Intermediate experience with Oracle database design—**4 points**
  - r_3 = Beginner experience with jQuery—**2 points**
- J2
  - r_1 = 4 years professional experience with Java EE—**4 points**
  - r_2 = Not found—**0 points**
  - r_3 = Intermediate experience with jQuery—**4 points**
- J3
  - r_1 = 1 year professional experience with Java EE—1 **point**
  - r_2 = Professional experience with Oracle database design—**6 points**
  - r_3 = Beginner experience with jQuery—**2 points**

Evaluation numbers for job applicants are:

- EN [J1] = 4.8 (2.4 + 2 + 0.4)
- EN [J2] = 3.8 (3.2 + 0 + 0.8)
- EN [J3] = 4.2 (0.8 + 3 + 0.4)

The output of this step is an ordered list of job applicants and their calculated evaluation number. The higher evaluation number, the better the job applicant matches the job requirements.

An ordered list from the previous example is:

- J1, EN [J1] = 4.8
- J3, EN [J3] = 4.2
- J2, EN [J2] = 3.8

3.4 Generating Questionnaire for Job Interview

The output of the previous step is the first part of specification of suitable job applicants. The second part is evaluation of the job interview and level of skills presented by each job applicant. Based on the job interview results and the outputs of the previous step, the HR manager may select the most suitable job applicant for a specific job position.

An auxiliary tool for the HR manager is a generated questionnaire to register important information during the job interview for each job applicant. For each job interview (connected with a specific job position), we can generate a different questionnaire for the HR manager.

An example of a few questions and answers in the questionnaire are:

- Knowledge of a foreign language? (possible answers: insufficient, sufficient, good, very good, excellent)
- Ability to communicate (possible answers: very low, low, medium, high, very high)
- Ability to work in team (possible answers: very low, low, medium, high, very high)
- Ability to work under time pressure (possible answers: very low, low, medium, high, very high)

After filling in the questionnaire for each job applicant, results are stored in the database and the HR manager may see the results of all job applicants.

3.5 Proposal and Visualization of Suitable Applicants

For the selected questionnaire from previous step, appropriate knowledge base of expert system is loaded. Knowledge base consists of IF-THEN rules connected with a question in a specific questionnaire. Based on the evaluation number of each job applicant and the results from the questionnaire, the expert system proposes suitable candidates for the job position:

An example of several IF-THEN rules is shown below:

```
IF (EVALUATION NUMBER IS VERY HIGH) AND
(KNOWLEDGE OF LANGUAGE IS GOOD) AND
(COMMUNICATION LEVEL IS HIGH) AND
(TEAM WORK LEVEL IS VERY HIGH) AND
(TIME PRESSURE WORK IS VERY HIGH) THEN
(DEGREE OF SUITABILITY IS VERY HIGH)

IF (EVALUATION NUMBER IS HIGH) AND
(KNOWLEDGE OF LANGUAGE IS GOOD) AND
(COMMUNICATION LEVEL IS MEDIUM) AND
(TEAM WORK LEVEL IS VERY HIGH) AND
(TIME PRESSURE WORK IS HIGH) THEN
(DEGREE OF SUITABILITY IS HIGH)
```

For the expert system knowledge base, the LFL Controller was used. Linguistic Fuzzy Logic Controller is more described in [8].

The degree of suitability of job applicants is then shown in an ordered list:

- J1—VERY HIGH
- J3—HIGH
- J2—HIGH

3.6 *Selection of the Most Suitable Applicant*

In the last step, the HR manager selects the most suitable candidate for the particular position. The expert system, evaluation numbers of job applicants and questionnaire results are only helpful tools for the human resources manager who will make the final selection of the most suitable candidate.

4 Conclusion

This article proposed a web application using an expert system for the selection of the most appropriate applicant for a job position within the company. The proposed application is connected with a database of job applicants and job positions. The article described the pre-selection mechanism for selecting appropriate job applicants in relation to job position requirements. Based on this information, a filled-in questionnaire and the knowledge base, the expert system suggests the most suitable candidates for employment.

The developed system has been compared with a previously realized selection procedure in a cooperating company. The previous system was based on a pointing system. Every applicant's skill was pointed, and the sum of points decided on candidate's invitation for the interview. A comparative analysis shows that a selection of 5 programmers out of 64 applicants needed an interview of 28 people based on the pointing system. The proposed system selected all 5 successfully selected applicants within the group out of 14 best ranked candidates.

Acknowledgment This work was supported by the project "LQ1602 IT4Innovations excellence in science" and during the completion of a Student Grant SGS02/UVAFM/2016 with student participation, supported by the Czech Ministry of Education, Youth and Sports.

References

1. Walek, B., Bartoš, J.: Expert system for selection of suitable job applicants. In: Proceedings of the 11th International FLINS Conference, pp. 68–73. World Scientific Publishing Co. Pte, Ltd., Singapore (2014)
2. Walek, B.: Fuzzy tool for selection of suitable job applicants. Global J. Technol. **2015**(8), 125–132 (2015)
3. Triantaphyllou, E.: Multi-criteria decision making methods: a comparative study, vol. 28, 288 p. Kluwer Academic Publishers, Boston, Mass (2000). ISBN: 0-7923-6607-7
4. Saaty, T.L.: Multi-decisions decision-making: in addition to wheeling and dealing, our national political bodies need a formal approach for prioritization. Math. Comput. Model. **46**(7–8), 1001–1016 (2007). ISSN: 0895-7177
5. Köksalan, M., Wallenius, J., Zionts, S.: Multiple Criteria Decision Making: From Early History to the 21st Century, vol. 11, 197 p. World Scientific, Hackensack, NJ (2011)

6. Ramík, J.: Pairwise comparison matrix with fuzzy elements. In: 32nd International Conference on Mathematical Methods in Economics (MME), Olomouc, pp. 849–854, 10–12 Sep 2014. ISBN: 978-80-244-4209-9
7. Ramík, J.: Isomorphisms between fuzzy pairwise comparison matrices. Fuzzy Optim. Decis. Making **14**(2), pp. 199–209 (2015). ISSN: 1568-4539
8. Habiballa, H., Novák, V., Dvořák, A., Pavliska, V.: Using software package LFLC 2000. In: 2nd International Conference Aplimat, pp. 355–358, Bratislava (2003)

Technique of Selecting Multiversion Software System Structure with Minimum Simultaneous Module Version Usage

Denis V. Gruzenkin, Roman Yu. Tsarev and Alexander N. Pupkov

Abstract Multiversion or N-version programming is well known as an effective approach, ensuring high level of software reliability. This approach is based on two fundamental strategies for enhancing the reliability of a software system—redundancy and diversity. Modules solving critical tasks are redundant and implemented in the form of functionally equivalent versions. In this connection versions can be developed by different programmer teams, in different languages, in different environment and can implement different methods and algorithms for solution of identical tasks in order to provide versions diversity. Complex software systems, as a rule, include a set of programs which can call the same modules for solving their target tasks, or to be more precise, versions of these modules. According to diversity concept call of different module versions allows to avoid identical failures. This article presents a technique of selecting optimal multiversion software system to minimize simultaneous usage of the same module versions.

Keywords Multiversion software · Reliability · Structure · Interface

1 Introduction

At present, different methods of designing highly reliable software are known [1–4]. One of the most perspective approaches is multiversion or N-version programming (NVP), first presented by Avizienis and Chen [5]. It says that several programming components (versions) duplicating each other are included in the

D.V. Gruzenkin (✉) · R.Yu.Tsarev · A.N. Pupkov
Siberian Federal University, Krasnoyarsk, Russia
e-mail: gruzenkin.denis@good-look.su

R.Yu.Tsarev
e-mail: tsarev.sfu@mail.ru

A.N. Pupkov
e-mail: alex007p@yandex.ru

R. Silhavy et al. (eds.), *Software Engineering Perspectives and Application in Intelligent Systems*, Advances in Intelligent Systems and Computing 465,
DOI 10.1007/978-3-319-33622-0_34

system. However, these versions are diverse, i.e. they implement different methods and algorithms to solve the same problem, they can be developed by different programmer teams, using different languages and different environment and so on [2, 6]. Multiversion implementation of program components provides system functioning regardless of hidden faults of some versions. The key advantage of N-version programming lies in the fact that system failure can occur only in case of considerable amount of versions failures [7]. Versions confirm each other's work which increases adequacy of results received [8].

N-version programming means that failures in functionally equivalent versions occur in different points, and thanks to this faults can be identified and resisted [9].

Use of module principle at a design stage is connected with the process of optimization of structure and correlations of independent multiversion software system components to achieve optimal parameters concerning development, debugging and exploitation of the system (see, for example, Kulyagin et al. [10]).

The set of tasks when selecting optimal structure of multiversion software systems includes the choice of optimal module set and data arrays, and system structure as a whole, formalized as a functional block scheme considering given technical and economical characteristics of system being developed.

Issues connected to optimization of the structure and components of multiversion software systems are contemplated in various works where different methods of this problem solution are presented. For example Kvasnica and Kvasnica in their work [11] use pseudoparallel optimization to form the structure of fault-tolerant software systems designed in line with the principle of the N-version programming approach. They propose several optimization procedures according to the features of the observed corresponding objective function.

Pham addresses the optimization issue for the cost of multiversion software systems and propose the solution for the minimum expected cost of multiversion software systems subject to the desired reliability level [12]. He also solves the problems of maximizing the reliability of the NVP subject to a constraint on expected system cost.

Redundancy can improve reliability, but increases the cost of software design and development. In [13] Rao et al. present a binary integer programming solution for multiversion software redundancy optimization.

Bhaskar and Kumar deal with the issues connected with criticality of the fault in multiversion program system and cost of its occurrence. Their work [14] suggests models for optimal release time under different constraints.

In [15] Kapur et al. propose a testing efficiency model incorporating the effect of imperfect fault debugging and error generation. Furthermore, they also formulate the optimal software release time problem for a 3-version software system under fuzzy environment and discuss a fuzzy optimization technique for solving the problem.

Probably, the greatest contribution to the solution of multiversion software optimization was made by Yamachi, Yamamoto and Tsujimura. They formulate the problem of multiversion software system optimal design as a bi-objective 0–1 nonlinear integer programming problem optimization model, maximizing the system reliability and minimizing the system cost [16–19]. They solve the optimization problem using a multi-objective genetic algorithm employing the random-key representation to provide effective genetic search ability and the elitism and Pareto-insertion based on distance between Pareto solutions in the selection proccss [16–19].

In their works [20, 21] Yamachi et al. formulate NVP design problem as the multi-objective optimization problem that seek Pareto solutions. In [20] they propose an algorithm that employs the breadth-first search method to find the Pareto solutions under practical computation time. Further they proposed employing the branch-and-bound method to find the Pareto solutions [21].

Besides the above, we can note Levitin's works, where he uses genetic algorithms to form optimal structure of a fault-tolerant software systems built according to N-version programming principle [22, 23].

Main criteria of synthesis of information processing module systems (multiversion software systems are undoubtedly such systems) alongside with reliability and cost at the stage of technical design can be: minimal intermodule interface complexity, minimum time exchange between operative and external computer memory when solving the task, minimal volume of unused data in exchange between operative and external computed memory, and maximum informational performance of the module system during solution of tasks. Values of these criteria, one way or another, depend upon structure of a multiversion software system, and particularly upon simultaneous use of versions of the same module by different multiversion software system programs.

This article considers the problem of maximization reliability with the limited cost of the system, along with minimization of simultaneous usage of the modules versions by different multiversion software system programs. It also offers a mathematical formalization of this problem. The problem solution is offered via recursive scheme of brute force (or exhaustive search), that allows to decrease the task solution time.

2 Problem's Statement

Let us consider the problem of selecting optimal multiversion software system structure which has high level of reliability, satisfies the given price constraints and provides minimal simultaneous usage of module versions.

This task appears at the stage of technical design which forms the common requirements to the software system, defines the system functions, procedures and data processing programs, including processing of input data, and getting intermediate and final results.

At module design, multiple versions which the module consists of are defined along with reliability of each version. On the basis of this data the module reliability can be calculated as follows [7, 10]:

$$R_i(X_{ij}) = 1 - \prod_{j=1}^{m_i} (1 - R_{ij})^{X_{ij}}, \quad i = 1, \ldots, n, \tag{1}$$

where

n number of modules;
m_i number of versions of module i;
R_{ij} reliability of version j of module i;
X_{ij} Boolean variable, equal to 1 if version j is used in module i, or 0—otherwise. According to N-version programming principle:

$$\sum_{j=1}^{m_i} X_{ij} \geq 2, \quad i = 1, \ldots, n.$$

Multiversion software module structure is shown in Fig. 1.

Processes of control or data processing, for which multiversion software is developed, often contain complex mathematical calculations and process big data. Therefore, usually, complex software systems, instead of independent programs are needed. Figure 2 Illustrates multiversion software system structure.

In Fig. 2, M (i) is module i. The problem of maximization of the multiversion software system reliability can be stated as follows:

$$\sum_{k=1}^{K} F_k \prod_{i \in S_k} R_i(X_{ij}) \rightarrow \max_{X_{ij}}, \quad j = 1, \ldots, m_i, \tag{2}$$

Fig. 1 Structure of multiversion module

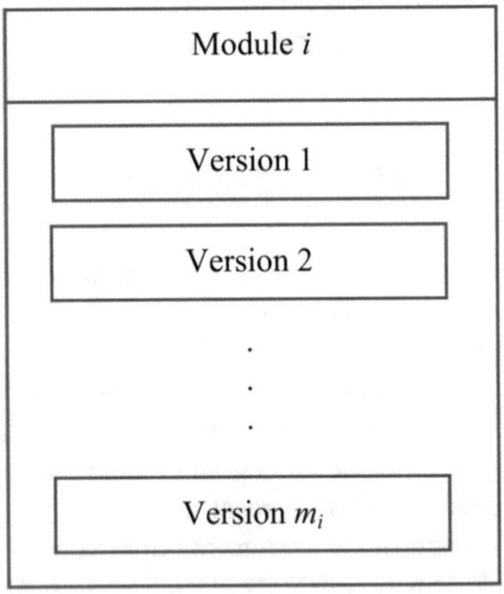

Module i

Version 1

Version 2

.
.
.

Version m_i

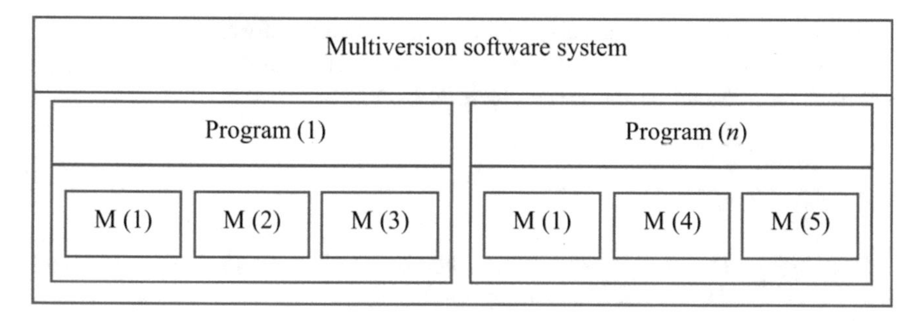

Fig. 2 Structure of multiversion software system

where

K number of programs;

S_k quantity of modules, corresponding to program k, $k = 1, \ldots, K$;

F_k frequency of program k usage, $k = 1, \ldots, K$.

When designing multiversion software system structure let us solve the problem of minimum simultaneous usage of a separate module versions in different programs (Fig. 3) i.e. minimize the number of a module versions which are used in different programs.

For formalization of this task let us define additional variables:

W_{ik} Boolean variable equals to 1 if module i is used by program k and 0 otherwise;

Y_{kj} Boolean variable equals 1 if version j of the module is used by program k and 0 otherwise:

$$Y_{kj}(W_{ik}, X_{ij}) = \begin{cases} 1, & \text{if } \sum_{i=1}^{n} W_{ik}X_{ij} \geq 1, \\ 0, & \text{if } \sum_{i=1}^{n} W_{ik}X_{ij} = 0. \end{cases} \tag{3}$$

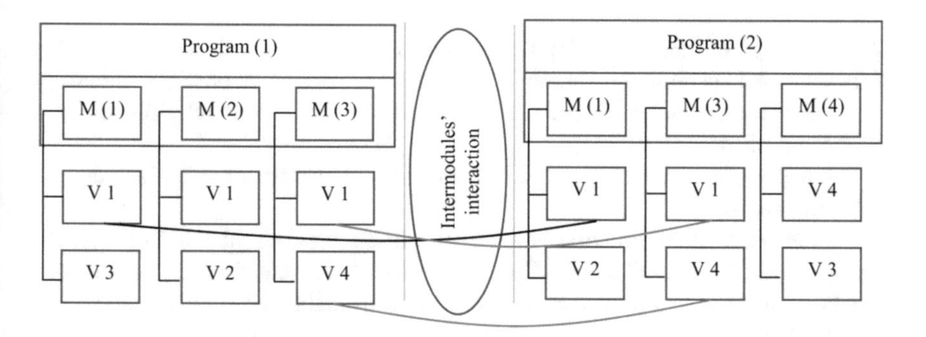

Fig. 3 The scheme of versions interaction in multiversion software system

Then the problem of selecting optimal multiversion software system structure with minimum simultaneous usage of the same modules by the multiverstion system programs can be formulated as follows:

$$\sum_{j=1}^{m_i} \sum_{k=1}^{K-1} \sum_{k'=k+1}^{K} Y_{kj}(W_{ik}, X_{ij}) \cdot Y_{k'j}(W_{ik'}, X_{ij}) \rightarrow \min_{W_{ik}, W_{ik'}, X_{ij}}, \quad i = 1, \ldots, n. \tag{4}$$

This task can be solved using brute force by considering all variants of matrix X and recalculating matrix Y according to (1).

The NVP design problem is constructing a multiversion software system with maximum system reliability and within a given budget and known as an NP-complete problem [21].

Yamachi et al. claim in [21], that the NVP design problem can be formulated as a problem for maximizing reliability under the constraint of budget limitations:

$$\sum_{i=1}^{n} \sum_{j=1}^{m_i} X_{ij} c_{ij} \leq b,$$

where

c_{ij} cost of version j of module i, $j = 1, \ldots, m_i$, $i = 1, \ldots, n$;

b budget available for system development.

Although, for such formulations, dynamic programming or genetic algorithms have been used, an algorithm that employs the branch-and-bound method can be applied as well [21].

In the example below the method of brute force has been used. This became possible due to limited number of programs, modules and versions of multiversion software system.

3 Algorithmic Basis for Method of Choosing a Multiversion Software System Structure

One of the approaches to defining an optimal structure of the designed multiversion software system is brute force method—exhaustive search of all possible variants of the modules versions usage in all programs of the system and selecting the best variant out of them.

The technique of choosing a multiversion software system structure includes the following steps.

On the basis of information about all available versions of the modules, the starting "basic" value of matrix X is set. Matrix X is created based on Boolean variables X_{ij}, equal to 1 in this case, if version j is available for use in module i, and zero if otherwise.

Brute force method allows to consider all possible values variants of designed multiversion software system structure. When doing so with all elements of Matrix X, the elements which starting value was equal to 1 change their value to zero and back, depending on the rate of duplicating the versions of a module in the structure of the multiversion software system given. Matrix X's elements which equal to zero originally, at brute force are not taken into consideration, and their values do not change.

At each stage of brute force method for a current variant of matrix X reliability of the modules (1) and multiversion software system as a whole are calculated (2).

Besides reliability analysis, each step of brute force method for each variant of matrix X involves calculation of matrix Y using the formula (3). After this, by using the criterion (4) we minimize simultaneous usage of the same versions of the modules by multiversion software system programs.

The problem of choosing a structure of multiversion software system can be considered as one of the following tasks:

- Maximization of reliability of a multiversion software system;
- Minimization of simultaneous usage of the modules version by multiversion software system programs;
- Maximization of reliability of multiversion software system with set level of modules versions usage by the programs;
- Minimization of simultaneous usage the modules versions with a set reliability of multiversion software system by the programs.

Naturally, solution of any of the abovementioned tasks corresponds to, a definite structure of multiversion software system which is formally introduced by means of matrix X.

However at brute force method a situation can arise when matrix X dimension will be big, and the amount of zeros in it will be drastically more than the amount of ones. In this case, calculation can take much time. For taking into account all possible combination of zeros and ones the recursion scheme is used, that is why a number of transitions grows exponentially with the increase of matrix X's dimension.

To avoid unnecessary steps of the algorithm in recursion we suggest introducing some additional arrays:

1. Flat Boolean array A, dimension of which equals to a number of modules. This array takes into account the rate of version duplication. Element i of this array equals to 1, if the amount of versions, used by the module is more than the rate of duplication, otherwise element i equals to zero. This allows to ignore the rows of matrix X where the original number of versions in the module equals to the rate of duplication.
2. Two-dimensional array B with the dimension of matrix X. Elements of this array are transition structures organized as follows: {number of row of matrix X, number of column of matrix X}. Each element of this array contains indexes of a row and a column of the following "basic element" (i.e. one), i.e. points to the

index of the element to be transferred to at the next step. If $X\ [i,\ j]$—the last "basic element", so the corresponding indexes in the structure of the given array will be equal to −1. If during recursion a transition to element $X\ [-1,\ -1]$ is encountered, this means that the final variant of matrix X has been obtained and we can move to calculations of the system parameters.

Array B is set as follows: we shall define indexes of a row and column in matrix X for each element $B\ [i, j]$ by moving from element to element along columns and rows until we find a "basic element". If the encountered "basic element" is in a row with index i and value of element i of array A equals to zero, then we move to the next row. If at reaching the end of the array the "basic element" has not been found, then the elements of array B which correspond to transition structures are set to −1.

The suggested approach efficiency is illustrated in Figs. 4 and 5. Figure 5 presents an example of transitions along matrix X with the rate of duplicating equal to 2.

Thus the suggested above approach lessen the number of recursive transition and decrease time needed for analysis of all possible variants of the multiversion software system structure.

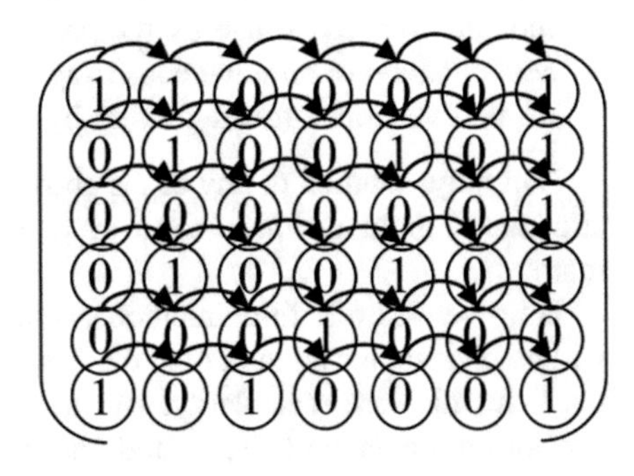

Fig. 4 Sequence of transitions at brute force method

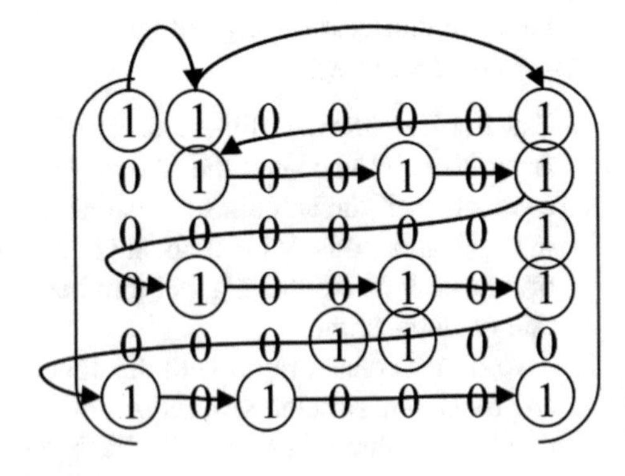

Fig. 5 Transitions by elements with the rate of duplicating equal to 2

4 Results and Discussion

Let us consider an example of selection an optimal structure for multiversion software system. The general structure of multiversion software system is given in Fig. 6. Here all versions available for each module are presented, the solid lines define the modules used in different programs of multiversion software system. In Table 1 values of reliability indices of the modules versions and frequency of usage by the programs are listed.

On the grounds of input data two tasks have been solved: maximization of multiversion software system reliability and minimization of simultaneous use of the same modules versions by multiversion software system programs.

Implementation of the above technique gave the following results: when solving the first task reliability of multiversion software system being designed constituted 0.99897; at the same time a number of simultaneous usage of the same versions of the modules in different programs is equal to 18.

When solving the second task reliability of multiversion software system being designed was 0.98837, while only six versions of the modules were used in different multiversion software system programs.

It is obvious that the given values of reliability and simultaneous usage of the versions by multiversion software system programs are threshold values for this example. With any other problem statement these values can not be excelled.

It should be noted that the size of the task allowed to implement the suggested technique based on a recursive scheme of brute force method. In case of big size of the task genetic algorithms or dynamic programming can be applied.

Moreover, we can consider the task of selecting an optimal structure of multiversion software system as a bi-objective one. This way the important things turn out to be the problem of balance between reliability and minimal level of the

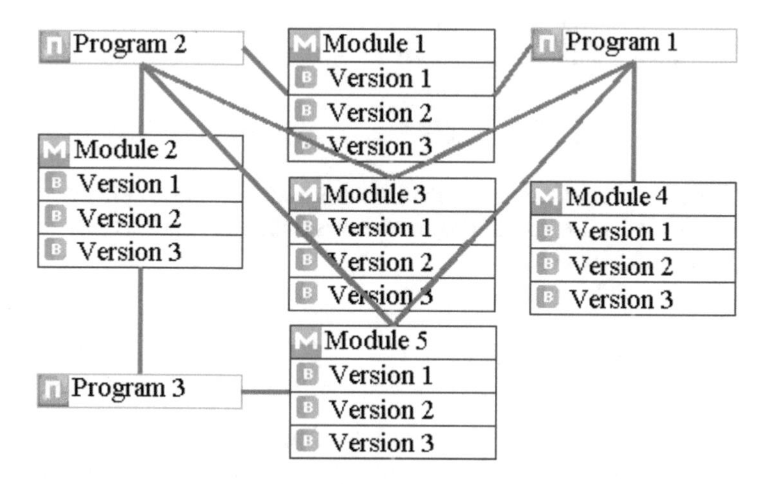

Fig. 6 Redundant structure of multiversion software system

Table 1 Input Data

Program	Module	Version	Version reliability	Frequency of program usage
P (1)	M (1)	V 1.1	0.950	0.6
		V 1.2	0.920	
		V 1.3	0.925	
	M (3)	V 3.1	0.960	
		V 3.2	0.860	
		V 3.3	0.930	
	M (4)	V 4.1	0.930	
		V 4.2	0.950	
		V 4.3	0.900	
	M (5)	V 5.1	0.960	
		V 5.2	0.910	
		V 5.3	0.950	
P (2)	M (1)	V 1.1	0.950	0.69
		V 1.2	0.920	
		V 1.3	0.925	
	M (2)	V 2.1	0.900	
		V 2.2	0.930	
		V 2.3	0.950	
	M (3)	V 3.1	0.960	
		V 3.2	0.860	
		V 3.3	0.930	
	M (5)	V 5.1	0.960	
		V 5.2	0.910	
		V 5.3	0.950	
P (3)	M (2)	V 2.1	0.900	0.5
		V 2.2	0.930	
		V 2.3	0.950	
	M (5)	V 5.1	0.960	
		V 5.2	0.910	
		V 5.3	0.950	

module versions usage by different multiversion software system programs. In this work such task was not considered, however solution of this task can become a subject for further research.

5 Conclusion

N-version programming allows reaching a maximum reliability of a software system. However, designing a redundant software system is not a trivial problem, demanding solving the set of tasks and considering different constraints.

This article represents the problem of choosing multiversion software system structure, which is defined with the usage of different versions of the same modules by system programs, is represented. The need for solving this problem is conditioned with the diversity principle realization to avoid potentially identical faults in different module versions.

The article also provides a technique allowing to solve the problem of selecting an optimal multiversion software system structure successfully considering requirements to its reliability, cost and structure constraints, structure complexity and simultaneous usage of module versions by multiversion system programs.

To solve the task of choosing multiversion software system structure involving minimal simultaneous use of the module versions the authors offered a modification of recursive scheme of exhaustive search, allowing to decrease the number of steps during search of the variants considered.

References

1. Carzaniga, A., Mattavelli, A., Pezze, M.: Measuring software redundancy. In: 37th IEEE International Conference on Software Engineering (ICSE), pp. 156–166. IEEE/ACM (2015)
2. Popov, P., Stankovic, V., Strigini, L.: An empirical study of the effectiveness of "Forcing" diversity based on a large population of diverse programs. In: 23rd IEEE International Symposium on Software Reliability Engineering (ISSRE), pp. 41–50 (2012)
3. Salewski, F., Kowalewski, S.: Achieving highly reliable embedded software: An empirical evaluation of different approaches. In: 26th International Conference on Computer Safety, Reliability, and Security, SAFECOMP, pp. 270–275, Nuremberg (2007)
4. Son, H.S., Koo, S.R.: Software reliability improvement techniques. Springer Ser. Reliab. Eng. **23**, 105–120 (2009)
5. Avizienis, A., Chen, L.: On the implementation of N-version programming for software fault-tolerance during program execution. In: Proceedings of IEEE Computer Society International Conference on Computers, Software and Applications Conference, COMPSAC, pp. 149–155 (1977)
6. Durmuş, M.S., Eriş, O., Yildirim, U., Söylemez, M.T.: A new bitwise voting strategy for safety-critical systems with binary decisions. Turk. J. Electr. Eng. Comput. Sci. **23**(5), 1507–1521 (2015)
7. Kapur, P.K., Pham, H., Gupta, A., Jha, P.C.: Software Reliability Assessment with OR Applications. Springer, London Limited (2011)
8. Latif-Shabgahi, G., Bass, J.M., Bennett, S.: A taxonomy for software voting algorithms used in safety-critical systems. IEEE Trans. Reliab. **53**, 319–328 (2004)
9. Sommerville, I.: Software Engineering, 9th edn. Pearson, Addison-Wesley (2011)
10. Kulyagin, V.A., Tsarev, R.Yu., Prokopenko, A.V., Nikiforov, A.Yu., Kovalev, I.V.: N-version design of fault-tolerant control software for communications satellite system. In: International Siberian Conference on Control and Communications (SIBCON), pp. 1–5 (2015)
11. Kvasnica, P., Kvasnica, I.: Parallel modelling of fault-tolerant software systems. Int. Rev. Comput. Softw. **7**(2), 621–625 (2012)
12. Pham, H.: On the optimal design of N-version software systems subject to constraints. J. Syst. Softw. **27**(1), 55–61 (1994)
13. Rao, N.M., Goura, V.M.K.P., Roy, D.S., Mohanta, D.K.: A binary integer programming solution for optimal reliability of computer relaying software incorporating redundancy. In:

Proceedings of IEEE Recent Advances in Intelligent Computational Systems, RAICS, pp. 524–527 (2011)
14. Bhaskar, T., Kumar, U.D.: A cost model for N-version programming with imperfect debugging. J. Oper. Res. Soc. **57**(8), 986–994 (2006)
15. Kapur, P.K., Gupta, A., Jha, P.C.: Reliability growth modeling and optimal release policy under fuzzy environment of an N-version programming system incorporating the effect of fault removal efficiency. Int. J. Autom. Comput. **4**(4), 369–379 (2007)
16. Yamachi, H., Tsujimura, Y., Yamamoto, H.: Pareto Distance-based MOGA for Solving bi-objective N-version Program Design Problem. Advances in Soft Computing (AISC), pp. 412–422 (2005)
17. Yamachi, H., Yamamoto, H., Tsujimura, Y.: Multiobjective evolutionary optimal design of N-version software system. In: Advances in Safety and Reliability—Proceedings of the European Safety and Reliability Conference, ESREL, vol. 2, pp. 2053–2060 (2005)
18. Yamachi, H., Tsujimura, Y., Yamamoto, H.: Evaluating the effectiveness of applying genetic algorithms for NVP system design. J. Jpn. Ind. Manage. Assoc. **57**(2), 112–119 (2006)
19. Yamachi, H., Tsujimura, Y., Kambayashi, Y., Yamamoto, H.: Multi-objective genetic algorithm for solving N-version program design problem. Reliab. Eng. Syst. Saf. **91**(9), 1083–1094 (2006)
20. Yamachi, H., Yamamoto, H., Tsujimura, Y., Kambayashi, Y.: Searching Pareto solutions of bi-objective NVP system design problem with breadth first search method. In: Proceedings of 5th IEEE/ACIS International Conference on Computer and Information Science, ICIS, pp. 252–258 (2006)
21. Yamachi, H., Yamamoto, H., Tsujimura, Y., Kambayashi, Y.: An algorithm employing the branch-and-bound method to search for Pareto solutions of Bi-objective NVP system design problems. J. Jpn. Ind. Manage. Assoc. **58**(1), 44–53 (2007)
22. Levitin, G.: Optimal structure of fault-tolerant software systems. Reliab. Eng. Syst. Saf. **89**(3), 286–295 (2005)
23. Levitin, G., Ben-Haim, H.: Genetic algorithm in optimization of fault-tolerant software. In: Advances in Safety and Reliability—Proceedings of the European Safety and Reliability Conference, ESREL, vol. 2, pp. 1259–1265 (2005)

Expressing Pre-, Post-conditions, Attributes and Business Constraints in Artifact-Centric Business Processes Using Object Role Modeling

Quân Nguyen-Le and Lam-Son Lê

Abstract Artifact-centric modeling is a good alternative to rigid modeling techniques (e.g., formalization) for business processes. First introduced by a research initiative of IBM, this approach has grown up considerably and drawn a lot of research interests from both academics and practitioners. Making context explicit by relating the control flow of business processes to business entities is one of many advantages of being artifact-centric in modeling business processes. Quite often, semi-formal techniques and diagrams are combined in artifact-centric modeling in order to capture both the behavior and data. We argue that the expressiveness of this approach could even be furthered by making a few modeling elements (e.g., attributes, rules, conditions) more intuitive in this approach. In this paper, we propose a modeling technique that represents these elements diagrammatically. We have chosen the Object Role Modeling language to represent pre-, post-conditions of tasks, artifact attributes and business constraints. The resulting models (of business processes) are purely diagrammatically described, as opposed to the hybrid nature of most work on artifact-centric business process modeling.

Keywords Artifact-centric process · Business process management · ORM

The original version of this chapter was revised.
An erratum to this chapter can be found at DOI 10.1007/978-3-319-33622-0_43.

Q. Nguyen-Le (✉) · L.-S. Lê
Faculty of Computer Science and Engineering, HCMC University of Technology, Ho Chi Minh, Vietnam
e-mail: nlquan.mis@gmail.com

L.-S. Lê
e-mail: lam-son.le@alumni.epfl.ch

R. Silhavy et al. (eds.), *Software Engineering Perspectives and Application in Intelligent Systems*, Advances in Intelligent Systems and Computing 465,
DOI 10.1007/978-3-319-33622-0_35

1 Introduction

Business process modeling is one of the main research concerns raised by the community of business process management (BPM) [1]. It has welcome considerable amount of work published in the BPM series and other venues. Majority of work in this realm is centered around activity-centric modeling, i.e., representation of the control flow in business processes. Artifact-centric business process modeling is an alternative [12]. First introduced by a research initiative of IBM, this approach has grown up considerably and drawn a lot of research interests from both academics and practitioners. Making context explicit by relating the control flow of business processes to business entities is one of many advantages of being artifact-centric in modeling business processes. Formal techniques in modeling might be employed in this approach, most notably on the representation of the pre- and post-conditions when it comes to the formalization of business tasks.

In this paper, we present how the Object-Role Modeling (ORM) language could be used for diagrammatically representing the main components of artifact-centric business process (ACP). Specifically, we propose a dual view modeling approach to the representation of the attributes of business artifacts, the pre- and post-conditions of business tasks and business constraints. Section 2 is dedicated to the preliminaries including the essence of being artifact-centric in business processes modeling and the ORM language. In Sect. 3, we present the motivation of our research by means of a running example. Section 4 is the core of our work where both the static view and the dynamic view are described in detail. Section 5 surveys the related work and Sect. 6 makes concluding remarks and outlines our further investigation.

2 Background

2.1 *Artifact-Centric Business Process*

Business process modeling traditionally put a lot of effort on the behavioral aspect. The so-called activity-centric modeling community became very crowded, resulting in a plenty of initiatives on both imperative and declarative modeling of the control-flow of business processes. Making an exhaustive list of them is not the goal of this paper. To name just a few, we have Business Process Execution Language (BPEL), Yet Another Workflow Language (YAWL), Unified Modeling Language (UML) activity diagram, Petri net, etc.

Artifact-centric modeling takes a different approach to modeling business processes [10, 13]. Conceptually, both business objects and the process control are taken into account in models, resulting in explicit representation of the transformation of business artifact in contract-like description[1] of business tasks [9]. This

[1]Each business task is described in terms of pre-conditions (i.e., conditions that must be held for the task to be invoked) and post-conditions (i.e., conditions that will be help upon the completion of the task).

approach has been further researched in connection to the UML [5] and Business Process Model and Notation (BPMN) [11].

One of the most cited work on ACP is called BALSA [9]. In the following, we summarize it:

- Business Artifact: Holding business information needed to complete business processes, a business artifact, or simply artifact, usually maps with a real world business entity. The representation of this element is also called information model, which includes attributes of an artifact, structure of artifacts, and relations between artifacts.
- (Artifact) Life-cycle: Included states of the artifact, (artifact) life-cycle demonstrates a sequence of states from initial to finish. One artifact have at least one initial state, at least one final state, and optional middle state(s). A current state of an artifact shows where it is and what have to be done next.
- Service: Contain conditions of business process. A service (or task) may change the state of an artifact by modifying its attributes. The representation of a service may include the notion of pre- and post-conditions.
- Association: Combination of business rules and business constraints.

2.2 *Object-Role Modeling Language*

Object-Role Modeling is a conceptual approach to information modeling using the concepts of objects (or things) and roles (or relations) [7]. For example, let us consider a `person` (as an object) `enrolls` (as a role) in a `class` (as another object). ORM includes both diagrammatic and textual representation. Providing rich graphical notations, ORM expresses the meaning of the conceptual design intuitively. Semantically, ORM representation is very close to database abstraction.

Instead of using attributes as a base construct, ORM uses *fact types* to express all relationships between objects, namely unary (e.g., `Person` smokes), binary (e.g., a `Car` has a `Registration Number`), ternary (e.g., a `Person` gets a `Grade` in a `Class`), and for generally, n-ary (n is the number of entities which involve in a relationship) [7].

An entity type is graphically represented under either a rectangle or a round rectangle or an eclipse. A value type is graphically represented under either a dash rectangle or a dash round rectangle or a dash eclipse. In this paper, we use a round rectangle to represent an entity type and a dash round rectangle to represent a value type.

A predicate is graphically represented under small rectangle(s). Notations of an unary predicate is a small rectangle, of a binary predicate is two small rectangles, of a ternary predicate is three small rectangles, and so on. A name of a predicate is placed below the symbol, meaning an attribute of an artifact that the symbol connected with. If there is a triangle on the left of the predicate text, it should be read in opposite direction. A purple line above the symbol shows whether its rule is unique. This one can represent the relation is $m : n$, $m : 1$, $1 : n$ or $1 : 1$. A purple dot in the connection visualizes the mandatory of the role. For example, a `person` (as an entity type) has exactly one (as a mandatory role) `personal ID` (as a value type). The dot maybe in either end of the connection.

3 Research Motivation

3.1 Example

In this subsection, we shall describe an example of a business process about car rental. The example is self-explanatory and in fact have been used as running examples in many scientific papers (e.g., Estañol et al. [5]) to illustrate ideas. The process involves several business participants, namely `Renter`, `Rental Company`. To start the process, a `Renter` either reserves a rental car (which she later comes for a pick-up), or simply walks in to select a rental car and picks it up at the same time. The reservation is secured or not depends on the *Renter*'s record maintained by the `Rental Company`. The process runs its course as the `Renter` takes the rental car with her while a rental agreement kicks in. The process ends when the `Renter` returns the rental car and the rental agreement is closed. If a reservation is made and the renter is about to pick up a car, the assigned car is the car that has the lowest class from the demand class above, which means its class must be a higher than or equal to the demand class.

Now let us represent this business process using the ACP approach (Fig. 1). This model is based on Car Rental process specification by Frias et al. [6] via Lohmann and Nyolt research [11]. There are a few points we need to consider. First, the diagram in Fig. 1 does not show attributes of artifact `Rental Agreement`, `Reservation` and `Car`. Second, pre- and post-conditions of service `reserve`, `pick-up`, `cancel`, `walk-in`, `return` of artifact `Rental Agreement`; `fill-in`, `non-guarantee`, `guarantee` of artifact `Reservation`, service `register payment`, `register`, `assign`, `use`, `return`, `make-available` and `sell` of artifact `Car` are not shown up in this Fig. 1. Third, we consider the business constraints could be represented in ACP approach or not.

3.2 Research Problems

Based on the analysis we have done for the example in Sect. 3.1, we come to generalization as follows:

- **Problem 1**: BPMN is well known for activity-centric modeling approach. BPMN was extended towards visual ACP modeling [11], but attributes of business objects are still not diagrammatically captured. We need diagramming techniques to visualize these attributes and how they are manipulated as the business process in question goes.

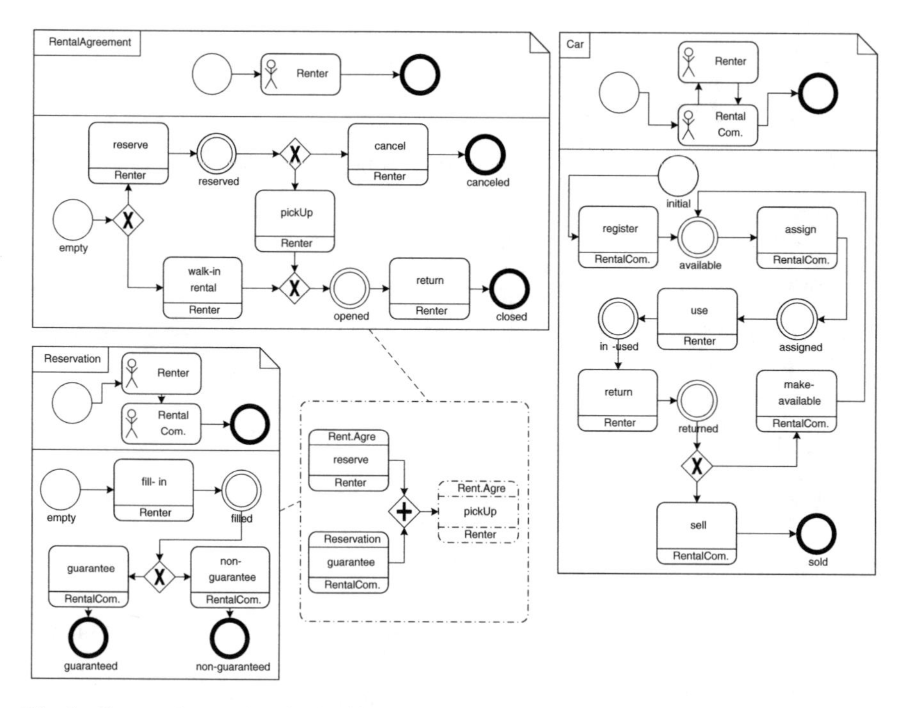

Fig. 1 Car rental example using in this paper

- **Problem 2**: We need multiple views to separate modeling concerns on artifacts and the pre- and post-conditions of business tasks individually. We consider ORM is an alternative to widely used modeling languages such as the UML [5] and used for making multi-view representation of artifact-centric modeling of business processes.
- **Problem 3**: Business constraints are usually used for enriching the semantics of business processes. We need to find a way to visually represent business constraints for artifact-centric modeling using the ORM language.

In this paper, we propose modeling framework to address the aforementioned research problems. The main idea behind our framework is that we consistently use the ORM language as an addition to traditional ACP approaches.

4 Modeling Framework

4.1 Static View

In this subsection we explicitly address our first research problem. We propose the notation of static view where business artifacts and their attributes are represented together in ORM diagrams.

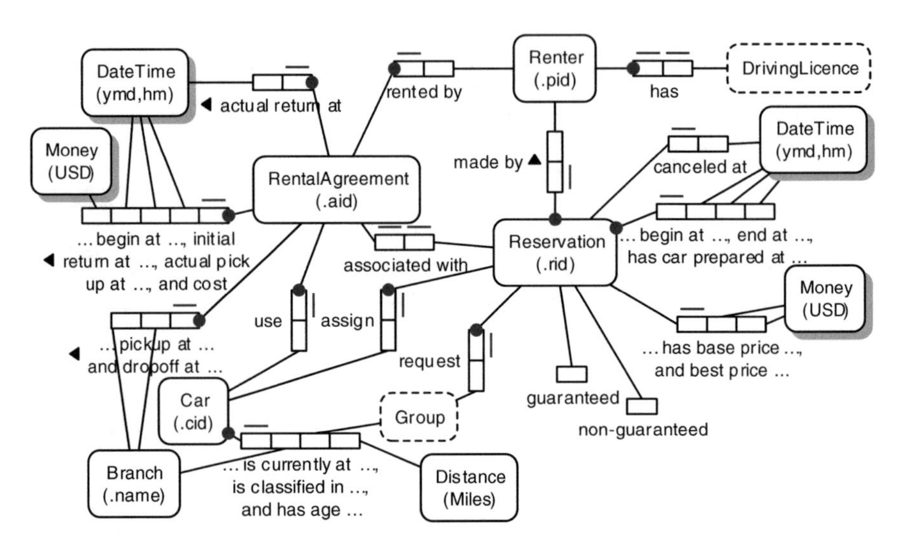

Fig. 2 Static view of artifact `Rental Agreement`

In our static view, we go with the ORM language following a few modeling principles. First, an entity type represents an artifact. Second, a predicate (relation) represents an attribute(s). Third, a value type represents a value-type attribute. Fourth, a connector represents a connection between artifacts, attributes and value types. Each artifact has a set of attributes that fall into two groups: value-type attributes and artifact-type attributes. We consider an artifact-type attribute is a binary predicate between artifacts, with mandatory role.

Figure 2 is a static view that shows artifact `Rental Agreement` in terms of: an identifier (`.aid`), and attributes, which can be verbalized as one artifact `Rental Agreement` is rented by exactly one artifact `EU_RentPerson` via an attribute `EU_RentPersonPid`, one artifact `Rental Agreement` is associated with at most one artifact `Reservation` identified by an attribute `ReservationRid`, one artifact `Rental Agreement` begins at exactly one attribute `DateTime` and initial returns at exactly one attribute `DateTime` and actual picks-up at exactly one attribute `DateTime` and cost exactly one attribute `Money` in `USD`, one artifact `Rental Agreement` actual returns at exactly one attribute `DateTime`, one artifact `Rental Agreement` picks up at exactly one artifact `Branch` identified by an attribute `BranchName` and drops off at exactly one artifact `Branch` identified by an attribute `BranchName`, one artifact `Rental Agreement` uses exactly one artifact `Car` via an attribute `CarCid`.

4.2 *Dynamic View*

In this subsection we explicitly address our second and third research problems. We propose the notation of dynamic view where preconditions, postconditions and business constraints are represented together.

Table 1 Mapping between the ACP terms and our ORM-based modeling

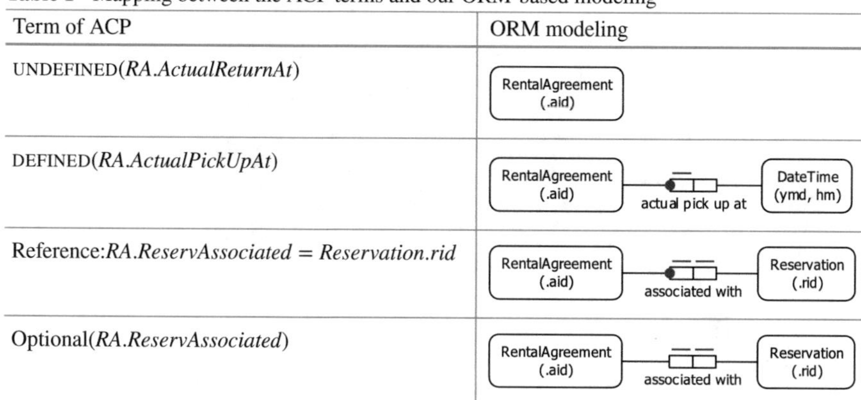

Term of ACP	ORM modeling
UNDEFINED(*RA.ActualReturnAt*)	RentalAgreement (.aid)
DEFINED(*RA.ActualPickUpAt*)	RentalAgreement (.aid) — actual pick up at — DateTime (ymd, hm)
Reference:*RA.ReservAssociated* = *Reservation.rid*	RentalAgreement (.aid) — associated with — Reservation (.rid)
Optional(*RA.ReservAssociated*)	RentalAgreement (.aid) — associated with — Reservation (.rid)

The pre- and post-conditions of a service are written either as a state proposition (INSTATE) or an attribute proposition (DEFINED, UNDEFINED and comparison operators). For each service of an artifact, user has to define some attributes in order to change the artifact into another state, i.e., each state requires a set of attributes which are defined. Therefore, we consider a state contains a set of DEFINED attributes. Additionally, there are some optional information required to complete a service, e.g., reservation details in returning car service. Attribute comparison from ACP has two different types, if there is no identifier of an artifact in either side of a comparison statement, it is a value comparison, which is considered as a parts of business constraints, otherwise, it is a reference from this artifact's attribute to other artifact's identifier.

In our visualization techniques, each dynamic view represents a service in terms of pre- and post-condition. The view is divided into an upper half (dedicated to the pre-condition) and a lower half (dedicated to the post-condition) that are separated by a line. Visually, UNDEFINED is represented as no relation between artifacts. DEFINED is represented as a relation of a tuple of artifact, mandatory rule, value-type. Each reference is represented as a relation of a tuple of artifact, mandatory rule, other artifact. Optional attributes of a service are represented under a relation of a tuple of artifact, rule, value type or other artifact. Table 1 summarizes the mapping between the ACP terms and our proposed modeling techniques using the ORM.

Figure 3a is a dynamic view for service `reserve`. Its upper half simply states that, as a precondition, no predicate is connected to artifact `Rental Agreement`. Its lower half represents the postcondition of service `reserve`. We partially verbalize the postconditions in the following. One artifact `Rental Agreement` is associated with exactly one artifact `Reservation` via an attribute `ReservationRid`, one artifact `Reservation` has exactly one attribute `ReservationRid`, one artifact `Reservation` is made by exactly one artifact `EU_RentPerson` via an attribute `EU_RentPersonPid`, one artifact `Reservation` has a base price in exactly one attribute `Money` in `USD`.

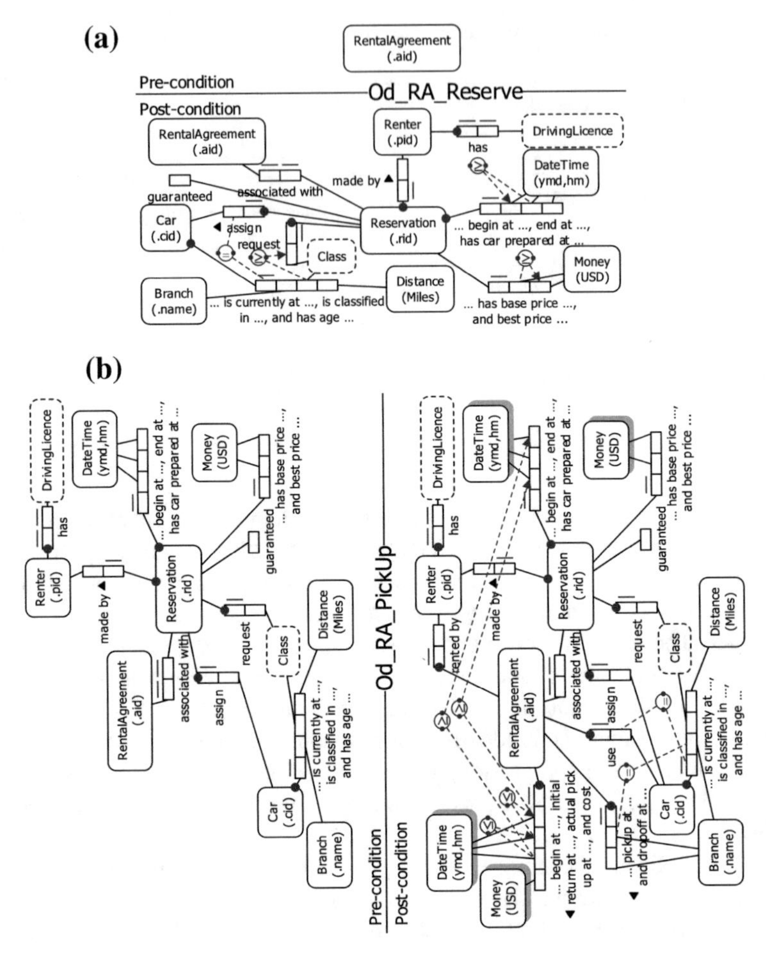

Fig. 3 Dynamic views of services of artifact `Rental Agreement`. **a** Service Reserve. **b** Service Pick-up

Figure 3b is another dynamic view that is dedicated to service `pick-up`. The left pane and the right pane of this figure represent the pre- and the post-condition of this service, respectively. The ORM language allows us to associate a comparison operator (e.g., $\geq$, $=$, $\leq$) with a role. If a business constraint cannot be represented by means of graphical notation, the ORM language allow us to put textual description instead [7]. In the aforementioned example, the assigned rental car must belong to a rental class that is higher than or equal to what the renter requested, as depicted in Fig. 3a. Note that we have the $\geq$ symbol that connects role `is classified in` to role `request`.

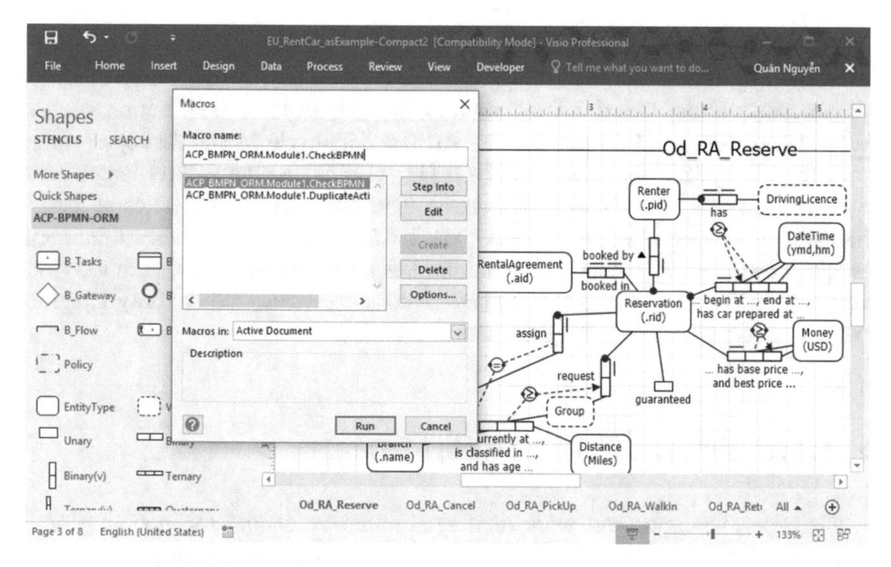

Fig. 4 A screenshot of our template prototype

4.3 *Implementation*

Based on this modeling technique, we had built a prototype as a template in Microsoft Visio.[2] This template features three types of pages: BPMN extended, static view and dynamic view. It also includes a stencil which allows users to quickly make BPMN-extended diagrams in [11] and ORM-based diagrams. Additionally, the prototype has advanced features such as: modeling wizard, checking consistency and views navigation. Figure 4 demonstrates our template prototype.

5 Related Work

UML-based modeling is popular in this line of research. De la Cruz et al. proposed the notion of visual contract [4] where the behavior of IT systems could be visualized in connection to its design concepts. Heckel and Sauer [8] also looked at UML diagrams integrating the state transformation with the structural and the interaction aspect. In our work, we leverage the advantage of visualizing conceptual attributes together with relation between concepts of the ORM language to express the behavioral aspect of business processes. The concepts and attributes in our work are formally described using the ACP approach.

[2]Microsoft Visio Homepage: https://products.office.com/en-us/visio/.

The ACP approach has been furthered with visual modeling techniques since first introduced by Bhattacharya et al. [2]. To name just a few, Bhattacharya et al. [3] introduced an Entity-Relation Diagram for capturing artifacts' information models. Yongchareon and Liu [13] provides an Artifact Life-cycle Model for representing artifacts' life-cycles. Lohmann and Nyolt [11] also extend the BPMN language to represent artifacts' life-cycles and services. The UML language defines at least 4 diagrams that match BALSA elements: UML Class Diagram matches artifacts and its attributes, UML Statechart Diagram matches artifact life-cycles, Operation Contracts in the Object Constraint Language matches services, and Activity Diagram matches associations; according to Estañol et al. [5].

6 Conclusion and Future Work

In this paper, we report our work on diagramming techniques for modeling ACP. The main idea behind this work is to make the representation of business artifacts, business tasks and business constraints fully diagrammatic, as opposed to hybrid modeling approaches whereby both text and graphical elements are put together in diagrams. We employ the ORM language to represent the information model of business artifacts, which is called static view. Another view is dedicated to the pre- and post-conditions of a business task, which is called dynamic view. In the dynamic view, we also explain how the ORM language can effectively be used for capturing business constraints. We have started to build a Microsoft Visio template as a tool associated with our modeling framework. Our future work includes a full implementation of the tool and investigation on the usability and applicability of our modeling approach.

References

1. van der Aalst, W.M.: Business process management: a comprehensive survey. ISRN Softw. Eng. (2013)
2. Bhattacharya, K., Gerede, C.E., Hull, R., Liu, R., Su, J.: Towards formal analysisof artifact-centric business process models. In: Business Process Management, 5th International Conference, pp. 288–304. Springer, Brisbane, Australia (Sept 2007)
3. Bhattacharya, K., Hull, R., Su, J.: A data-centric design methodology for business processes. In: Handbook of Research on Business Process Modeling, pp. 503–531 (2009)
4. De la Cruz, J.D., Lê, L., Wegmann, A.: VISUAL CONTRACTS—A way to reason about states and cardinalities in IT system specifications. In: ICEIS 2006—Proceedings of the Eighth International Conference on Enterprise Information Systems: Databases and Information Systems Integration, pp. 298–303. Paphos, Cyprus (May 2006)
5. Estañol, M., Queralt, A., Sancho, M., Teniente, E.: Artifact-centric business process models in UML. In: Business Process Management Workshops, pp. 292–303. Springer, Tallinn, Estonia (May 2013)
6. Frias, L., Queralt, A., Olivé, A.: EU-rent car rentals specification. Research report, Universitat Politècnica de Catalunya (2003)

7. Halpin, T., Morgan, T.: Information Modeling and Relational Databases, 2nd edn. Morgan Kaufmann (2008)
8. Heckel, R., Sauer, S.: Strengthening UML collaboration diagrams by state transformations. In: 4th InternationalConference Fundamental Approaches to Software Engineering, pp. 109–123. Genova, Italy (Apr 2001)
9. Hull, R.: Artifact-centric business process models: brief survey of research results and challenges. In: On the Move to Meaningful Internet Systems: OTM 2008, pp. 1152–1163. Springer, Monterrey, Mexico (Nov 2008)
10. Kunchala, J., Yu, J., Yongchareon, S.: A survey on approaches to modeling artifact-centric business processes. In: Web Information Systems Engineering—WISE 2014 Workshops, pp. 117–132. Springer, Thessaloniki, Greece (Oct 2014)
11. Lohmann, N., Nyolt, M.: Artifact-centric modeling using BPMN. In: Service-Oriented Computing—ICSOC 2011 Workshops, pp. 54–65. Springer, Paphos, Cyprus (Dec 2011)
12. Nigam, A., Caswell, N.S.: Business artifacts: an approach to operational specification. IBM Syst. J. **42**(3), 428–445 (2003)
13. Yongchareon, S., Liu, C.: A process view framework for artifact-centric business processes. In: On the Move to Meaningful Internet Systems: OTM 2010, pp. 26–43. Springer, Crete, Greece (Oct 2010)

SPAM Detection: Naïve Bayesian Classification and RPN Expression-Based LGP Approaches Compared

Clyde Meli and Zuzana Kominkova Oplatkova

Abstract An investigation is performed of a machine learning algorithm and the Bayesian classifier in the spam-filtering context. The paper shows the advantage of the use of Reverse Polish Notation (RPN) expressions with feature extraction compared to the traditional Naïve Bayesian classifier used for spam detection assuming the same features. The performance of the two is investigated using a public corpus and a recent private spam collection, concluding that the system based on RPN LGP (Linear Genetic Programming) gave better results compared to two popularly used open source Bayesian spam filters.

Keywords Reverse polish notation (RPN) · Naïve bayesian classifier · Spam detection · Linear genetic programming (LGP) · Genetic programming (GP)

1 Introduction

1.1 Overview

The paper deals with the spam detection based on the Reverse Polish Notation (RPN) [1, 2] expression based Linear Genetic Programming (LGP) [3–5] approach compared to Naïve Bayesian classifier [6–8].

Traditionally spam was fought using blacklists and heuristics [9]. In recent years, the spam war has had the help of machine learning and text classification methods.

C. Meli (✉)
CIS Department, Faculty of ICT, University of Malta, Msida, Malta
e-mail: Clyde.meli@um.edu.mt

Z.K. Oplatkova
Department of Informatics and Artificial Intelligence,
Faculty of Applied Informatics, Tomas Bata University in Zlin,
Nam. T.G. Masaryka 5555, Zlín, Czech Republic
e-mail: oplatkova@fai.utb.cz

R. Silhavy et al. (eds.), *Software Engineering Perspectives and Application in Intelligent Systems*, Advances in Intelligent Systems and Computing 465,
DOI 10.1007/978-3-319-33622-0_36

The war has not been won yet. In 2012, a record number of 27 million new malware were found, according to Panda Security [10]. A study [11] reported in 1998 that 10 % of incoming messages on a LAN were spams. Spam is an economic expense and it slows down legitimate traffic.

One main frontline against spam has recently been the statistical approach using the Naive Bayesian classifier [6–8]. Paul Graham in “A Plan for Spam” [12] and “Better Bayesian Filtering” [13] presented his algorithm and implementation of the Naïve Bayesian classifier, the latter with a more complicated tokenizer. Academic implementations include Pantel and Lin’s SpamCop [14] and Sahami et al’s Bayesian filter [15]. SpamAssassin [16] is a well known open source implementation of this algorithm. Günter Bayler [17] discusses a number of attacks on Bayesian spam filters in his book, which exploit their weakness.

For the Bayesian classifier to be effective, it requires a large number of records. Secondly where a predictor category is not present in the training data, Naïve Bayes takes the assumption that a new record with that category would have 0 probability. If such a predictor category is important, this can be a problem. As a result when the goal is to determine class membership probability, Bayes is very biased, and as a result rarely used in Credit Scoring [18].

1.2 Related Work

Different techniques were developed for the fight with spam detection from various field of machine learning.

Sangeetha et al. [19] proposed a local concentration (LC) based feature extraction approach for spam detection. Goweder et al. [20] developed a system to detect spam based upon a neural network, used as a classifier, whose weights were evolved by a GA. Artificial immune systems such as described by Khorsi [21] have been used to fight spam and computer viruses by imitating biological immune systems.

Katirai [22] proposed a method for junk email classification which used the classical form of Genetic Programming (GP) to automatically evolve a Bayesian filter. The Bayesian filter program was represented by a syntactic tree whose nodes are numbers representing word frequencies, operations on numbers, words and operations on words. The program would evolve the better filter program according to a fitness function developed to minimise misclassifications, i.e. the sum of squared errors over all documents.

Hirsch et al. [23] implemented a GP system to evolve rules based on n-grams (strings n character long) for document classification. GP has been used by Shengen et al. [24] to generate new features used as inputs to Support Vector Machines (SVM) and GP classifiers to detect link webspam.

Earlier systems include Magi [25] which used decision trees to automatically route new messages to the relevant folders, RIPPER [1] which automatically learnt rules to classify email into categories (spam was not mentioned), and Genetic

Document Classifier [2] which using a classical GP routed inbound documents to interested research groups within a large organisation.

Davenport et al. [26] implemented a GP system which evolved a Reverse Polish Notation postfix representation rather than the classical trees. The operators included logical operators.

Various spam corpora exist such as SpamBase [27], SpamAssassin and Untroubled.Org. SpamBase was not used in this study because it required the use of specific features, the raw emails were not available and use of more fresh emails was preferred.

2 Methods

2.1 Bayesian Approach

The Bayesian Network is defined as a directed acyclic graph compactly representing a probability distribution [28]. Nodes represent a random variable F_i. A directed edge between two such nodes indicates a dependency or probabilistic influence. It is assumed by the network structure that each node Fi in the network is conditionally independent of its non-descendants. Thus each node Fi in the network is associated with a conditional probability table which specifies the distribution over Fi.

A Bayesian classifier is a Bayesian network applied to classification, containing a node C representing the class variable and a node Fi for every feature.

From Bayes Theorem, given feature variables $f_1, f_2, \ldots f_n$, the Bayesian network is used to compute the probability that a document k with vector $F = \langle f_1, f_2, \ldots f_n \rangle$ belongs to category c is (1):

$$P(C=c|F=f) = \frac{P(C=c)P(Fi=fi|C=c)}{\sum_{k\in\{spam,ham\}} P(C=k)P(Fi=fi|C=k)} \quad (1)$$

The Naïve Bayesian Classifier assumes that $f_1, f_2, \ldots f_n$ are conditionally independent, which gives (2):

$$P(C=c|F=f) = \frac{P(C=c)\prod_{i=1}^{n} P(Fi=fi|C=c)}{\sum_{k\in\{spam,ham\}} P(C=k)\prod_{i=1}^{n} P(Fi=fi|C=k)} \quad (2)$$

where both P(Fi|C) and P(C) are relative frequencies which can be calculated from the training corpus.

It is possible to define a criterion in (3) and further in (4) by development of the idea that mistaking a ham message (legitimate) as being spam is much more severe as a mistake than the opposite (mistaking spam as ham). If it is assumed that misclassification of ham as spam is *L* more times costly than misclassification of spam as ham, that the independence criterion holds and that the probabilities have been estimated correctly. Thus a message can be classified as being a spam if it complies with (3).

$$\frac{P(C=spam|F-f)}{P(C=ham|F-f)} > L \tag{3}$$

Following Sahami et al's formulation [15] using P(C = spam|F = f) = 1 − P (C = ham|F = f), the criterion can be rewritten as (4):

$$P(C=spam|F=f) > t, where\, t = \frac{L}{1+L}, L = \frac{t}{1-t} \tag{4}$$

Sahami et al. [15] set the threshold t to 0.999 and L to 999, meaning that blocking a ham message as spam is as bad as letting 999 spam messages through the filter. There are cases as indicated by Sahami et al. where it would be feasible to use lower threshold, such as in the case of having a different action instead of blocking detected spams; e.g. sending a challenge to the sender in the case of a potential spam. Sahami has noted the independence assumptions are often violated in practice, resulting in poor performance. As a threshold, Graham [13] uses 0.9 with the argument that few probabilities end up in the midrange.

2.2 Genetic Programming

Genetic Programming (GP) [29–31] is a technique used to solve problems by methods based upon evolutionary mechanisms inspired from biology applied to the evolution of computer programs. Using Genetic Algorithms (GA), every individual represents a program and the fitness function depends on how good it solves the task represented by the fitness function.

Linear Genetic Programming systems (LGP) [3–5] which are extensions of the classical GP take a different approach. The program which is evolved is represented using a sequence of instructions in either machine language or imperative programming language. This makes it easier to use classical genetic operators like bit mutation and crossover as opposed to the classical GP as well as avoiding the standard GP's interpretation stage in the case of a machine opcode implementation. It also has been found superior over a set of benchmark problems (to the classic GP) [3, 32, 33]. A typical LGP program would have registers and constants from pre-defined sets.

Various applications of LGP's have been implemented, including web usage mining [34], formulation of the compressive strength of concrete cylinders [35], time-series modelling [36], and intrusion detection [37, 38].

2.3 RPN Expression

For the purpose of comparing Naïve Bayesian classification and Reverse Polish Notation Expressions, where the latter are made up of operators (such as +, −, *, /, sine, cosine, AND, OR and XOR) and operands (constants and evaluated feature values Fi), we assume that Fi are the same for the RPN Expression as for the Naïve Bayesian classifier.

It is known that the Naïve Bayesian classifier is incapable of expressing propositional operators such as XOR [39]. However, the attributes or features are independent, they act independently to classify and they do not interact. Thus, they cannot capture concepts like XOR (similar to the limitation of single-layer perceptrons).

On the other hand, a Reverse Polish Notation [40, 41] expression can easily be used to express propositional logic. Reverse Polish Notation was proposed by Burks, Warren and Wright [42], based on Polish Notation which was invented by logician Jan Lukasiewicz in the 1920's. One can express propositional logic with an RPN expression by using specific AND, OR and XOR operators or by using integer arithmetic operators (e.g. XOR can be built from modulus, sum and multiply operators, and modulus itself can be built from division, multiplication and subtraction operators). Galculator [43] is an example implementation of an algebraic and RPN calculator with propositional operators.

One advantage of Reverse Polish Notation is that there is no need to use parentheses which are sometimes needed. 3 − x + y would be written as 3 x − y + using RPN. 4 + x * y also can be written as 4 + (x * y). However changing the location of the parenthesis would result in something very different, i.e. (4 + x)*y. The parentheses are used to force order. On the other hand, the RPN postfix equivalent of the former is 4 × y * +. Calculations with RPN are performed faster using an iterative interpreter than the equivalent infix.

Evidence has been found [44] that Naïve Bayes produces poor probability estimates (Bennett writes that it tends to produce uncalibrated probability estimates). Monti and Cooper [45] similarly provided evidence that the Naïve Bayes model is poorly calibrated.

Thus based on this, it is to be expected that Reverse Polish Notation Expressions are more expressive than Naïve Bayesian classification. Implementation of a system based on RPN expressions should be as good as one based on Naïve Bayes.

2.4 Implementation Overview

Based on the preceding, it was expected that RPN Expressions might give better results in a practical implementation than the use of an equivalent Naïve Bayesian classifier. The idea was to build an LGP [4] system which evolves a program in the form of an RPN Expression (using machine code format), which would be able to

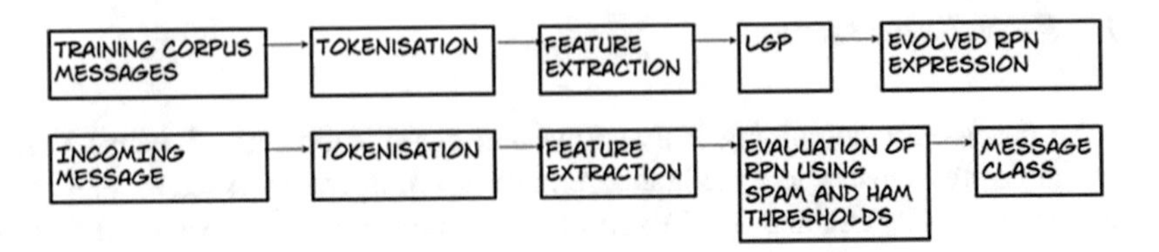

Fig. 1 Training and classification phases of the RPN/LGP model

be used as a detector of spam in emails. Once a sufficiently good RPN expression would be evolved, it would be able to be used for classification apart from the LGP system.

The system was implemented by the author. It utilises the feature extraction approach using RPN's and uses multi-threading. It was assumed that alphanumeric characters (including non-English characters), dashes, apostrophes, euro and dollar symbols are to be considered part of tokens, and any left-over symbol is taken to be a token separator. No stopword filtering was performed.

Figure 1 shows the training and classification phases of the RPN/LGP model. After the tokens are recognised, a number of features are extracted and then used within an RPN expression made up of feature values as represented by the current LGP chromosome. A stack is used to evaluate the RPN expression.

The features used are grouped in four. (1) A group of features are evaluated on the subject line; (2) a group on the Priority and Content-Type headers; (3) a group on the whole message body; (4) a group of features are evaluated on any URL found within the body. The latter features include an evaluation of Internet browser services WOT [46] and usage of the Google Safe Browsing API [47] as well as DNS RBL [48] checks. Some of the other features included Yule's Measure [49], Hapax Legomena and Simpson's Measure [50], plus some new measures developed and implemented by the author inspired from text measures used with author attribution. Characteristics of texts have been examined for obtaining their mathematical characteristics and for document clustering [51, 52]. One measure evaluates Zipf's Law [51] and the other a metric similar to Zipf's Law. Extra mechanisms like transcription/repair of chromosomes inside the optimization algorithm were not needed. When evaluating an expression, attempts to pop an empty stack return UNDEF (undefined), meaning that operators requiring two operands would evaluate to UNDEF. Selection pressure removed such expressions from the population. More details on features was not included due to space constraints in this paper.

The implementation was coded under Windows and Linux operating systems in C++ using the Boost library for the bit string implementation. The LGP's GA was initialised using the Mersenne Twister [53]. The fitness value is stored to speed up evaluation of fitness values for an unchanged chromosome. Changes to a chromosome will trigger reevaluation. Under Linux a DNS caching server [54] was used to speed up the system.

In the LGP implementation, single point crossover is performed with Elitism, so the best chromosome always survives. The implemented selection operator uses Holland's Proportion Fitness Selection. "Per locus" (per bit) mutation is specified rather than "per chromosome", as in De Jong and Spears [55]. Linear chromosomes rather than trees were implemented using binary opcodes.

Fitness determines how good an RPN expression (chromosome) detects spam and ham correctly using this algorithm (5) and (6):

$$\begin{aligned} \textit{Incorrectly classed} = {} & \text{number of incorrectly recognised ham from ham corpus} \\ & + \text{number of incorrectly recognised spam from spam corpus.} \end{aligned} \tag{5}$$

$$\text{Fitness value} = \text{MAXSPAMHAM} - \text{Incorrectly Classed} \tag{6}$$

MAXSPAMHAM is a constant which is equal to the total number of spams and hams in the corpus. The objective was to maximize fitness.

For every email, multiple features are evaluated and the values are cached avoiding unnecessary computation. The evolved RPN expression is evaluated with the results of the features and other constants. Two static thresholds are utilised with the result from the RPN expression for determining whether the email is a spam or a ham, in a similar way to Zhang and Cieselksi's GP object classification system [56]. It is also reminiscent of Paul Graham's two probability thresholds for spam and ham [13]. The thresholds were chosen arbitrarily initially and justified by tests, which included testing of different threshold algorithms.

Finally, the final or best thresholds used were 0 for spam and 1000 for ham. Test results here used the 40 value for ham which was as good as using the 1000 value.

The following was the final threshold algorithm used:

if we KNOW this is supposed to be a spam (supervised learning),
if evaluation < SPAM THRESHOLD then increment wronglyclassed;
else (in the case of ham) if (evaluation <HAM THRESHOLD && evaluation> SPAM THRESHOLD) then increment wronglyclassed

LRU (Least Recently Used) Caching was used to speed up the system, and caches can be saved and reloaded. The caches should be cleared from time to time because feature values based on Internet services like dns or WOT will change in time (a legitimate domain may expire and become a malicious website).

The Labelled Training Set used during various testing phases was the following:

(a) 280 Personally collected spam and ham, from personal mailboxes and mailing lists
(b) 610 Spams from the public untroubled.org spam corpus (which only contains spam)

An RPN expression is made up of machine code opcodes, each representing an operand or an operator. Operators used were +, −, *, /, sine and cosine. Registers

RPN Example 1: f[1] f[2] f[3] f[4] f[5] f[6] * * * * *
RPN Example 2: f[1] f[2] * f[3] + 10 * 5 –

Fig. 2 Examples of RPN expressions used in GAGENES/LGP, where f[n] is the value of the feature function no n, evaluated on the current email being tested

are Feature detector values evaluated on the email. Operands are registers or constant values.

The LGP is run with a training collection of emails already classified as spam or ham. This did not include attachments. Emails were stored in separate folders for spam and ham (Fig. 2).

The opcode bits determine whether it represents an operator, register or constant. If the top three bits are equal to '000', then the rest of the opcode specifies a register number, if it is '001' then the rest is a constant value, otherwise it is taken to be an operator (from 010 to 111).

3 Results

This study was looking for an effective method for email spam detection. The feature extraction approach with evolved RPN expressions was used.

During simulations, the following settings was applied: Population size in LGP was set to 15, Mutate probability to 0.5, Crossover probability to 0.8 and Copy probability to 0.001. Based on results, it can be stated that this settings—high mutation rates like 0.5 and a small population—performed good solutions. It is corroborating what was found in the GA literature [57] about high mutation rates.

Every test was repeated 5 times with the same parameters. Test platform ran Windows 7 and Gentoo Linux, with an Athlon 64-bit Phenom II X6 2.86 Ghz. The number of fitness evaluations per run was equal to 5 * 15; evaluation resulted in working out the feature values for every email which were then cached, speeding things up. Originally, the average population fitness tended to decrease throughout generations, however it tended to remain stable. A good chromosome tended to dominate early in the population. Thus the Selection algorithm was modified to select only non-zero fitness chromosomes, which helped increase population average fitness, though some bloat was exhibited in at least one test.

A test corpus of 3657 emails which included 280 hams was used to test the evolved RPN expression. The test corpus was built as follows:

(1) 280 Hand-chosen Hams taken (from the Training Corpus)
(2) Another 32 Hand-chosen Hams
(3) 607 Hand-chosen Spams
(4) 2763 Spams from the Untroubled.org archive (dated 2012)

LGP was compared with two popularly used and capable open source spam filters, SpamAssassin and BogoFilter. It is to be noted that SpamAssassin includes a Bayes filter derived from Graham and Robinson's proposals (it is a hybrid of Bayes plus a Rule-based system), whereas BogoFilter is a pure bayesian filter [58].

The same training set was used with BogoFilter. The three filters used the same testing set.

BogoFilter was not able to work with small training sets as was noted above. SpamAssassin was used with its default configuration. All emails were stored with full headers and did not have attachments. A bash script was used to call filters, and keep count of spams and hams detected.

In 2004, Graham-Cumming [59] proposed an alternative way of counting spam filter accuracy by working out the following individual hit rates (7) and (8):

$$spam\ hit\ rate = total\ number\ of\ correctly\ detected\ spams/total\ number\ of\ spams\ received \quad (7)$$

$$ham\ hit\ rate = total\ number\ of\ incorrectly\ detected\ hams/total\ number\ of\ spams\ received \quad (8)$$

However the industry still works out spam filter accuracy by working out total number of correctly detected spams and hams divided by the total number of messages (9).

$$Accuracy = (correctly\ detected\ spams + correctly\ detected\ hams)/number\ of\ messages \quad (9)$$

Accuracy (using data from Table 1 for the number of emails, false positives, etc.) using (9) was found to be as follows and in graphically form in Fig. 3.:

SpamAssassin: 82 % ((3682–663)/3682); BogoFilter: 26.67 % ((3682–2700)/3682); RPN LGP: 100 % ((3682–0)/3682).

Unsure spam emails are assumed to be hams, so the False Positives (FP) for Bogofilter amount to 568 (366 + 202 unsure) for TESTSPAM and 2132

Table 1 Summary of SpamAssassin (SA), Bogofilter (BF) and GAGENES/LGP (LGP) performance on spam and ham

Mailbox	# Emails	SA # FP	SA % FP (%)	BF # FP	BF % FP (%)	LGP # FP	LGP % FP (%)
HAM	280	6	2.1	0	0.0	0	0
TESTHAM	32	0	0.0	0	0.0	0	0
ALL HAM	312	6	1.9	0	0.0	0	0
TESTSPAM	607	481	79.2	568	93.6	0	0
UNTROUB	2763	176	6.4	2132	77.2	0	0
ALL SPAM	3370	657	19.5	2700	80.1	0	0

FP stands for false positives

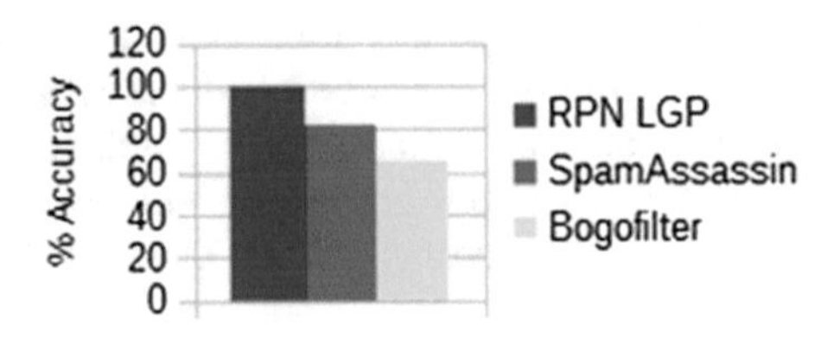

Fig. 3 Spam filter testing

(75 + 2057 unsure) for UNTROUB (This refers to spam taken from the untroubled.org corpus).

False Positives on ham were higher with SpamAssassin than with BogoFilter, whereas False Positives on spam were much higher with BogoFilter than with SpamAssassin. This is consistent with spam filter testing done by Claypool and O'Brien [60]. In the words of Paul Graham, "email is not just text; it has structure" [13], which is one reason why the LGP system uses different features per email structure. With the author's LGP system there were no misclassifications with the testing set. GAGENES is the author's base C++ genetic algorithm library.

4 Discussion and Conclusion

This paper gives a brief introduction to the RPN expression-based LGP system of GAGENES/LGP applied to spam detection and the test results performed. It has been determined that the RPN representation utilised in this LGP system is a potentially good alternative to the classical Bayesian classifier. This study has found the feature extraction approach with evolved RPN expressions to be effective in email spam detection. The simulation proved that the proposed system has no misclassification within the used testing set of hams and spams.

To speed up fitness evaluation, feature results are cached and groups of features used multi-threading.

In future, the spam and ham thresholds may be evolved themselves. Also RPN expressions will be represented in a way which will eliminate invalid expressions. Future testing will involve k-fold testing and principal component analysis. The system may be extended to detect web spam, network intrusions and other malware.

Acknowledgement Acknowledgements go to my Ph.D. Supervisors Dr Vitezlav Nezval. Thanks also to Tom Fawcett who answered my email query about the subject of Bayesian classifiers and RPN. This work was supported by Grant Agency of the Czech Republic—GACR P103/15/06700S, further by financial support of research project NPU I No. MSMT-7778/2014 by the Ministry of Education of the Czech Republic and also by the European Regional Development Fund under the Project CEBIA-Tech No. CZ.1.05/2.1.00/03.0089.

References

1. Cohen, W.: Learning rules that classify e-mail. In: Papers from the AAAI Spring Symposium on Machine Learning in Information Access, pp. 18–25. AAAI Press
2. Clack, C., Farringdon, J., Lidwell, P., Yu, T.: Autonomous document classification for business. In: Proceedings of the first international conference on Autonomous Agents, pp. 201–208. ACM, New York, NY, USA (1997)
3. Brameier, M.: On linear genetic programming (2004). https://eldorado.tu-dortmund.de/handle/2003/20098
4. Brameier, M.F., Banzhaf, W.: Linear Genetic Programming. Springer (2006)
5. M. Brameier, W. Banzhaf, A comparison of linear genetic programming and neural networks in medical data mining. IEEE Trans. Evol. Comput. **5**, 17–26 (2001)
6. Androutsopoulos, I., Koutsias, J., Chandrinos, K.V., Paliouras, G., Spyropoulos, C.D.: An evaluation of Naive Bayesian anti-spam filtering (2000). arXiv:cs/0006013
7. Duda, R.O., Hart, P.E., Nilsson, N.J.: Subjective bayesian methods for rule-based inference systems. In: Proceedings of the June 7–10, 1976, National Computer Conference and Exposition, pp. 1075–1082. ACM, New York, NY, USA (1976)
8. Mitchell, T.M.: Machine Learning. McGraw-Hill Science/Engineering/Math (1997)
9. Zdziarski, J.: Ending Spam: Bayesian Content Filtering and the Art of Statistical Language Classification. No Starch Press (2005)
10. Reports| Press Panda Security. http://press.pandasecurity.com/press-room/reports/
11. Cranor, L.F., LaMacchia, B.A.: Spam! Commun. ACM **41**, 74–83 (1998)
12. Graham, Paul: A Plan for Spam. http://www.paulgraham.com/spam.html
13. Graham, P.: Better Bayesian Filtering. http://www.paulgraham.com/better.html
14. Pantel, P., Lin, D.: SpamCop: A spam classification & organization program. In: Learning for Text Categorization: Papers from the 1998 Workshop, pp. 95–98 (1998)
15. Sahami, M., Dumais, S., Heckerman, D., Horvitz, E.: A bayesian approach to filtering junk e-mail. In: Proceedings of AAAI-98 Workshop Learn. Text Categ. (1998)
16. SpamAssassin Homepage. http://spamassassin.apache.org/
17. Bayler, G.: Penetrating Bayesian Spam Filters: Exploiting Redundancy in Natural Language to Disguise Spam Emails. Vdm Verlag Dr. Müller (2008)
18. Shmueli, G., Patel, N.R., Bruce, P.C.: Data Mining for Business Intelligence: Concepts, Techniques, and Applications in Microsoft Office Excel with XLMiner. Wiley (2011)
19. C. Sangeetha, P. Amudha, S. Sivakumari, Feature extraction approach for spam filtering. Int. J. Adv. Res. Technol. **2**, 89–93 (2012)
20. Goweder, A.M., Rashed, T.E., Ali, S., Alhammi, H.A.: An Anti-spam system using artificial neural networks and genetic algorithms. Proc. 2008 Int. Arab Conf. Inf. Technol. 1–8 (2008)
21. A. Khorsi, An overview of content-based spam filtering techniques. Inform. Slov. **31**, 269–277 (2007)
22. Katirai, H.: Filtering Junk E-Mail: A Performance Comparison Between Genetic Programming and Naive Bayes (1999). http://citeseer.ist.psu.edu/310632.html
23. L. Hirsch, M. Saeedi, R. Hirsch, Evolving rules for document classification, in *Genetic Programming*, ed. by M. Keijzer, A. Tettamanzi, P. Collet, J. van Hemert, M. Tomassini (Springer, Berlin, 2005), pp. 85–95
24. Shengen, L., Xiaofei, N., Peiqi, L., Lin, W.: Generating new features using genetic programming to detect link spam. In: Proceedings of the 2011 Fourth International Conference on Intelligent Computation Technology and Automation, vol. 01. pp. 135–138. IEEE Computer Society, Washington, DC, USA (2011)
25. Payne, T., Payne, T.: Learning Email Filtering Rules with Magi A Mail Agent Interface. Presented at the Department of Computing Science, University of Aberdeen (1994)
26. Davenport, G.F., Ryan, M.D., Rayward-Smith, V.J.: Rule induction using a reverse polish representation. In: GECCO, pp. 990–995 (1999)

27. Lichman, M.: UCI Machine Learning Repository, Irvine, CA, University of California, School of Information and Computer Science (2013). http://archive.ics.uci.edu/ml
28. J. Pearl, *Probabilistic Reasoning in Intelligent Systems: Networks of Plausible Inference* (Morgan Kaufmann Publishers Inc., San Francisco, CA, USA, 1988)
29. Koza J.R.: Genetic Programming: On the Programming of Computers by Means of Natural Selection. A Bradford Book (1992)
30. Koza J.R.: Genetic evolution and co-evolution of computer programs. In: Artificial Life II, pp. 603–629. Addison-Wesley Publishing Company (1990)
31. Koza J.R., K.M.A.: Genetic Programming IV. Kluwer Academic Publishers (2003)
32. Downey, C.: Explorations in Parallel Linear Genetic Programming: A Thesis Submitted to the Victoria University of Wellington in Fulfilment of the Requirements for the Degree of Master of Science in Computer Science. Victoria University of Wellington (2011)
33. Downey, C., Zhang, M.: Parallel linear genetic programming. In: Proceedings of the 14th European conference on Genetic programming, pp. 178–189. Springer, Berlin (2011)
34. Abraham, A., Ramos, V.: Web usage mining using artificial ant colony clustering and linear genetic programming. In: The 2003 Congress on Evolutionary Computation, 2003. CEC'03, vol. 2, pp. 1384–1391 (2003)
35. A.H. Gandomi, A.H. Alavi, M.G. Sahab, New formulation for compressive strength of CFRP confined concrete cylinders using linear genetic programming. Mater. Struct. **43**, 963–983 (2009)
36. A. Guven, Linear genetic programming for time-series modelling of daily flow rate. J. Earth Syst. Sci. **118**, 137–146 (2009)
37. Song, D., Heywood, M.I., Zincir-Heywood, A.N.: A linear genetic programming approach to intrusion detection. In: Genetic and Evolutionary Computation—GECCO 2003, pp. 2325–2336. Springer, Berlin (2003)
38. S. Mukkamala, A.H. Sung, A. Abraham, Modeling Intrusion Detection Systems Using Linear Genetic Programming Approach, in *Innovations in Applied Artificial Intelligence*, ed. by B. Orchard, C. Yang, M. Ali (Springer, Berlin, 2004), pp. 633–642
39. I. Kononenko, Semi-naive bayesian classifier, in *Machine Learning—EWSL-91*, ed. by Y. Kodratoff (Springer, Berlin, 1991), pp. 206–219
40. C.L. Hamblin, Translation to and from polish notation. Comput. J. **5**, 210–213 (1962)
41. RPN.: An Introduction To Reverse Polish Notation. http://h41111.www4.hp.com/calculators/uk/en/articles/rpn.html
42. A.W. Burks, Don W. Warren, J.B. Wright, An analysis of a logical machine using parenthesis-free notation. Math. Tables Aids Comput. **8**, 53–57 (1954)
43. galculator—a GTK 2/GTK 3 algebraic and RPN calculator. http://galculator.sourceforge.net/
44. Bennett, P.N.: Assessing the Calibration of Naive Bayes' Posterior Estimates. School of Computer Science, Carnegie Mellon University (2000)
45. Monti, S., Cooper, G.F.: A Bayesian Network Classifier that Combines a Finite Mixture Model and a Naive Bayes Model (2013). arXiv:1301.6723
46. Safe Browsing Tooll WOT (Web of Trust). http://www.mywot.com/
47. Safe Browsing API—Google Developers. https://developers.google.com/safe-browsing/
48. Damodaram, R., Valarmathi, D.M.L.: RBL Global Toolbar with Clustering Algorithm for Fake Website Detection
49. P.E. Bennett, The statistical measurement of a stylistic trait in julius caesar and as you like it. Shakespeare Q. **8**, 33–50 (1957)
50. E. Stamatatos, N. Fakotakis, G. Kokkinakis, Computer-based authorship attribution without lexical measures. Comput. Humanit. **35**, 193–214 (2001)
51. V.A. Yatsko, Automatic text classification method based on Zipf's law. Autom. Doc. Math. Linguist. **49**, 83–88 (2015)
52. M. Basavaraju, D.R. Prabhakar, A novel method of spam mail detection using text based clustering approach. Int. J. Comput. Appl. **5**, 15–25 (2010)

53. M. Matsumoto, T. Nishimura, Mersenne Twister: A 623-dimensionally equidistributed uniform pseudorandom number generator. ACM Trans. Model. Comput. Simul. **8**, 3–30 (1998)
54. Pdnsd: pdnsd homepage. http://members.home.nl/p.a.rombouts/pdnsd/
55. Jong, K.A.D., Spears, W.M.: An analysis of the interacting roles of population size and crossover in genetic algorithms. In: Proceedings of the 1st Workshop on Parallel Problem Solving from Nature, pp. 38–47. Springer, London, UK (1991)
56. M. Zhang, V. Ciesielski, Genetic programming for multiple class object detection, in *Advanced Topics in Artificial Intelligence*, ed. by N. Foo (Springer, Berlin, 1999), pp. 180–192
57. Piszcz, A., Soule, T.: Genetic programming: analysis of optimal mutation rates in a problem with varying difficulty. In: FLAIRS Conference, pp. 451–456 (2006)
58. G.V. Cormack, T.R. Lynam, Online supervised spam filter evaluation. ACM Trans. Inf. Syst. **25**, 11 (2007)
59. Graham-Cumming, John: Understanding Spam Filter Accuracy (Newsletter). http://www.jgc.org/antispam/11162004-baafcd719ec31936296c1fb3d74d2cbd.pdf
60. Mark, C., O'Brien, J.: An Analysis of Spam Filters. Computer Science Department, WPI (2003)

An Agile Approach to Improve Process-Oriented Software Development

Adriana Herden, Pedro Porfirio Muniz Farias and Adriano Bessa Albuquerque

Abstract This article describes a software development process called AgilePDD, specially focused on workflow systems. This process gives priority for defining the scope of the targeted system use cases, immediate prototyping and continuous delivery of working releases of the system, which are produced by a BPMS tool. To reduce rework, it is proposed to unite BPMN and BPMS with agile practices in a software development process. This paper summarizes two case studies using the AgilePDD, which were assessed in quantitative and qualitative ways. Lessons learned from these experiments show that some UML diagrams can be replaced by BPMN diagrams in the development of systems, without damage the documentation and modeling. Moreover, the stakeholders can get benefits from the possibility of execution of diagrams for process engines.

Keywords Business Process Management · Agile practices · Service Oriented Architecture · BPM · BPMS

1 Introduction

Software development is a complex and also creative and critical activity, which is carried out by people who use methods and technologies supported by software engineering. On the other hand, software engineering is a subject that is concerned with all the aspects of software production, from the early stages of system specification to maintenance [6, 7, 15].

A. Herden (✉) · P.P.M. Farias · A.B. Albuquerque
University of Fortaleza and Federal University of Technology of Parana,
Av: Washington Soares 1321, Fortaleza, CE, Brazil
e-mail: aherden@gmail.com

P.P.M. Farias
e-mail: porfirio@unifor.br

A.B. Albuquerque
e-mail: adrianoba@unifor.br

R. Silhavy et al. (eds.), *Software Engineering Perspectives and Application in Intelligent Systems*, Advances in Intelligent Systems and Computing 465,
DOI 10.1007/978-3-319-33622-0_37

Initially, the specification and development of the system are made from the definition of the functional and nonfunctional requirements. Thereafter, validation and verification are performed in order to ensure the product quality. Finally, evolution and software changes are managed to meet user needs [15].

Errors may occur in each stage of system development, generally related to requirements change. These requirements change cause rework for the team, which usually add costs associated with its activities. There are software development models, where changes are welcome, as in agile methods.

From 2000, with the Agile Manifesto and its approaches, social and human aspects have been incorporated into the activities of software development. In short, the values and the principles of the Agile Manifesto broke previous paradigms, emphasizing the interaction and collaboration between developers and users. Furthermore, the focus was on the continuous delivery of software versions via the incremental development and in rapidly responding to changes, especially in requirements [4].

At the same time companies get acquainted with a new approach to organizational management focused on the client, called Business Process Management (BPM). Such an approach is defined as “Supporting business processes using methods, techniques, and software to design, enact, control, and analyze operational processes involving humans, organizations, applications, documents and other sources of information” [17].

So BPM begins to be used as a viable alternative for companies carry out the alignment between IT and business, solving problems such as communication between participants of business processes, as well as the delay in forwarding documents between departments. This goal was achieved through the use of Business Process Management Systems (BPMS). Such a tool is defined as “A system that defines, creates and manages the execution of workflows through the use of software, running on one or more workflow engines, which is able to interpret the process definition, interact with workflow participants and, if required, invoke the use of IT tools and applications” [17].

As occurred with the standardization of the Unified Modeling Language (UML) [version 1.3 in 2000] [12] to improve the modeling and documentation of systems, also occurred with the standardization of the Business Process Model and Notation (BPMN) [version 1.1 in 2008] [11], to facilitate business process modeling, both held by the Object Management Group (OMG).

The BPMN version 2.0 was standardized by the OMG in 2011. Since then, the acceptance is growing and been supported by a large number of modeling and software development tools. The BPMN has been used both to model business processes and to specify the integration of systems in the Service Oriented Architecture (SOA) as also to model the flow of activities in information systems. One of BPMN advantages against the sequence diagram and the activity diagram, present in the UML, is that the business process written in BPMN can be performed using a BPMS tool.

The BPMS, beyond facilitate to design the processes graphically, can associate screens to every task building interfaces to users, may indicate which data will be persisted or transmitted between activities enabling the rapid construction of executable prototypes that automate the processes.

Therefore this work investigates how the diagrams and tools that relate to BPMN could be introduced in the context of system development. Which diagrams could be complemented by BPMN, or which diagrams could be replaced by the new standard. Due to the prototyping alternative offered by the BPMS was considered that the BPMN was more friend to agile methodologies.

This article describes the development methodology called Agile Process Driven Development (AgilePDD), which emphasizes the integration of BPMN in the context of agile methodologies. Moreover shows how two government companies used the proposed methodology, addressing the problem of rework. These case studies aimed to investigate how to effectively incorporate agile practices to AgilePDD.

The paper is organized as follows: Sect. 2 presents some methods used; Sect. 3 presents the obtained results; Finally, Sect. 4 presents the discussions.

2 Methods

2.1 Agile Process Driven Development

The AgilePDD process is characterized by the values and principles postulated by the Agile Manifesto [4], especially in their agile practices, and adopts the concepts of BPM and SOA in their life cycle. The Table 1 show the agile values adopted by AgilePDD.

2.1.1 AgilePDD Phases

The main phases of AgilePDD are: Scope Definition, System Prototyping, Sprint Production, Deployment, Monitoring and Optimization. The Fig. 1 shows the overview of the AgilePDD, represented in Software and System Process Engineering Meta-Model v.2 (SPEM) [10].

Table 1 Agile manifesto values incorporated in AgilePDD

Agile manifesto values	AgilePDD
Individuals and interactions over processes and tools	The BPMN-based prototypes are iteratively constructed with better details
Working software over comprehensive documentation	The BPMN-based prototypes are executable software more than simple documentations
Customer collaboration over contract negotiation	The business analysts and the users, together, specify the process. The informal communication is valued
Responding to change over following a plan	The process can be quickly revised

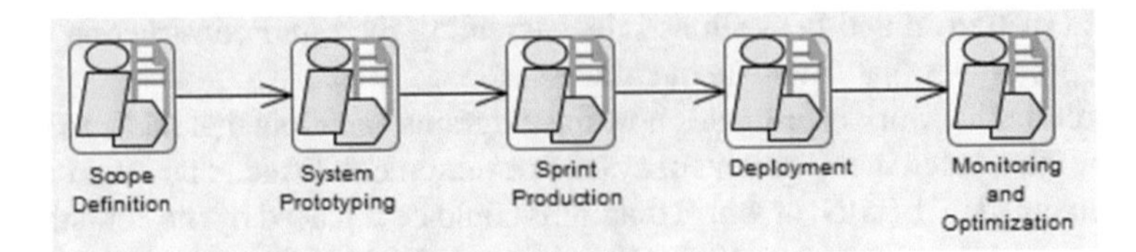

Fig. 1 AgilePDD overview

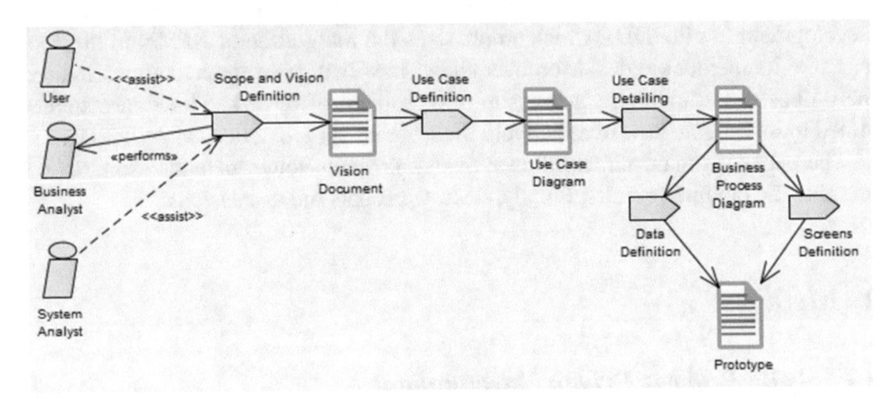

Fig. 2 AgilePDD—scope definition and system prototyping

The software development starts with defining the scope and identifying the requirements using use cases that are detailed in BPMN diagrams. It is noteworthy that the list of requirements is not contained in user stories as in the Scrum, nor the use cases are detailed in textual descriptions as in the Rational Unified Process (RUP).

In compliance to the agile principle related to preferring executable code than documentation, are adopted the BPMN diagrams to make possible the generation of prototypes rather than simply to produce the requirements documentation of requirements, adopting user stories or textual representations of the use cases.

The prototypes produced through BPMS tools are refined with the design of data and user interfaces design. Finally, the productions of sprints is performed iteratively, incorporating the integration of existing data and web services that already exist and the implementation of activities where the coding is still required.

The Fig. 2 presents an Activity Diagram of SPEM, representing the first two phases of AgilePDD, the Scope Definition and System Prototyping. In this figure it can be observed the changes and dependences that occur in artifacts generated at each stage, allowing also to perform the requirements traceability.

The Fig. 3 presents an Activity Diagram of SPEM which represents the transformation of the artifacts generated at Sprint Production phase.

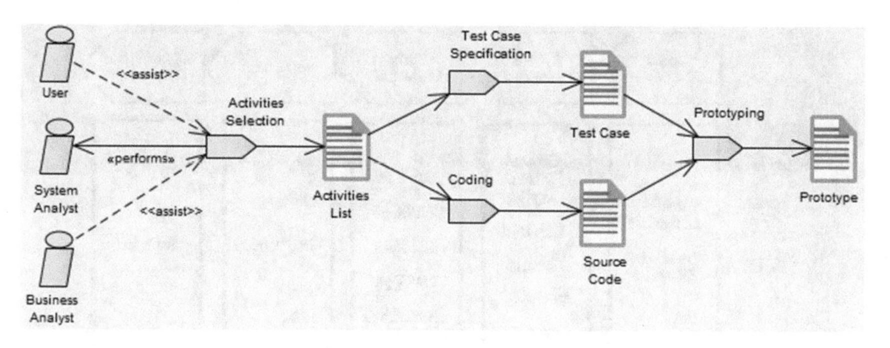

Fig. 3 AgilePDD—sprint production

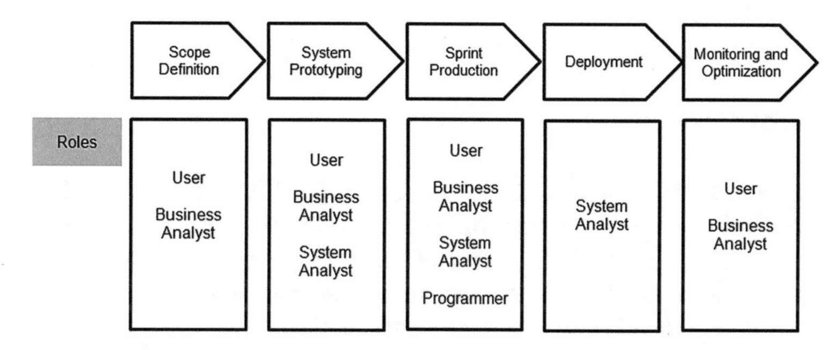

Fig. 4 AgilePDD roles

2.1.2 AgilePDD Roles

The AgilePDD assumes the existence of only three roles: the Business Analysts, the Systems Analysts and the Programmers. Besides the constant presence of the User during all the software life cycle. The Fig. 4 shows the roles related to each phase of AgilePDD.

The Business Analysts are responsible for the scope definition, for eliciting requirements, for capturing, enhancing and modeling the business processes. The technical aspects, for example, to refine the data model, are of responsibility of the Systems Analysts. The Programmers are involved mainly in coding. The same person can play more than one role.

The definition of the roles in the AgilePDD is minimalist as in other agile methodologies, but is more detailed, for example, than the Scrum one [5] (Product Owner, Scrum Master, Development Team). The users and the business analysts hold the vision for the product, the system analyst or the business analyst can plays as a Scrum Master and the business analyst, the system analyst and the programmers are the development team. So, the AgilePDD roles reflect a very common division of responsibilities in IT careers. The Fig. 4 displays the roles related to each phase of AgilePDD.

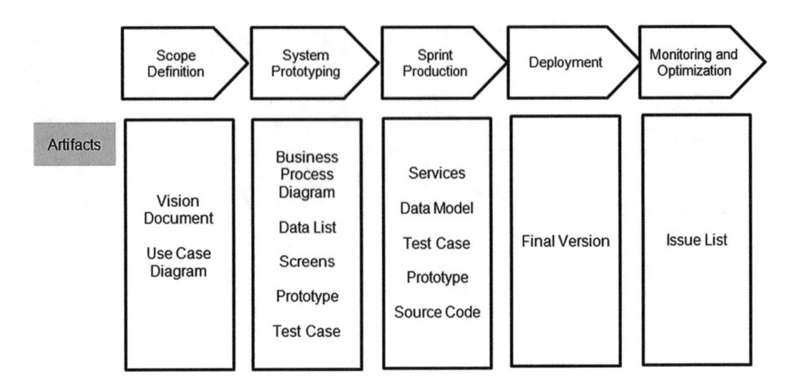

Fig. 5 AgilePDD artifacts

2.1.3 AgilePDD Artifacts

The following artifacts were identified in AgilePDD: Vision Document, Use Case Diagram, Business Process Diagram, Data List, Screens, Prototypes, Test Cases, Services, Integrated Data Model, Source Code, Final Version and Issue List. Such artifacts serve to gather important information generated in each phase. The Fig. 5 displays artifacts related to each phase of AgilePDD.

The vision document is prepared in the Scope Definition phase and presents the essential system requirements. The vision document, generally, identify the actors and contains a first version of a simple use case diagram. The AgilePDD assumes that the Scope Definition is very important and can be modified but is necessary for the initial decision of allocate resources and initiate the software development.

Each of the use cases is detailed as a Business Process in the System Prototyping phase. In AgilePDD the use cases are not detailed in a textual way but can be draw in BPMN diagrams to represent the action sequences, the flow controls and alternative scenarios for each use case. It is used a BPMS tool for graphical specification of processes. Many BPMS allow turn a Business Process Diagram (BPD) into an executable prototype that can be used to help the user to view and validate the business process. Details of this approach can be seen in [9].

Each activity of a business process can be refined by specifying data to be gathered in lists subsequently manipulated and exchanged between these activities. The layout of screens that present the data users can be set after the development of the data list. Both screens as data lists tore not mere documentation artifacts. They are a refinement of the code that is immediately incorporated in the new generated prototypes. At this phase can be specified integration test cases.

The AgilePDD predicts the generation of short cycles of prototyping to each business process, screen or data list specified. As soon as specified inputs and outputs of each activity and, if necessary, an early version of the screen can be initiated in the Sprint Production phase.

The sprints purpose is coding, in languages such as Java, the internal logic of each activity. Firstly, the inputs and outputs of the activity are associated with an integrated data model of all the application. Web Services and sub-processes previously developed are identified and can be reused. The source code is the main artifact of this phase and unit tests are added.

When a final version of the software is obtained, it can be deployed in the Deployment phase. The phase of Monitoring and Optimization occurs when the software is being producted. Issues are identified specifying fixes and improvements to be introduced in future releases. The fixes and improvements to be introduced in future releases can be identified and grouped in an artifact called issue list.

2.2 Related Work

The data modeling and the activities flow have been pillars of the systems analysis for decades. In structured analysis while the Entity Relationship Model was centered on the data specification, the Data Flow Diagrams described the activities flow.

The object-oriented analysis evolved to encapsulate in the same abstraction the data structure of the objects and their behavior. With the advent of UML, the Activity Diagram began to model the activities flow necessary for the functioning of the systems.

After the emergence of web services and Service Oriented Architecture (SOA), several methodologies have been developed focusing on integrating applications through services. It was given greater emphasis in the process view as opposed to traditional systemic view. The processes have become orchestrated with Business Process Execution Language (BPEL) and, more recently, with BPMN.

Explicitly assuming that the various systems of the company would be accessible through services, [13] proposes an Architecture-Centric Processes (PCA) for the development of information systems. The central element of the PCA is a BPMS that runs the processes in the Process Execution Layer and is also in charge to persist the processes in a process repository. The BPMS takes on a role of integrator receiving calls and calling other enterprise systems through an API layer.

In [14] is proposed a methodology, called Integrated BPM Project Methodology (IBPM), for performing BPM projects through a traditional approach to development. In this methodology are inserted the Business Design and Implementation Design activities in the classic software life cycle, emphasizing the perspectives of processes and services at different levels of detail.

The authors of [16] propose an approach for the development of BPM projects based on principles of agile methodologies, called Agile BPM. This approach uses generic parts of IBPM within a methodology based on Scrum, focusing on high frequency feedback cycles that take place between the IT teams and business teams as well as in the daily synchronization of the software development progress.

Although there are proposals such as PCA using the BPMS as an orchestrator element of the information system, and Agile BPM using the agile approach in the context of BPM projects, the work presented here considers that the development of modern BPMS able to generate executable prototypes processes, enable a new approach to developing workflow systems. This proposal suggests the immediate prototyping, coming from the use cases represented in BPMN.

2.3 Case Studies

In this study were chosen two government institutions to observe if the use of a process as AgilePDD is efficient for developers. There was training in AgilePDD in both organizations before the process has being used.

The AgilePDD was applied in the development of systems in order to assess if the presence of agile characteristics and practices together with BPMN and BPMS assist the developer. The characteristics and agile practices have been identified from the studies found in [2] e [3]. In the Project A the system developed was the "allocation of vehicles", and Project B has developed an "app for control of user's account".

In both studies was used the experimental evaluation method as a case study, in order to assess the degree of agility in the AgilePDD, from the point of view of developers, observing the teams and using qualitative and quantitative analysis to interpretation of results. According [18] before presenting the results is important to assess the validity of the case study in order to identify factors that may affect results. The following criteria are met in relation to methodological validity of both experiments.

- **Internal validity**: It is assumed that the team is representative of the population of developers with experience in BPMS and BPMN. A training was carried out put the developers at same knowledge level and reduce the influence of other methodologies used by developers. The number of participants in each project was: (i) 4 in the Project A, (ii) 5 in the Project B.
- **External validity**: Both systems were real, fruit of necessity of the organizations. So the methodology is applicable to practical situations.

2.4 Projects

One of the selected institutions, here called Company A [is related to Project A], is committed to serve the public interest through regulation, planning, monitoring, control and supervision of concessions and permits, promoting and ensuring technical efficiency of public services.

The other institution here called Company B [is related to Project B], is committed to contribute to improved health and quality of life by providing solutions in sanitation, with economic, social and environmental sustainability.

Project A In this case study a system to control the allocation of vehicles was developed using the BPMS BonitaSoft® v. 6.22, the DBMS Postgres® as database server and the email server VMware Zimbra®. The team involved in developing the system consisted of two analyst (who performed the system analyst role and business analyst role) and two programmers. Both were accompanied by a user that controls the allocations of the vehicles. The time spent for the development was about two months.

Project B In this case study a mobile application available for smartphones and tablets that use Android or iOS, was developed. The team involved in developing the system consisted of three analyst (who performed the system analyst role and business analyst role) and two programmer. Such implementation prioritized the development of the version 2 of the application compared to version 1 (without using the methodology AgilePDD). The following variables Were observed quantitatively: productivity, transparency, facility and interaction. The time spent to develop the software was about four months.

3 Results

This section presents some results obtained from the interview with the target population after the deployment of the systems.

3.1 Report—Project A

The methodology was used to develop the system and later was qualitatively assessed by the researcher. Through reports the presence of typical characteristics and practices of agile methodologies during project execution was evidenced.

3.2 Interview—Project B

An interview was conducted with three of the five participants of the Project B,with the objective of discuss some positive and negative aspects of this experience. Some of the questions used in the interview are presented below, along with a summary of participants' responses.

Question 1: **What aspects characterize AgilePDD as an agile method?** The agile methods produce code more directly instead of documentation or models. Using a BPMS, the BPMN is not only a model but also a specification that executes and that can be linked with screens and data allowing the construction of funcional prototypes that can be improved and validated by the stakeholders.

Question 2: **In the development of APP the AgilePDD brought agility to develop the software?** Yes, mainly because the Prototypation facilitated the validation of the activities flow , the fields and the screens. Three prototypes of the business process were designed until the stakeholders agreed with the version that would be automated.

Question 3: **What benefits the automatic generation of screens bring to prototyping?** The main benefit is to facilitate the communication reducing the rework.

Question 4: **What gains were obtained from AgilePDD in the application development?** The AgilePDD avoided rework and saved time.

3.3 Lessons Learned

According [1] to classify a method as agile method is necessary that it has a basic set of features such as adaptability, incrementality, iterative, colaborativity, cooperation, people oriented, leanness and restriction of time.

In this context, the case studies served to classify AgilePDD as an agile method. During the project implementation the participants were observed in their daily activities and it was found that some agile characteristics and practices were present at the execution of the projects. Among the characteristics identified by [2, 3] the most frequently observed were: adaptability, modularity; being incremental; being iterative; constant testing.

The adoption of automatically prototypes generated from BPMN facilitates *adaptability* because a simple change in the diagrams already allows the execution of the modified system. The automation of each activity separately from the rest enable high *modularity* and the sprints production cycle makes the methodology *incremental* and *iterative*. *Testing* and validation with the user are performed in each prototype generation. Such prototypes are generated in short time windows.

As well as some agility characteristics were identified during the implementation of projects, some agile practices were also identified on behavior of the development team, such as: on site customer; test driven development; simple design; game planning; refactoring; releases short; daily meetings.

The overall objective of both projects was to assess whether the elements and guidelines proposed by AgilePDD facilitate the development of workflow systems. The Project A was considered a pilot project in which the aim was to apply the methodology and observe the emergence of problems and benefits that were emerging during the application development. The results showed that prototypes generated by BPMS facilitate the validation requirements. After, the AgilePDD was applied on a new project (Project B) in which the goal this time would be to address the problem of staff rework. The results showed that to produce prototypes in each phase of the life cycle served to prevent and/or minimize the rework. Furthermore the visualization of BPMN diagrams before the system implementation phase facilitated the understanding of the functions that should be implemented.

Therefore, the lessons learned from Project A corroborated with the execution of the Project B because the pilot project results were favorable to the use of AgilePDD in the development of workflow systems. After the case study it was observed that there was a reduction in system development time, comparing with similar projects related to vehicle allocation. The methodology was used for the development of the system and was subsequently assessed by the researcher. The AgilePDD was observed in a qualitative way, typical features of agile methodologies were found, and a causal link in relation to the proposed methodology was identified.

The implementation of Project B has prioritized the development of the application version 2 compared to version 1 (without using the methodology). After the case study the proposed process was performed in a quantitative way of the following variables: productivity, transparency, facility and interaction. The results showed a 20 percent reduction of time for development, and 40 percent reduction in the number of errors found in use cases. More details of quantitative evaluation can be seen in [8].

In the Project B, the reduction in rework was one of the topics discussed during the interview, because the modeling and prototyping workflow system from the early stages decreased uncertainty requirements also facilitated the system acceptance tests. Therefore, from these two case studies it was observed that the elements and guidelines of AgilePDD are favorable for the business process oriented software development.

However, the limitations of the case studies showed common problems that occur in several software development companies that were: (i) resistance to change in development paradigm on the part of directors, especially in Company B, (ii) small team involved in parallel demands. Moreover, in the Project B the application was developed for the IOS operating system. This fact restricted the use of BPMN only for application modeling and prototyping.

4 Discussions

This paper presented the phases, roles and artifacts of a process oriented software development methodology called AgilePDD. Moreover, there were two case studies that identified in the AgilePDD the benefits of the separation of abstraction processes and business logic of each particular activity, as well as, the benefits of union of architecture centric processes features with agile methodologies features.

The ability to generate executable prototypes quickly from the graphical representation of business processes opens the possibility of using BPMN in the context of agile systems development, especially for workflow systems.

Another conclusion is that the production of prototypes reduced the need for textual specifications and documentation artifacts, resulting, more directly, in the production of executable code. Textual descriptions of the data or screens were not used. Instead, the prototypes were refined with the specifications of the data and screens in an iterative way.

Finally the production of sprints, in the spirit of agile methodologies, allowed integrating and coding the necessary activities to produce executable releases with progressive degree of refinement in less time.

The future work include the specification of a metric based on counting of activities to evaluate the software size and estimate the effort required to develop it. Moreover, there will be a comparative analysis between the AgilePDD and other process oriented software development methodologies.

References

1. Abrantes, J.F., Travassos, G.H.: Characterization of agile method software development. In: VI Brazilian Symposium on Software Quality (2007)
2. Abrantes, J.F.: Experimental studies on Agility in the Software Development and its use in the Test Process. Doctoral Thesis. Program Systems Engineering and Computer—COPPE/UFRJ. http://objdig.ufrj.br/60/teses/coppe_d/JoseFortunaAbrantes.pdf (2012). Accessed Jan 2015
3. Abrantes, J.F., Travassos, G.H.: Towards pertinent characteristics of agility and agile practices for software processes. CLEI 2013—Electron. J. **16**(1) (2013)
4. AGILE ALLIANCE.: http://www.agilealliance.org/ (2001). Accessed Jan 2015
5. Cohn, M.: Software Development with Scrum. Bookman (2011)
6. Fuggetta, A.: Software process: a roadmap. In: Proceedings of the Conference on the Future of Software Engineering, pp. 25–34. ACM (2000)
7. Fuggetta, A., Di Nitto, E.: Software process. In: Proceedings of the on Future of Software Engineering, pp. 1–12. ACM (2014)
8. Herden, A., et. al.: Agile PDD: One approach to software development using BPMN. In: 11th International Conference Applied Computing (IADIS 2014), vol. 11, pp. 214–221. IADIS press, Lisboa, Portugal (2014)
9. Herden, A., Farias, P.P.M., Albuquerque, A.B.: An approach based on BPMN to detail use cases. In: New Trends in Networking, Computing, E-learning, Systems Sciences, and Engineering. Springer International (2013)
10. OMG Object Management Group.: Software and Systems Process Engineering Metamodel Specification (SPEM version 2.0). (Technical report—OMG document number formal/2008-04-01). http://www.omg.org/spec/SPEM/2.0/ (2008). Accessed Jan 2015
11. OMG Object Management Group.: Business Process Model and Notation (BPMN version 2.0). (Technical report—OMG document number formal/2011-01-03). http://www.omg.org/spec/BPMN/2.0 (2011). Accessed Jan 2015
12. OMG Object Management Group.: Unified Modeling Language (UML version 2.5). (Technical report—OMG document number formal pct 13-09-05). http://www.omg.org/spec/UML/2.5/Beta2/PDF/ (2013). Accessed Jan 2015
13. Seshan, P.: Process-Centric Architecture for Enterprise Software Systems. CRC Press (2010)
14. Slama, D., Nelius, R., Breitkreuz, D.: Enterprise BPM. In: Erfolgsrezepte für unternehmensweites Prozessmanagement. dpunkt-Verlag, Heidelberg (2011)
15. Sommerville, I.: Software Engineering. Pearson Brazil (2011)
16. Thiemich, C., Puhlmann, F.: An agile BPM project methodology In: 11th International Conference on Business Process Management, Beijing, China (2013)
17. van der Aalst, W.M., Ter Hofstede, A.H., Weske, M.: Business process management: a survey. In: Business process management, pp. 1–12. Springer, Berlin (2003)
18. Wohlin, C., Höst, M., Hemmingsson, K.: Empirical research methods in software engineering. In: Empirical methods and studies in software engineering, pp. 7–23 Springer, Berlin (2003)

Ad-Hoc Routing Protocols Comparison Using Open-Source Simulation Tool

Tomas Sochor and Tomas Gatek

Abstract Despite their slow practical adoption, MANETs will play much more important role in the future Internet, especially the emerging Internet of Things. Routing in such network is much different comparing traditional networks. So far a variety of protocols exists with no clear leader. The paper tries to add some more information among existing ad-hoc routing protocol performance comparison, namely amongst AODV. DSR and OLSR. The open-source software tool OMNET++ has been used for simulation and some results are presented showing slight precedence of OLSR as the proactive protocol in the simulated cases.

Keywords Ad-hoc routing · MANET · AODV · DSR · OLSR · OMNET++ · Open-source

1 Introduction

Wireless communication has been omnipresent nowadays. While its older issues with insufficient securing of wireless networks seem to be overcome nowadays, its widening applications pose new challenges that research should face. Wireless connections are used not only for intercommunication of computers but more and more among various sensors, actuators and other components of emerging Internet of Things in so-called Mobile Ad-hoc NETworks (MANET) [1].

Because the radio signal emitted by small low-energy sensors and similar items often cannot reach to recipients directly, various concepts of multipath transmission in MANET have been developed. On of the most critical issues in MANETs is finding a feasible path between transmitter and recipient(s) as well as possibly keeping

T. Sochor (✉) · T. Gatek
Department of Informatics and Computers, University of Ostrava,
30. dubna 22, 70103 Ostrava, Czech Republic
e-mail: tomas.sochor@osu.cz
URL: http://prf.osu.eu/kip

R. Silhavy et al. (eds.), *Software Engineering Perspectives and Application in Intelligent Systems*, Advances in Intelligent Systems and Computing 465,
DOI 10.1007/978-3-319-33622-0_38

such information up-to-date (at least during a single transmission period) that is usually done by so-called ad-hoc routing.

There are multiple standard protocols of ad-hoc routing. They are usually classified among [2]

- reactive protocols that look for the path just at the moment of request for data transit (e.g. AODV, DSR, TORA),
- proactive protocols keeping the routing table up-to-date continuously (e.g. OLSR, DSDV, GSR), and
- hybrid protocols having come properties of both above groups (e.g. ZRP, HARP).

While reactive protocols are expected to be better in MANETs with scarce traffic because of its low overhead, proactive protocols could excel when traffic is frequent. So far just limited experience with hybrid protocols prevent from formulating any expectations.

One of the main goals of the paper is to identify differences among selected representatives of reactive and proactive protocols and to confirm the expectations above. Among the protocols available, namely AODV [3], DSR [4], and OLSR [5] were studied in details.

2 Current State of Research in the Field of MANET Routing Protocol

There are numerous studies trying to compare various MANET routing protocols [6–8]. The recently prevailing approach is a simulation using a suitable software tool.

There are some common limitations of such studies. The first is given by a truly endless set of pattern of MANET nodes behavior (mainly motion). When reviewing recent papers focusing this topic, you can see that almost every paper use its own topology resulting in worse comparability of results with others.

The other aspects is associated with the simulation tool used. While majority of recent paper seem to use commercial tools like OPNet [9], it is clear that this approach could be a bit problematic because of the lack of information about the real implementation of studied protocols in the simulation software.

Therefore the paper focuses to perform the simulation using OMNET++[1] open-source simulation software tool together with INET package.

3 Simulation Setup

One of the issues to solve when preparing simulations was what application communication process to choose for traffic initiation. The selection seems to be important because the application protocol and/or transport protocol used could influence the

[1] Available at http://www.onmetpp.org.

communication patterns of lower layers and it could reflect in the measured (or rather simulated) results.

For the purpose of the simulation, two different application were chosen, namely reliable FTP, and on the other hand, a videostream distributed via UDP. The main metrics that were investigated in simulation results were round-trip time (RTT) and efficient transmission rate. While focusing to transmission rate is obvious, the selection of RTT requires a comment. In spite of the fact that in simulated environment even one-trip delay (latency) could be easily measured, RTT was chosen primarily because it seems to reflect better the communication properties especially in the case of connection-oriented (TCP) connections used for FTP.

Simulation parameters The complete set of simulation parameters is summarized in Table 1. The meaning of the Motion model parameter values is the following While TurtleMobility is applied in the case of static nodes for the purpose of random node distribution, MassMobility is used in the case of moving nodes for simulation of non-zero inertia of each node.

3.1 Simulation Topology

The topology used in the results was composed of four nodes named Client, Server, Bridge 1 and Bridge 2. All nodes are located in the same line. While the Client was moving along the line defined by the positions of the other nodes, the other nodes were static. The layout of the topology is shown in Fig. 1.

Thanks to the Client movement, first the direct connection between Client and server was used for test transmission as described further. Next, when the Client

Table 1 Main simulation parameters

Node number	10
Simulation period length	1,000 s
Simulation area dimensions	$1{,}000 \times 1{,}000$ m
Motion model	TurtleMobility, MassMobility
Motion speed	0 m/s, 3 m/s
Application protocols	FTP, UDP video
Application transmission rate	2 Mb/s

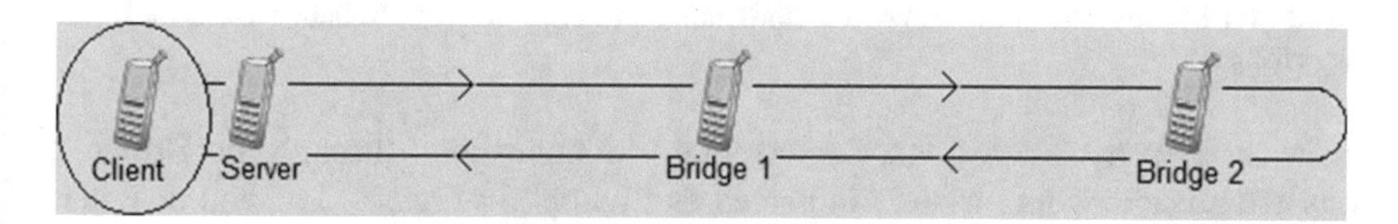

Fig. 1 Layout of the simulation topology. *Arrows* indicate the motion of the Client node

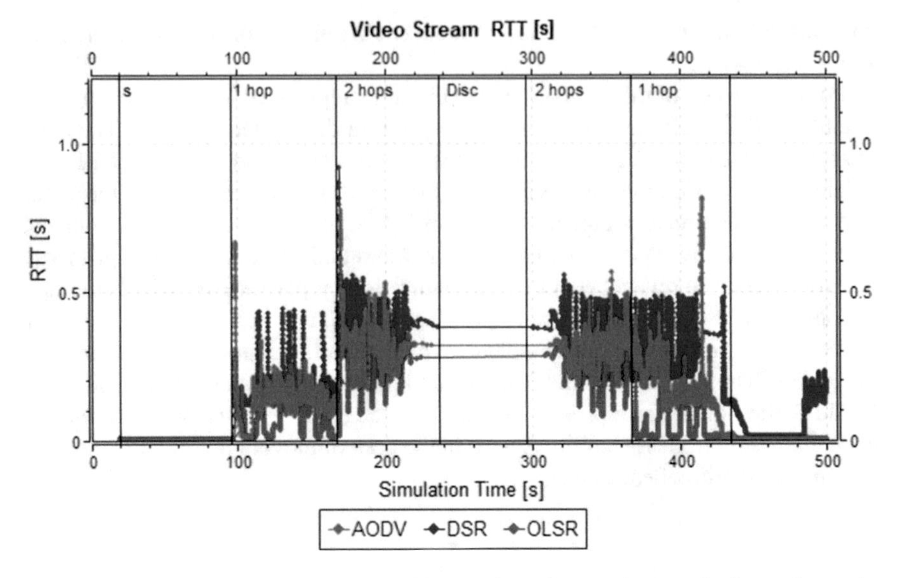

Fig. 2 RTT for UDP videostream. *Vertical lines* indicate interruptions (path change due to the topology change)

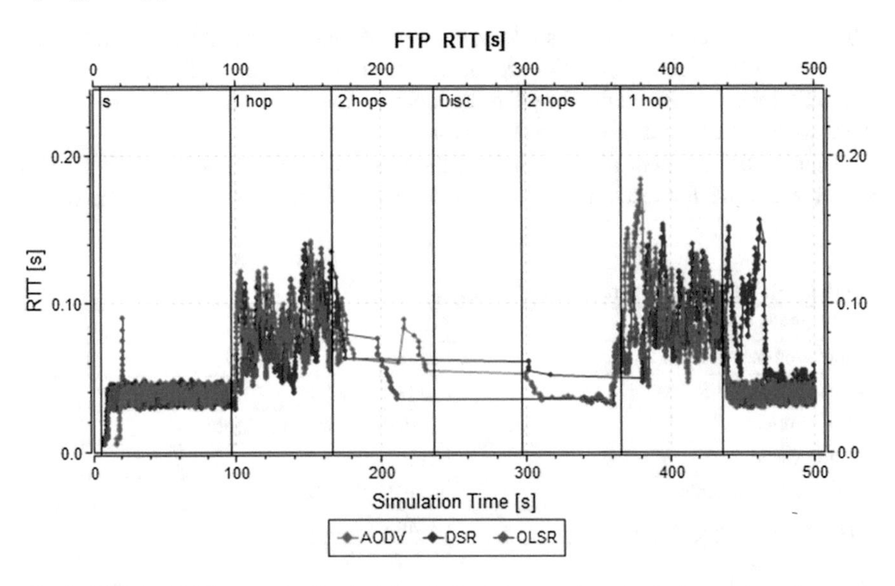

Fig. 3 RTT for FTP file transfer. *Vertical lines* indicate interruptions (path change due to the topology change)

distance from the Server became too long, the connection through the Bridge 1 was established. This phase is indicated as "1 hop" in Figs. 2, 3, 4 and 5. Later, when even the Client distance to the Bridge 1 became too long, the connection through Bridge 2 and Bridge 1 was established. This phase is indicated as "2 hops"

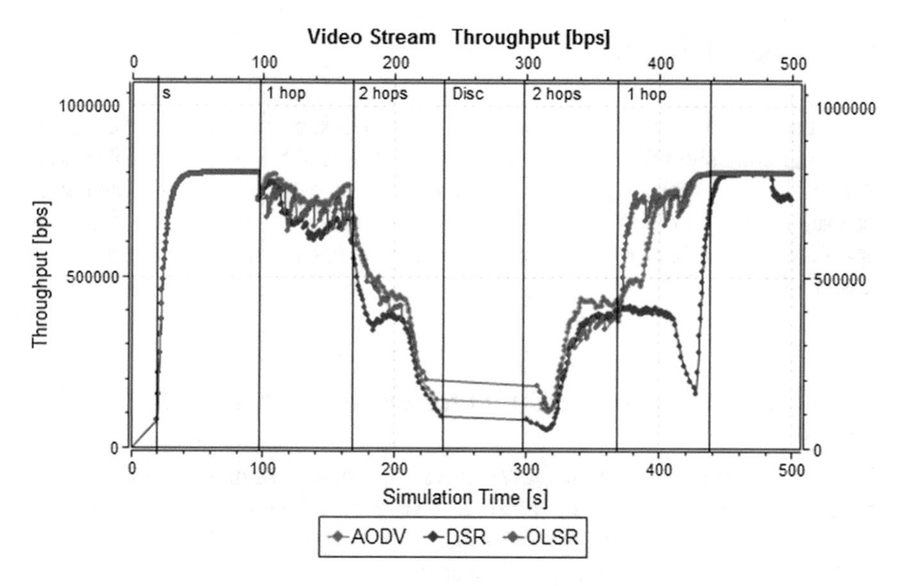

Fig. 4 Efficient transmission rate for UDP videostream. *Vertical lines* indicate interruptions (path change due to the topology change)

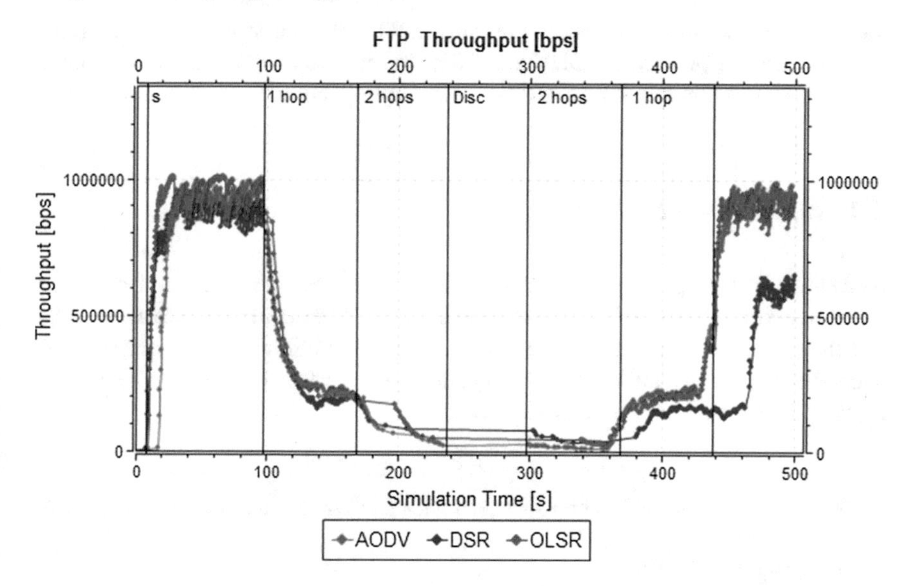

Fig. 5 Efficient transmission rate for FTP file transfer. *Vertical lines* indicate interruptions (path change due to the topology change)

in Figs. 2, 3, 4 and 5. Even later, the connection cannot be established at all because the excessive Client distance from all the remaining nodes and then the Client started to return and the cycle described above is repeated in the opposite sequence.

4 Simulation Results

Several sets of simulation experiments were run. Three of them are presented here. While the first simulation set was focused primarily to application protocol characteristics and its influence to the MANET routing behavior, the second simulation set was devoted to the investigation of the static node scenario and the third one was done in order to verify the obtained results by comparing the with previously published data.

4.1 Connectionless Versus Connection-Oriented Results

The first set focused to finding differences in MANET routing protocol behavior under different application transfers. FTP protocol was chosen here to represent connection-oriented transfers using TCP transport protocol while a UDP videostream represents a connectionless data transfer.

The averaged results are summarized in the Table 2. As one can see, the results were just slightly influenced by the application protocols because OSLR provided the highest transmission rate as well as the shortest RTT. Not surprisingly, the difference between reactive protocols AODV and DSR are much less significant in both cases but the transmission rate for AODV is higher slightly, but still significantly.

4.2 Static Node Scenario

As averaged results presented in Table 3 show, the removal of movement from nodes in the network did not influence the results much. What could be surprising is the fact that all simulation show better averaged results (both higher transmission rate and lower RTT) that static case. This results will be further investigated because so far there is no acceptable explanation for it.

Table 2 Results for two different types of application protocols and for three different MANET routing protocols

Routing protocol	FTP efficient transmission rate (kbps)	FTP RTT (s)	UDP efficient transmission rate (kbps)	UDP RTT (s)
OLSR	741.8	0.049	707.8	0.054
DSR	627.5	0.053	621.6	0.111
AODV	681.7	0.055	697.1	0.106

Table 3 Results for static nodes compared to slowly moving nodes for three different MANET routing protocols

Routing protocol	Static efficient transmission rate (kbps)	Static RTT (s)	Moving efficient transmission rate (kbps)	Moving RTT (s)
OLSR	661.6	0.057	868.3	0.0408
DSR	622.6	0.055	797.2	0.0422
AODV	674.2	0.060	901.1	0.0409

4.3 Result Verification

As already mentioned, the latter set of simulation experiments was focused to the result verification (i.e. the comparison of data obtained from authors' simulation with older published results). There is an obvious limitation of the comparison consisting in different simulation tool used here (OM?NET++) and in the work [7] with which the results are compared (OPNET Modeler). There are differences in simulation scenarios as well.

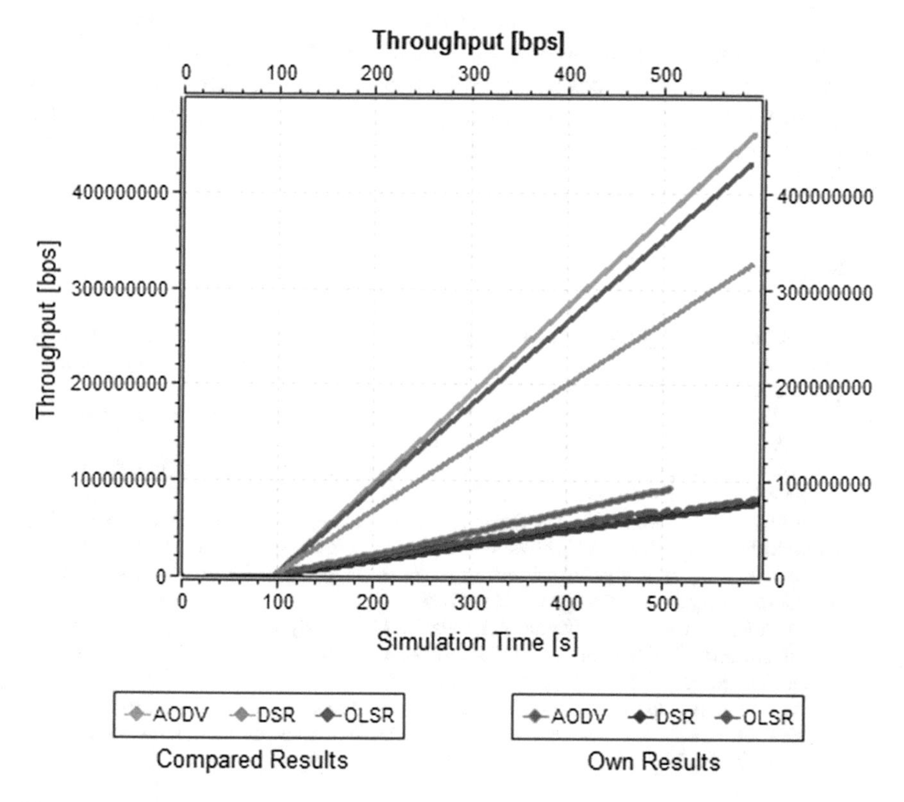

Fig. 6 Comparison of our results (transmission rate) with the results published in [7]

Therefore the comparison was limited just to compare the trends and not exact comparison of measured data (moreover, such data are almost never available from other publications). As it is shown in the Fig. 6, the obtained results (namely transmission rate) show the same trend as published in [7]. Also the order of routing protocols (AODV first, OLSR last) is the same.

5 Conclusions

The previous sections has demonstrated that the presented simulation model has been successfully verified by comparing the result data with previous work (details in Sect. 4.3). Therefore the open-source simulation approach could be considered reliable while keeping better control above the code of MANET routing protocols in comparison to commercial simulation tools thus obtaining better repeatability of obtained results.

The main conclusion from the presented results is that proactive protocols (here OLSR) show better behavior (faster transmission and lower latency) comparing reactive protocols (here AODV and DSR). This is true (and somehow could be expected) for longer transmissions that were simulated (FTP file transfer or UDP videostream). The situation for short transmissions that could be relatively frequent especially in Internet of Things applications could be different. A lot of unresolved issues regarding carious other MANET aspects remain, however. Among them the one of the ultimate significance seems to be MANET security [10].

References

1. Ramanathan, R. Redi, J.: A brief overview of ad hoc networks: challenges and directions. IEEE Commun. Mag. **40**, 20–22 (2002). http://www.ir.bbn.com/~ramanath/pdf/commmag-manets.pdf. [quot. 2015-04-06]
2. Royer, E.M., Toh, C.K.: A review of current routing 'protocols for ad hoc mobile wireless networks. IEEE Pers. Commun. **6**(2), 46–55 (1999)
3. Perkins, C., Belding-Royer, E., Das, S.: Ad hoc on-demand distance vector (AODV) routing. IETF (2003). https://www.ietf.org/rfc/rfc3561.txt. [quot. 2015-05-28]
4. Johnson, D., Hu, Y., Maltz, D.: The dynamic source routing protocol (DSR) for mobile ad hoc networks for IPv4. IETF (2007). https://www.ietf.org/rfc/rfc4728.txt. [quot. 2015-05-28]
5. Clausen, T., Jacquet, P.: Optimized link state routing protocol (OLSR). IETF (2003). https://www.ietf.org/rfc/rfc3626.txt. [quot. 2015-10-05]
6. Gupta, A.K., Sadawarti, H., Verma, A.K.: Performance analysis of AODV, DSR & TORA routing protocols. IACSIT Int. J. Eng. Technol. **2**(2), 226–231 (2010)
7. Aujla, G.S., Kang, S.S.: Comprehensive evaluation of AODV, DSR, GRP, OLSR and TORA routing protocols with varying number of nodes and traffic applications over MANETs. IOSR J. Comput. Eng. **9**(3), 8 (2013). ISSN: 2278-0661. http://www.iosrjournals.org/iosr-jce/papers/Vol9-Issue3/J0935461.pdf. [quot. 2015-11-06]

8. Manvi, S.S., Kakkasageri, M.S., Mahapurush, C.V.: Performance analysis of AODV, DSR, and swarm intelligence routing protocols in vehicular ad hoc network environment. In: Proceedings of International Conference on Future Computer and Communications, pp. 21–25. IEEE (2009)
9. Guo, L., et al.: Performance evaluation for on-demand routing protocols based on OPNET modules in wireless mesh networks. Comput. Electric. Eng. **37**(1), 106–114 (2011)
10. Zacek, J., Hunka, F.: Reusable object-oriented model. e-Inform. Softw. Eng. J. **7**, 35–44 (2013)

Discrete Event Simulation of Loading Unloading Operations in a Specific Intermodal Transportation Context

Ezzeddine Fatnassi and Jouhaina Chaouachi

Abstract With the development of economy, the movement of goods toward urban areas has already become an important issue. For that purpose, we have seen an increasing interest for developing Intermodal transportation mode. In this work, we focus on the loading/unloading operations in a specific context of an Intermodal transportation mode. We aim to develop a sophisticated discrete event simulation tool in order to simulate these operations. The paper describes the different steps and objects implemented in the study to develop our simulation model. The structure of the simulation program is also explained. Preliminaries computational results to validate our model are presented and discussed.

Keywords Discrete event simulation · Intermodal transportation · Freight transportation · Loading · Unloading · Logistics

1 Introduction

1.1 The Background of the Paper

The boom that has been observed in the size of urban areas and urban goods consumption has spurred tremendous growth in goods transportation to urban areas. This latter have led to the appearance of new ways of moving goods and persons to urban areas. Among these new practices one could note the Intermodal transportation [20]. Intermodal transportation is a new tendency in transportation logistics. Intermodal transportation is defined as the process of transporting goods, loads or

E. Fatnassi (✉)
Institut Supérieur de Gestion de Tunis, Université de Tunis,
41, Rue de la Liberté - Bouchoucha, 2000 Bardo, Tunisia
e-mail: ezzeddine.fatnassi@gmail.com

J. Chaouachi
Institut des Hautes Etudes Commerciales de Carthage,
Université de Carthage - IHEC Carthage Présidence, 2016 Tunis, Tunisia

R. Silhavy et al. (eds.), *Software Engineering Perspectives and Application in Intelligent Systems*, Advances in Intelligent Systems and Computing 465,
DOI 10.1007/978-3-319-33622-0_39

persons from an origin location to a destination location through at least two tools of transportation. Generally in Intermodal transportation [22], transfer between the two means of transportation is performed at a specific intermodal terminal. In the rest of this paper, we focus on Intermodal freight transportation movements.

The concept of Intermodal transportation is rich and include various examples. We could note for instance the transportation of palette of goods by combining trucks rail and ocean shipping, moving persons or goods over long distances via dedicated and different rails service, etc. [23].

Generally, intermodal freight transportation consists on a set of moving successively containers or loads of freights through a specific chain of container-transportation services. This chain is generally used to link the departure location to the destination location of the container.

Today, the tremendous increasing rate of goods movement especially in urban areas has put pressure on the urban intermodal transportation to make facilities more efficient in order to cope with these dynamic challenges [2].

We should note also that container and palette loading/unloading at intermodal terminal is a major component of intermodal transportation facilities [17].

Among the different performance measure used to test the quality of intermodal transportation facilities one could note the intermodal terminal turnaround time. The intermodal terminal turnaround time represents the average period of time that container or palette of goods stays waiting in an intermodal terminal [21]. This latter represents an important aspect as the turnaround time has a related economic cost that need to be minimized. We should note also that a large portion of the terminal turnaround time is spent on container, palette loading/unloading [15].

There exists different tools and types of palettes and containers transportation modes that could be involved in intermodal transportation. However, we limit the scope of this paper to a specific context where container and palette of goods are moved through two tools of transportation: train and specific electric on-demand freight transportation tool called Freight Rapid Transit (FRT). FRT represents an efficient and sustainable tool of moving goods in urban areas. We focus on this context as it represents a new trend in Intermodal and shared transportation means [1].

1.2 Objectives of the Paper

The aim of this study is to evaluate loading/unloading operations by means of a simulation in the context of train/FRT connection in an intermodal transportation context.

In order to develop a simulation model which is capable of fulfilling the needs of this study, it will be essential to present the current context of our simulation model. To achieve this, this paper draws on the FRT transportation tool, its advantages and its possible connections with train in order to develop an intermodal transportation tool. Next, the developed simulation model is presented and explained in this paper. The simulation model is used in order to evaluate the efficiency of loading/unloading operations in our specific context.

1.3 Contributions of the Paper

The contribution of this paper is two folds. We **first** develop a specific simulation model to model loading/unloading operation between trains and FRT vehicles in an intermodal terminal. **Second**, we evaluate this operation through the new developed simulation model.

1.4 Outlines of the Paper

The rest of this paper is organized as follow. Section 2 presents our specific studied context. Section 3 presents our developed simulation model. Preliminaries results are exposed in Sect. 4. Section 5 concludes the paper and present our future works.

2 The Studied Context

FRT system has been in development since the 1950s. FRT is used to move goods in urban areas. It is considered as an extension of the classical Personal Rapid Transit system (PRT) which is mainly used to move persons in urban areas. FRT has the potential to solve various problems related to the goods movement in urban areas such as congestion, pollution, carbon emission and so on. FRT is a public transportation tool that presents a specific alternative to the use of small trucks in urban areas.

As FRT is a new transportation mode, it offers a new way to move palette and container of goods. In fact, FRT includes five main features: (i) On demand service, (ii) Exclusive guideways (iii) Fully automated electric vehicles and control system, (iv) Offline stations, (v) Low emission and electric power usage.

In addition to these characteristics, various studies showed that the FRT could offer a very high reliability level of 98 % [19]. From these different features, we should note that FRT has a full potential to solve many of the problems related to goods movement in urban areas and therefore deals with the inefficiency of classic public transportation such as congestion, high energy profile, etc.

However, the developing of such an intelligent and enhanced system is still at its beginning. In a recent study called EDICT[1] conducted by the European commission, it was concluded that such an intelligent system has the full potential to solve the different problems related to urban movements. This study concluded also that for such a system the only obstacle is the hesitant local community.

[1] http://www.transport-research.info/web/projects/project_details.cfm?ID=4381.

In the literature, several studies related to FRT and its passengers counterpart PRT were proposed. One could note for example static routing of PRT vehicles [4, 10, 13], dynamic routing [3, 8, 12], minimization of waiting time of passengers [16] and so on.

According to us, the FRT system deserves a better positioning into its urban context in a way to integrate it with the existent urban transportation tools. This integration could be done in a smart interconnected way in order to move more efficiently goods into urban areas. Furthermore, the feasibility of such an action must be argued through **various optimization and simulation tools** in order to assert the economic, ecological and societal performance of such a system. Following this way of thinking, the next section provides an illustrative example of the inter-connectivity between FRT and the train transportation mode.

2.1 The Intermodal FRT/train Transportation Option

As presented in the previous section, an interconnected option for the FRT with the existent urban transportation tools is crucial for its success. In our point of view, an interconnected way of integrating FRT with existent transportation modes helps to move efficiently goods in urban areas. The FRT interconnected transportation option that we seek to present in this paper aims to integrate it into the global urban transportation system. Our proposal aims to consider the FRT transportation option as a complementary transportation tool for existent transportation tool such as train. Therefore and in order to better explain the intermodal FRT/train transportation option proposed in this paper, let us consider the following context. In a specific urban area, we propose that goods flows coming to a specific urban area are transferred via mass transit system such as train. Then in a specific offline Intermodal terminal, load of goods are transferred from train to FRT vehicles. This intermodal terminal offers the possibility to make the loading/unloading operations via dedicated specific facilities. Then, palette of goods are transferred via FRT vehicles to their final destination in urban areas. In recent study [9], the integration of PRT and FRT with other transportation service was studied from the operational point of view. However, the feasibility of its related intermodal facilities was not studied.

The development of such an intermodal transportation option relies on the mutualisation of the facilities and material sources in order to offer a specific joint usage. This option is also concerned with the specific inter-connectivity between the various transportation tools in order to moves efficiently goods in urban areas. This is essential, since it is well established that in city centers the operations of loading/unloading are atypical ones.

The rest of the paper will focus on the operational level of decisions related to our intermodal transportation option. In fact, we focus on **simulating and evaluating the loading and unloading operation movements** between train and FRT vehicles in the specific offline intermodal facility.

3 The Developed Simulation Model

In this section, we present our specific developed simulation model to simulate the loading/unloading operations in a FRT/Train intermodal transportation system.

3.1 The Used Simulation Software

The presented simulation model was developed inside the Xjtek's AnyLogic simulation software. Anylogic is a multi paradigm Java based simulation tool that offers to use System Dynamics, Discrete Event (aka Process-centric) and Agent Based approaches in the same simulation tool.[2] AnyLogic also offers the possibility to simulate the behavior of pedestrians within a physical environment. Anylogic is a powerful tool that combines also the use of various statistical objects that enables to collect useful information about the simulated system. Finally, Anylogic proposes to develop animation of any developed simulation model. This enables to observe the evolution of the system over the time. In the literature, Anylogic was used to simulate various complex systems [7, 14, 18]. Therefore and based on all of these advantages that Anylogic offers, our choice was turned to use it to develop our simulation model.

3.2 Modeling Loading/Unloading Operations

Generally, palette and container of goods are not a homogeneous group. Palette of goods differs mainly in their size, weight, and so on. Each single palette of goods represents an individual objects in our system that need to be simulated over both time and space. Figure 1 presents the life cycle as well as the different operations related to a specific palette of goods in the specific intermodal facility. In our approach, palette of goods are classified into various types according to their size and type of loads. Each palette of goods belongs to a specific type and is modeled as specific object instantiation of a Java class denoted "Palette". Our simulation model has been developed to fulfil the following characteristics:

- To implement related complex decisions for the loading/unloading operations of palette of goods.
- To simulate many different types of palette of goods interacting with the various complex objects that characterize their environment (i.e. the train, the FRT vehicles, etc.).
- To include significant probabilistic elements.

[2]More details could be found in http://www.xjtek.com/anylogic/why_anylogic.

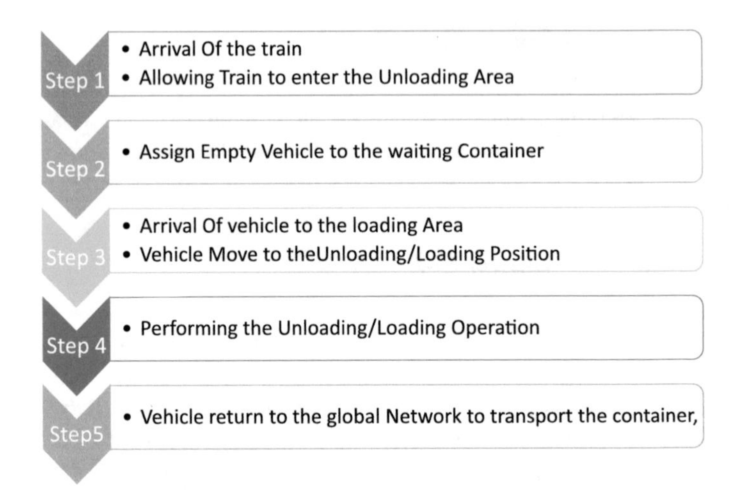

Fig. 1 Simulation process

Fig. 2 Overview of our network object

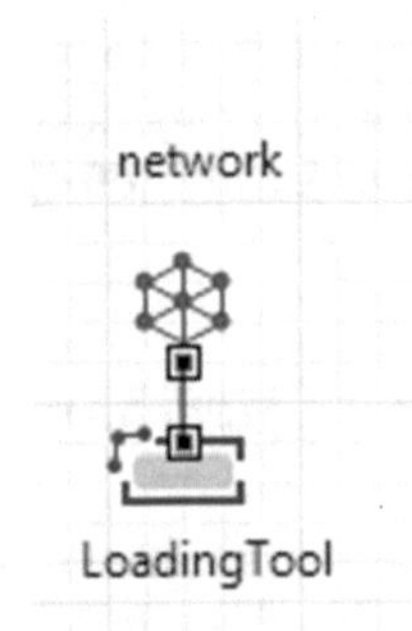

According to our simulation approach, the loading/unloading simulation model should be able to describe efficiently its environment, to operate without the intervention of the user and to exhibit goal-directed behavior of the simulation model.

The Intermodal facility is simulated using a global network object (see Fig. 2).

The train behavior was modeled using a specific railyard object(see Fig. 3). This object is used to simulate the arrival of trains as well as the loading of the goods into specific train container. The railyard object is modular and can be parameterized in order to englobe several specific parameters to our context. The arrival of trains was modeled using a specific Anylogic event object (the trainArrival Object shown in Fig. 3). This event is used to trigger the arrival of trains.

Several functions was used in our simulator (see Fig. 4). The callFRTFromFarStation function was used to call FRT vehicles from the network and to assign them to the Intermodal facility. The AffectFRTToContainer function assigns empty FRT vehicles to waiting palette in the Intermodal Facility. The CheckPalette function will periodically check if there is waiting palette in the area.

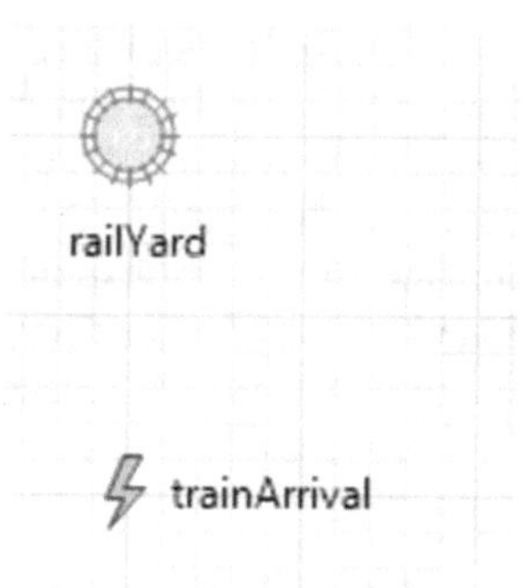

Fig. 3 Overview of the developed train process

Fig. 4 Overview of the distinctive developed functions

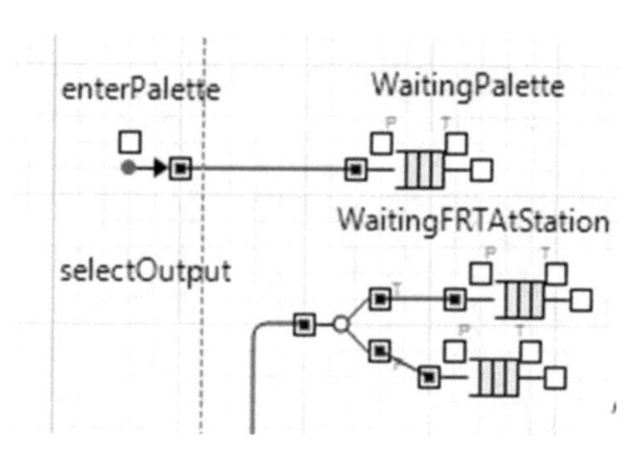

Fig. 5 Overview of initial treatment process of palette and vehicles

All the waiting palettes are grouped in a specific queue object (see Fig. 5). Also all the new arrived FRT vehicles are first put in a queue object while waiting to be affected to a specific palette. Then, if a palette of goods have been affected to a FRT vehicle, both the palette and the vehicle are processed. They enter the specific network using the two network enter objects EnterPalette and EnterFRT (see Fig. 5). FRT vehicles moving process to the specific position of unloading palette of goods from the train and loading it into the vehicle is modeled using a NetworkMoveTo object.

Finally, the specific unloading and loading operation is modeled using a entreprise library object. This object is used to simulate the specific process of unloading the palette and loading it in the vehicle.

Using the combination of these different simulated process allows us to model carefully the loading and unloading operations in our specific context. This model also allows us to model specific algorithms for managing the whole process. In fact using the different implemented functions, several procedures could be tested and implemented to observe their impact on the whole system. In this work and in order to test and validate our developed simulation model, we used a simple First in First out (FIFO) strategy to assign vehicles to palettes.

4 Preliminaries Computational Results

The simulation model was developed using the Anylogic 6.4 version. Tests were performed a computer with a 3.2 GHZ CPU and 8 GB of RAM. In order to test our proposed intermodal simulation model, we developed it as an integrated part of a PRT simulation model [6, 11] using Anylogic in order to test its feasibility from an operational level. Results of our preliminaries tests are shown in Table 1. We used simulation'use case and scenarios adapted from [5]. Based on these assumptions, we generated 15 test'use case. In Table 1, we present the average waiting time for the different palette to be loaded into a FRT vehicle. We should note that from these tests

Table 1 Preliminaries obtained results

Use case	Average waiting time in minutes
1	20.986
2	21.099
3	21.413
4	22.998
5	19.869
6	20.063
7	23.959
8	21.776
9	22.546
10	24.78
11	21.944
12	21.313
13	25.835
14	22.577
15	21.556
Average	22.225
Minimum	19.869
Maximum	25.835

the simulation model was validated. In fact, our simulation model was able to model our complex Intermodal context. We could conclude also that having an average waiting time of 22.225 min could be considered as a satisfactory preliminaries results for our context.

5 Conclusion

This paper presented a new simulation model in order to simulate the loading/unloading operation in an intermodal transportation tool. Details of our simulation model was provided and implemented in order to mimic the real behavior of palette of goods in intermodal facilities. Simulation results proved that considering the specific intermodal transportation mode is an interesting option for urban areas. In fact, our performance measure proved that palette of goods don't stay more than 25 min waiting in order to be moved. This latter represents an interesting finding. Several extensions are possible to this work. We are currently in the stage of developing a whole simulation model which integrates the loading/unloading operations in order to simulate the movement of palette of goods in a specific urban areas. We are also developing more enhanced algorithms to optimize the movement of palette of goods in the loading/unloading facilities.

References

1. Abbott, D., Marinov, M.V.: An event based simulation model to evaluate the design of a rail interchange yard, which provides service to high speed and conventional railways. Simul. Model. Pract. Theor. **52**, 15–39 (2015)
2. Assadipour, G., Ke, G.Y., Verma, M.: Planning and managing intermodal transportation of hazardous materials with capacity selection and congestion. Transp. Res. Part E: Logist. Transp. Rev. **76**, 45–57 (2015)
3. Chebbi, O., Chaouachi, J.: Modeling on-demand transit transportation system using an agent-based approach. In: Computer Information Systems and Industrial Management, pp. 316–326. Springer (2015)
4. Chebbi, O., Chaouachi, J.: Optimal fleet sizing of personal rapid transit system. In: Computer Information Systems and Industrial Management, pp. 327–338. Springer (2015)
5. Chebbi, O., Chaouachi, J.: Reducing the wasted transportation capacity of personal rapid transit systems: an integrated model and multi-objective optimization approach. Transp. Res. Part E: Logist. Transp. Rev. (2015)
6. Chebbi, O., Siala, J.C.: Priority rule strategy for managing a personal rapid transit system. In: 2014 IEEE 17th International Conference on Intelligent Transportation Systems (ITSC), pp. 2852–2857. IEEE (2014)
7. Choe, P., Tew, J.D., Tong, S.: Effect of cognitive automation in a material handling system on manufacturing flexibility. Int. J. Prod. Econ. **170**, 981–899 (2015)
8. Daszczuk, W.B., Mieścicki, J., Grabski, W.: Distributed algorithm for empty vehicles management in personal rapid transit (PRT) network. J. Adv. Transp. (2016)

9. Fatnassi, E., Chaouachi, J., Klibi, W.: Planning and operating a shared goods and passengers on-demand rapid transit system for sustainable city-logistics. Transp. Res. Part B: Methodol. **81**, 440–460 (2015)
10. Fatnassi, E., Chebbi, O., Chaouachi, J.: Discrete honeybee mating optimization algorithm for the routing of battery-operated automated guidance electric vehicles in personal rapid transit systems. Swarm Evol. Comput. (2015)
11. Fatnassi, E., Chebbi, O., Siala, J.C.: Evaluation of different vehicle management strategies for the personal rapid transit system. In: 2013 5th International Conference on Modeling, Simulation and Applied Optimization (ICMSAO), pp. 1–5. IEEE (2013)
12. Fatnassi, E., Chebbi, O., Siala, J.C.: Two strategies for real time empty vehicle redistribution for the personal rapid transit system. In: 2013 16th International IEEE Conference on Intelligent Transportation Systems-(ITSC), pp. 1888–1893. IEEE (2013)
13. Fatnassi, E., Chebbi, O., Siala, J.C.: Comparison of two mathematical formulations for the offline routing of personal rapid transit system vehicles. In: The International Conference on Methods and Models in Automation and Robotics (2014)
14. Fu-gui, D., Hui-mei, L., Bing-de, L.: Agent-based simulation model of single point inventory system. Syst. Eng. Proc. **4**, 298–304 (2012)
15. Huang, Y., Liang, C., Yang, Y.: The optimum route problem by genetic algorithm for loading/unloading of yard crane. Comput. Ind. Eng. **56**(3), 993–1001 (2009)
16. Lees-Miller, J.D.: Minimising average passenger waiting time in personal rapid transit systems. Ann. Oper. Res. **236**(2), 405–424 (2016)
17. Lehnfeld, J., Knust, S.: Loading, unloading and premarshalling of stacks in storage areas: survey and classification. Eur. J. Oper. Res. **239**(2), 297–312 (2014)
18. Li, J., Daaboul, J., Tong, S., Bosch-Mauchand, M., Eynard, B.: A design pattern for industrial robot: user-customized configuration engineering. Robot. Comput. Integr. Manuf. **31**, 30–39 (2015)
19. Lichtenberg, A., Guimarães, P., Podsedkowska, H.: Planning for sustainable mobility with personal rapid transit in small European cities. In: Highway and Urban Environment, pp. 3–14. Springer, The Netherlands (2010)
20. Monios, J., Bergqvist, R.: Intermodal terminal concessions: lessons from the port sector. Res. Transp. Bus. Manag. **14**, 90–96 (2015)
21. Monios, J., Bergqvist, R.: Operational constraints on effective governance of intermodal transport. Res. Transp. Bus. Manag. **14**, 1–3 (2015)
22. Pallme, D., Lambert, B., Miller, C., Lipinski, M.: A review of public and private intermodal railroad development in the memphis region. Res. Transp. Bus. Manag. **14**, 44–55 (2015)
23. Santos, B.F., Limbourg, S., Carreira, J.S.: The impact of transport policies on railroad intermodal freight competitiveness—the case of Belgium. Transp. Res. Part D: Transp. Environ. **34**, 230–244 (2015)

Parallelization of Fuzzy Logic Analysis for Pattern Recognition

M. Hires, H. Habiballa and V. Novak

Abstract Fuzzy logic analysis in pattern recognition is a unique method developed on Institute for Research and Applications of Fuzzy Modeling, University of Ostrava. Its efficiency was also proved on an industrial application for automated symbolic recognition of signatures on metal ingots. In this article we briefly recall main notions of Fuzzy Logic Analysis and we especially take into account parallel implementation of the algorithm.

Keywords Parallelization · Fuzzy logic · Pattern recognition · OCR

1 Introduction

The main obstacle in the task of character recognition lies in damaged or incomplete graphical information. The input of the task includes an image with the presence of a symbolic element. Our goal is to recognize this symbolic element (character or other pattern) from a picture, which we can assume as a pixel matrix. Best solver of such a task still remains human brain. Human beings are capable of character recognition in highly uncertain situations as well as badly damaged symbols. Computer science and Artificial Intelligence as its branch study automatization of recognition for a long time. We have many methods how to recognize patterns, e.g. neural networks are efficient for these purposes.

In this paper we will present parallelization of software tool based on mathematical fuzzy logic calculus with evaluated syntax. This calculus was initiated in [7] in propositional version and further developed in first order version in [4] and crowned in the book [5]. The pattern recognition method was originally described in [6]. The

M. Hires (✉) · H. Habiballa · V. Novak
Department of Informatics and Computers, University of Ostrava,
30. dubna 22, Ostrava 1, Czech Republic
e-mail: matej.hires@osu.cz

R. Silhavy et al. (eds.), *Software Engineering Perspectives and Application in Intelligent Systems*, Advances in Intelligent Systems and Computing 465,
DOI 10.1007/978-3-319-33622-0_40

core of the method and its efficiency is already published in detail in [3], so we will only recall necessary formalisms and algorithms for further parallel implementation of the method.

2 Fuzzy Logic Analysis

The proposed method is based on the theory of mathematical fuzzy logic with evaluated syntax $Ev_Ł$. The idea is to split the image into a grid of parts, each of which can consists of one or more pixels.

First, we consider a special language J of first-order $Ev_Ł$. We suppose that it contains a sufficient number of terms (constants) $t_{i,j}$ which will represent *locations* in the two-dimensional space (i.e., selected parts of the image). Each location can be whatever part of the image, including a single pixel or a larger region of the image.

The two-dimensional space will be represented by matrices of terms taken from the set of closed terms M_V:

$$M = \left(t_{i,j}\right)_{\substack{i\in I\\ j\in J}} = \begin{pmatrix} t_{11} & \dots & t_{1n} \\ \vdots & \vdots & \vdots \\ t_{m1} & \dots & t_{mn} \end{pmatrix} \tag{1}$$

where $I = \{1, \dots, m\}$ and $J = \{1, \dots, n\}$ are some index sets. The matrix (1) will be called the *frame of the pattern*. In other words, the frame of the pattern is the underlying grid of parts of the given image. The pattern itself is the letter which we suppose to be contained in the image and which is to be recognized. A vector $t_i^L = \left(t_{i1}, \dots, t_{in}\right)$ is a *line* of the frame M and $t_j^C = \left(t_{1j}, \dots, t_{mj}\right)$ is a *column* of the frame M.

The simplest content of the location is the *pixel* since pixels are points of which images are formed. A *pixel* is represented by a certain designated (and fixed) atomic formula $P(x)$ where the variable x can be replaced by terms from (1), i.e., it runs over the locations. Another special designated formula is $N(x)$. It will represent "nothing" or also "empty space". We put $N(x) := \mathbf{0}$.

The algebra of truth values is the standard Łukasiwicz MV-algebra

$$\mathcal{L} = \langle [0, 1], \vee, \wedge, \otimes, \rightarrow, 0, 1 \rangle \tag{2}$$

where

$$\wedge = \text{minimum}, \qquad \vee = \text{maximum},$$
$$a \otimes b = \max(0, a + b - 1), \qquad a \rightarrow b = \min(1, 1 - a + b),$$
$$\neg a = a \rightarrow 0 = 1 - a.$$

Formulas of the language J are *properties of the given location* (its content) in the space. They can represent whatever shape, e.g., circles, rectangles, hand-drawn curves, etc. As mentioned, the main concept in the formal theory is that of *evaluated formula*. It is a couple ${}^{a}/A$ where A is a formula and $a \in [0, 1]$ is a syntactic truth value. In connection with the analysis of images, we will usually call a *intensity* of the formula A. Note that ${}^{0}/\mathbf{0}$, i.e., "nothing" has always the intensity 0.

Let M_Γ be a frame. The *pattern* Γ is a matrix of evaluated formulas

$$\Gamma = \left({}^{a_{ij}}/A_x[t_{ij}]\right)_{\substack{i \in I_{M_\Gamma} \\ j \in J_{M_\Gamma}}} \tag{3}$$

where $A(x) \in \Sigma(x)$, $t_{ij} \in M_\Gamma$ and I_{M_Γ}, J_{M_Γ} are index sets of terms taken from the frame M_Γ.

A *horizontal component* of the pattern Γ is

$$\Lambda_i^H = \left({}^{a_{ij}}/A_x[t_{ij}] \in \Gamma \mid t_{ij} \in t_i^L\right), \qquad j \in J_{M_\Gamma} \tag{4}$$

where t_i^L is a line of M_Γ. Similarly, a *vertical component* of the pattern Γ is

$$\Lambda_j^V = \left({}^{a_{ij}}/A_x[t_{ij}] \in \Gamma \mid t_{ij} \in t_j^C\right), \qquad i \in I_{M_\Gamma} \tag{5}$$

where t_i^C is a column of M_Γ. When the direction does not matter, we will simply talk about *component*. The *empty component* is

$$E = ({}^{0}/\mathbf{0}, \ldots, {}^{0}/\mathbf{0}).$$

Hence, a component is a vertical or horizontal line selected in the picture which consists of some well defined elements represented by formulas.

Dimension of the component $\Lambda = ({}^{a_1}/A_1, \ldots, {}^{a_n}/A_n)$ is $\dim(\Lambda) = k^{max} - k^{min} + 1$, $1 \leq k^{min}, k^{max} \leq n$, where

$$k^{min} = \min_{i=1,\ldots,n} \{i \mid \vdash_a \neg(A_i(x) \Leftrightarrow \mathbf{0}), a > 0\}, \tag{6}$$

$$k^{max} = \max_{i=1,\ldots,n} \{i \mid \vdash_a \neg(A_i(x) \Leftrightarrow \mathbf{0}), a > 0\} \tag{7}$$

Intensity of a pattern is the matrix

$$Y_\Gamma = \left(a_{ij}\right)_{\substack{i \in I_{M_\Gamma} \\ j \in J_{M_\Gamma}}}. \tag{8}$$

The intensity $Y_{\Lambda_i^H}$ ($Y_{\Lambda_j^V}$) of a horizontal (vertical) component is defined analogously.

Maximal intensity of a pattern Γ is

$$\check{H}_\Gamma = \max\{a_{ij} \mid a_{ij} \in Y_\Gamma \text{ and } \vdash_{a_{ij}} \neg(A_x[t_{ij}] \Leftrightarrow \mathbf{0})\}.$$

Minimal intensity $\hat{H}_\Gamma$ of a pattern Γ is defined analogously. Intensity of a pattern Γ is said to be *normal* if $\check{H}_\Gamma = 1$. The pattern with normal intensity is called *normal*.

2.1 Comparison of patterns

To be able to *compare* two patterns

$$\Gamma = \left(a_{ij}/A_x[t_{ij}]\right)_{\substack{i \in I_{M_\Gamma} \\ j \in J_{M_\Gamma}}}$$

and

$$\Gamma' = \left(a'_{ij}/A'_x[t_{ij}]\right)_{\substack{i \in I_{M_{\Gamma'}} \\ j \in J_{M_{\Gamma'}}}},$$

their frames must have the same dimension. This can always be assured because we can embed two given frames in bigger frames of equal size. Thus, without loss of generality, we will assume in the sequel that $I_{M_\Gamma} = I_{M_{\Gamma'}}$ and $J_{M_\Gamma} = J_{M_{\Gamma'}}$.

The patterns will be compared both according to the content as well as intensity of the corresponding locations. Hence, we will consider a bijection $f : M_\Gamma \longrightarrow M_{\Gamma'}$ $f(t_{ij}) = t'_{ij}$, $i \in I, j \in J$ between the frames M_Γ and $M_{\Gamma'}$.

Let two components, $\Lambda = (a_1/A_1, \dots, a_n/A_n)$ and $\Lambda' = (a'_1/A'_1, \dots, a'_n/A'_n)$ be given. Put $K_1 = \min(k^{min}, k'^{min})$ and $K_2 = \max(k^{max}, k'^{max})$ where k^{min}, k'^{min} are the corresponding indices defined in (6) and k^{max}, k'^{max} are those defined in (7), respectively. Note that K_1 and K_2 are the left-most and right-most indices of some non-empty place which occurs in either of the two compared patterns in the direction of the given components. Furthermore, we put

$$n^C = \sum \{b_i \Vdash_{b_i} A_i(x) \Leftrightarrow A'_i(x), a/A_{i,x}[t] \in \Lambda, a'/A'_{i,x}\}, \tag{9}$$

$$n^I = \sum \{b_i = a_i \leftrightarrow a'_i \mid a_i/A_{i,x}[t] \in \Lambda, a'_i/A'_{i,x}[f(t)]\}. \tag{10}$$

The number n^C represents the total degree in which the corresponding places in both patterns tally *in the content*. This extends the power of the procedure as from the formal point of view, A'_i may differ from A_i but they still may represent the same object; at least to some degree b_i. The number n^I is similar but it reflects the compared intensity of the objects residing in the respective locations.

The components Λ and Λ' are said to *tally in the degree* q if

$$q = \begin{cases} \frac{n^C + n^I}{2(K_2 - K_1 + 1)} & \text{if } K_2 - K_1 + 1 > 0 \\ 1 & \text{otherwise.} \end{cases} \tag{11}$$

We will write

$$\Lambda \approx_q \Lambda'$$

to denote that two components Λ and Λ' tally in the degree q. When $q = 1$ then the subscript q will be omitted.

It can be demonstrated that if all formulas A in $^{a}/A$, for which it holds that $\vdash \neg(A \Leftrightarrow \mathbf{0})$ and which occur in Λ and Λ', are the same then we can compute q using the following formula:

$$q = 1 - \frac{1}{2} \frac{\sum_{K_1 \leq i \leq K_2} |a_i - a'_i|}{K_2 - K_1 + 1}. \tag{12}$$

The pattern Γ can be viewed in two ways:

(a) From the *horizontal view*, i.e., as consisting of horizontal components

$$\Gamma = \left(\Lambda^H_i\right)_{i \in I} = (\Lambda^H_1, \ldots, \Lambda^H_m). \tag{13}$$

(b) From the *vertical view*, i.e., as consisting of vertical components

$$\Gamma = \left(\Lambda^V_j\right)_{j \in J} = (\Lambda^V_1, \ldots, \Lambda^V_n). \tag{14}$$

If the distinction between horizontal and vertical view of the pattern is inessential, we will simply use the term *pattern* in the sequel.

A *subpattern* (horizontal or vertical) $\Delta \subseteq \Gamma$ of $\Gamma = (\Lambda_1, \ldots, \Lambda_p)$ is any connected sequence

$$\Delta = (\Lambda_{j_1}, \ldots, \Lambda_{j_k}), \qquad 1 \leq j_1, j_k \leq p \tag{15}$$

of components (horizontal or vertical, respectively) from Γ. If $\Lambda_{j_1} \neq E$ and $\Lambda_{j_k} \neq E$ then Δ is a *bare subpattern* of Γ and the number k is its *dimension*. Δ is a *maximal bare subpattern* of Γ if $\bar{\Delta} \subseteq \Delta$ for every bare subpattern $\bar{\Delta}$. The dimension of a maximal bare subpattern of Γ is the *dimension of* Γ and it will be denoted by $\dim(\Gamma)$.

Recall from our previous agreement that, in fact, we distinguish horizontal ($\dim_H(\Gamma)$) or a vertical dimensions ($\dim_V(\Gamma)$) of the pattern depending on whether a pattern is viewed horizontally or vertically. Note that both dimensions are, in general, different.

Suppose now that two patterns Γ and Γ' are given and let $\Delta \subseteq \Gamma$ have a dimension k. The following concepts of maximal common subpattern and the degree of matching of patterns are inspired by the paper [8].

Let q_0 be some threshold value of the the degree (11) (according to experiments, it is useful to set $q_0 \approx 0.7$). We say that Δ *occurs in* Γ' with the degree q_0 (is a q_0-*common subpattern* of both Γ and Γ') if there is a subpattern $\Delta' \subseteq \Gamma'$ of dimension k for which the property

$$\Lambda_i \approx_q \Lambda'_i, \qquad q \geq q_0 \tag{16}$$

holds for every pair of components from Δ and Δ', $i = 1, \ldots, k$, respectively. If $q_0 = 1$, then Δ is a common subpattern of Γ and Γ'.

A q_0-common subpattern of Γ and Γ' is *maximal* if every subpattern $\bar{\Delta} \subseteq \Gamma$ such that $\bar{\Delta} \supseteq \Delta$ is not a q_0-common subpattern of Γ and Γ'.

Let $\Delta_1, \ldots, \Delta_r$ be all maximal q_0-common subpatterns of Γ and Γ'. The *degree of matching* of Γ and Γ' is the number

$$\eta(q_0) = \frac{1}{2}\left(\frac{\sum_{j=1}^{r} \dim(\Delta_j)}{\dim(\Gamma)} + \frac{\sum_{j=1}^{r} \dim(\Delta_j)}{\dim(\Gamma')}\right) \tag{17}$$

where $\dim(\Delta_j) \geq 2$ for all $j = 1, \ldots, r$.

We must compute (17) separately for horizontal and vertical view so that we obtain horizontal η_H as well as vertical η_V degrees of matching, respectively. Then the *total degree of matching* of the patterns Γ and Γ' is the number

$$\bar{\eta}(q_0) = \frac{\eta_H(q_0) + \eta_V(q_0)}{2}. \tag{18}$$

3 Design

The parallel procedure for pattern recognition is derived from its sequential version. We can assume one input picture containing damaged character. This character will be compared with all sample patterns. There are 36 standard alphanumeric characters in our example. Let's consider the suitable cluster with certain number of nodes. Every particular node could be assigned with certain part of patterns from the whole pattern set. The desired effect would be the enhanced time efficiency of the parallel recognition procedure.

The first step of the algorithm is the loading of the necessary data, i.e. pattern set and test picture or even a batch of tested pictures. The load procedure is done in parent process that is also responsible of the data distribution over the subprocesses.

The ideal distribution should be proportional with respect to the whole pattern set. Since the ideal solution is not assured we have to adopt the procedure to handle non-proportional pattern subsets. The idea of such adaptation could be seen in Fig. 1. Further the algorithm would be distributing the test picture (damaged character) over all subprocesses (Fig. 2). Then the test picture is compared with all patterns of the particular subprocess. This procedure is equivalent to sequential version of the method. The result of comparison is then stored in suitable structure after all subprocesses are finished and aggregated. Note that the non-proportionality mentioned above should be also preserved in the result aggregation phase. This phase is illustrated in Fig. 3.

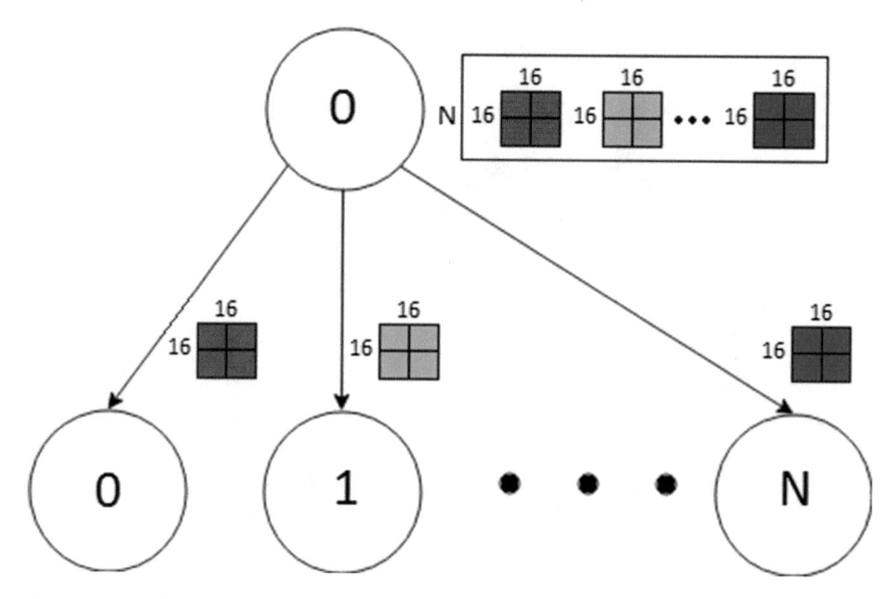

Fig. 1 The distribution of patterns to subprocesses

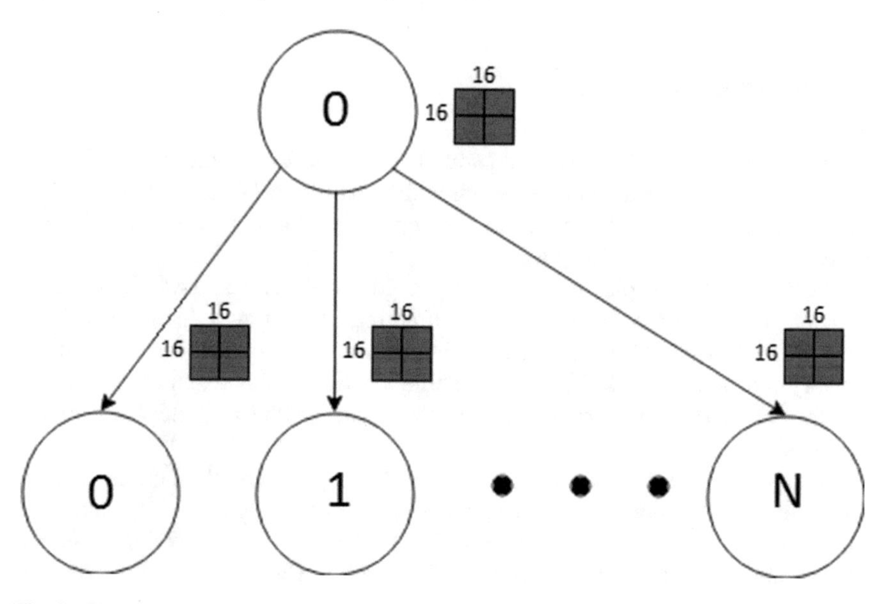

Fig. 2 The test picture (damaged) distribution to subprocesses

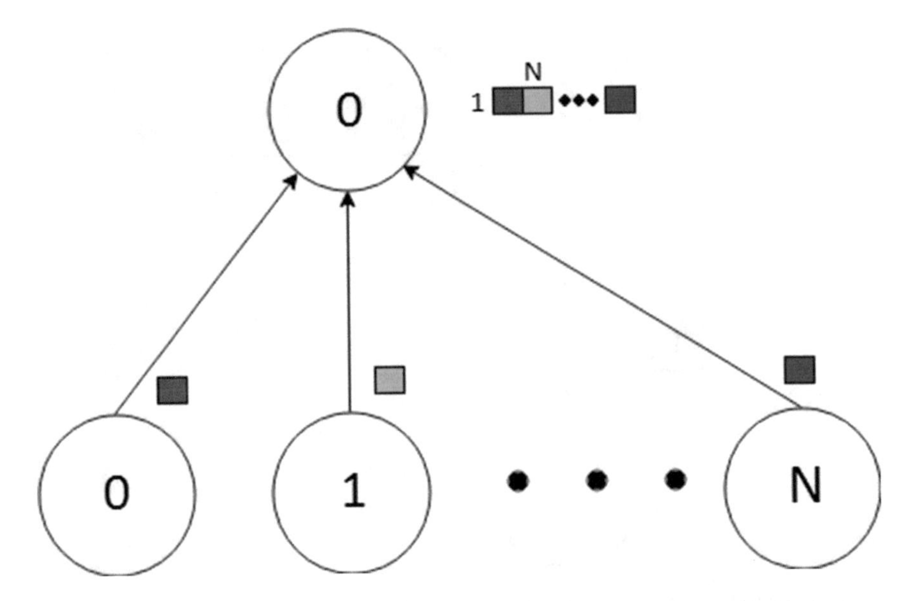

Fig. 3 The aggregation of results from subprocesses

4 Implementation

The implementation of the designed parallel version of fuzzy logic analysis is done with the help of MS-MPI library, which is the realization of the MPI standard by Microsoft corporation. The testing of the implementation is realized on Windows Server 2008 R2 HPC Edition platform.

Every character and picture respectively could be perceived as a two-dimensional array of pixels and the character set could be represented by a three-dimensional array. Nevertheless the MPI standard requires the transferred data to form continuous memory block. That's why the most suitable solution to this requirement was the transformation to the conform one-dimensional array (both for single character and character sets).

It can been seen from the example, how to allocate and to fill the pattern set array. There are several important constants, which are used globally in the algorithm. The constant **ROOT** defines parent process. The **ROW** and **COLUMN** define picture dimensions. The last key value **NUMBER OF PATTERNS** defines the extent of the pattern set.

```
int* pattern = (int)malloc(sizeof(int) * ROW*  COLUMN);
int* all_patterns =
  (int*)malloc(sizeof(int) * NUMBEROFPATTERNS * ROW * COLUMN);
if (rank == ROOT)
{
```

```
    for (int i = 0; i < NUMBEROFPATTERNS; i++)
    {
      pattern = ReadImage(PATTERNS, i);
      for (int j = 0; j < ROW; j++)
      for (int k = 0; k < COLUMN; k++)
        all_patterns[( i * COLUMN * ROW) + (( j * COLUMN) + k)] =
          pattern[k * COLUMN + j];
    }
  }
```

The distribution to subprocesses is done by the method **scatterv**, which is similar to the method **scatter** enabling distribution of non-uniform data blocks. In order to use the method it has to be defined two arrays and they have to be passed as parameters. The first array **scounts** defines the extents of data blocks and the second array **sdispls** contains starting positions in data array (data to be passed to the subprocesses).

```
  int* GetScounts(int* rcounts, int size)
  {
    int* result = InitArray(size);
    for (int i = 0; i < size ; i++)
      result[i] = rcounts[i] * (ROW * COLUMN);
    return result;
  }

  int* GetDispls(int* arr , int size)
  {
    int* result = InitArray(size);
    for (int i = 1; i < size ; i++)
      result[i] = (arr[i - 1] + result[i - 1]);
    return result;
  }
```

The usage for this method is the following.

```
MPI_Scatterv(all_patterns, scounts, sdispls, MPI_INT,
  block_patterns, scounts[rank], MPI_INT, ROOT, comm);
```

The broadcast method

```
MPI_Bcast(picture, ROW COLUMN, MPI INT, ROOT, comm);
```

The parallel version of fuzzy logic analysis is equivalent to the sequential version with one difference —the subprocesses have only a portion of all patterns from pattern set. The results are stored in a particular process into one-dimensional array and then after comparison of a picture with all patterns the partial results are aggregated in the parent process. The method **gatherv** (similar to **gather**) serves for this purpose.

```
MPI_Gatherv(block_results, rcounts[rank], MPIDOUBLE,
  results, rcounts, rdispls, MPIDOUBLE, ROOT, comm);
```

5 Conclusions

We have extended the former application PREPIC [9], which utilizes standard sequential version of fuzzy logic analysis. Several sample recognition with PREPIC is given on the following example pictures (Fig. 4).

The preliminary results of experiments show the enhanced efficiency even only on limited number of subprocesses. We also plan to make experiments with heterogeneous computer networks on multiplatform nodes. The simple testing environment was the following (Table 1):

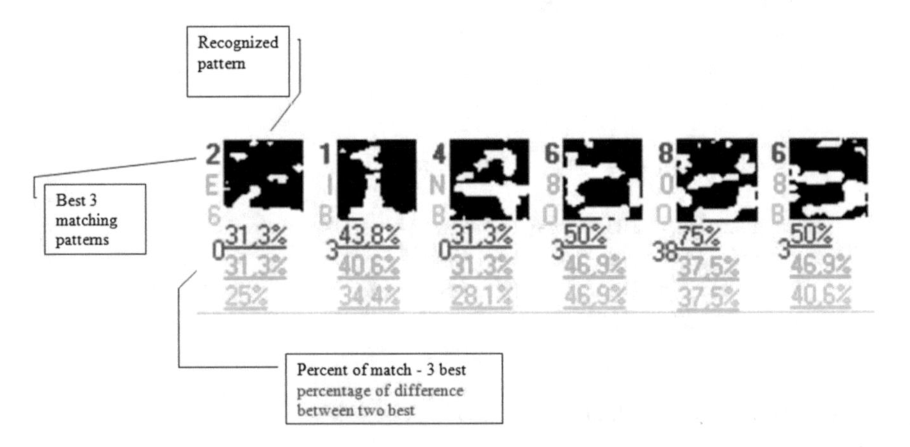

Fig. 4 Batch process by PREPIC

Table 1 Parallel version of fuzzy logic analysis—simple testing

Test pct number	1 process (ms)	2 processes (ms)	3 processes (ms)	4 processes (ms)
1	24	24	24	23
10	66	42	36	34
100	458	393	374	339
1000	3310	2460	2210	2100

- Processor: Intel(R) Core(TM) i5-4670K
- RAM: 16 GB
- OS: Windows Server 2008 R2 HPC Edition

References

1. Hireš, M.: Paralelizace Fuzzy Logick Analzy. University of Ostrava (2015)
2. Novák, V., Habiballa, H.: Recognition of damaged letters based on mathematical fuzzy logic analysis. In: International Joint Conference CISIS 12—SOCO 12 Special Sessions, pp. 497–506. Springer, Berlin (2012) ISBN: 978-3-642-33017-9
3. Novák, V., Hurtik, P., Habiballa, H., Stepnicka, M.: Recognition of damaged letters based on mathematical fuzzy logic analysis. J. Appl. Logic **13**, 94–104 (2015)
4. Novák, V.: On the syntactico-semantical completeness of first-order fuzzy logic I. II. Kybernetika **26**(47–66), 134–154 (1990)
5. Novák, V., Perfilieva, I., Močkoř, J.: Mathematical Principles of Fuzzy Logic. Kluwer, Boston (1999)
6. Novák, V., Zorat, A., Fedrizzi, M.: A simple procedure for pattern prerecognition based on fuzzy logic analysis. Int. J. Uncertain. Fuzz. Knowl.-Based Syst. **5**, 31–45 (1997)
7. Pavelka, J.: On fuzzy logic I, II, III. Z. Math. Logik Grundlagen der Math. **25**, 45–52, 119–134, 447–464 (1979)
8. Schek, H.J.: Tolerating fuzziness in keywords searching. Kybernetes **6**, 175–184 (1977)
9. Habiballa, H., et al.: PRErognition of PICtures (PREPIC)—Win32 software application for pattern recognition, University of Ostrava (2013). http://irafm.osu.cz/en/c149_prepic/

Performance Management Using Autonomous Control-Based Distributed Coordination Approach in a Volunteer Grid Computing Environment

Saddaf Rubab, Mohd Fadzil Hassan, Ahmad Kamil Mahmood and Syed Nasir Mehmood Shah

Abstract In volunteer grid environment, it is difficult to fulfill the requirements of all jobs due to increasing demands of resources. A resource requester submits a job, require resources for the job to be completed within deadline and budget if specified any. Whereas resource provider makes use of available resources and wants to utilize resources to maximum. Therefore, satisfying the requirements of both i.e., jobs and resources makes it difficult to manage the performance of a volunteer grid. In performance management, the main objectives include maintaining service level agreements, maximization of resource utilization, meeting job deadline/budget and minimizing the job transfer. In this paper, only the maximization of resource utilization and meeting job deadlines will be addressed for managing the performance of a volunteer grid computing environment. An autonomous approach is introduced that provides dynamic resource allocation for submitted jobs in a volunteer grid environment depending on the availability and demand of resources. Grid resource brokers are considered third party organizations that work as intermediaries between volunteer resource provider and requester. Proposed autonomous approach is developed by utilizing distributed coordination approach for interactive assignment of volunteer resources. The proposed approach is applying distributed coordination approach and giving priority to maximization of volunteer resource usage while completing jobs within deadline.

S. Rubab (✉) · M.F. Hassan · A.K. Mahmood
Department of Computer and Information Sciences, University Teknologi PETRONAS, 31750 Bandar Seri Iskandar, Tronoh, Perak, Malaysia
e-mail: saddaf_g02754@utp.edu.my

M.F. Hassan
e-mail: mfadzil_hassan@petronas.com.my

A.K. Mahmood
e-mail: kamilmh@petronas.com.my

S.N.M. Shah
Department of Computer Sciences, Dr. A. Q. Khan Institute of Computer Sciences & Information Technology, Kahuta, Pakistan
e-mail: dr.shah@kicsit.edu.pk

R. Silhavy et al. (eds.), *Software Engineering Perspectives and Application in Intelligent Systems*, Advances in Intelligent Systems and Computing 465,
DOI 10.1007/978-3-319-33622-0_41

Keywords Resource provider • Resource requester • Resource broker • Volunteer grid computing • Distributed coordination

1 Introduction

Volunteer computing is a form of distributed computing which allows different participants around the world to contribute their idle resources or unused CPU cycles in a grid system [1, 2]. In distributed networks, the resources volunteered to be used in grid system for storing large data and processing computations are termed as volunteer grid resources. With the recent interests of researchers in large computations for scientific purposes, a good response of resource volunteers has been observed. This helps in having large number of resources that can be used anytime and anywhere through an access to the volunteer grid environment. It also aids in minimizing the need of installing physical infrastructure of large computing resources.

In a volunteer grid, all the submitted tasks or applications are subdivided into small dependent or independent jobs. Jobs regardless of dependent or independent in nature will be considered in proposed approach. These jobs need additional computing resources. The resource requester holds all the jobs and asks the resource provider for the matching resource of a job. In recent works reported by researchers, an intermediary between resource provider and requester has been explored for selection of resources and assignment of jobs, called resource broker. It selects a resource from the pool of resource providers and assigns it to the job submitted to resource requester and waiting for resource assignment.

A volunteer resource has to be made available by the resource provider on receiving a request. This includes an agreement of complying with the negotiated Service Level Agreements (SLAs) between resource provider and requester. There are multi-objectives SLAs between resource provider and requester to satisfy their constraints. These may be budget, utilization or availability of resource and jobs deadline. In this paper, SLAs will only study the resource availability aspects of a volunteer grid environment. In order to avoid any SLA violations, the job requirements must be studied by resource broker before finding any matching resource and availability of resource should also be premeditated. If there is an underestimation of resource requirement for a job, the resource inadequacy will occur that will eventually violate the SLA of job. If resource requirement is overestimated, the job assignment may not be efficiently done and will increase the computational costs and budget that will violate SLA of resource.

Moreover, due the unpredictability of volunteer resources, makes it unavoidable to reserve additional resources from resource providers' pool. There can be a situation that most of the jobs request for additional resources simultaneously; this will lead to a competition among the jobs for obtaining resources, which may be limited.

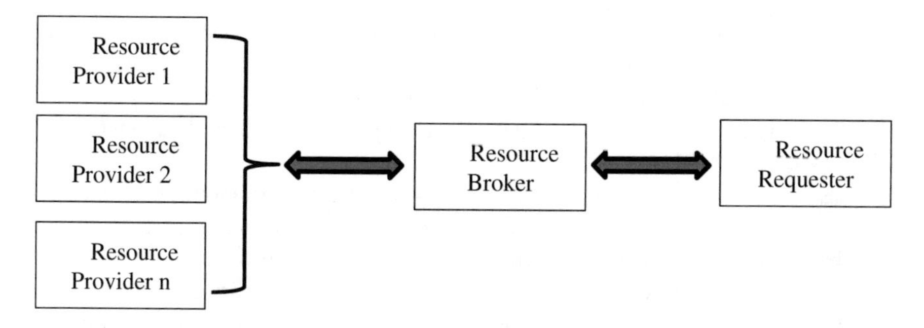

Fig. 1 Integrated scheme

This paper proposed an autonomous control-based distributed coordination approach for performance management in volunteer computing environment. The proposed work utilizes control-based approach for solving the coordination between the distributed resource providers, resource broker and resource requester to get an optimized allocation of volunteer resources. This approach will serve the requirements of jobs computing in volunteer grid environment and maximize the resource utilization. The proposed approach will consider integrated resource providers in a volunteer grid environment, where one resource provider can acquire resource from another provider using resource broker to serve jobs submitted to resource requester (Fig. 1). The following two primary objectives will be addressed:

1. To improve the maximum utilization of resources to fulfill the purpose of volunteer computing
2. To satisfy the job requirements to reduce the communication cost and job transfer

These objectives are contradiction to each other and make performance management more challenging. A presentation of deployment of an autonomous approach in volunteer grid environment is also presented.

The outline of rest of the paper is as follows. The overview of the related literature is explained in Sect. 2, which is followed by Sect. 3 give details of proposed autonomous control-based distributed coordination performance management approach for volunteer grid computing environment. Section 4 presents the simulation results and the scope of work presented is concluded in Sect. 5.

2 Related Work

The proposed approach is addressing the performance management of volunteer grid computing environment using control-based technique. The control-based techniques help to design performance management framework for a system if accurately estimated and designed. In any framework presented using control-based technique, it can state various performance management issues and can also assist in studying feasibility solution of system prior to its deployment. Control-based techniques have been applied in task scheduling [3, 4], energy management [5], load balancing [6, 7] and QoS issues [8].

This section will review the brief literature of different techniques presented for performance management to study different methods for managing performance in any distributed computing environment like P2P, cluster, grid etc.

In [9], a performance management system using agents for distributed computing environment is proposed, in which users need not to know the hierarchy of resource arrangement. The reported work balances the load to manage the performance of overall distributed resources. The monitoring agents are used to update the current state of resources and this information is broadcasted to overall distributed system using brokering agents.

Hierarchical control framework for distributed computing systems has been presented in [10] to manage the system itself and satisfy QoS while different operating systems running on resources/machines. The temporal and control decomposition along with function decomposition has been used to achieve system control. A three-level hierarchical structure was used to manage performance and save energy of a computer cluster. The proposed algorithm [10] choose control inputs depending on the future states predicted from the current states of resources. For evaluation of the performance, time varying workload from WC'98 [11] was tested.

In [12], authors proposed a new decentralized resource management framework for exploiting multi-core nodes in a P2P grid system. The key innovation is to use distinct logical nodes to represent the static and dynamic aspects of node utilization. The original Content-Addressable Network (CAN) does not allow two nodes to have identical coordinates, but multiple nodes with the same resource capabilities can exist in the CAN. A dimension to the CAN is presented that has randomly generated values for both nodes and jobs [12]. This multi-core environment makes use of match making and execute job using FIFO.

Highly Available Job Execution Service (HA-JES) are described in [13], which dynamically and transparently virtualizes underlying low-level computational resources to meet imbalanced and unpredictable resource usage requirements. From the grid user's perspectives, HA-JES is the same as the ordinary job execution service; it takes a job description and requirements from the user and executes the job if it can meet the user's requirements. From the architectural point of view, HA-JES is similar to a typical resource broker as it acts as a mediator between grid users and grid resources. However, instead of merely brokering resources which meet the user's requirements, HA-JES actively composes underlying underutilized low-quality resources to build a high quality resource satisfying the user's requirements. In particular, the process of virtualization in HA-JES occurs in a market-driven efficient way; underutilized and therefore cheap resources are exploited to build a high quality resource and hence foster balanced resource usage.

3 Proposed Autonomous Control-Based Distributed Coordination Approach for Performance Management

In this paper, a single resource broker can coordinate and communicate with multiple resource providers underlying in one volunteer grid environment. The communication is done for negotiating on the available volunteer resources. The resource broker firstly assigns a priority c_{ini} to each of the resource depending on the available CPU cycles that can be used by volunteer grid. To solve the control problem and maximize the resource utilization, the resource broker will update the priority c_{ini} depending on the amount of CPU cycles of volunteer resource has been requested. If resource requester is making request of large number of CPU cycles, resource broker will decrease the priority so that the interest of job in the particular resource can be minimized if there is a great competition on resource to reduce the communication load also. In other case, to promote one particular resource the priority of resource will be increased to get attraction of jobs waiting. After updating, resource broker must send the c_{ini} of each resource to resource requester. Resource requester will again compute the optimal value depending on the required amount of resource. This coordination and communication cycle will continue until there is no more resource left in any of resource provider pool or no more jobs is waiting for resource assignment. There is a possibility that for any job, no consensus has been made or there was no resource available. The priority value assigned to each resource has upper bound c_{max} and lower bound $c_{min.}$ Here

$$c_{min} \le c_{ini} \le c_{(t)} \le c_{max} \tag{1}$$

where c (t) is priority of a volunteer grid resource at time 't'.

In proposed approach multiple jobs are considered who want to acquire resources from volunteer resources pooled in 'N' resource providers. The details are illustrated in Fig. 2. The proposed approach utilizes the concept of coordination for maximization of volunteered resource usage accessed via resource broker and submitted tasks/applications divided into 'N' jobs can be assigned to any of the available resource. The jobs compete for volunteer resources 'R(t)' at a particular time 't'. If a large number of resources are already assigned to job 'i', the other 'N' jobs will compete hard and generate a large communication time overhead. The state dynamics at each resource provider 'i' can be described by following equations:

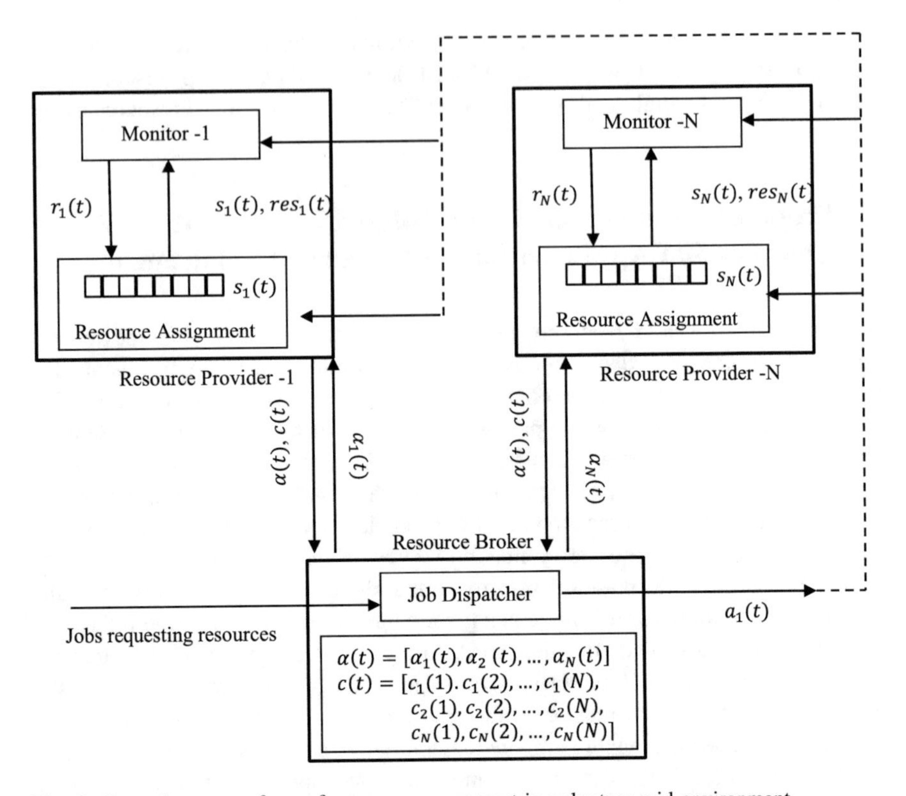

Fig. 2 Control structure for performance management in volunteer grid environment

$$S_i(t+1) = \left[S_i(t) + a_i(t) - \frac{r_i(t)T}{p_i(t)} \right] \tag{2}$$

$$res_i(t+1) = [1 + S_i(t+1)] \frac{p_i(t)}{r_i(t+1)} \tag{3}$$

$$r_i(t) = \alpha_i(t) R(t) \tag{4}$$

All of the variable notations are defined in Table 1.

The performance behavior of resource providers with respect to the available volunteer resources $r_i(t)$, computational jobs $a_i(t)$ arrival rate, expected response time $res_i(t+1)$, size of queue $S_i(t)$ is represented in Eqs. 2, 3 and 4. The problem of assignment of desired resources and mechanism of selecting desired resources is out of scope of the addressed approach. It is only giving importance to solving a

Table 1 Variable glossary

Variable name	Description
c_{ini}	Priority assigned initially to each resource depending on the CPU cycles available
c_{min}	Minimum priority value of resource
c_{max}	Maximum priority value of resource
c_t	Priority value of resource at time t
T	Sampling time
t	Time interval
$R(t)$	Total volunteer grid resources at time t
$r_i(t)$	Amount of computing resource of resource provider i assigned from the total available resource $R(t)$
$r_i(t+1)$	Expected required resource i
$S_i(t)$	Size of queue at resource provider i at time t
$S_i(t+1)$	Queue size of expected resource i
$a_i(t)$	Arrival rate of expected required resource request
$p_i(t)$	Predicted average resource time per request
$\alpha_i(t)$	Fraction of resource in use
$res_i(t+1)$	Expected response time

problem of controlling communication and fulfilling SLAs using a group of volunteer resources. The goal is to allocate the maximum available resources. In this process resource broker may have to increase or decrease the priority initially assigned c_{ini} to each resource. Therefore, the volunteer resource optimization problem is to find the optimal priority value c_t at time 't' and fraction of resource $\alpha_i(t)$ that can be assigned, such that resource broker can maximize the resource utilization while not having any job starving and also not violating SLAs.

3.1 Problem Statement

The resource broker has 'N' resource providers having volunteer resource $r_i(t)$. To have efficient coordination and communication between resource providers and resource broker, a variable $comm_i(t)$ must be defined at each resource provider to observe the effect of resource provider i on the overall volunteer grid system.

$comm_i(t)$ is sum of resources i assigned from each resource provider by resource broker to other resource provider j.

$$comm_i(t) = \sum_{j \neq i}^{N} \alpha_j(t) \tag{5}$$

3.2 Control at Resource Provider

The resource usage is to maximized using Eqs. 3 and 4 with volunteer resource as control input $r_i(t)$. To find the optimal value of $r_i(t)$, calculate $c_i^l(t)$ and $\alpha_i^l(t)$ given by resource broker over a set of t ∈ [1, T] by using Algorithm 1. Here '*l*' indicates coordination instance between resource provider and monitor within time t. Broker should be notified about values of $\alpha_i^l(t)$ over t ∈ [1, T].

3.3 Control at Resource Broker

At resource broker, the aim is minimize the communication error '**e**'.

$$e_i(t) = 1 - \sum_{j=i}^{N} \alpha_j^l(t) \tag{6}$$

The resource provider is notified about c_{ini}. The α_i will be calculated and forwarded to resource broker to calculate the error rate using Eq. 6. If e_{i2}^l is less or equal to error rate 'e', then assign the volunteer resource. Otherwise change c_{ini} satisfying Eq. 1, send the updated c_{ini} to resource broker. Increment '*l*' and solve for error rate until all resources assigned to maximum and no job left waiting for resource.

3.4 Expected Arrival Rate of Jobs

To predict number of jobs and their arrival rate in a volunteer grid environment is difficult, where all available resources at time 't' are also dynamic. There is a possible pattern of arrival rate of some regular jobs requesting resources during the 24 h [14]. The arrival rates might also vary slightly for such regular jobs. There are different methods used to predict the arrival rate like Kalman filters [15, 16], decision trees [17], ARIMA [18], and some techniques using automatic code instrumentation [19].

Algorithm 1: RPControl $_i$ (t) Control at Resource Provider

```
Inputs: Size of queue S_i(t)
          Time scale T
          c_ini and α^l_j(t) from resource broker
          a_i(t) expected arrival rate of resource request
          α^l_i at each RP_i
          F_i = [f_i1, f_i2, ..., f_iN] set of resource fraction at re-
     source provider 'i'

  1. Resource provider state state_i(t) = [S_i(t) * res_i(t)]
  2. for all f ∈ F_i do
  3.          α^l_i(1) = α^l_i(2) = α^l_i(3) = ··· α^l_i(T)
  4.          for all t within time scale T do
  5.                  Compute S_i(t + 1)
  6.                  Compute res_i(t + 1)
  7.                  Compute comm_i(t)
  8.                  Compute e_i(t)
  9.                  state_i(t) = [S_i(t + 1) * res_i(t + 1)]
  10.                      t = t + 1
  11.             end for
  12.       end for
  13.       Select maximum state_i(t) ∈ state_T
  14.       Select f with minimum comm_i(t)
  15.       Return α^l_i
```

4 Simulation Results

The coordination and communication approach for managing the performance of volunteer grid resources has been programmed in C ++. In the experimental setup, five resource providers are considered only to test the performance. Each of the resource providers is having 10 resources (5 resource providers * 10 resources = 50 resources). The LCG1 (Large Hadron Collider Computing Grid) [20] workload dataset has been used for jobs submitted to RPs (Resource Providers).

To experiment under the above mentioned simulation criterion, the CPU cycles requested by jobs and available CPU cycles at each resource is required. Synthetic datasets have been generated for jobs to evaluate the proposed approach and observe the behavior to run it in real volunteer grid environment. In this process, CPU cycles available at each resource and requested CPU cycles by each job are artificially created. With this process the testing data depending on the simulation needs can be generated quickly. For instance, if performance management of resources is to be evaluated, the resource CPU cycles available and allocated total must be equal to 1 or 100 % and not less than 0. To assign jobs to resources there

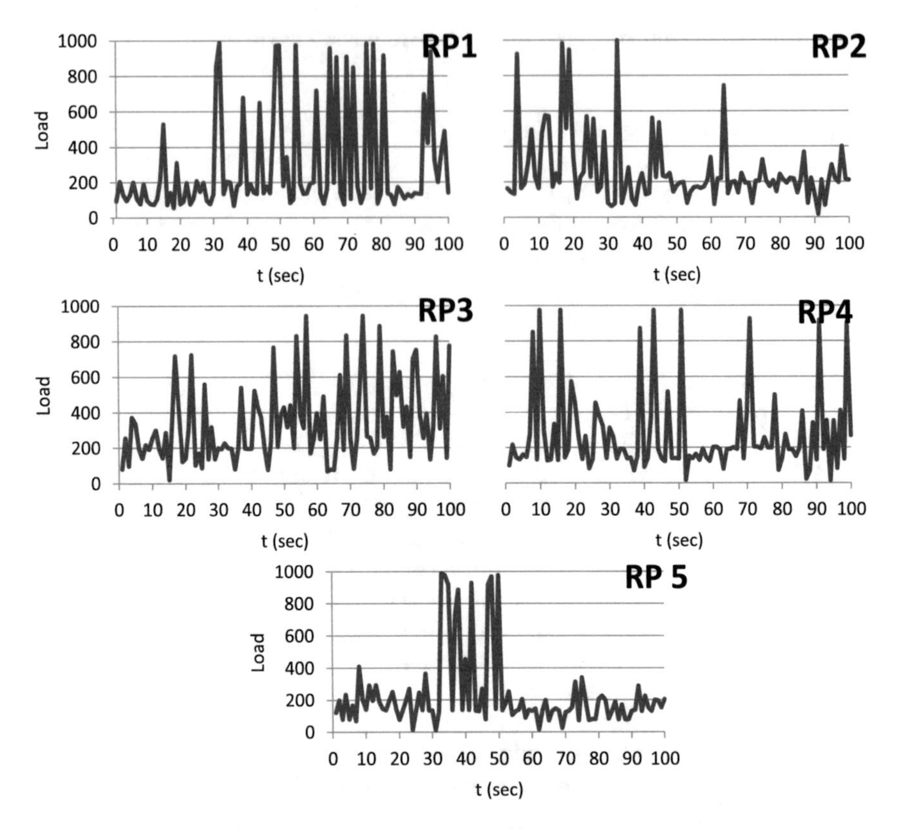

Fig. 3 Expected arrival of load at each resource provider

must be some scheduling policy according to which jobs are being submitted for execution on resources. In our proposed approach, the scheduling scheme is not discussed, so for experimental setup simple First Come First Serve (FCFS) is used.

The resource providers RP2 and RP5 have lower queue size and response time, therefore RP2 and RP5 are expected to perform fewer amounts of jobs depending on the available queue size and CPU cycles requested. In comparison, the resource providers RP1, RP3 and RP4 have higher queue size and response time, therefore will perform more amounts of jobs. The expected jobs load arriving at each RPs is shown in Fig. 3.

The comparison of the proposed distributed control based approach with the typical centralized approach has been made in Fig. 4. According to the proposed approach the jobs are assigned to the available resources equally depending on the deadline and the CPU cycles available at each resource of different RPs. To make a comparison, only two resources has been considered from random RPs. Using the centralized approach, the overall resource utilization is reduced and it will in turn minimize the overall volunteer grid performance. The proposed approach in comparison to the centralized approach balance the job workload distribution whereas,

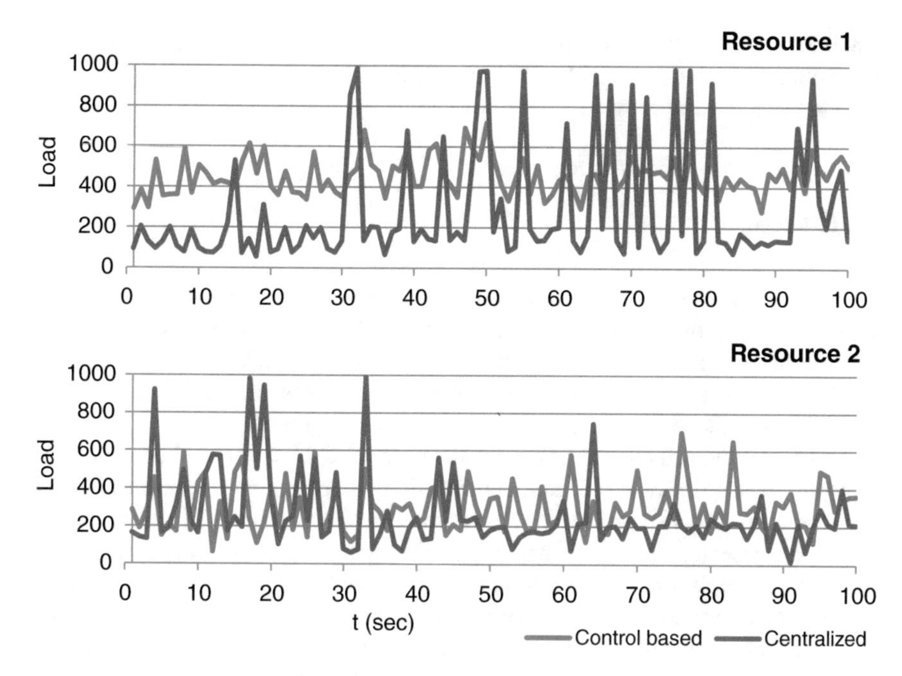

Fig. 4 Job submission rate comparison

in the centralized approach a large proportion of jobs are assigned to few resources and other resource go in starvation state.

Arriving rate of jobs load for RP2 and RP5 is more at some time samples. This is obvious to note that queue size of RP1, RP3 and RP4 are higher, so the resources of these resource providers will receive more jobs load. The share of all resources from each resource providers allocated to jobs is also dependent on the expected arrival rate of jobs. Figure 5 presents the average resources allocated and available from each resource provider with respect to time t (sec). The resources of all the resource providers being utilized practicing, the proposed approach during a time 't'

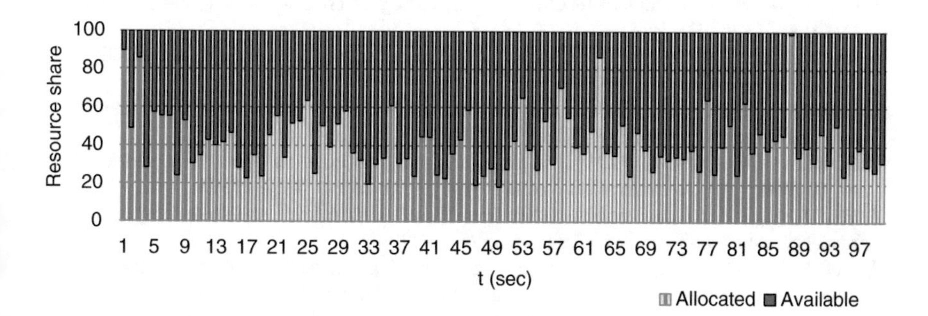

Fig. 5 Average resources allocated and available during time sample 't'

is maximized depending on the jobs load allocated to each resource. If the resources are grouped based on available CPU cycles, it will be easy to study the effect of applying proposed approach.

5 Conclusion

In this paper, a distributed coordination control-based approach has been presented for performance management to efficiently use the available resources and also balance the load meanwhile satisfying the job need in a volunteer grid computing environment. The proposed approach is adaptive to the increasing rate of jobs towards the volunteer resources and changes the job allocation dynamically depending on the arrival rate of expected future jobs. The proposed control-based approach makes use of distributed coordination and communication to satisfy the SLAs of both jobs and resources. Framework for the experimental volunteer grid has been presented and evaluated using the synthetic workloads generated. It provides an insight of volunteer computing, and depicts a picture what will be the behavior of true volunteer grid environment if the proposed approach is applied.

The proposed approach can be extended to apply in the real volunteer grid computing environment for managing the performance of resource providers. In future, the proposed approach will be used to reduce the interactions time and allocating the most matched resources at the first assignment to complete the jobs within deadline specified.

References

1. Watanabe, K., Fukushi, M., Kameyama, M.: Adaptive group-based job scheduling for high performance and reliable volunteer computing. J. Inf. Process. **19**, 39–51 (2011)
2. Nouman Durrani, M., Shamsi, J.A.: Volunteer computing: requirements, challenges, and solutions. J. Netw. Comput. Appl. **39**, 369–380 (2014)
3. Cervin, A., et al.: Feedback–feedforward scheduling of control tasks. Real-Time Syst. **23** (1–2), 25–53 (2002)
4. Tabuada, P.: Event-triggered real-time scheduling of stabilizing control tasks. IEEE Trans. Autom. Control **52**(9), 1680–1685 (2007)
5. Sharma, V., et al. Power-aware QoS management in web servers. In: 24th IEEE on Real-Time Systems Symposium, 2003. RTSS 2003. IEEE (2003)
6. Bertsekas, D.P., Tsitsiklis, J.N.: Neuro-dynamic programming: an overview. In: Proceedings of the 34th IEEE Conference on Decision and Control, 1995. IEEE (1995)
7. Parekh, S., et al.: Using control theory to achieve service level objectives in performance management. Real-Time Syst. **23**(1–2), 127–141 (2002)
8. Abdelzaher, T.F., Shin, K.G., Bhatti, N.: Performance guarantees for web server end-systems: A control-theoretical approach. IEEE Trans. Parallel Distrib. Syst. **13**(1), 80–96 (2002)
9. Haring, G., et al.: A transparent architecture for agent based resource management. In: Proceedings of IEEE International Conference on Intelligent Engineering. Citeseer (1998)

10. Kandasamy, N., Abdelwahed, S., Khandekar, M.: A hierarchical optimization framework for autonomic performance management of distributed computing systems. In: 26th IEEE International Conference on Distributed Computing Systems, 2006. ICDCS 2006. IEEE (2006)
11. Arlitt, M., Jin, T.: A workload characterization study of the 1998 world cup web site. Netw. IEEE **14**(3), 30–37 (2000)
12. Lee, J., Keleher, P., Sussman, A.: Decentralized resource management for multi-core desktop grids. In: IEEE International Symposium on Parallel & Distributed Processing (IPDPS), 2010. IEEE (2010)
13. Kang, W., Huang, H.H., Grimshaw, A.: Achieving high job execution reliability using underutilized resources in a computational economy. Future Gener. Comput. Syst. **29**(3), 763–775 (2013)
14. Foster, I. Kesselman, C.: The Grid 2: Blueprint for a New Computing Infrastructure. Elsevier (2003)
15. Kalman, R.E.: A new approach to linear filtering and prediction problems. J. Fluids Eng. **82** (1), 35–45 (1960)
16. Chapman, C., et al.: Predictive resource scheduling in computational grids. In: IEEE International on Parallel and Distributed Processing Symposium, 2007. IPDPS 2007. IEEE (2007)
17. Quinlan, J.R.: Induction of decision trees. Mach. Learn. **1**(1), 81–106 (1986)
18. Bowerman, B.L., O'Connell, R.T., Koehler, A.B.: Forecasting, Time Series, and Regression: An Applied Approach. Thomson Brooks/Cole (2005)
19. Taylor, V., et al.: Prophesy: Automating the modeling process. In: Third Annual International Workshop on Active Middleware Services, 2001. IEEE (2001)
20. Worldwide LHC Computing Grid. http://lcg.web.cern.ch/lcg/ Accessed 24 Oct 2011

Intelligent Speech Interaction of Devices and Human Operators

Maciej Majewski and Wojciech Kacalak

Abstract The article describes research and development of intelligent speech interaction systems between mobile lifting devices and their human operators. A general processing scheme, using several functional modules, for the interaction has been presented. The paper proposes an intelligent subsystem for assessment of human operator's ability for efficient processing of information streams from many sources. The paper also formulates rigorously developed concepts of process information mining and assessment of task execution and correlated working conditions using hybrid neural networks. These methods enable a variety of analyzes of task parameters and working conditions and allow us to classify different process quality classes and safety condition types. Furthermore, the methodology allows us to determine regression of task execution quality assessment. The presented research offers the possibility of motivating and inspiring further development of the intelligent speech interaction system and methods that have been elaborated in this paper.

Keywords Human-machine interface · Intelligent interaction · Speech communication · Natural language processing · Hybrid neural networks

1 Introduction

The aim of the experimental research is to design a prototype of an innovative system for controlling a mobile crane, equipped with a vision and sensorial system, interactive manipulators with force feedback, as well as a system for bi-directional voice communication through speech and natural language between an operator and the controlled lifting device. The designed structure of the innovative system for

M. Majewski (✉) · W. Kacalak
Faculty of Mechanical Engineering, Koszalin University of Technology,
Raclawicka 15-17, 75-620 Koszalin, Poland
e-mail: maciej.majewski@tu.koszalin.pl

W. Kacalak
e-mail: wojciech.kacalak@tu.koszalin.pl

R. Silhavy et al. (eds.), *Software Engineering Perspectives and Application in Intelligent Systems*, Advances in Intelligent Systems and Computing 465,
DOI 10.1007/978-3-319-33622-0_42

interaction is presented, implementing augmented reality and interactive systems. The design provides versatility in terms of application of the system when used for controlling and supervising modern machines and devices in conditions of difficulty or increased risk.

1.1 *Intelligent Speech Interaction of Loader Cranes and Operators*

In the new concept (Fig. 1), the intelligent speech interaction system between mobile lifting devices and their human operators includes intelligent spoken command and sentence interfaces. It is capable of adaptation to the human operator through an assessment system that evaluates human ability of information processing. The system also allows for adjustment of the level of automated supervision of cargo manipulation processes.

The proposed speech interaction system allows for optimal control of mobile lifting devices using spoken natural language commands. The commands are processed

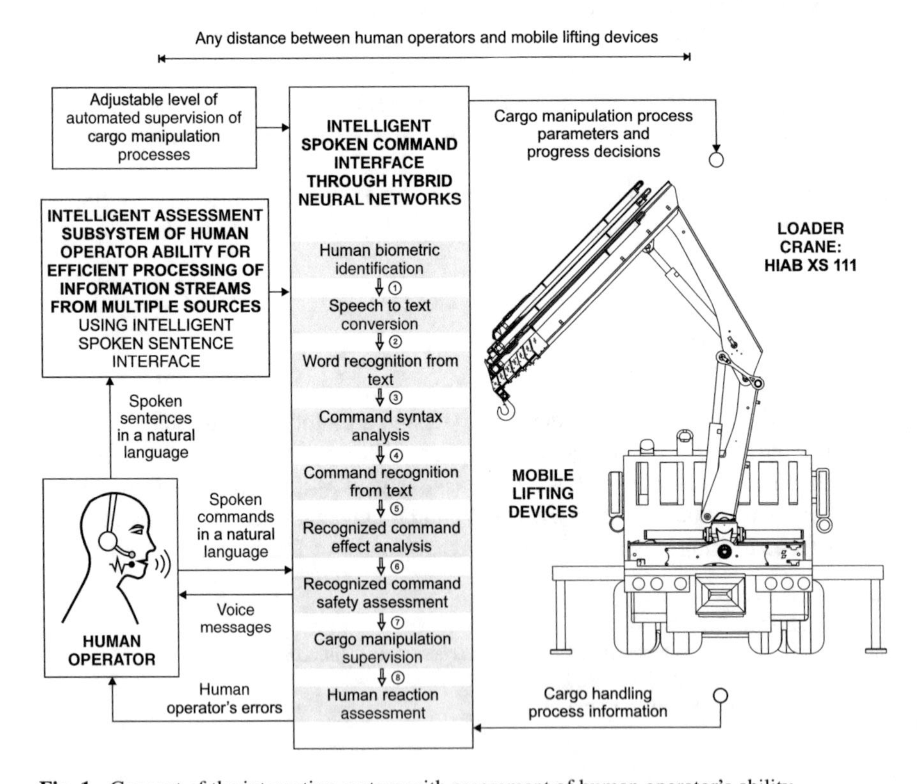

Fig. 1 Concept of the interaction system with assessment of human operator's ability

using artificial intelligence methods. The processing involves meaning analysis of words, commands and sentences. It includes command effect analysis and safety assessment. Therefore the system is capable of determination of optimal cargo manipulation process parameters and progress decisions with the aim of supporting the human operator. The novelty of the system also consists of inclusion of several adaptive layers in the spoken command and sentence interface for human biometric identification, speech recognition, word recognition, command and sentence analysis and recognition, sentence meaning analysis, command effect analysis and safety assessment, process supervision as well as human reaction assessment.

1.2 Intelligent Assessment of Human Operators

In the new concept, the intelligent human-device speech interaction system is supported by the assessment subsystem of human operator's ability for efficient processing of information streams from multiple sources. The subsystem is based on an intelligent spoken sentence interface between the assessment subsystem and the human operator. It has been implemented as shown in Fig. 2. The assessment subsystem allows for intelligent adaptation of the speech interaction system by determination of its parameters.

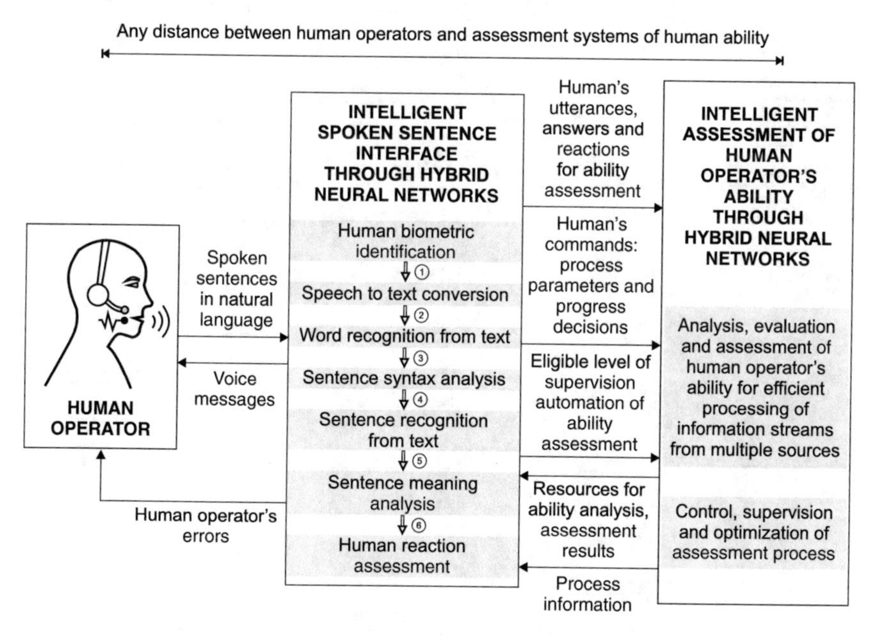

Fig. 2 Concept of the intelligent assessment subsystem of human operator's ability

1.3 Scientific and Application Significance

The proposed design can be considered as an attempt to create universal interaction systems for execution, control, supervision and optimization of cargo manipulation processes using communication by speech and natural language. It is very significant for development of new effective and flexible cargo manipulation methods. It can also contribute for increase of efficiency and decrease of costs of cargo positioning processes. This system provides an innovative solution allowing for more complete advantages of applied automated cargo handling processes nowadays.

Safe communication between human operators and mobile cranes in the automated cargo positioning processes requires analysis of the state of the lifting device and processing that state before the human operator issues a command, and so it is a necessity that artificial intelligence be applied in order to perform prediction of the command's effect as well as safety assessment of the possible outcome. Human-machine interfaces using natural languages [1–3] and aided by artificial intelligence methods lead to quality improvement. This especially applies when the expert assistance is needed regarding distant effects and complex decisions, heuristic circumstances of decisions, and sudden changes of conditions. The utilization of this knowledge can take place via remote communication.

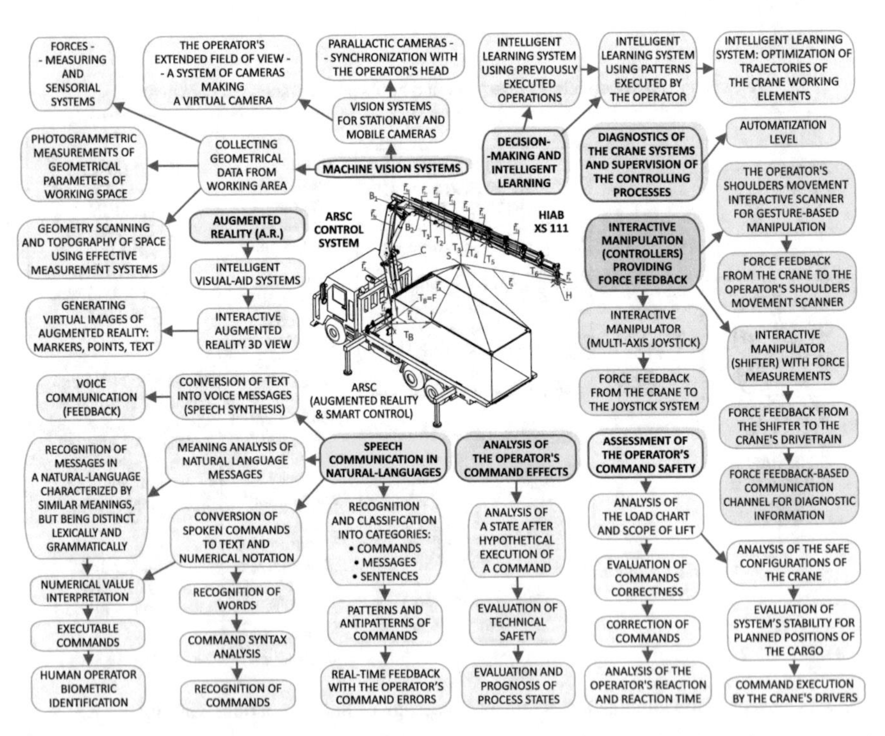

Fig. 3 Map of the technical and scientific problem space of the augmented reality and smart control systems for mobile cranes

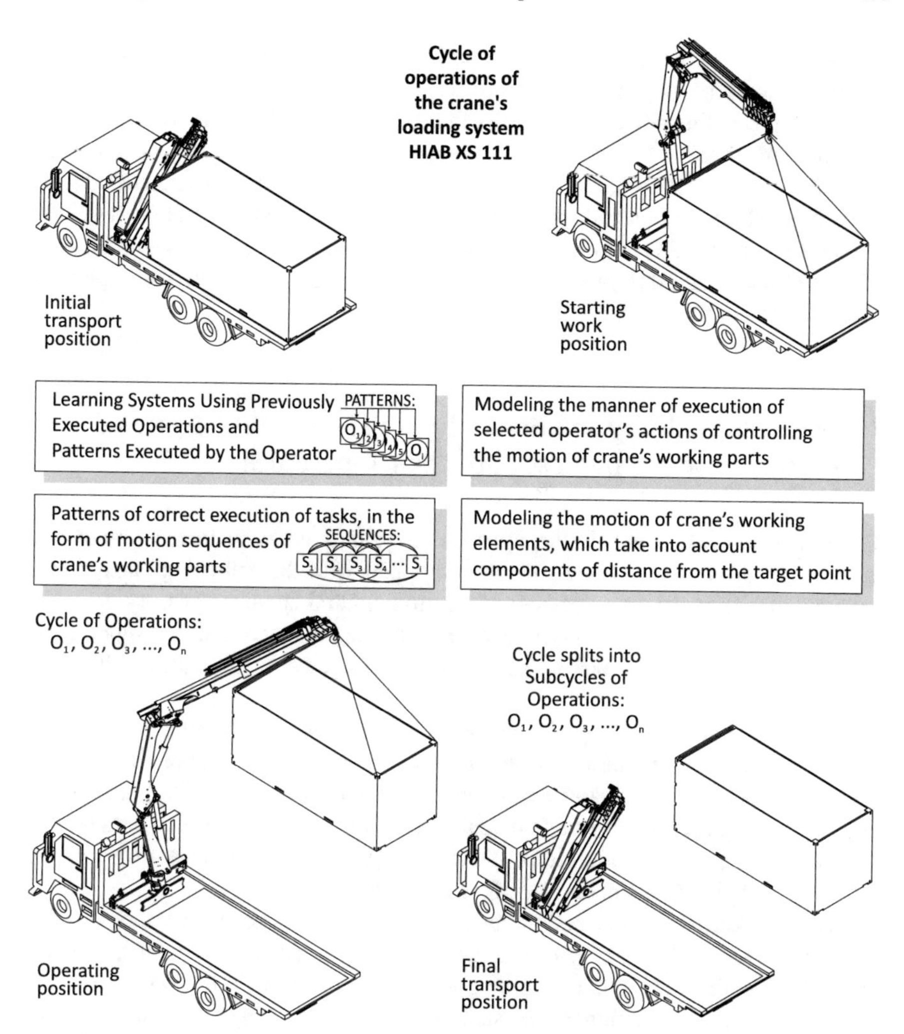

Fig. 4 Modeling the manner of execution of selected operator's actions of controlling the motion of crane's working parts. Cycles of operations of the crane's loading system HIAB XS 111

The aim of the presented research is to design an ARSC (Augmented Reality and Smart Control) system which uses: intelligent visual-aid systems based on augmented reality, interactive manipulation systems providing force feedback, as well as natural-language voice communication techniques. A sketch of the technical and scientific problem space facing a scalable and universal realization of the ARSC systems is shown in Fig. 3.

As part of the simulations and experiments with the chosen crane, an analysis of the configuration system of the crane's loading system was carried out in order to ensure collision-free motion of parts, as well as kinematic models for different tasks were developed. Moreover a set of tasks for the crane were devised, motion components for movable elements were determined, as well as ranges of motion and allowed trajectories depending on characteristics of executed tasks. In addition the patterns representing correct execution of tasks were devised, in the form of motion sequences of crane's working parts. The analyzes were used to enable rigorous development of algorithms and software for modeling the manner of execution of selected operator's actions of controlling the motion of crane's working parts (Fig. 4). The research also allowed for rigorous development of algorithms and software for modeling the motion of crane's working elements, which take into account components of distance from the target point.

2 Towards Building Intelligent Speech Interaction Systems Between Lifting Devices and Their Operators

The fundamental concept of the interaction systems between loader cranes and human operators assumes that they are equipped with the following subsystems: augmented reality vision, voice communication, natural language processing, command effect analysis, command safety assessment, command execution, supervision and diagnostics, decision-making and learning, interactive manipulation with force feedback.

The subsystem of visually aiding loader crane control with augmented reality generates virtual images of augmented reality (including markers, points), and also projects images from the vision system on a monitor (or a monitor setup) or inside of 3D-vision googles. The operator's extended field of view contains a camera system in configurations: parallactic setup—synchronization with the operator's head, and a system of stationary cameras making a virtual camera.

The subsystem for speech communication from the operator to the crane is used to perform the following tasks: processing the operator's spoken commands, operator biometric identification and authentication, converting voice commands to text and numerical notation, handling errors, analysis of character-strings, analysis of words, recognition of words, analysis of commands' syntax, analysis of commands' segments, recognition of commands, meaning analysis of natural-language messages, as well as converting text into voice messages (speech synthesis). The voice communication subsystem also provides voice feedback to the operator including reporting on the crane's working conditions' safety and expert information for exploitation and controlling. It is also communicated with a subsystem of interactive manipulators with force feedback.

The subsystem of effect analysis and evaluation of the operator's commands is designed for the following tasks: analysis of a state after hypothetical execution of a command, evaluation of technical safety, evaluation of crane systems' and process's states, evaluation of crane working conditions' safety, forecasting and signaling of process states' causes, evaluation of commands' correctness, as well as detection and signaling of possible operator's errors. The commands' safety assessment subsystem is assigned to evaluate the membership of commands to the correct commands category, and correct them.

The subsystem of the operator's commands execution is capable of signaling of a state of the process, as well as commands approved for execution. It also analyzes and evaluate the operator's reaction and its time. The execution of a command involves determination of process's parameters and its manner of execution for the configuration of the loader crane.

The subsystem for supervision and diagnostics implements crane diagnostics, supervision of the controlling process, remote supervision with mobile technologies, determination of supervision automatization level. It also includes the tasks related to measurements of the crane's working space and collection of geometrical data using photogrammetric techniques. The decision-making and learning subsystem is composed of expert systems for exploitation and controlling the crane, and an intelligent learning kernel integrated with augmented reality.

In the system, there is also a linkage with the interactive manipulators providing force feedback, which include the operator's shoulders' movement interactive scanner for gesture-based manipulation, an interactive manipulator (a shifter) with a forces-measuring system, an interactive manipulator (a multi-axis joystick). It is a connection to a force feedback-based communication channel (crane's working conditions diagnostic information) containing force feedback from the crane to the operator's shoulders' movement scanner system, force feedback from the shifter to the crane's drivetrain, as well as force feedback from the crane to the joystick system.

The intelligent speech interaction system with intelligent assessment of human operator's ability is presented in abbreviated form in Fig. 5. The numbers in the cycle represent the successive phases of information processing. The system first performs biometric identification of the human operator. Commands in the form of continuous speech are then converted to text and numeric values. The recognized text is processed by the meaning analysis subsystem performing recognition of words and commands. The result is sent to the subsystem of effect analysis and safety assessment which diagnoses the process conditions, analyzes the conditions for hypothetical command execution, and consecutively evaluates the process and crane parameters. The subsystem also evaluates the technical safety and process state, and models the process quality. It also assesses the command correctness and estimates the grades of command safety. Then commands are sent to the command execution subsystem that corrects commands and analyzes human reactions, and defines new parameters and progress decisions of the process. The crane kernel subsystem is composed of mobile cranes, as well as process automated supervision and optimization. The subsystem for speech communication converts text back to voice messages for the human operator.

The intelligent assessment subsystem of human operator's ability performs adaptation of the intelligent speech interaction system in a cycle shown in Fig. 6. After the sentence meaning analysis [4] of the human operator's utterance, response or command, the recognized meaningful sentences are subject to analysis, evaluation and assessment of the human operator's ability. The assessment subsystem analyses

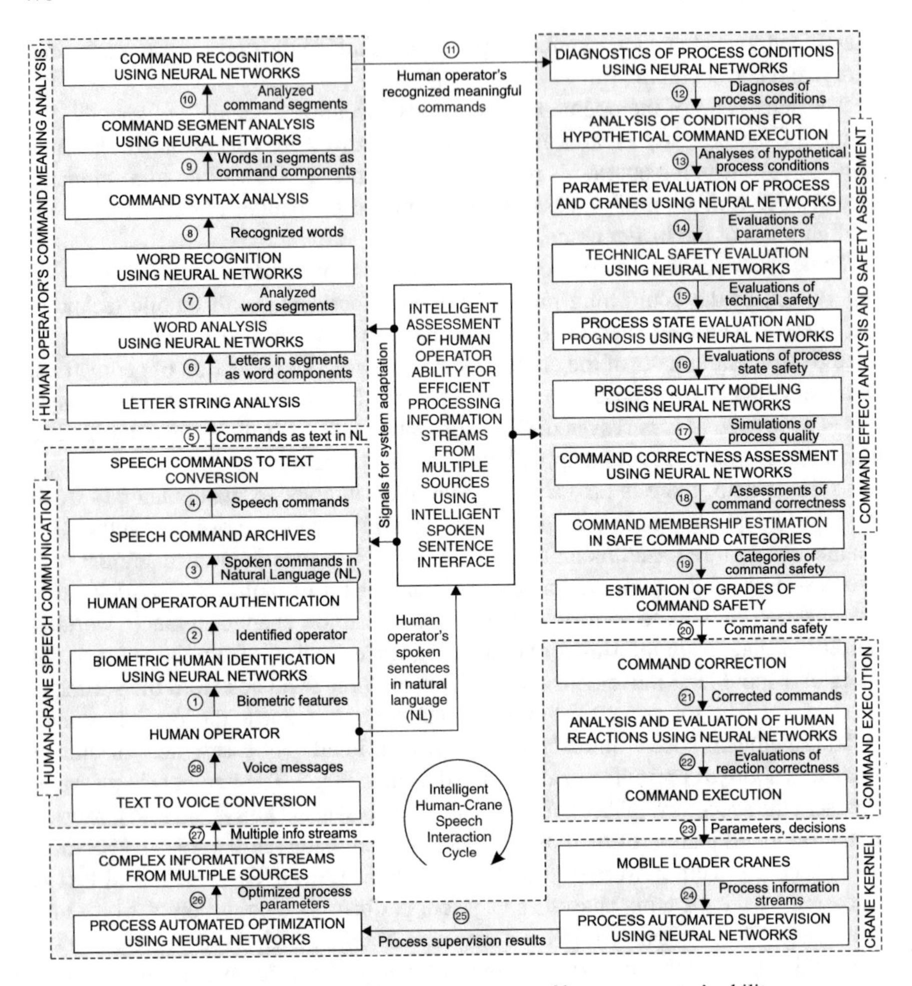

Fig. 5 Intelligent interaction with intelligent assessment of human operator's ability

the user's utterances or responses, and the level of information perception. It also analysis the level of information analyzing and reasoning.

The subsystem evaluates the information memorizing state and information comprehension level. It prognoses the information processing state. It also diagnoses the command execution state, and analyses commands and reactions. The subsystem investigates errors in responses and reactions, and assesses the human operator's actions and reactions. It assesses the correctness of the human operator's responses and evaluates the information utilization effectiveness. It also classifies the human operator's abilities and estimates the membership of human operator's ability in ability categories. The subsystem estimates the grades of the human operator's ability, and consecutively adjusts the requirement range, criterion and standards for the

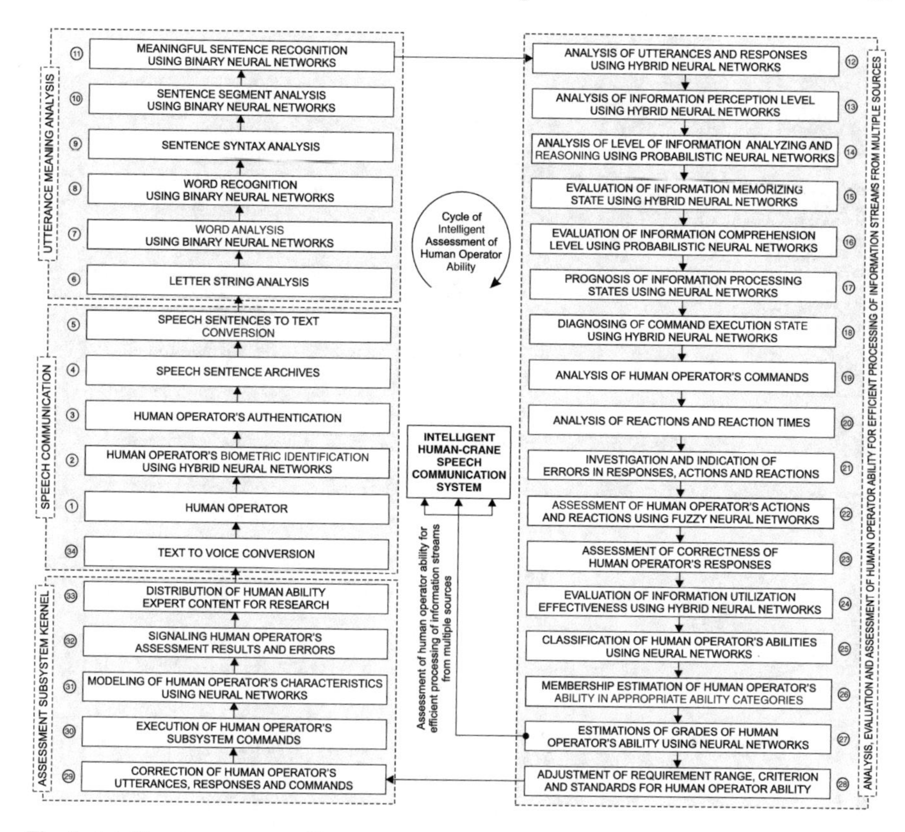

Fig. 6 Intelligent assessment of human operator's ability

human operator's ability. The kernel of the assessment subsystem corrects utterances, responses and commands, and executes commands. It models of the human operator's characteristics, and signals the assessment results and errors. The kernel also distributes expert content for the human operator's ability assessment.

3 Process Information Mining and Assessment

The intelligent speech interaction system also consists of a proposed hybrid method (Fig. 7) for information mining for task execution quality assessment. It involves extraction of hidden information through uncovering implicit relations for various types of entities in the assessment of task executions. The developed method for information mining includes the extraction method for hidden information about the cargo manipulation process and a methodology based on a regression neural network [5]. It allows to determine linear regressions of task execution quality assessment. The fuzzy assessment result processor is equipped with fuzzy sets and corresponding

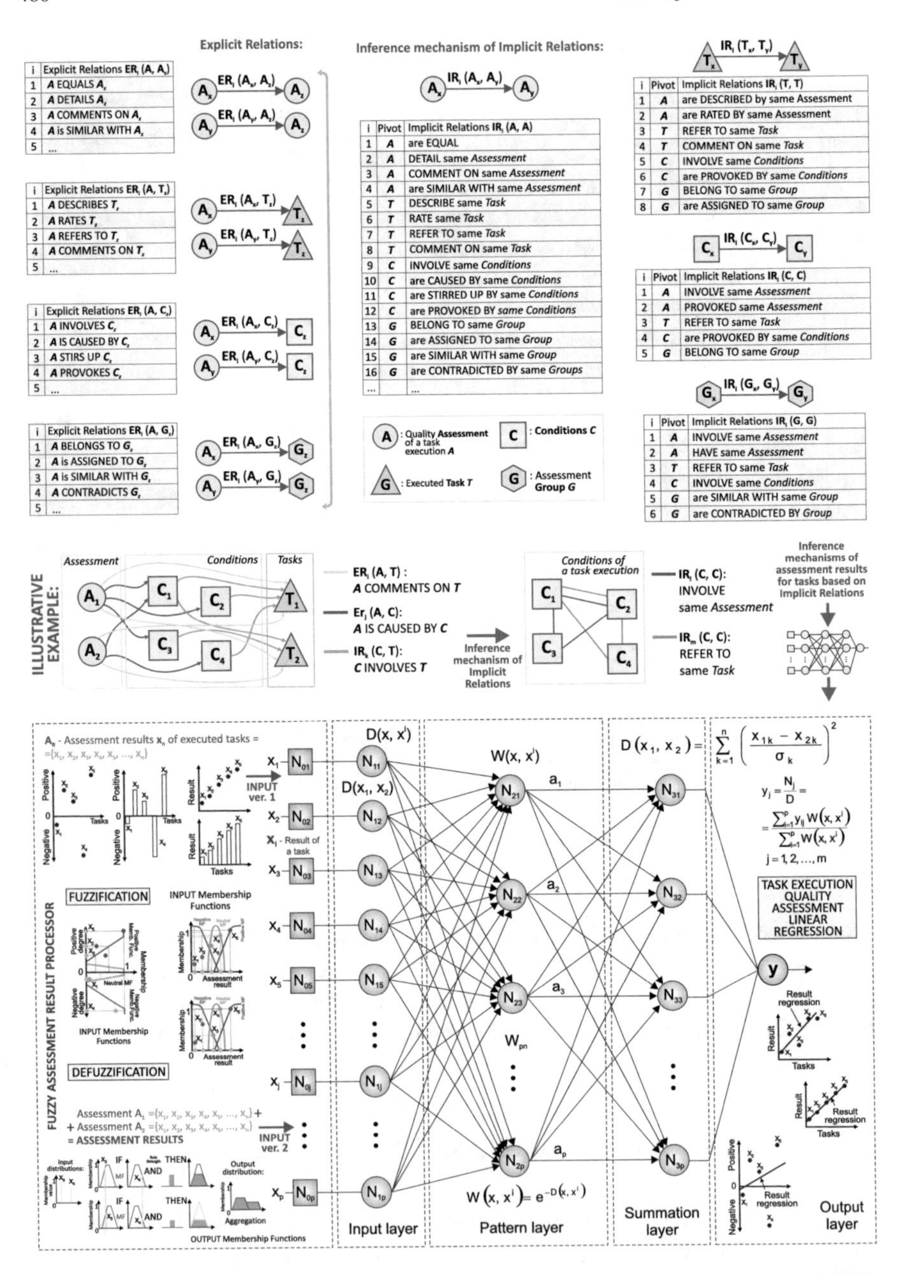

Fig. 7 Method for information mining for task execution quality assessment: extraction of hidden information through uncovering implicit relations for various types of entities in quality assessment of task executions

membership degrees. It produces quantifiable results of the task assessment. Basing on developed rules, it transforms a vector of results into fuzzy results, and the results are described in terms of membership in fuzzy sets. The fuzzy processor also designs aggregated rules in the form of output distribution depending on the patterns from the intelligent learning system. The defuzzification process interprets the membership degrees of the fuzzy sets into specific real values for assessment regressions. Inputs of the network comprise assessment results in a vector for executed tasks or joint vectors after defuzzification as concluding results obtained using fuzzy assessment result processor. The network's output produces regressions of task execution quality assessments. The analysis is modeled with a linear combination of the assessment results to produce their approximation for the cargo manipulation tasks.

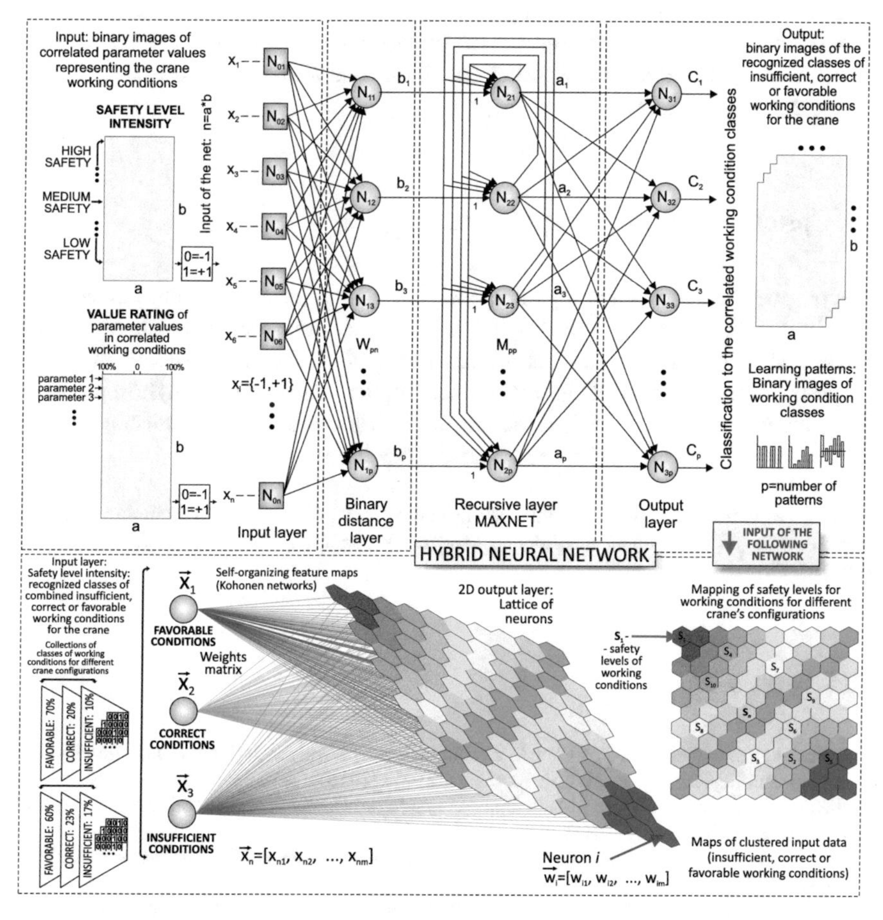

Fig. 8 Hybrid neural networks for safety assessment of the crane's correlated working conditions for different crane configurations based on the positive, neutral or negative value rating of process parameter values

Another rigorously developed hybrid method proposes a classification process of the crane's correlated working conditions being examined based on the positive, neutral or negative value rating of parameter values of signals (Fig. 8). The inputs of the hybrid neural network comprise binary images of correlated parameter values representing the crane working conditions. The outputs provide binary images of the recognized classes of insufficient, correct or favorable working conditions for the crane. The Hamming neural network [4] is chosen for the recognition of normalized parameter values and their ratings. The hybrid method also comprises self-organizing feature maps (Kohonen networks [6]). Inputs of the network consist of the classes of working conditions for different crane configurations (positive and negative ratings of parameter values). The output represents safety levels of working conditions through mapping of clustered input data (collection of classes of insufficient, correct or favorable working conditions), which are examined based on the rating of correlated parameter values. The proposed methods enable a variety of analyzes of task parameters and working conditions and allow us to classify technical safety and process states.

4 Conclusions and Perspectives

The proposed design of intelligent speech interaction between lifting devices and their human operators, equipped with intelligent decision support mechanisms and assessment of human operator's ability, carries a lot of meaning to effectiveness and progress in development of these systems. Application of various hybrid methods based on neural network architectures allows for various analyzes of the crane's configurations, process parameters and working conditions.

Acknowledgments This project is financed by the National Centre for Research and Development, Poland (NCBiR), under the Applied Research Programme—Grant agreement No. PBS3/A6/28/2015.

References

1. Kacalak, W., Majewski, M., Budniak, Z.: Interactive systems for designing machine elements and assemblies. Manag. Prod. Eng. Rev. **6**(3), 21–34 (2015). De Gruyter Open
2. Kumar, A., Metze, F., Kam, M.: Enabling the rapid development and adoption of speech-user interfaces. IEEE Comput. **47**(1), 40–47 (2014)
3. Ortiz, C.L.: The road to natural conversational speech interfaces. IEEE Internet Comput. **18**(2), 74–78 (2014)
4. Majewski, M., Zurada, J.M.: Sentence recognition using artificial neural networks. Knowl.-Based Syst. **21**(7), 629–635 (2008). Elsevier
5. Specht, D.F.: A general regression neural network. IEEE Trans. Neural Netw. **2**(6), 568–576 (1991)
6. Kohonen, T.: Self-Organization and Associative Memory. Springer (1984)

Erratum to: Expressing Pre-, Post-conditions, Attributes and Business Constraints in Artifact-Centric Business Processes Using Object Role Modeling

Quân Nguyen-Le and Lam-Son Lê

Erratum to: Expressing Pre-, Post-conditions, Attributes and Business Constraints in Artifact-Centric Business Processes Using Object Role Modeling: R. Silhavy et al. (eds.), *Software Engineering Perspectives and Application in Intelligent Systems*, Advances in Intelligent Systems and Computing 465, DOI 10.1007/978-3-319-33622-0_35

The original version of this chapter was inadvertently published with an incorrect author name. The name should read Quân Nguyen-Le.

The updated online version of the original chapter can be found at
DOI 10.1007/978-3-319-33622-0_35

Q. Nguyen-Le (✉) · L.-S. Lê
Faculty of Computer Science and Engineering, HCMC University of Technology, Ho Chi Minh, Vietnam
e-mail: nlquan.mis@gmail.com

L.-S. Lê
e-mail: lam-son.le@alumni.epfl.ch

R. Silhavy et al. (eds.), *Software Engineering Perspectives and Application in Intelligent Systems*, Advances in Intelligent Systems and Computing 465,
DOI 10.1007/978-3-319-33622-0_43

Author Index

R. Silhavy et al. (eds.), *Software Engineering Perspectives and Application in Intelligent Systems*, Advances in Intelligent Systems and Computing 465,
DOI 10.1007/978-3-319-33622-0

Printed in the United States
by Baker & Taylor Publisher Services